Functional Analysis

Joseph Muscat

Functional Analysis

An Introduction to Metric Spaces,
Hilbert Spaces, and Banach Algebras

Second Edition

 Springer

Joseph Muscat ⓘ
Mathematics Department
University of Malta
Msida, Malta

ISBN 978-3-031-27536-4   ISBN 978-3-031-27537-1   (eBook)
https://doi.org/10.1007/978-3-031-27537-1

Mathematics Subject Classification: 46, 47

This Springer imprint is published by the registered company Springer Nature Switzerland AG
The registered company address is: Gewerbestrasse 11, 6330 Cham, Switzerland

Paper in this product is recyclable.

To Jennifer-Claire and Bernadette

Preface

Originally, functional analysis was the study of functions. It is now considered to be a unifying subject that generalizes much of linear algebra and real/complex analysis, with emphasis on infinite dimensional spaces. This book introduces this vast topic from these elementary preliminaries and develops both the abstract theory and its applications in three parts: (I) metric spaces, (II) Banach and Hilbert spaces, and (III) Banach algebras.

Especially with the digital revolution at the turn of the millennium, Hilbert spaces and least squares approximation have become necessary and fundamental topics for a mathematical education, not only just for mathematicians, but also for engineers, physicists, and statisticians interested in signal processing, data analysis, regression, quantum mechanics, etc. Banach spaces, in particular L^1 and L^∞ methods, have gained popularity in applications and are complementing or even supplanting the classical least squares approach to many optimization problems.

Aim of This Book

The main aim of this book is to provide the reader with an introductory textbook that starts from elementary linear algebra and real analysis and develops the theory sufficiently to understand how various applications, including least squares approximation, etc., are all part of a single framework. A textbook must try to achieve a balance between rigor and understanding: not being too elementary by omitting 'hard' proofs, but neither too advanced by using too strict a language for the average reader and treating theorems as mere stepping stones to yet other theorems. Despite the multitude of books in this area, there is still a perceived gap in learning difficulty between undergraduate and graduate textbooks. This book aims to be in the middle: it covers much material and has many exercises of varying difficulty, yet the emphasis is for the student to remember the theory clearly using intuitive language. For example, real analysis is redeveloped from the broader picture of metric spaces (including a construction of the real number space), rather than through the even more abstract topological spaces.

Audience

This book is meant for the undergraduate who is interested in mathematical analysis and its applications, or the research engineer/statistician who would like a more rigorous approach to fundamental mathematical concepts and techniques. It can also serve as a reference or for self-study of a subject that occupies a central place in modern mathematics, opening up many avenues for further study.

The basic requirements are mainly the introductory topics of mathematics: set and logic notation, vector spaces, and real analysis (calculus). Apart from these, it would be helpful, but not necessary, to have taken elementary courses in Fourier series, Lebesgue integration, and complex analysis. Reviews of vector spaces and measurable sets are included in this book, while the other two mentioned subjects are developed only to the extent needed.

Examples are included from many areas of mathematics that touch upon functional analysis. It would be helpful at the appropriate places, for the reader to have encountered these other subjects, but this is not essential. The aim is to make connections and describe them from the viewpoint of functional analysis. With the modern facilities of searching over the Internet, anyone interested in following up a specific topic can easily do so.

The sections follow each other in a linear fashion, with the three parts fitting into three one-semester courses, although Part II is twice as long as the others. The following sections may be omitted without much effect on subsequent topics:

Section 6.4 $C(X, Y)$

Section 9.2 Function Spaces

Section 11.5 Pointwise and Weak Convergence

Sections 12.1 and 12.2 Differentiation and Integration

Sections 14.4 and 14.5 Functional Calculus and the Gelfand Transform

Section 15.6 Representation Theorems

Acknowledgments

This book grew out of lecture notes that were used in a course on functional analysis for 20 years. I wish to thank my students over the years for their feedback, but mostly my colleagues, Emanuel Chetcuti and Josef Lauri for their encouragement and very helpful suggestions when reading the first draft, as well as the anonymous reviewers who, with their invaluable comments and constructive critiques, helped shape the contents of this book. For the second edition, I wish to thank the several readers who generously pointed out errors in the first edition, especially Konstantin Koifman and

Nicholas Stevenson. I am sincerely grateful to the Springer editors who made these editions a reality, for their care in preparing the manuscripts and their unwavering support throughout the process, in particular Joerg Sixt and Catherine Waite for the first edition, and Remi Lodh for the second.

Msida, Malta   Joseph Muscat

Contents

Chapter 1
Introduction

Modern mathematics depends to a considerable degree upon extending the finite to the infinite. In this regard, imagine extending the geometric vectors that we are familiar with to an infinite number of components. That is, consider

$$v = a_1 e_1 + a_2 e_2 + \cdots = (a_1, a_2, a_3, \ldots)$$

where e_i are unit independent vectors just like $\mathbf{i}$, $\mathbf{j}$ and $\mathbf{k}$ in Cartesian geometry. It is not at all clear that we can do so—for starters, what do those three dots "$\cdots$" on the right-hand side mean? Surely they signify that as more terms are taken one gets better approximations of v. This immediately suggests that not every such "infinite" vector is allowed; for example, it might be objected that the vector

$$v = e_1 + e_2 + e_3 + \cdots$$

cannot be approximated by a finite number of these unit vectors, as the remainder $e_n + \cdots$ looks as large as v. Instead we might allow the infinite vector

$$v = e_1 + \tfrac{1}{2} e_2 + \tfrac{1}{3} e_3 + \cdots$$

although even here, it is unclear whether this may also grow large, just as

$$1 + \frac{1}{2} + \frac{1}{3} + \cdots = \infty.$$

To continue with our experiment, let us just say that the coefficients become zero rapidly enough. There are all sorts of operations we can attempt to do with

© The Author(s), under exclusive license to Springer Nature Switzerland AG 2024
J. Muscat, *Functional Analysis*, https://doi.org/10.1007/978-3-031-27537-1_1

these "infinite" vectors, by analogy with the usual vectors: addition of vectors and multiplication by a number are easily accomplished,

$$(1, \tfrac{1}{2}, \tfrac{1}{3}, \ldots) + (1, \tfrac{1}{2}, \tfrac{1}{4}, \ldots) = (2, 1, \tfrac{7}{12}, \ldots),$$

$$2 \times (0, 1, -\tfrac{1}{2}, \ldots) = (0, 2, -1, \ldots).$$

One can even generalize the "dot product"

$$(a_1, a_2, \ldots) \cdot (b_1, b_2, \ldots) = a_1 b_1 + a_2 b_2 + \cdots,$$

assuming the series converges—and we have no guarantee that it always does. For example, if x is equal to $(1, \tfrac{1}{\sqrt{2}}, \tfrac{1}{\sqrt{3}}, \ldots)$, then $x \cdot x = \sum_{n=1}^{\infty} \tfrac{1}{n}$ is infinite. Again let us remedy this situation by insisting that vectors have coefficients that decrease to 0 fast enough.

Having done this, we may go on to see what infinite matrices would look like. They would take an infinite vector and return another infinite one, as follows,

$$\begin{pmatrix} a_{11} & a_{12} & \cdots \\ a_{21} & a_{22} & \\ \vdots & & \ddots \end{pmatrix} \begin{pmatrix} x_1 \\ x_2 \\ \vdots \end{pmatrix} = \begin{pmatrix} y_1 \\ y_2 \\ \vdots \end{pmatrix},$$

where $y_1 = a_{11}x_1 + a_{12}x_2 + \cdots = \sum_n a_{1n}x_n$, etc. Perhaps we may need to have the rows of the matrix vanish sufficiently rapidly as we go down and to the right of the matrix.

Once again, many familiar ideas from finite matrices seem to generalize to this infinite setting. Not only is it possible to add and multiply these matrices without any inherent difficulty, but methods such as Gaussian elimination can also be applied in principle. There seems to be no intrinsic problem to working with infinite-dimensional linear algebra.

It may come as a slight surprise to the reader that in fact they have already encountered these infinite vectors before! When a function is expanded as a MacLaurin series

$$f(t) = f(0) + f'(0)t + \tfrac{1}{2}f''(0)t^2 + \cdots,$$

it is in effect written as an infinite sum of the basis vectors (or functions) $1, t, t^2, \ldots$, each with the numerical coefficients $f(0), f'(0), \tfrac{1}{2}f''(0), \ldots$, respectively. Adding two functions is the same as adding the two infinite vectors of coefficients; and multiplying by a number is equivalent to multiplying each coefficient by the same number. What about infinite matrices? Take a look at the following form of differentiation, here written in matrix form,

$$f' = \begin{pmatrix} 0 & 1 & 0 & \\ & 0 & 2 & 0 \\ & & 0 & 3 & \ddots \\ & & & & \ddots \\ & & & & & \ddots \end{pmatrix} \begin{pmatrix} f(0) \\ f'(0) \\ \frac{1}{2}f''(0) \\ \vdots \end{pmatrix}.$$

And just as there are various bases that can be used in geometry, so there are different ways to expand functions, the most celebrated being the Fourier series

$$f(t) = a_0 + a_1 \cos t + b_1 \sin t + a_2 \cos 2t + b_2 \sin 2t + \cdots.$$

The basis vectors are now $1, \cos t, \sin t, \cos 2t$, etc. What matrix does differentiation take with respect to this basis?

If we accept that all this is possible and makes sense, we are suddenly made aware of a new unification of mathematics: functions can be thought of as 'points' in a space of infinite vectors, certain differential equations *are* matrix equations, the Fourier and Laplace transforms can be thought of as generalized "matrices" mapping a function (vector) to another function, etc. Solving a linear differential equation, and finding the inverse Fourier transform, are equivalent to finding the inverses of their "matrices".

Do we gain anything by converting to a matrix picture? Apart from the practical matter that there are many known algorithms that deal with matrices, a deeper reason is that linear algebra and geometry give insights to the subject of functions that we may not have had before. Euclid's theorems may possibly still be valid for functions if we think of them as 'points' in an infinite-dimensional vector space. We wake up to the possibility of a function being perpendicular to another, for example, and that a function may have a closest function in a "plane" of functions.

Conversely, ideas from classical analysis may be transferred to linear algebra. Since square matrices can be multiplied with themselves, can the geometric series $1 + A + A^2 + \cdots$ make sense for matrices? Perhaps one can take the exponential of a matrix $e^A := 1 + A + A^2/2! + A^3/3! + \cdots$. There's no better way than to take the plunge and try it out, say on the differentiation 'matrix' D,

$$e^D f(t) = (1 + D + D^2/2 + \cdots)f(t) = f(t) + f'(t) + f''(t)/2 + \cdots = f(t+1)$$

(by a Taylor expansion around t). The "matrix" e^D certainly has meaning: it performs an unexpected, if mundane, operation—it *shifts* the function f one step to the left! Again, suppose we have the equation $y' - 2y = e^t$; manipulating the derivative blindly as if it were a number gives a correct solution (but not the general solution)

$$y = (D-2)^{-1}e^t = -\tfrac{1}{2}(1 + D/2 + D^2/4 + \cdots)e^t = -e^t.$$

(Yet repeating for the equation $y' - 2y = e^{2t}$ fails to give a meaningful solution.)

In fact, historically, the subject of functional analysis as we know it started in the nineteenth century when mathematicians began to notice the connections between differential equations and matrices. For example, the equation

$$y'(t) = a(t)y(t) + g(t)$$

can be written in equivalent form as

$$y(t) = \int_{t_0}^t a(s)y(s)\,\mathrm{d}s + \tilde{g}(t). \tag{1.1}$$

The integral $\int_{t_0}^t a(s)y(s)\,\mathrm{d}s$ is an infinitesimal version of $\sum_{n=1}^N a_n y_n$ and can be thought of as a transformation of $y(t)$. Equation (1.1) is akin to a matrix equation $y = Ay + b$, and we are tempted to try out the solution $y = (1-A)^{-1}b = (1 + A + A^2 + \cdots)b$.

Nonetheless, technical problems in carrying out this generalization arise immediately: are the components of an infinite vector unique? They would be if the vectors e_n are in some sense 'perpendicular' to each other. But what is this supposed to mean, say for the MacLaurin series? After all, there do exist non-zero functions whose MacLaurin coefficients are all zero. The question of whether the Fourier coefficients are unique took almost a century to answer! And extra care must be taken to handle infinite vectors. For example, let

$$
\begin{aligned}
v_1 &:= (\ \ 1, \ \ \ 0, \ \ \ 0, \ \ \ 0, \ \dots\,) \\
v_2 &:= (-1, \ \ \ 1, \ \ \ 0, \ \ \ 0, \ \dots\,) \\
v_3 &:= (\ \ 0, -1, \ \ \ 1, \ \ \ 0, \ \dots\,) \\
v_4 &:= (\ \ 0, \ \ \ 0, -1, \ \ \ 1, \ \dots\,) \\
&\ \cdots
\end{aligned}
$$

It seems clear that

$$v_1 + v_2 + v_3 + \cdots = 0,$$

yet the size of the sum of the first n vectors never diminishes:

$$v := v_1 + \cdots + v_n = (0, \dots, 0, 1, 0, \dots) \Rightarrow v \cdot v = 1.$$

Here's another seeming paradox: Consider the infinite number of equations $x_1 = x_2$, $x_2 = x_3$, ..., $x_n = x_{n+1}$, ... Clearly it has the solutions $x_1 = x_2 = \cdots = \lambda$ for any $\lambda \in \mathbb{R}$. But let us try to use infinite matrices to solve this problem. The equations in matrix form become $Ax = 0$,

$$\begin{pmatrix} 1 & -1 & 0 \\ 0 & 1 & -1 \\ & & & \ddots \end{pmatrix} \begin{pmatrix} x_1 \\ x_2 \\ \vdots \end{pmatrix} = \begin{pmatrix} 0 \\ 0 \\ \vdots \end{pmatrix}$$

Its inverse can be calculated using Gaussian elimination to get

$$A^{-1} = \begin{pmatrix} 1 & 1 & 1 \\ 0 & 1 & 1 \\ \vdots & & & \ddots \end{pmatrix}$$

One can verify that $AA^{-1} = I = A^{-1}A$. But doesn't this then imply that $x = A^{-1}0 = 0$? What happens to the solutions we obtained above? On the other hand, if the non-zero vector $(1, 1, \ldots)$ satisfies $Ax = 0$, then 0 would be an eigenvalue, but then shouldn't it appear on the main diagonal of A?

Because of these trapfalls, we need to proceed with extra caution. It turns out that many of the equations written above are capable of different interpretations and so cannot be taken to be literally true.

These considerations force us to consider the meaning of convergence. The reader may already be familiar with the real line $\mathbb{R}$, in which one can speak about convergence of sequences of numbers, and continuity of functions. Some of the main results in real analysis are:

(i) Cauchy sequences converge,
(ii) For continuous functions, if $t_n \to t$ then $f(t_n) \to f(t)$,
(iii) Continuous real functions are bounded on intervals of type $[a, b]$ and have the intermediate value property, that is, they map intervals to intervals.

We seek generalizations of these to $\mathbb{R}^n$ and possibly to infinite dimensional spaces. We do seem to have an intuitive sense of what it means for vectors to converge $x_n \to x$, but can it be made rigorous? Is it true that if $x_n \to x$ and $y_n \to y$ then $f(x_n, y_n) \to f(x, y)$ when f is a continuous function? Are continuous real functions bounded on "rectangles" $[a, b] \times [c, d]$, and is the latter the correct analogue of an "interval"? Since vector functions are common in applications, it is important to show how these theorems apply in a much more general setting than $\mathbb{R}$, and this can be achieved by stripping off any inessential structure, such as its order ($\leqslant$). As we proceed to answer these questions, we will see that the real line is very special indeed. Intervals play several roles in real analysis, roles that are distinguished apart in $\mathbb{R}^n$, where we speak instead of connected sets, balls, etc.

The book is divided into three parts: the first considers *convergence*, continuity, and related concepts, the second part treats infinite vectors and their matrices, and the third part tackles infinite series of matrices and more.

Functional analysis is a rich subject because it combines two large branches of mathematics: the *topological* branch concerns itself with convergence, continuity, connectivity, boundedness, etc.; the *algebraic* branch concerns itself with operations, groups, rings, vectors, etc. Problems from such different fields as matrix algebras, differential equations, and approximation theory, can be unified in one framework. As in most of mathematics, there are two streams of study: the *abstract theory* deduces the general results, starting from axioms, while the *concrete examples* are shown to be part of this theory. Inevitably, the former appears elegant and powerful, and the latter full of detail and perhaps daunting. Nonetheless, both pedagogically and historically, it is often by examples that one understands the abstract, and by the theory that one makes headway with concrete problems.

Most sections contain a number of worked out examples, notes, and exercises: it is suggested that a section is first read in full, including its propositions and exercises. These exercises are an essential part of the book; they should be worked out before moving to the next section (some hints and answers are provided in the appendix, and many worked solutions can be found in the Instructor's solutions manual). To prevent the exercises from becoming a litany of "Show ..." and "Prove ...", these terms have frequently been omitted, partly to instil an attitude of critical reading. As a guide, the notes and exercises have been marked as follows:

▶   refers to important notes and results;
⋆   more advanced or difficult exercises that can be skipped on a first reading;
◇   side remarks that can be skipped without losing any essential ideas.

1.1   Preliminaries

Familiarity with the following mathematical notions and notation is assumed:

Logic and Sets
The basic logical symbols are $\Rightarrow$ (implies), NOT, AND, OR, as well as the quantifiers $\exists$ (there exists) and $\forall$ (for all). The reader should be familiar with the basic proof strategies, such as proving $\phi \Rightarrow \psi$ by its contrapositive (NOT ψ) $\Rightarrow$ (NOT ϕ), and proofs by contradiction. The negation of $\forall x \; \phi_x$ is $\exists x$ (NOT ϕ_x); and NOT ($\exists x \; \phi_x$) is the same as $\forall x$ (NOT ϕ_x). The symbol $:=$ is used to *define* the left-hand symbol as the right-hand expression, e.g., $e := \sum_{n=0}^{\infty} \frac{1}{n!}$.

A *set* consists of *elements*, and $x \in A$ denotes that x is an element of the set A. The empty set $\varnothing$ contains no elements, so $x \in \varnothing$ is a contradiction.

The following sets of numbers are the foundational cornerstones of mathematics: the natural numbers $\mathbb{N} = \{0, 1, \dots\}$, the integers $\mathbb{Z}$, the rational numbers $\mathbb{Q}$, the real numbers $\mathbb{R}$, and the complex numbers $\mathbb{C}$, the last containing the *imaginary* number i. The *induction principle* applies for $\mathbb{N}$,

> If $A \subseteq \mathbb{N}$ AND $0 \in A$ AND $\forall n$ ($n \in A \Rightarrow n + 1 \in A$) then $A = \mathbb{N}$.

Although variables should be quantified to make sense of statements, as in $\forall a \in \mathbb{Q}, a^2 \neq 2$, in practice one often takes shortcuts to avoid repeating the obvious. This book uses the convention that if a statement mentions variables without accompanying quantifiers, say, $\|x + y\| \leqslant \|x\| + \|y\|$, these are assumed to be $\forall x$, $\forall y$, etc., in the space under consideration. Natural numbers are usually, but not exclusively, denoted by the variables m, n, N, ..., real numbers by t, a, b, ..., and complex numbers by z, w, ... An unspecified X or Y refers to a metric space, a normed space, or a Banach algebra, depending on the chapter.

Sets are often defined in terms of a property, $A := \{ x \in X : \phi_x \}$, where X is a given 'universal set' and ϕ_x a 'well-formed' statement about x. For example, $\mathbb{R}^+ := \{ x \in \mathbb{R} : x \geqslant 0 \}$.

$A \subseteq B$ denotes that A is a *subset* of B, i.e., $x \in A \Rightarrow x \in B$; $A \subset B$ means $A \subseteq B$ but $A \neq B$. A "non-trivial" or "proper" subset of X is one which is not $\varnothing$ or X. "Nested sets" are contained in each other as in $A_1 \subseteq A_2 \subseteq A_3 \subseteq \ldots$ or $A_1 \supseteq A_2 \supseteq \ldots$.

The *complement* of a set A is denoted by $X \smallsetminus A$, or by A^c for short; $A^{cc} = A$, and $A \subseteq B \Leftrightarrow B^c \subseteq A^c$. $A \cap B$ and $A \cup B$ are the *intersection* and *union* of two sets, respectively. Two sets are "disjoint" when $A \cap B = \varnothing$, while we say "A intersects B" to mean $A \cap B \neq \varnothing$. De Morgan's laws state that $(A \cup B)^c = A^c \cap B^c$ and $(A \cap B)^c = A^c \cup B^c$. The 'symmetric difference' of two subsets is $A \triangle B := (A \cup B) \smallsetminus (A \cap B)$. In general, the union and intersection of a number of sets are denoted by $\bigcup_i A_i$ and $\bigcap_i A_i$ (where the range of the index i is understood by the context). A "cover" of A is a collection of sets $\{ B_i : i \in I \}$ whose union includes A, i.e., $A \subseteq \bigcup_i B_i$; a "partition" of X is a cover by disjoint subsets of X.

Pairs of elements are denoted by (x, y), or as $\binom{x}{y}$, generalized to finite ordered lists $(x_1, \ldots, x_n)$. The *product* of two sets is the set of pairs

$$X \times Y := \{ (x, y) : x \in X, y \in Y \}$$

in particular $X^2 := X \times X = \{ (x, y) : x, y \in X \}$, and by analogy

$$X^n := \{ (x_1, \ldots, x_n) : x_i \in X, i = 1, \ldots, n \}.$$

An important example is the *plane* $\mathbb{R}^2$, whose points are pairs of real numbers (called "coordinates"). The unit *disk* is $\{ (x, y) \in \mathbb{R}^2 : x^2 + y^2 \leqslant 1 \}$; its perimeter is the unit *circle*, $\mathbb{S}^1 := \{ (x, y) \in \mathbb{R}^2 : x^2 + y^2 = 1 \}$.

Functions

A *function* $f : X \to Y$, $x \mapsto f(x)$, assigns, for every input $x \in X$, a unique output element $f(x) \in Y$. (It need not be an explicit procedure.) X is called the "domain" of f and Y its "codomain". Functions are also commonly referred to as "maps" or "transformations". To avoid being too pedantic, we sometimes say, for example, "the function $x \mapsto e^x$" without reference to the domain and codomain, when these are obvious from the context.

The "image" of a subset $A \subseteq X$, and the "pre-image" of a subset $B \subseteq Y$ are

$$fA := \{ f(a) \in Y : a \in A \}, \quad f^{-1}B := \{a \in X : f(a) \in B \},$$

also denoted by $f[A]$ and $f^{-1}[B]$ for clarity. The *image* of f is im $f := fX$. It is easy to show that for any number of sets A_i,

$$f[\bigcup_i A_i] = \bigcup_i f[A_i], \quad f[\bigcap_i A_i] \subseteq \bigcap_i f[A_i],$$

$$f^{-1}[\bigcup_i A_i] = \bigcup_i f^{-1}[A_i], \quad f^{-1}[\bigcap_i A_i] = \bigcap_i f^{-1}[A_i].$$

The set of functions $f : X \to Y$ is denoted by Y^X.

Some functions can be *composed* together $(f \circ g)(x) := f(g(x))$ whenever the image of g lies in the domain of f. Composing with the trivial *identity* function $I : X \to X, x \mapsto x$ (one for each set X), has no effect, $f \circ I = f$ and $I \circ f = f$.

The *restriction* of a function $f : X \to Y$ to a subset $M \subseteq X$ is the function $f|_M : M \to Y$ which agrees with f on M, i.e., $f|_M(x) = f(x)$ whenever $x \in M$. Conversely, f is said to be an *extension* of $f|_M$.

The reader should be familiar with the functions $t \mapsto -t$, t^n, $|t|$, for $t \in \mathbb{R}$ or $\mathbb{C}$; $(x, y) \mapsto x + y$, xy, with domain $\mathbb{R}^2$ or $\mathbb{C}^2$; $(x, y) \mapsto x/y$ for $y \neq 0$; and $(x_1, \dots, x_n) \mapsto \max(x_1, \dots, x_n)$ for real numbers x_i. In particular, the *absolute value* function satisfies

$$|a + b| \leqslant |a| + |b|, \quad |a| \geqslant 0, \quad |a| = 0 \Leftrightarrow a = 0, \quad |ab| = |a||b|.$$

Conjugation is the function $\overline{} : \mathbb{C} \to \mathbb{C}, a + ib \mapsto a - ib$; its properties are

$$\overline{z + w} = \bar{z} + \bar{w}, \; \overline{zw} = \bar{z}\bar{w}, \; \bar{\bar{z}} = z, \; \bar{z}z = |z|^2.$$

The Kronecker *delta* function is $\delta_{ij} := \begin{cases} 1 & i = j \\ 0 & i \neq j \end{cases}$. The *exponential* function $t \mapsto e^t$, $\mathbb{R} \to \mathbb{R}$, may be defined by $e^t := \sum_{n=0}^{\infty} \frac{t^n}{n!}$; it satisfies $e^0 = 1$ and $e^t > 0$.

Sequences are functions $x : \mathbb{N} \to X$, but $x(n)$ is usually written as x_n, and the whole sequence x is referred to by $(x_n)_{n \in \mathbb{N}}$ or $(x_0, x_1, \dots)$ or even just (x_n); real or complex-valued sequences are denoted by bold symbols, $\boldsymbol{x}$. For example $(1/2^n)$ is the sequence $(1, \frac{1}{2}, \frac{1}{4}, \dots)$, which is shorthand for $0 \mapsto 1, 1 \mapsto \frac{1}{2}$, etc. It is important to realize that $(x_n)_{n \in \mathbb{N}}$ is a function and not a set of values, e.g. $(1, -1, 1, -1, \dots)$ is quite different from $(-1, 1, 1, 1, \dots)$ and $(-1, 1, -1, 1, \dots)$, even if they have the same set of values. The set of real-valued sequences is denoted by $\mathbb{R}^{\mathbb{N}} := \{ x : \mathbb{N} \to \mathbb{R} \}$, and of the complex-valued sequences by $\mathbb{C}^{\mathbb{N}}$. Functions $x : \mathbb{Z} \to X$ are also sometimes called sequences and are denoted by $(x_n)_{n \in \mathbb{Z}}$.

Polynomials (of one variable) are functions $p : \mathbb{C} \to \mathbb{C}$ that are a finite number of compositions of additions and multiplications only; every polynomial can be written in the standard form $p(z) = a_n z^n + \cdots + a_1 z + a_0$ ($a_i \in \mathbb{C}$, $a_n \neq 0$ unless $p = 0$); n is called the *degree* of p. The set of all polynomials in the variable z, with complex coefficients, is denoted $\mathbb{C}[z]$.

A function $f : X \to Y$ is *1–1* ("one-to-one") or *injective* when $f(x) = f(y) \Rightarrow x = y$; it is *onto* or *surjective* when $fX = Y$. A *bijection* is a function which is both 1–1 and onto; every bijection has an *inverse* function f^{-1}, whereby $f^{-1} \circ f(x) = x$, $f \circ f^{-1}(y) = y$.

Sets may be *finite*, *countably infinite*, or *uncountable*, depending on whether there exists a bijection from the set to, respectively, (i) a set $\{ 1, \ldots, n \}$ for some natural number n, or (ii) $\mathbb{N}$, or (iii) otherwise. In simple terms, a set is countable when its elements can be listed, and finite when the list terminates. If A, B are countable sets then so is $A \times B$; more generally, the union of the countable sets A_n, $n = 0, 1, 2, \ldots$, is again countable:

$$A_0 = \{ \quad a_{00}, \quad a_{01}, \quad a_{02}, \quad \ldots \quad \}$$
$$A_1 = \{ \quad a_{10}, \quad a_{11}, \quad a_{12}, \quad \ldots \quad \}$$
$$A_2 = \{ \quad a_{20}, \quad a_{21}, \quad a_{22}, \quad \ldots \quad \}$$
$$\ldots$$

$$\bigcup_{n=0}^{\infty} A_n = \{ a_{00}, a_{01}, a_{10}, a_{02}, \ldots \}$$

Relations and Orders

A *relation* is a statement about pairs of elements taken from $X \times Y$, e.g. $x = y^2 + 1$ for $(x, y) \in \mathbb{R}^2$. An *equivalence* relation $\approx$ on a set X is a relation on X^2 which is

$$\text{reflexive}: \qquad x \approx x,$$
$$\text{symmetric}: \qquad x \approx y \Leftrightarrow y \approx x,$$
$$\text{transitive}: \qquad x \approx y \approx z \Rightarrow x \approx z.$$

An equivalence relation induces a partition of the set X into *equivalence classes* $[a] := \{ x \in X : x \approx a \}$.

An *order* $\leqslant$ is a relation which is reflexive, transitive, and anti-symmetric:

$$x \leqslant y \leqslant x \Rightarrow x = y.$$

One writes $x < y$ when $x \leqslant y$ but $x \neq y$. A *linear order* is one which also satisfies $x \leqslant y$ OR $y \leqslant x$. A number x is *positive* when $x > 0$, whereas "non-negative" means $x \geqslant 0$. An *upper bound* of a set A is a number b which is larger than any $a \in A$. A "least upper bound", denoted sup A, is the smallest such upper bound (if it exists), i.e., every upper bound of A is greater than or equal to sup A. There are analogous definitions of lower bound and greatest lower bound, denoted inf A.

Groups and Fields

A *group* is a set G with an *associative* operation and an *identity* element 1, such that each element $x \in G$ has an *inverse* element x^{-1},

$$x(yz) = (xy)z, \quad 1x = x = x1, \quad xx^{-1} = 1 = x^{-1}x.$$

A *subgroup* is a subset of G which is itself a group with the same operation and identity. A *normal* subgroup is a subgroup H such that $x^{-1}Hx \subseteq H$ for all $x \in G$. An example of a group is the set $\mathbb{C} \setminus \{0\}$ with the operation of multiplication; the set $S := \{e^{i\theta} : \theta \in \mathbb{R}\}$ is a subgroup since $e^{i\theta}e^{i\phi} = e^{i(\theta+\phi)}$, $1 = e^{i0}$, $(e^{i\theta})^{-1} = e^{-i\theta}$ are all in S.

A *field* $\mathbb{F}$ is a set of numbers, such as $\mathbb{Q}$, $\mathbb{R}$, or $\mathbb{C}$, whose elements can be *added* and *multiplied* together associatively, commutatively, and distributively, that is, for all $a, b, c \in \mathbb{F}$,

$$(a+b)+c = a+(b+c), \qquad (ab)c = a(bc),$$
$$a+b = b+a, \qquad\qquad ab = ba,$$
$$(a+b)c = ac+bc,$$

there is a *zero* 0 and an *identity* 1, every element a has an additive inverse, or *negative*, $-a$, and every $a \neq 0$ has a multiplicative inverse, or *reciprocal*, $1/a$:

$$0+a = a, \qquad 1a = a,$$
$$a+(-a) = 0, \qquad a\tfrac{1}{a} = 1 \ (a \neq 0).$$

The real number space $\mathbb{R}$ is that unique field which has a linear order $\leqslant$ such that

(a) $a \leqslant b \Rightarrow a+c \leqslant b+c$, and $0 \leqslant a, b \Rightarrow 0 \leqslant ab$,
(b) Every non-empty subset with an upper bound has a least upper bound.

The *intervals* are the subsets

$$[a, b] := \{x \in \mathbb{R} : a \leqslant x \leqslant b\}, \quad]a, b] := \{x \in \mathbb{R} : a < x \leqslant b\},$$
$$[a, b[:= \{x \in \mathbb{R} : a \leqslant x < b\}, \quad]a, b[:= \{x \in \mathbb{R} : a < x < b\},$$
$$[a, \infty[:= \{x \in \mathbb{R} : a \leqslant x\}, \qquad]a, \infty[:= \{x \in \mathbb{R} : a < x\},$$
$$]-\infty, a] := \{x \in \mathbb{R} : x \leqslant a\}, \qquad]-\infty, a[:= \{x \in \mathbb{R} : x < a\},$$
$$]-\infty, \infty[:= \mathbb{R}$$

where $a < b$ are fixed real numbers. The real numbers satisfy the *Archimedean property*

$$\forall x > 0, \exists n \in \mathbb{N}, \quad x < n.$$

The proof is simple: If the set $\mathbb{N}$ had an upper bound in $\mathbb{R}$ then it would have a least upper bound α; by definition, this implies that $\alpha - 1$ is not an upper bound, meaning there is a number $n \in \mathbb{N}$ such that $n > \alpha - 1$; yet α is an upper bound, so $n + 1 \leqslant \alpha$. This contradiction shows that no $x \in \mathbb{R}$ is an upper bound of $\mathbb{N}$: there is an $n \in \mathbb{N}$ such that $n > x$.

The Axiom of Choice

There is an important set principle, called the *axiom of choice*, that is not usually covered in elementary mathematics textbooks:

If $\mathcal{A} = \{ A_i : i \in I \}$ is a collection of non-empty subsets of a set X, then there is a function $f : I \to X$ such that $f(i) \in A_i$.

That is, this 'choice' function selects an element from each of the sets A_i, where the index i ranges over some set I. The Axiom of Choice is often used to create a sequence $(x_n)_{n \in \mathbb{N}}$ from a given list of non-empty sets A_n, with $x_n \in A_n$. It seems obvious that if a set is non-empty then an element of it can be selected, but the existence of such a procedure for arbitrary collections of sets cannot be proved from the other standard set axioms.

Part I
Metric Spaces

Chapter 2
Distance

Metric spaces can be thought of as very basic spaces, with only a few axioms, where the ideas of *convergence* and *continuity* exist. We wish to understand what it means in general for x_n to converge to x, whether they are real numbers, vectors, matrices or functions. One fundamental ingredient that makes these concepts rigorous is that of a *distance*, also called a *metric*, which is a measure of how close elements are to each other.

Definition 2.1

A **distance** (or **metric**) on a **metric space** X is a function

$$d : X^2 \to \mathbb{R}^+$$
$$(x, y) \mapsto d(x, y)$$

such that the following properties (called the *axioms* of a metric space) hold for all $x, y, z \in X$,

(i) $d(x, y) \leqslant d(x, z) + d(z, y)$ Triangle Inequality,

(ii) $d(y, x) = d(x, y)$ Symmetry,

(iii) $d(x, y) = 0 \Leftrightarrow x = y$ Distinguishability.

A metric space is not just a set, in which the elements have no relation to each other, but a set X equipped with a particular structure, its distance function d. One can emphasize this by denoting the metric space by the pair (X, d), although it is more convenient to denote different metric spaces by different symbols such as X, Y, etc.

Maurice Fréchet (1878–1973) Fréchet studied under Hadamard (who had proved the prime number theorem and had succeeded Poincaré) and Borel at the University of Paris (École Normale Supérieure); his 1906 thesis developed "abstract analysis", an axiomatic approach to abstract functions that allows the Euclidean concepts of convergence and distance, as well as the usual algebraic operations, to be applied to functions. Many terms, such as metric space, completeness, compactness etc., are due to him.

In what follows, X will denote any abstract set with a distance, not necessarily $\mathbb{R}$ or $\mathbb{R}^n$, although these are of the most immediate interest. We still call its elements "points", even if in reality they are geometric points, sequences, or functions. What matters, as far as metric spaces are concerned, is not the internal structure of its points, but their outward relation to other points.

Although most distance functions treated in this book are of the type $d(x, y) = |x - y|$, as for $\mathbb{R}$, the point of studying metric spaces in more generality is not only that there are some exceptions that don't fit this type, but also to emphasize that addition/subtraction is not essential, as well as to prepare the groundwork for even more general spaces, called *topological spaces*, in which pure convergence is studied without reference to distances (but which are not covered in this book).

There are two additional axioms satisfied by some metric spaces that merit particular attention: *complete* metrics, which guarantee that their Cauchy sequences converge, and *separable* metric spaces whose elements can be handled by approximations. Both properties are possessed by *compact* metric spaces, which is what is often meant when the term "finite" is applied in a geometric sense. These are considered in later chapters.

Easy Consequences

1. $d(x, z) \geqslant |d(x, y) - d(z, y)|$.
2. If $x_1, \ldots, x_n$ are points in X, then by induction on n,

$$d(x_1, x_n) \leqslant d(x_1, x_2) + \cdots + d(x_{n-1}, x_n).$$

Examples 2.2

1. The spaces $\mathbb{N}$, $\mathbb{Z}$, $\mathbb{Q}$, $\mathbb{R}$, and $\mathbb{C}$ have the standard distance $d(a, b) := |a - b|$. Check that the three axioms for a distance are satisfied, making use of the in/equalities $|s + t| \leqslant |s| + |t|$, $|-s| = |s|$, and $|s| = 0 \Leftrightarrow s = 0$.
2. ▶ The vector spaces $\mathbb{R}^n$ and $\mathbb{C}^n$ have the standard *Euclidean* distance defined by $d(x, y) = \|x - y\| := \sqrt{\sum_{i=1}^n |a_i - b_i|^2}$ for $x = (a_1, \ldots, a_n)$, $y = (b_1, \ldots, b_n)$ (prove this for $n = 2$).

3. One can define distances on more general spaces. For example, we will later show that the space of real continuous functions with domain $[0, 1]$ has a distance defined by $d(f, g) := \max_{x \in [0,1]} |f(x) - g(x)|$.

4. ◇ The space of 'shapes' in $\mathbb{R}^2$ (roughly speaking, subsets that have an area) have a metric $d(A, B)$ defined as the area of the symmetric difference $A \triangle B$.

5. ▶ Any subset of a metric space is itself a metric space (with the 'inherited' or 'induced' distance). (The three axioms are such that they remain valid for points in a subset of a metric space.)

6. ▶ The product of two metric spaces, $X \times Y$, can be given several distances, none of which have a natural preference. Two of them are the following

$$D_1 \left(\left(\begin{smallmatrix} x_1 \\ y_1 \end{smallmatrix} \right), \left(\begin{smallmatrix} x_2 \\ y_2 \end{smallmatrix} \right) \right) := d_X(x_1, x_2) + d_Y(y_1, y_2),$$

$$D_\infty \left(\left(\begin{smallmatrix} x_1 \\ y_1 \end{smallmatrix} \right), \left(\begin{smallmatrix} x_2 \\ y_2 \end{smallmatrix} \right) \right) := \max(d_X(x_1, x_2), d_Y(y_1, y_2)).$$

For convenience, we choose D_1 as our standard metric for $X \times Y$, except for $\mathbb{R}^n$ and $\mathbb{C}^n$, for which we take the Euclidean distance.

Proof for D_1: Positivity of D_1 and axiom (ii) are obvious. To prove axiom (iii), $D_1(x_1, x_2) = 0$ implies $d_X(x_1, x_2) = 0 = d_Y(y_1, y_2)$, so $x_1 = x_2$, $y_1 = y_2$, and $x_1 = \left(\begin{smallmatrix} x_1 \\ y_1 \end{smallmatrix} \right) = \left(\begin{smallmatrix} x_2 \\ y_2 \end{smallmatrix} \right) = x_2$. As for the triangle inequality,

$$D_1(x_1, x_2) = d_X(x_1, x_2) + d_Y(y_1, y_2)$$
$$\leqslant d_X(x_1, x_3) + d_X(x_3, x_2) + d_Y(y_1, y_3) + d_Y(y_3, y_2)$$
$$= D_1(x_1, x_3) + D_1(x_3, x_2).$$

Exercises 2.3

1. Show that

(a) If $d(x, z) > d(z, y)$ then $x \neq y$;

(b) For any z, $\frac{1}{2}d(x, y) \leqslant \max(d(x, z), d(y, z))$;

(c) If $d(u, v) = d(u, x) + d(x, v)$ then either $d(x, u) \leqslant d(y, u)$ or $d(x, v) \leqslant d(y, v)$.

2. Write in mathematical language,

(a) The subsets A, B are close to within 2 distance units;

(b) A and B are arbitrarily close.

3. The set of bytes, i.e., sequences of 0s and 1s (bits) of length 8 (or any length), has a "Hamming distance" defined as the number of bits where two bytes differ; e.g. the Hamming distance between 10010111 and 11001101 is 4.

4. Any non-empty set can be given a distance function. The simplest is the *discrete* metric $d(x, y) := \begin{cases} 1, & x \neq y \\ 0, & x = y \end{cases}$. Indeed, there are infinitely many other metrics

on the same set (except when there is only one point!); for example, if d is a distance function then so are $2d$ and $d/(1 + d)$.

($\star$ Not every function of d will do though! The function d^2 is not generally a metric; what properties does $f : \mathrm{im}\, d \to \mathbb{R}^+$ need to have in order that $f \circ d$ also be a metric?)

5. A set may have several distances defined on it, but each has to be considered as a different metric space. For example, the set of positive natural numbers has a distance defined by $d(m, n) := |1/m - 1/n|$ (prove!); the metric space associated with it has very different properties from $\mathbb{N}$ with the standard Euclidean distance. For example, in this space, distinct natural numbers come arbitrarily close to each other.

6. Let $n = \pm 2^k 3^r \cdots$ be the prime decomposition of any integer $n \in \mathbb{Z}$ and define $|n|_2 := 1/2^k$, $|0|_2 := 0$. Show that $|\cdot|_2$ satisfies the same properties as the standard absolute value and hence that $d(m, n) := |m - n|_2$ is a distance on $\mathbb{Z}$ (called the 2-adic metric).

7. $\star$ Given the distances between n points in $\mathbb{R}^n$, can their positions be recovered? Can their relative positions be recovered?

2.1 Balls and Open Sets

The distance function provides an idea of the "surroundings" of a point. Given a point a and a number $r > 0$, we can distinguish between those points 'near' to it, satisfying $d(x, a) < r$, and those that are not.

Definition 2.4

An (open) **ball**, with *center a* and *radius r > 0*, is the set

$$B_r(a) := \{ x \in X : d(x, a) < r \}.$$

Despite the name, we should lay aside any preconception we may have of it being "round" or symmetric. We are now ready for our first, simple, proposition:

Proposition 2.5

Distinct points of a metric space can be separated by disjoint balls,

$$x \neq y \quad \Rightarrow \quad \exists r > 0 \quad B_r(x) \cap B_r(y) = \emptyset.$$

Proof If $x \neq y$ then $d(x, y) > 0$ by the distinguishability axiom (iii). Letting $r := d(x, y)/2$, then $B_r(x)$ is disjoint from $B_r(y)$ else we get a contradiction,

$$z \in B_r(x) \cap B_r(y) \Rightarrow d(x, z) < r \text{ AND } d(y, z) < r$$
$$\Rightarrow d(x, y) \leqslant d(x, z) + d(y, z)$$
$$< 2r = d(x, y).$$

$\square$

Examples 2.6

1. In $\mathbb{R}$, a ball is an *open* interval

$$B_r(a) = \{x \in \mathbb{R} : |x - a| < r\} = \,]a - r, a + r[.$$

Conversely, any open interval of type $]a, b[$ is a ball in $\mathbb{R}$, namely $B_{|b-a|/2}(\frac{a+b}{2})$.

2. In $\mathbb{R}^2$, the ball $B_r(a)$ is the disk with center a and radius r without the circular perimeter.
3. In $\mathbb{Z}$, $B_2(m) = \{n \in \mathbb{Z} : |n - m| < 2\} = \{m - 1, m, m + 1\}$ and $B_1(m) = \{m\}$.
4. It is clear that balls differ depending on the context of the metric space; thus $B_{1/2}(0) = \,]-\frac{1}{2}, \frac{1}{2}[$ in $\mathbb{R}$, but $B_{1/2}(0) = \{0\}$ in $\mathbb{Z}$.

Open Sets

We can use balls to explore the relation between a point x and a given subset A. As the radius of the ball $B_r(x)$ is increased, it is certain to include some points which are in A and some points which are not, unless $A = X$ or $A = \varnothing$. So it is more interesting to investigate what can happen when the radius is small. There are three possibilities as r is decreased: either $B_r(x)$ eventually contains (i) *only* points of A, or (ii) *only* points in its complement A^c, or (iii) points of *both* A and A^c, no matter how small we take r.

Definition 2.7

A point x of a set A is called an **interior** point of A when it can be "surrounded completely" by points of A, i.e.,

$$\exists r > 0, \quad B_r(x) \subseteq A.$$

In this case, A is also said to be a *neighborhood* of x.
A point x (not in A) is an **exterior** point of A when

$$\exists r > 0, \quad B_r(x) \subseteq X \setminus A.$$

(continued)

All other points of X are called **boundary** points of A (see Fig. 2.1), that is when

$$\forall r > 0, \ \exists a, b \in B_r(x), \quad a \in A \text{ AND } b \in A^c.$$

Accordingly, the set A partitions X into three parts: its *interior* $A°$, its *exterior* $(\bar{A})^c$, and its *boundary* ∂A. The set of interior and boundary points of A is called the *closure* of A and denoted by $\bar{A} := A° \cup \partial A$.

A subset A is **open** in X when all its points are interior points of it, i.e., $A = A°$.

Examples 2.8

1. In $\mathbb{R}$, the intervals $]a, b[$, $[a, b[$, $]a, b]$, and $[a, b]$ have the same interior $]a, b[$, exterior, and boundary $\{a, b\}$; their closure is $\overline{]a, b[} = [a, b]$.

   Proof: For any $a < x < b$, let $0 < \epsilon < \min(x - a, b - x)$, then $a < x - \epsilon$ and $x + \epsilon < b$, that is, $B_\epsilon(x) =]x - \epsilon, x + \epsilon[\subset]a, b[$; this makes x an interior point of the interval.

   For $x < a$, take any $\epsilon < a - x$ so that $x \in B_\epsilon(x) \subset]-\infty, a[\subset \mathbb{R} \setminus [a, b]$. Similarly, any $x > b$ is an exterior point of the interval.

   For $x = a$, any small interval $B_\epsilon(a)$ contains points such as $a + \epsilon/2$, that are inside the interval, and points outside it, such as $a - \epsilon/2$, making a (and similarly b) a boundary point.

2. ▶ The following subsets are open in *any* metric space X:

   (a) $X \setminus \{x\}$ for any point x. The reason is that any other point $y \neq x$ is separated from x by disjoint balls (our first proposition); this makes y an interior point of $X \setminus \{x\}$.

   (b) The empty set is open by default, because it does not contain any point. The whole space X is also open because $B_r(x) \subseteq X$ for any $r > 0$ and $x \in X$.

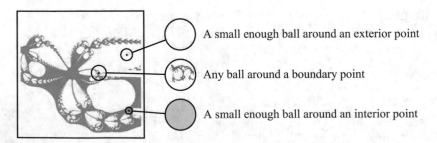

Fig. 2.1 The distinction between interior, boundary, and exterior points

(c) Balls are open sets in any metric space.

> *Proof*: Let $x \in B_r(a)$ be any point in the given ball, meaning $d(x, a) < r$.
> Let $\epsilon := r - d(x, a) > 0$; then $B_\epsilon(x) \subseteq B_r(a)$ since for any $y \in B_\epsilon(x)$,
>
> $$d(y, a) \leqslant d(y, x) + d(x, a) < \epsilon + d(x, a) = r.$$

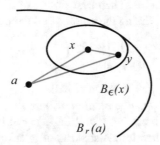

3. ▶ The least upper bound of a set A in $\mathbb{R}$ is a boundary point of it.
   Proof: Let α be the least upper bound of A. For any $\epsilon > 0$, $\alpha + \epsilon/2$ is an upper
   bound of A but does not belong to it (else α would not be an upper bound).
   Even if $\alpha \notin A$, then the interval $]\alpha - \epsilon/2, \alpha[$ cannot be devoid of elements of
   A, otherwise α would not be the *least* upper bound. So the neighborhood $B_\epsilon(\alpha)$
   contains elements of both A and A^c.

Proposition 2.9

The set of interior points A° is the largest open set inside A.

Proof If $B \subseteq A$ then the interior points of B are obviously interior points of A,
so $B^\circ \subseteq A^\circ$. In particular every open subset of A lies inside A° (because then
$B = B^\circ$), and every (open) ball in A lies in A°. This implies that if $B_r(x) \subseteq A$ then
$B_r(x) \subseteq A^\circ$, so that every interior point of A is surrounded by other interior points,
and A° is open.

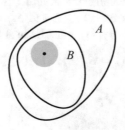

□

Proposition 2.10

> **A set A is open $\Leftrightarrow$ A is a union of balls.**

Proof Let A be an open set. Then every point of it is interior, and can be covered by a ball $B_{r(x)}(x) \subseteq A$. Taking the union of all the points of A gives

$$A = \bigcup_{x \in A} \{x\} \subseteq \bigcup_{x \in A} B_{r(x)}(x) \subseteq A,$$

forcing $A = \bigcup_{x \in A} B_{r(x)}(x)$, a union of balls.

Now let $A := \bigcup_i B_{r_i}(a_i)$ be a union of balls, and let x be any point in A. Then x is in at least one of these balls, say, $B_r(a)$. But balls are open and hence $x \in B_\epsilon(x) \subseteq B_r(a) \subseteq A$. Therefore A consists of interior points and so is open. □

The early years of research in metric spaces have shown that most of the basic theorems about metric spaces can be deduced from the following characteristic properties of open sets:

Theorem 2.11

> **Any union of open sets is open.**
> **The finite intersection of open sets is open.**

Proof (i) Consider the union of open sets, $\bigcup_i A_i$. Any $x \in \bigcup_i A_i$ must lie in at least one of the open sets, say A_j. Therefore,

$$x \in B_r(x) \subseteq A_j \subseteq \bigcup_i A_i$$

shows that it must be an interior point of the union.

(ii) It is enough, using induction (show!), to consider the intersection of two open sets $A \cap B$. Let $x \in A \cap B$, meaning $x \in A$ and $x \in B$, with both sets being open. Therefore there are open balls $B_{r_1}(x) \subseteq A$ and $B_{r_2}(x) \subseteq B$. The smaller of these two balls, with radius $r := \min(r_1, r_2)$, must lie in $A \cap B$,

$$x \in B_r(x) = B_{r_1}(x) \cap B_{r_2}(x) \subseteq A \cap B.$$

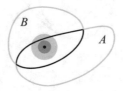

□

Relatively Open Sets

The interior, boundary, and exterior of a subset depend on the metric space under consideration. Changing the space may change whether a subset is open or not, even though its elements remain the same. We thus need to specify that a subset A is open *in* X. For example, the singleton $\{0\}$ is not open in $\mathbb{R}$ nor in $\mathbb{Q}$ but it is open in $\mathbb{Z}$; in other words, if one takes any ball of small enough radius around 0, one is sure to find non-zero real or rational numbers, but no other integers. Thus, when the space is $\mathbb{R}$, the interior of $\{0\}$ is empty and the boundary is $\{0\}$; when the space is $\mathbb{Z}$, the interior is $\{0\}$ and there is no boundary. Similarly, the interval $]a, b[$ is open in $\mathbb{R}$, but not open when considered as a subset of the x-axis in $\mathbb{R}^2$.

More tricky examples arise if the metric space is already a subspace. Fortunately, if one is familiar with the open sets in a bigger space such as $\mathbb{R}$ or $\mathbb{R}^2$, then the following proposition gives an immediate criterion for deducing the open subsets of a subspace of it. In words, if $Y \subseteq X$ then a subset A is open in Y when it can be extended to an open set U in X.

Proposition 2.12

> **Let Y be a subspace of X. Then a subset A is open in Y if, and only if, $A = U \cap Y$ for some subset U open in X.**

Proof Let $Y \subseteq X$ inherit X's distance. Care must be taken to distinguish balls in Y, considered as a metric space in its own right, from those in X

$$B_r^Y(x) = \{ y \in Y : d(y, x) < r \} = \{ y \in X : d(y, x) < r \text{ AND } y \in Y \}$$
$$= B_r^X(x) \cap Y.$$

If A is open in Y, then by Proposition 2.10, it is the union of balls of Y,

$$A = \bigcup_{a \in A} B_{r(a)}^Y(a) = \bigcup_{a \in A} B_{r(a)}^X(a) \cap Y = U \cap Y,$$

where $U := \bigcup_{a \in A} B_{r(a)}^X(a)$ is open in X.

For the converse, points in A are those points of an open set $U \subseteq X$ which happen to be in Y, and so are interior points of A as a subset of Y,

$$y \in B_r^X(y) \subseteq U \implies y \in B_r^X(y) \cap Y \subseteq U \cap Y = A,$$

that is, $U \cap Y$ is open in Y. □

The same considerations apply to exterior points and to boundary points. One has to be careful to interpret the definitions correctly, in particular substituting $Y \setminus A$

instead of $X \smallsetminus A$. A point is exterior to A in Y when there is an $r > 0$ such that $B_r(x) \subseteq Y \smallsetminus A$. The boundary of a subset A relative to a subspace Y is often denoted by $\partial_Y(A)$, and may differ from its boundary relative to the parent space X.

$$\partial_Y(A) = \{ x \in Y : \forall r > 0, \ \exists a \in A, \exists b \in Y \smallsetminus A, \ a, b \in B_r(x) \cap Y \}$$

Note that this boundary $\partial_Y(A)$ is not necessarily $\partial_X(A) \cap Y$. All that can be said in this regard is that, in general, $\partial_Y(A) \subseteq \partial_X(A)$.

Examples 2.13

1. ▶ The exterior $(\bar{A})^{\complement} = (A^{\complement})^{\circ}$ of a subset A is open in X.
2. $A^{\circ} = A \smallsetminus \partial A$. So a set is open iff it does not contain any of its boundary points.
3. (a) Let $X := \mathbb{R}$ and $Y := [0, 2]$; then the subset $A := [0, 1[$ is open in Y: a ball about $x = 0$ in the space $[0, 2]$ has the form $[0, r[$, and this is therefore open by Example 2.8.(2c). Alternatively, and simpler, $A = U \cap Y$ where $U =]-1, 1[$ is open in $\mathbb{R}$. Its boundary in Y is $\partial_Y(A) = \{1\}$ whereas $\partial_X(A) = \{0, 1\}$.
   (b) In the space $Y := [0, 1] \cup [2, 3] \subset \mathbb{R}$, the subset $[0, 1]$ is open in Y and has no boundary in it. Why isn't 1 a boundary point of it? Because the ball $B_{\epsilon}^{Y}(1) =]1 - \epsilon, 1]$ surrounds it completely in Y; the other points of $\mathbb{R}$ are nonexistent as far as Y is concerned. Alternatively, $[0, 1] =]-\frac{1}{2}, 1\frac{1}{2}[\cap Y$.
   (c) For a more extreme example, consider the subspace $\mathbb{Q}$ of $\mathbb{R}$. Then $\partial_{\mathbb{Q}}(\mathbb{Q}) = \varnothing$, whereas $\partial_{\mathbb{R}}(\mathbb{Q}) = \mathbb{R}$. If our world were rational, then each rational point would be surrounded by rationals only, but in the real world, each rational is close to irrationals.
4. In the unit square $Y := [0, 1]^2 \subset \mathbb{R}^2$, the open balls are either disks themselves (when the radius is small enough) or they are the intersection of a disk with Y. The subset $A := [0, \frac{1}{2}]^2$ has interior $[0, \frac{1}{2}[^2$ and a boundary consisting of the lines $\{ (t, \frac{1}{2}) : 0 \leqslant t \leqslant \frac{1}{2} \} \cup \{ (\frac{1}{2}, t) : 0 \leqslant t \leqslant \frac{1}{2} \}$. The origin is not a boundary point but an interior point of A in Y.

Exercises 2.14

1. In $\mathbb{R}$, the set $\{a\}$ has no interior points, a single boundary point a, and all other points are exterior. It is not an open set in $\mathbb{R}$. There are ever smaller open sets that contain a, but there is no smallest one.
2. In $\mathbb{R}$, $\overline{\{ 1/n : n \in \mathbb{N} \}} = \{ 1/n : n \in \mathbb{N} \} \cup \{0\}$.
3. The set $\mathbb{Q}$, and also its complement $\mathbb{Q}^{\complement}$, the set of irrational numbers, do not have interior (or exterior) points in $\mathbb{R}$. Every real number is a boundary point of $\mathbb{Q}$.
   Similarly every complex number is a boundary point of $\mathbb{Q} + i\mathbb{Q}$.
4. The set $\{m\}$ does not have any boundary points in $\mathbb{Z}$; it is an open set in $\mathbb{Z}$ ($B_{1/2}(m) = \{m\}$).
5. Of the proper intervals in $\mathbb{R}$, only $]a, b[$, $]a, \infty[$, and $]-\infty, a[$ are open.
6. In $\mathbb{R}^2$, the half-plane $\{ (x, y) \in \mathbb{R}^2 : y > 0 \}$ and the rectangles $]a, b[\times]c, d[:= \{ (x, y) \in \mathbb{R}^2 : a < x < b, c < y < d \}$ are open sets.

7. Describe the interior, boundary and exterior of the sets

$$\{(x, y) \in \mathbb{R}^2 : |x| + |y| \leqslant 1\}, \qquad \{(x, y) \in \mathbb{R}^2 : \tfrac{1}{2} < \max(|x|, |y|) \leqslant 1\},$$

   in (i) $\mathbb{R}^2$, (ii) the first quadrant $(\mathbb{R}^+)^2$.

8. ▶ A^c has the same boundary as A; its interior is the exterior of A, that is, $(\bar{A})^c = (A^c)^\circ$ (and $\bar{A} = A^{c\circ c}$); so $\partial A = \bar{A} \cap \overline{A^c}$.

9. Find an open subset of $\mathbb{R}$, apart from $\mathbb{R}$ itself, without an exterior.
   So, the exterior of the exterior of A need not be the interior of A. Similarly, the boundary of $\bar{A}$ or A° need not equal the boundary of A.

10. ▶ *An infinite intersection of open sets need not be open.* For example, in $\mathbb{R}$, the open intervals $]-1/n, 1/n[$ are nested one inside another. Their intersection is the non-open set $\{0\}$. Find another example in $\mathbb{R}^2$.

11. Deduce from Theorem 2.11 that if every singleton $\{x\}$ is open in X, then *every* subset of X is open in X. This 'extreme' property is satisfied by $\mathbb{N}$, $\mathbb{Z}$, and any discrete metric space.

12. Any point x with $d(x, a) > r$ is in the exterior of the open ball $B_r(a)$. But the boundary of $B_r(a)$ need not be the set $\{x : d(x, a) = r\}$. Illustrate this by an example in $\mathbb{Z}$.

13. ⋆ Every open set in $\mathbb{R}$ is a countable disjoint union of open intervals. (Hint: An open set in $\mathbb{R}$ is the disjoint union of open intervals; take a rational interior point for each.)
   In contrast to this simple case, the open sets in $\mathbb{R}^2$, say, can be much more complicated—there is no simple characterization of them, apart from the definition.

2.2 Closed Sets

An open set is one that does not contain its boundary points; the dual concept is that of a closed set, one that contains all its boundary points. Logically speaking, the terms "open" and "closed" are not mutually exclusive because a set may possibly not have any boundary points; in a sense, they are misnomers, but they have stuck in the literature, being derived from the earlier use of "open/closed intervals".

Definition 2.15

A subset F is **closed** in a space X when $X \smallsetminus F$ is open in X.

Proposition 2.16

A subset F is closed $\Leftrightarrow$ F contains its boundary $\Leftrightarrow$ $\bar{F} = F$.

Proof We have already seen that the boundary of a set F and of its complement F^c are the same (because the interior of F^c is the exterior of F). So F is closed, and F^c open, precisely when this common boundary does not belong to F^c, but belongs instead to $F^{cc} = F$.   □

Examples 2.17

1. In $\mathbb{R}$, the set $[a, b]$ is closed, since $\mathbb{R} \setminus [a, b] = \,]-\infty, a[\, \cup \,]b, \infty[$ is the union of two open sets, hence itself open. Similarly $[a, \infty[$ and $]-\infty, a]$ are closed in $\mathbb{R}$.
2. $\mathbb{N}$ and $\mathbb{Z}$ are closed in $\mathbb{R}$, but $\mathbb{Q}$ is not.
3. ▶ In any metric space X, the following sets are closed in X (by inspecting their complements):

   (a) the singleton sets $\{x\}$,
   (b) the 'closed balls' $B_r[a] = \{x \in X : d(x, a) \leqslant r\}$; it contains, but need be equal to, $\overline{B_r(a)}$;
   (c) X and $\varnothing$,
   (d) the boundary of any subset (the complement of ∂A is $A^\circ \cup (A^c)^\circ$).

4. ▶ The complement of an open set is closed. More generally, if U is an open set and F a closed set in X, then $U \setminus F$ is open and $F \setminus U$ is closed. The reasons are that $U \setminus F = U \cap F^c$ and $(F \setminus U)^c = F^c \cup U$.

Closed sets are complements of open sets, and their properties reflect this:

Proposition 2.18

The finite union of closed sets is closed.
Any intersection of closed sets is closed.

Proof These are the complementary results for open sets (Theorem 2.11). For F, G closed sets in X, the subsets F^c, G^c are open, so the result follows from

$$(F \cup G)^c = F^c \cap G^c, \quad \left(\bigcap_i F_i\right)^c = \bigcup_i F_i^c,$$

and the definition that the complement of a closed set is open.   □

Theorem 2.19 (Kuratowski's Closure 'Operator')

The closure of a subset, $\bar{A}$, is the smallest closed set containing A.

$$A \subseteq B \Rightarrow \bar{A} \subseteq \bar{B}, \quad \bar{\bar{A}} = \bar{A}, \quad \overline{A \cup B} = \bar{A} \cup \bar{B}.$$

Proof The complement of $\bar{A}$ is the exterior of A, which is an open set, so $\bar{A}$ is closed. This implies $\bar{\bar{A}} = \bar{A}$ by Proposition 2.16.

If $A \subseteq B$, then an exterior point of B is obviously an exterior point of A, that is $(\bar{B})^c \subseteq (\bar{A})^c$; so $\bar{A} \subseteq \bar{B}$. It follows that if F is any closed set that contains A, then $\bar{A} \subseteq \bar{F} = F$, and this shows that $\bar{A}$ is the smallest closed set containing A. (Alternatively, Proposition 2.9 can be used: how?)

Of course, $\bar{A} \subseteq \overline{A \cup B}$ follows from $A \subseteq A \cup B$; combined with $\bar{B} \subseteq \overline{A \cup B}$, it gives $\bar{A} \cup \bar{B} \subseteq \overline{A \cup B}$. Moreover, $\bar{A} \cup \bar{B}$ is a closed set which contains $A \cup B$, and so must contain its closure $\overline{A \cup B}$. □

Exercises 2.20

1. It is easy to find sets in $\mathbb{R}$ which are neither open nor closed (so contain only part of their boundary). Can you find any that are both open and closed?
2. The set $\{x \in \mathbb{Q} : x^2 < 2\}$ is closed, and open, in $\mathbb{Q}$.
3. In any metric space, a finite collection of points $\{a_1, \ldots, a_n\}$ is a closed set.
4. The following sets are closed in $\mathbb{R}$: $[0, 1] \cup \{5\}$, $\bigcup_{n=0}^{\infty}[n, n + \frac{1}{2}]$.
5. The infinite union of closed sets may, but need not, be closed. For example, the set $\bigcup_{n=1}^{\infty}\{\frac{1}{n}\}$ is not closed in $\mathbb{R}$; which boundary point is not contained in it?
6. Find two disjoint closed sets (in $\mathbb{R}^2$ or $\mathbb{Q}$, say) that are arbitrarily close to each other.
7. Start with the closed interval $[0, 1]$; remove the open middle interval $]\frac{1}{3}, \frac{2}{3}[$ to get two closed intervals $[0, \frac{1}{3}] \cup [\frac{2}{3}, 1]$. Remove the middle interval of each of these intervals to obtain four closed intervals $[0, \frac{1}{9}] \cup [\frac{2}{9}, \frac{1}{3}] \cup [\frac{2}{3}, \frac{7}{9}] \cup [\frac{8}{9}, 1]$. If we continue this process indefinitely we end up with the *Cantor set*. Show it is a closed subset of $\mathbb{R}$.

8. Denote the decimal expansion of any number in $[0,1]$ by $0.n_1 n_2 n_3 \ldots$. Show that the set

$$\{x \in [0, 1] : x = 0.n_1 n_2 n_3 \ldots \Rightarrow \forall k, \frac{n_1 + \cdots + n_k}{k} \leqslant 5\}$$

is closed in $\mathbb{R}$.
9. ▶ One can define the "distance" between a *point* and a *subset* of a metric space by $d(x, A) := \inf_{a \in A} d(x, a)$. Then $x \in \bar{A}$ exactly when $d(x, A) = 0$.
10. ⋆ More generally, the Hausdorff distance between two subsets is defined to be

$$d(A, B) := \sup_{a \in A} d(a, B) + \sup_{b \in B} d(b, A).$$

11. Let x be an exterior point of A, and let $y \in \bar{A}$ have the least distance between x and $\bar{A}$. Do you think that y is unique? or that it must be on the boundary of A? Prove or disprove. For starters, take the metric space to be $\mathbb{R}^2$.

12. Show $\overline{A \cap B} \subseteq \bar{A} \cap \bar{B}$ and prove that equality need not hold. Indeed, two disjoint sets may 'touch' at a common boundary point.
13. Show the complementary results of the theorem: $A° \cap B° = (A \cap B)°$, $A°° = A°$.
14. If $A \subseteq \bar{B}$, does it follow that $A° \subseteq B$?

Limit Points and Dense Subsets

When dealing with subsets, we can see that some of their points are clustered within the rest, while other single points are separated or unattached to the rest; some are evenly spread out in space, others are sparse. Think of the difference between the integers and the rational numbers in $\mathbb{R}$. The concepts introduced before, of balls, open sets, and closure, allow us to formulate these ideas rigorously.

Definition 2.21

> A point a in a set A is an **isolated point** when there is a ball which contains no points of A other than itself,
>
> $$\exists \epsilon > 0, \quad B_\epsilon(a) \cap A = \{a\}.$$
>
> A point b (not necessarily in A) is a **limit point** of a set A when every ball around it contains other points of A,
>
> $$\forall \epsilon > 0, \quad \exists a \neq b, \quad a \in A \cap B_\epsilon(b).$$

Thus a limit point cannot be isolated from the rest of A. Every point of $\bar{A}$ is either a limit point or an isolated point of A, so a set that contains its limit points is closed.

We often need to approximate an element $x \in X$ to within some small distance ϵ by an element from some special subset $A \subseteq X$. The elements of A may be simpler to describe, or more practical to work with, or may have nicer theoretical qualities. For example, computers cannot handle arbitrary real numbers and must approximate them by rational ones; polynomials are easier to work with than general continuous functions. The property that elements of a set A can be used to approximate elements of X to within any ϵ, namely,

$$\forall x \in X, \ \forall \epsilon > 0, \ \exists a \in A, \ d(x, a) < \epsilon,$$

is equivalent to saying that any ball $B_\epsilon(x)$ contains elements of A, in other words there are no points exterior to A, i.e., $\bar{A} = X$.

Definition 2.22

A set A is **dense** in X when $\bar{A} = X$ (so $\bar{A}$ contains all balls).
A set A is **nowhere dense** in X when $\bar{A}$ contains no balls.

Exercises 2.23

1. Can a set *not* have limit points? Can an infinite set not have limit points?
2. In $\mathbb{R}$, the set of integers $\mathbb{Z}$ has no limit points, but all real numbers are limit points of $\mathbb{Q}$.
3. (a) 1 is an interior isolated point of $\{1, 2\}$ in $\mathbb{Z}$;
   (b) 1 is a boundary isolated point of $\{1, 2\}$ in $\mathbb{R}$;
   (c) 1 is an interior limit point of $[0, 2]$ in $\mathbb{R}$;
   (d) 1 is a boundary limit point of $[0, 1]$ in $\mathbb{R}$.
4. In $\mathbb{R}$ and $\mathbb{Q}$, an isolated point of a subset must be a boundary point, or, equivalently, an interior point is a limit point.
5. ▶ $\mathbb{Q}$ is dense in $\mathbb{R}$. (This is equivalent to the Archimedean property of $\mathbb{R}$.) More generally, a set A is dense in $\mathbb{R}$ when for any two distinct real numbers $x < y$, there is an element $a \in A$ between them $x < a < y$.
6. The intersection of two open dense sets is again open and dense.
7. A finite union of straight lines in $\mathbb{R}^2$ is nowhere dense. $\mathbb{Z}$ and the Cantor set are nowhere dense in $\mathbb{R}$.
8. Nowhere dense sets have no interior points.
9. The complement of a nowhere dense set is dense. But the complement of a dense set need not be nowhere dense.
10. A is nowhere dense in X $\Leftrightarrow$ $X \smallsetminus \bar{A}$ is dense in X $\Leftrightarrow$ $\bar{A}$ is the boundary of an open set.

Remarks 2.24

1. If $d(x, y) = 0$ does not guarantee $x = y$, but d satisfies the other two axioms, then it is called a *pseudo-distance*. In this case, let us say that points x and y are *indistinguishable* when $d(x, y) = 0$ ($\Leftrightarrow$ $\forall z, d(x, z) = d(y, z)$). This is an equivalence relation, which induces a partition of the space into equivalence classes $[x]$. The function $D([x], [y]) := d(x, y)$ is then a legitimate well-defined metric.

   In a similar vein, if d satisfies the triangle inequality, but is not symmetric, then $D(x, y) := d(x, y) + d(y, x)$ is symmetric and still satisfies the triangle inequality.

   Positivity of d follows from axioms (i) and (ii) (unless $X = \{x, y\}$),

$$d(x, y) \geqslant |d(x, z) - d(y, z)| \geqslant 0.$$

2. The axioms for a distance can be re-phrased as axioms for balls:

   (a) $B_0(x) = \varnothing, \bigcap_{r>0} B_r(x) = \{x\}, \bigcup_{r>0} B_r(x) = X$,
   (b) $\{y : x \in B_r(y)\} = B_r(x)$,
   (c) $B_s \circ B_r(x) \subseteq B_{r+s}(x)$, i.e., if $y \in B_s(z)$ where $z \in B_r(x)$ then $y \in B_{r+s}(x)$.

3. The concept of open sets is more basic than that of distance. One can give a set X a collection of open subsets satisfying the properties listed in Theorem 2.11 (taken as axioms), and study them without any reference to distances. It is then called a *topological space*; most theorems about metric spaces have generalizations that hold for topological spaces. There are some important topological spaces that are not metric spaces, e.g. the arbitrary product of metric spaces $\prod_i X_i$, and spaces of functions $X^Y := \{f : Y \to X\}$.

Chapter 3
Convergence and Continuity

3.1 Convergence

The previous chapter was primarily intended to expand our vocabulary of mathematical terms in order to better describe and clarify the concepts that we will need. Our first task is to define *convergence*.

Definition 3.1

A sequence $(x_n)_{n \in \mathbb{N}}$ in a metric space X **converges** to a **limit** x, written

$$x_n \to x \text{ as } n \to \infty,$$

when

$$\forall \epsilon > 0, \quad \exists N, \quad n \geqslant N \Rightarrow x_n \in B_\epsilon(x).$$

A sequence which does not converge is said to *diverge* .

One may express this as "any neighborhood of x contains all the sequence from some point onwards," or "eventually, the sequence points get arbitrarily close to the limit".

Felix Hausdorff (1868–1942) Hausdorff studied atmospheric refraction in Bessel's school at Leipzig in 1891. In 1914, at 46 years of age in the University of Bonn, he published his major work on set theory, with chapters on partially ordered sets, measure spaces, topology and metric spaces, where he built upon Fréchet's abstract spaces, using open sets and neighborhoods. Later, in 1919, he introduced fractional dimensions. But in the late 1930s, increasing Nazi persecution made life impossible for him.

Proposition 3.2

> In a metric space, a sequence $(x_n)_{n \in \mathbb{N}}$ can only converge to one limit, denoted $\lim_{n \to \infty} x_n$.

Proof Suppose $x_n \to x$ and $x_n \to y$ as $n \to \infty$, with $x \neq y$. Then x and y can be separated by two disjoint balls $B_r(x)$ and $B_r(y)$ (Proposition 2.5). But convergence means

$$\exists N_1 \quad n \geqslant N_1 \Rightarrow x_n \in B_r(x),$$

$$\exists N_2 \quad n \geqslant N_2 \Rightarrow x_n \in B_r(y).$$

For $n \geqslant \max(N_1, N_2)$ this would result in $x_n \in B_r(x) \cap B_r(y) = \varnothing$, a contradiction.

$\square$

Examples 3.3

1. In $\mathbb{R}$, the definition of $a_n \to a$ reduces to $\forall \epsilon, \exists N, n \geqslant N \Rightarrow |a_n - a| < \epsilon$.
2. In any metric space, $x_n \to x \Leftrightarrow d(x_n, x) \to 0$ as $n \to \infty$ (because $x_n \in B_\epsilon(x) \Leftrightarrow d(x_n, x) < \epsilon$). For example, $x_n \to x$ when $d(x_n, x) \leqslant 1/n$ holds.
3. In $\mathbb{R}$, $n/(n+1) \to 1$ as $n \to \infty$, since for any ϵ, there is an N such that $1/N < \epsilon$ (Archimedean property of $\mathbb{R}$), so

$$n \geqslant N \Rightarrow \left| 1 - \frac{n}{n+1} \right| = \frac{1}{n+1} < \frac{1}{N} < \epsilon.$$

4. Given two convergent real sequences $a_n \to a$ and $b_n \to b$, then $a_n + b_n \to a + b$. *Proof*: For any $\epsilon > 0$, there are N_1, N_2, such that

$$n \geqslant N_1 \Rightarrow |a_n - a| < \epsilon, \qquad n \geqslant N_2 \Rightarrow |b_n - b| < \epsilon.$$

Thus for $n \geqslant \max(N_1, N_2)$,

$$|(a_n + b_n) - (a + b)| \leqslant |a_n - a| + |b_n - b| < 2\epsilon.$$

5. ► A sequence $\binom{x_n}{y_n}$ in $X \times Y$ converges to $\binom{x}{y}$ if, and only if, $x_n \to x$ and $y_n \to y$. *Proof*: Any distance in Example 2.2(6) can be used, but we will use the standard metric here. The distance between $\binom{x_n}{y_n}$ and $\binom{x}{y}$ is

$$\delta := d\left(\binom{x_n}{y_n}, \binom{x}{y}\right) = d(x_n, x) + d(y_n, y) \to 0, \text{ as } n \to \infty.$$

As both $d(x_n, x)$ and $d(y_n, y)$ are less than δ, the converse follows.

6. Consider a composition of functions $\mathbb{N} \to \mathbb{N} \to X$ where the first function is 1–1, and the second is a sequence. A *subsequence* is the case when the first function is strictly increasing, and a *rearrangement* is the case when it is 1–1 and onto. For example, $1, 1/4, 1/9, \ldots$ is a subsequence of $(1/n)$, and $1/2, 1, 1/4, 1/3, \ldots$ is a rearrangement. Any such 'sub-selection' of a convergent sequence also converges, to the same limit.

   Proof: Suppose $n \geqslant N \Rightarrow d(x_n, x) < \epsilon$. Let (x_{n_i}) be a sub-selection of (x_n). As $n_i \leqslant N$ can only be true of a finite number of indices i, with the largest being, say, M, it follows that

$$i > M \Rightarrow n_i > N \Rightarrow d(x_{n_i}, x) < \epsilon.$$

7. A sequence converges *fast* (or 'linearly') when $d(x_n, x) \leqslant Ac^n$ for some real constants $A > 0, 0 < c < 1$. *Quadratic* convergence, $d(x_n, x) \leqslant Ac^{2^n}$, is even faster. Instead, $1/n$ and $\sqrt[n]{2}$ converge *slowly*.

There are many questions in analysis of the type: If x_n has a property A, and $x_n \to x$, does x still have this property? For example, if a convergent sequence of vectors in the plane lies on a circle, will its limit also lie on the same circle? Or, can continuous functions (or differentiable, or integrable, etc.) converge to a discontinuous function? The following proposition answers this question in a general setting: the 'property' A needs to be closed in the metric space.

Proposition 3.4

If $x_n \in A$ and $x_n \to x$, then $x \in \bar{A}$.

Conversely, in a metric space, for any $x \in \bar{A}$ there is a sequence $x_n \in A$ which converges to x.

In particular, closed sets are "closed" under the process of taking the limit (since $\bar{A} = A$ by Proposition 2.16).

Proof Take any ball $B_\epsilon(x)$ about x. If x_n converges to x, then all the sequence points will be in the ball for n large enough. Since $x_n \in A$, x cannot be an exterior point, and so $\lim_{n\to\infty} x_n = x \in \bar{A}$.

For the converse, let $B_{1/n}(x)$ be a decreasing sequence of nested balls around $x \in \bar{A}$; whether x is a boundary or interior point of A, $B_{1/n}(x)$ contains at least a point a_n in A (which could be x itself). So $d(a_n, x) < 1/n \to 0$ as $n \to \infty$, and $a_n \to x$.   □

Examples 3.5

1. If $x_n \to x$ in a metric space X, and $x_n \neq x$ for all n, then x is a limit point of the set $\{x_1, x_2, x_3, \dots\}$. However, despite the name, the limit of a converging sequence need not be a limit point of its set of values. If x_n is eventually constant $(n \geqslant N \implies x_n = x)$, then $x_n \to x$ with x being an isolated point, not a limit point, of $\{x_n : n \in \mathbb{N}\}$. The confusion is ultimately caused by the fact that a sequence is a function, not the set of its values.

2. Several sequences appear to get close to more than one limit, e.g., $(-1)^n$ or e^{in}. These are not truly convergent sequences, by the proposition, but one can introduce a new concept, a *cluster point* of a sequence, to denote a point which the sequence gets arbitrarily close to infinitely many times, that is,

$$\forall \epsilon > 0, \ \forall n \in \mathbb{N}, \ \exists m \geqslant n, \ d(x_m, x) < \epsilon.$$

   Given any cluster point of a sequence, one can find a subsequence which converges to it. In general, any limit point of the values of a sequence is a cluster point of the sequence; an isolated point of the values is a cluster point only if it is visited infinitely many times.

3. If one were to list the rational numbers as a sequence, the result would have every point of $\mathbb{R}$ as a cluster point. At the other extreme, the sequence $(1, 2, 3, \dots)$ has no cluster points at all.

   If a real sequence $(a_n)_{n\in\mathbb{N}}$ has several cluster points, then the largest one, if it exists, is called its *limit superior*, denoted $\limsup_{n\to\infty} a_n$, and the smallest $\liminf_{n\to\infty} x_n$, the *limit inferior*.

Exercises 3.6

1. ▶ In $\mathbb{R}$,

   (a) $1/n \to 0$ (this is a rewording of the Archimedean property of the real numbers: for every $a > 0$, there is an $n \in \mathbb{N}$ such that $n > a$).
   (b) $a^n \to 0$ when $0 < a < 1$, but diverges for $a > 1$. (Hint: When $1 < a = 1 + \delta$, then $a^n = 1 + n\delta + \cdots > n\delta$; otherwise consider $1/a$.)
   (c) $n/a^n \to 0$ when $a > 1$, hence $n^k/a^n = (n/b^n)^k \to 0$.
   (d) $\sqrt[n]{a} \to 1$ for any $a > 0$, and $n^{1/n} \to 1$ (so $(\log n)/n \to 0$). (Hint: Assuming $a > 1$, expand $a^{1/n} =: 1 + a_n$ using the binomial theorem to show that $a_n < a/n \to 0$; similarly show $a_n^2 < 2/(n-1)$ for the second sequence.)

(e) $\star$ $(1 + 1/n)^n$ converges to a number denoted e. This is too hard to show for the moment. Show at least that the sequence is increasing but bounded by 3, using the binomial theorem. (This highlights the need of "convergence tests": how can one know that a sequence converges when the limit is unknown?)

(f) $\sqrt[n]{n!} \to \infty$ (what should this mean?)

2. What do the sequences $2 + \sqrt{2 + \sqrt{2 + \cdots}}$ and $1 + \dfrac{1}{1 + \frac{1}{1 + \cdots}}$ converge to, assuming they do?

3. In $\mathbb{R}$, if $a_n \to 0$ then $a_n^n \to 0$; find examples where (i) $a_n \to 0$ but $a_n^{1/n} \not\to 0$, (ii) $a_n \to 1$ but $a_n^n \not\to 1$.

4. ▶ If $a_n \leqslant b_n$ for two convergent real sequences then $\lim\limits_{n \to \infty} a_n \leqslant \lim\limits_{n \to \infty} b_n$ (Hint: $[0, \infty[$ is closed). In particular, if a_n converges and $a_n < a$, then $\lim\limits_{n \to \infty} a_n \leqslant a$.

5. *Squeezing principle*: In $\mathbb{R}$, if $a_n \leqslant x_n \leqslant b_n$ and $\lim\limits_{n \to \infty} a_n = a = \lim\limits_{n \to \infty} b_n$, then x_n converges (to a).

6. It is possible for a divergent sequence to have a convergent subsequence. Find one in the sequence $(1, -1, 1, -1, \ldots)$. But any rearrangement must diverge. If a sequence has only one cluster point, need it converge to it?

7. ▶ We may occasionally encounter 'sequences' with two indices $(a_{m,n})$ (they are more properly called *nets*). The example $n/(n + m)$ shows that in general

$$\lim_{m \to \infty} \lim_{n \to \infty} a_{m,n} \neq \lim_{n \to \infty} \lim_{m \to \infty} a_{m,n}.$$

The same example shows that, in $\mathbb{R}$, generally, $\sup_n \inf_m a_{n,m} \neq \inf_m \sup_n a_{n,m}$. But the following are true:

(a) $\sup_n \sup_m a_{n,m} = \sup_m \sup_n a_{n,m}$,
(b) $\sup_n (a_n + b_n) \leqslant \sup_n a_n + \sup_n b_n$.

3.2 Continuity

One is often not particularly interested in the actual values of the distances between points: no new theorems will result by substituting metres with feet. What matters more, in most cases, is the relation of points to each other captured by the concept of convergence. Accordingly, functions that preserve convergence (rather than distance) take on a central importance.

Definition 3.7

> A function $f : X \to Y$ between metric spaces is **continuous** when it preserves convergence,
>
> $$x_n \to x \text{ in } X \implies f(x_n) \to f(x) \text{ in } Y.$$

In this case therefore, $f(\lim_{n \to \infty} x_n) = \lim_{n \to \infty} f(x_n)$. Before we see any examples of continuous functions, let us prove that the following two statements are equivalent formulations of continuity in metric spaces, so any of them can be taken as the definition of continuity.

Theorem 3.8

> **A function $f : X \to Y$ between metric spaces is continuous if, and only if, any of the following statements holds:**
>
> (i) $\forall x \in X,\ \forall \epsilon > 0,\ \exists \delta > 0,\ \forall x' \in X,$
>
> $$d_X(x, x') < \delta \implies d_Y(f(x), f(x')) < \epsilon,$$
>
> (ii) **For every open set V in Y, $f^{-1}V$ is open in X.**

Statement (i) is often written as $\lim_{x' \to x} f(x') = f(x)$ for all x.

Proof Let (d) denote the defining statement that f is continuous.
(d) $\Rightarrow$ (i): Suppose statement (i) is false; then there is a point $x \in X$ and an $\epsilon > 0$ such that arbitrarily small changes to x can lead to sudden variations in $f(x)$,

$$\forall \delta > 0, \quad \exists x', \quad d_X(x, x') < \delta \text{ AND } d_Y(f(x), f(x')) \geqslant \epsilon$$

In particular, letting $\delta = 1/n$, there is a sequence[1] $x_n \in X$ satisfying $d_X(x, x_n) < 1/n$ but $d_Y(f(x), f(x_n)) \geqslant \epsilon$. This means that $x_n \to x$, but $f(x_n) \nrightarrow f(x)$, contradicting statement (d).
(i) $\Rightarrow$ (ii): Note that (i) can be rewritten as

$$\forall x \in X, \quad \forall \epsilon > 0, \quad \exists \delta > 0, \quad x' \in B_\delta(x) \implies f(x') \in B_\epsilon(f(x))$$

or even as

$$\forall x \in X, \quad \forall \epsilon > 0, \quad \exists \delta > 0, \quad f[B_\delta(x)] \subseteq B_\epsilon(f(x)).$$

---

[1] This selection of points x_n needs the Axiom of Choice for justification.

Let V be an open set in Y. To show that $U := f^{-1}V = \{x \in X : f(x) \in V\}$ is open in X, let x be any point of U; then $f(x) \in V$ and V is open. Hence

$$f(x) \in B_\epsilon(f(x)) \subseteq V,$$

and so

$$\exists \delta > 0, \quad f[B_\delta(x)] \subseteq B_\epsilon(f(x)) \subseteq V.$$

In other words, x is an interior point of U:

$$\exists \delta > 0, \quad B_\delta(x) \subseteq f^{-1}V = U.$$

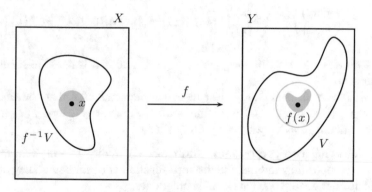

(ii) $\Rightarrow$ (d): Let $(x_n)_{n \in \mathbb{N}}$ be a sequence converging to x. Consider any open neighborhood $B_\epsilon(f(x))$ of $f(x)$. Then $f^{-1}[B_\epsilon(f(x))]$ contains x, and is an open set by (ii), so

$$\exists \delta > 0, \quad x \in B_\delta(x) \subseteq f^{-1}[B_\epsilon(f(x))],$$

$$\Rightarrow \exists \delta > 0, \quad f[B_\delta(x)] \subseteq B_\epsilon(f(x)).$$

But eventually all the points x_n are inside $B_\delta(x)$,

$$\exists N > 0, \quad n > N \Rightarrow x_n \in B_\delta(x)$$

$$\Rightarrow f(x_n) \in f[B_\delta(x)] \subseteq B_\epsilon(f(x))$$

$$\Rightarrow d_Y(f(x_n), f(x)) < \epsilon.$$

This shows that $f(x_n) \to f(x)$ as $n \to \infty$. $\qquad\qquad\qquad\qquad\qquad\square$

Examples 3.9

1. The square root function on $\mathbb{R}^+$ is continuous.
   Proof: Let $x, \epsilon > 0$, and $\delta := \epsilon\sqrt{x}$ (for $x = 0$, choose $\delta = \epsilon^2$), then

$$|x - y| < \delta \Rightarrow |\sqrt{x} - \sqrt{y}| < \frac{\delta}{\sqrt{x} + \sqrt{y}} = \frac{\epsilon}{1 + \sqrt{y/x}} < \epsilon.$$

2. Let X, Y, Z be metric spaces, then the function $h : X \to Y \times Z$ defined by $h(x) := (f(x), g(x))$ is continuous if, and only if, f, g are continuous. For example, the circle path $\theta \mapsto (\cos\theta, \sin\theta)$ is a continuous map $\mathbb{R} \to \mathbb{R}^2$.
   Proof: The statement follows directly from Example 3.3(5),

$$\begin{pmatrix} f(x_n) \\ g(x_n) \end{pmatrix} \to \begin{pmatrix} f(x) \\ g(x) \end{pmatrix} \Leftrightarrow f(x_n) \to f(x) \text{ AND } g(x_n) \to g(x)$$

3. ▶ If $f : X \to Y$ is continuous, then $f\bar{A} \subseteq \overline{fA}$. So if A is dense in X, then fA is dense in fX.
   Proof: If $x \in \bar{A}$, then there is a sequence of elements of A that converge to x, $x_n \to x$ (Proposition 3.4). By continuity of f, $f(x_n) \to f(x)$, so $f(x) \in \overline{fA}$. It follows that if $\bar{A} = X$ then $fX \subseteq \overline{fA} \cap fX$.

The following two propositions affirm that continuity is well-behaved with respect to composition and that the distance function is continuous. They allow us to build up continuous functions from simpler ones.

Proposition 3.10

If $f : X \to Y$ and $g : Y \to Z$ are continuous, so is $g \circ f : X \to Z$.

Proof Let $x_n \to x$ in X. Then by continuity of f, $f(x_n) \to f(x)$ in Y, and by continuity of g,

$$g \circ f(x_n) = g(f(x_n)) \to g(f(x)) = g \circ f(x) \text{ in } Z.$$

Alternatively, let W be any open set in Z. Then $g^{-1}W$ is an open set in Y, and so $f^{-1}[g^{-1}W]$ is an open set in X. But this set is precisely $(g \circ f)^{-1}W$.   □

Proposition 3.11

The distance function $d : X^2 \to \mathbb{R}$ is continuous.

Proof Let $x_n \to x$ and $y_n \to y$ in X. Then, by the triangle inequality,

$$|d(x_n, y_n) - d(x, y)| \leqslant |d(x_n, y_n) - d(x, y_n)| + |d(x, y_n) - d(x, y)|$$
$$\leqslant d(x_n, x) + d(y_n, y) \to 0,$$

which gives $d(x_n, y_n) \to d(x, y)$ as $n \to \infty$. $\qquad\square$

Homeomorphisms

Continuous functions preserve *convergence*, a central concept in metric spaces; in this sense, they correspond to the *morphisms* of groups and rings, which preserve the group and ring operations. The analogue of an *isomorphism* is called a *homeomorphism*:

Definition 3.12

A **homeomorphism** between metric spaces X and Y is a mapping $J : X \to Y$ such that

$$J \text{ is bijective (1–1 and onto)},$$

$$J \text{ is continuous},$$

$$J^{-1} \text{ is continuous}.$$

X is *homeomorphic* to Y when there exists a homeomorphism between them.
A metric space X is said to be **embedded** in another space Y, when there is a subset $Z \subseteq Y$ such that X is homeomorphic to Z.

Like all other isomorphisms, "X is homeomorphic to Y" is an equivalence relation on metric spaces. When X and Y are homeomorphic, they are not only the same as sets (the bijection part) but also with respect to convergence:

$$x_n \to x \iff J(x_n) \to J(x),$$

$$A \text{ is open in } X \iff JA \text{ is open in } Y.$$

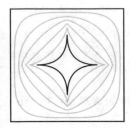

The elements of Y are those of X in different clothing, as far as convergence is concerned. The most vivid picture is that of "deforming" one space continuously and reversibly from the other. The by-now classic example is that a 'teacup' is homeomorphic to a 'doughnut'.

Exercises 3.13

1. Any constant function $f : x \mapsto y_0 \in Y$ is continuous. The identity function $I : X \to X, x \mapsto x$, is always continuous.
2. The functions that map the real number t to $t + 1$, $2t$, t^n ($n \in \mathbb{N}$), a^t ($a > 0$), and $|t|$ are all continuous.
3. In $\mathbb{R}$, addition and multiplication are continuous, i.e., if $x_n \to x$ and $y_n \to y$ then $x_n + y_n \to x + y$ and $x_n y_n \to xy$. Deduce that if $f, g : X \to \mathbb{R}$ are continuous functions, then so are $f + g$ and fg. For example, the polynomials on $\mathbb{R}$ are continuous. The function max $: \mathbb{R}^2 \to \mathbb{R}$ is also continuous, i.e., $\max(x_n, y_n) \to \max(x, y)$.
4. The function $f :]0, \infty[\to]0, \infty[$, defined by $f(t) := 1/t$ is continuous.
5. Conjugation in $\mathbb{C}$, $z \mapsto \bar{z}$, is continuous.
6. In $\mathbb{R}$, the *characteristic* function $1_A(x) = \begin{cases} 1, & x \in A \\ 0, & x \notin A \end{cases}$ is always discontinuous except when $A = \varnothing$ or $A = \mathbb{R}$. Is this true for all metric spaces?
7. When $f : X \to \mathbb{R}$ is a continuous function, the set $\{ x \in X : f(x) > 0 \}$ is open in X.
8. Any function $f : \mathbb{N} \to X$ is continuous, where X is any metric space.
9. The *graph* of a continuous function $f : X \to Y$, namely

$$\{ (x, f(x)) : x \in X \},$$

is closed in $X \times Y$ (with the D_1 metric).
10. Find examples of continuous functions f (e.g., $\mathbb{R} \to \mathbb{R}^n$), such that

    (a) f is invertible but f^{-1} is not continuous.
    (b) $f(x_n) \to f(x)$ in Y but $(x_n)_{n \in \mathbb{N}}$ does not converge at all.
    (c) U is open in X but fU is not open in Y. However functions which map open sets to open sets do exist (find one) and are called *open mappings*.

11. If F is a closed set in Y and $f : X \to Y$ is a continuous function, then $f^{-1}F$ is closed in X. But f may map a closed set to a non-closed set (even if f is an open mapping).

12. It is not enough that $f(x, y)$ is continuous in x and y separately in order that f be continuous. For example, show that the function

$$f(x, y) := \frac{xy}{x^2 + y^2}, \quad f(0, 0) := 0,$$

is discontinuous at $(0, 0)$ even though $f(x_n, 0) \to 0$, $f(0, y_n) \to 0$, when $x_n \to 0$, $y_n \to 0$. It needs to be "jointly continuous" in the sense that $f(x_n, y_n) \to f(x, y)$ for any $(x_n, y_n) \to (x, y)$.

13. Show carefully that the function $f(t) := t^2 + 5t^{3/2} - 3t + 4t^{1/2} - 1$ on the domain $\mathbb{R}^+$ is continuous. (Hint: Use Proposition 3.10.)

14. The roots of a quadratic equation $ax^2 + bx + c = 0$ vary continuously as the coefficients change (but maintaining $b^2 \geqslant 4ac$), except at $a = 0$.

15. ▶ Find a short proof that the *sphere* $S_r := \{ y : d(x, y) = r \}$ is closed using the continuity of d.

16. Given a set $A \subseteq X$, the map $x \mapsto d(x, A)$ is continuous. (Hint: $d(y, A) \leqslant d(y, x) + d(x, A)$.)

17. Given disjoint non-empty closed subsets $A, B \subseteq X$, find a continuous function $f : X \to [0, 1]$ such that $fA = 0$, $fB = 1$ (Hint: use $d(x, A)$ and $d(x, B)$).

18. Every interval in $\mathbb{R}$ is homeomorphic to $[0, 1]$, $[0, \infty[$, or $\mathbb{R}$.

19. $\mathbb{N}$ is homeomorphic to the discrete metric space on a countable set, but $\mathbb{Q}$ is not. (Hint: The convergent sequence $1/n \to 0$ must correspond to a divergent sequence in $\mathbb{N}$.)

20. ◇ A bent line in the plane, consisting of two straight line segments meeting at their ends, is homeomorphic to the unbent line. Thus angles are meaningless as far as homeomorphisms are concerned; triangles, squares and circles are homeomorphic.

Chapter 4
Completeness and Separability

4.1 Completeness

Our task of rigorously defining convergence in a general space has been achieved, but there seems to be something circular about it, because convergence is defined in terms of a *limit*. For example, take a convergent sequence $x_n \to x$ in a metric space X, and "artificially" remove the point x to form $X \smallsetminus \{x\}$ (assume $\forall n, \ x_n \neq x$). The other points x_n still form a sequence in this subspace, but it no longer converges (otherwise it would have converged to two points in X)—its limit is "missing". The sequence $(x_n)_{n \in \mathbb{N}}$ is convergent in X but divergent in $X \smallsetminus \{x\}$. How are we to know whether a metric space has "missing" points? And if it has, is it possible to *create* them when the bigger space X is unknown?

To be more concrete, let us take a look at the rational numbers: consider the sequences $(1, 2, 3, \ldots)$, $(1, -1, 1, -1, \ldots)$, and $(1, 1.5, 1.417, 1.414, 1.414, \ldots)$, the last one defined iteratively by $a_0 := 1$, $a_{n+1} := \frac{a_n}{2} + \frac{1}{a_n}$. It is easy to show that the first two do not converge, but, contrary to appearances, neither does the third, the reason being that were it to converge to $a \in \mathbb{Q}$, then $a = a/2 + 1/a$, implying $a^2 = 2$, which we know cannot be satisfied by any rational number. This sequence seems a good candidate of one which converges to a "missing" number not found in $\mathbb{Q}$. Having found one missing point, there are an infinite number of them: $(2, 2.5, 2.417, 2.414, \ldots)$ and $(2, 3, 2.834, 2.828, \ldots)$ cannot converge in $\mathbb{Q}$.

But could it be that the first two sequences also converge to "missing" numbers? How are we to distinguish between sequences that "truly" diverge from those that converge to "missing" points? There is a property that characterizes *intrinsic convergence*: suppose that $(x_n)_{n \in \mathbb{N}}$ is divergent in the metric space Y, but converges $x_n \to a$ in a bigger space X. Then the points get close to each other (in Y),

$$d_Y(x_n, x_m) = d_X(x_n, x_m) \leqslant d_X(x_n, a) + d_X(a, x_m) \to 0, \quad \text{as } n, m \to \infty.$$

© The Author(s), under exclusive license to Springer Nature Switzerland AG 2024
J. Muscat, *Functional Analysis*, https://doi.org/10.1007/978-3-031-27537-1_4

Definition 4.1

A **Cauchy sequence** is one such that $d(x_n, x_m) \to 0$ as $n, m \to \infty$, that is,

$$\forall \epsilon > 0, \quad \exists N, \quad n, m \geqslant N \Rightarrow d(x_n, x_m) < \epsilon.$$

To clarify this idea further, we prove:

Proposition 4.2

Two sequences $(x_n)_{n\in\mathbb{N}}$, $(y_n)_{n\in\mathbb{N}}$ are defined to be *asymptotic* when $d(x_n, y_n) \to 0$ as $n \to \infty$.

(i) **Being asymptotic is an equivalence relation.**
(ii) **For $(x_n)_{n\in\mathbb{N}}$ asymptotic to $(y_n)_{n\in\mathbb{N}}$,**

    (a) **if $(x_n)_{n\in\mathbb{N}}$ is Cauchy then so is $(y_n)_{n\in\mathbb{N}}$,**
    (b) **if $(x_n)_{n\in\mathbb{N}}$ converges to x then so does $(y_n)_{n\in\mathbb{N}}$.**

(iii) **A sequence $(x_n)_{n\in\mathbb{N}}$ is Cauchy if, and only if, every subsequence of $(x_n)_{n\in\mathbb{N}}$ is asymptotic to $(x_n)_{n\in\mathbb{N}}$.**

Proof (i) Let $(x_n)_{n\in\mathbb{N}} \sim (y_n)_{n\in\mathbb{N}}$ signify $d(x_n, y_n) \to 0$ as $n \to \infty$. Reflexivity and symmetry of $\sim$ are obvious. If $(x_n)_{n\in\mathbb{N}} \sim (y_n)_{n\in\mathbb{N}} \sim (z_n)_{n\in\mathbb{N}}$ then transitivity holds:

$$d(x_n, z_n) \leqslant d(x_n, y_n) + d(y_n, z_n) \to 0 \quad \text{as } n \to \infty.$$

(ii) If $d(x_n, y_n) \to 0$ and $d(x_n, x_m) \to 0$ as $n, m \to \infty$, then

$$d(y_n, y_m) \leqslant d(y_n, x_n) + d(x_n, x_m) + d(x_m, y_m) \to 0.$$

Similarly, if $d(x_n, x) \to 0$ then $d(y_n, x) \leqslant d(y_n, x_n) + d(x_n, x) \to 0$.

(iii) A Cauchy sequence satisfies

$$\forall \epsilon > 0, \ \exists N, \ n, m \geqslant N \Rightarrow d(x_n, x_m) < \epsilon.$$

Given a subsequence (x_{n_i}), its indices satisfy $n_i \geqslant i$ (by induction on i: $n_1 \geqslant 1$, $n_2 > n_1 \geqslant 1$ so $n_2 \geqslant 2$, etc.). Thus

$$i \geqslant N \Rightarrow n_i, i \geqslant N \Rightarrow d(x_{n_i}, x_i) < \epsilon$$

and $d(x_{n_i}, x_i) \to 0$ as $i \to \infty$.

Conversely, suppose $(x_n)_{n \in \mathbb{N}}$ is not Cauchy. Then

$$\exists \epsilon > 0, \ \forall i, \ \exists n_i, m_i \geqslant i, \ d(x_{n_i}, x_{m_i}) \geqslant \epsilon,$$

from which we can create the subsequences $(x_{n_1}, x_{n_2}, \ldots)$ and $(x_{m_1}, x_{m_2}, \ldots)$. If both these subsequences were asymptotic to $(x_n)_{n \in \mathbb{N}}$ then there would exist an N such that $i > N$ implies $d(x_i, x_{n_i}) < \epsilon/2$ as well as $d(x_i, x_{m_i}) < \epsilon/2$. Combining these two then gives a contradiction

$$d(x_{n_i}, x_{m_i}) \leqslant d(x_i, x_{n_i}) + d(x_i, x_{m_i}) < \epsilon,$$

so one of the two subsequences is not asymptotic to $(x_n)_{n \in \mathbb{N}}$. □

Examples 4.3

1. Convergent sequences are always Cauchy, since if $x_n \to x$ then $d(x_n, x_m) \to d(x, x) = 0$ by continuity of the distance function. But the discussion above gives examples of Cauchy sequences which do not converge.
2. In $\mathbb{R}$ or $\mathbb{Q}$, any increasing sequence that is bounded above, $a_n \leqslant b$, is Cauchy.
   Proof: Split the interval $[a_0, b]$ into subintervals of length ϵ. Let I be the last subinterval which contains a point, say a_N. As the sequence is increasing, I contains all of the sequence from N onward, proving the statement.
3. $\mathbb{R}$ and $\mathbb{Q}$ have the *bisection property*:
   Let $[a_0, b_0]$ be an interval in $\mathbb{R}$ or $\mathbb{Q}$, and divide it into halves, $[a_0, c]$ and $[c, b_0]$, where $c := (a_0 + b_0)/2$ is the midpoint. Choose $[a_1, b_1]$ to be either $[a_0, c]$ or $[c, b_0]$, randomly or according to some criterion; continue taking midpoints to get a nested sequence of intervals $[a_n, b_n]$, whose lengths are

$$b_n - a_n = (b_0 - a_0)/2^n \to 0.$$

So, for any $\epsilon > 0$, there is an $N > 0$ such that $b_N - a_N < \epsilon$, and for any $n \geqslant N$, $a_n, b_n \in [a_N, b_N]$. Hence $(a_n)_{n \in \mathbb{N}}$ and $(b_n)_{n \in \mathbb{N}}$ are asymptotic Cauchy sequences.
4. Let B_{r_n} be a nested sequence of balls, $B_{r_{n+1}} \subseteq B_{r_n}$, with $r_n \to 0$. Then choosing any points $x_n \in B_{r_n}$ gives a Cauchy sequence.
   Proof: For any $m \geqslant n$,

$$x_m \in B_{r_m} \subseteq B_{r_{m-1}} \subseteq \cdots \subseteq B_{r_n}$$

so that $d(x_m, x_n) < 2r_n \to 0$ as $n, m \to \infty$.

5. ▶ In any metric space, if $d(x_{n+1}, x_n) \leqslant ac^n$ with $c < 1$ then x_n is Cauchy.
Moreover, if $x_n \to x$ then $d(x_n, x) \leqslant \frac{ac^n}{1-c}$.
Proof: Taking $n \leqslant m$, without loss of generality,

$$d(x_n, x_m) \leqslant d(x_n, x_{n+1}) + \cdots + d(x_{m-1}, x_m)$$

$$\leqslant a(c^n + \cdots + c^{m-1})$$

$$\leqslant \frac{ac^n}{1-c} \to 0 \quad \text{as } m, n \to \infty.$$

For the second part, take $m \to \infty$ in the above.

6. A Cauchy sequence cannot stray too far in the sense that $d(x_0, x_n) \leqslant R$ for all n, for some $R \geqslant 0$. Hence Cauchy sequences are "bounded".
Proof: By the definition of a Cauchy sequence for $\epsilon := 1$ say, there is an N such that $n, m \geqslant N \Rightarrow d(x_n, x_m) < \epsilon$. Therefore

$$d(x_0, x_n) \leqslant d(x_0, x_N) + d(x_N, x_n) < d(x_0, x_N) + \epsilon.$$

7. A Cauchy sequence in $\mathbb{Q}$ either converges to 0, or is eventually greater than some $\epsilon > 0$ or less than some $-\epsilon < 0$. In each case, an asymptotic sequence behaves in the same manner.
Proof: If $a_n \not\to 0$ yet is Cauchy, then

$$\exists \epsilon > 0, \ \forall M, \ \exists m \geqslant M, \quad |a_m| \geqslant \epsilon,$$

$$\exists N, \quad m, n \geqslant N \Rightarrow |a_n - a_m| < \epsilon/2.$$

Assuming, for example, $a_m \geqslant \epsilon$ for some $m \geqslant N$,

$$n \geqslant N \Rightarrow a_n - a_m \geqslant -|a_n - a_m|$$

$$\Rightarrow a_n \geqslant a_m - |a_n - a_m| > \epsilon/2.$$

If $(b_n)_{n \in \mathbb{N}}$ is an asymptotic sequence to $(a_n)_{n \in \mathbb{N}}$, then there is an M such that $|a_n - b_n| < \epsilon/2$ whenever $n \geqslant M$, and so

$$n \geqslant \max(N, M) \Rightarrow b_n \geqslant a_n - |a_n - b_n| \geqslant \epsilon/2.$$

When $a_m \leqslant -\epsilon$ for $m \geqslant N$, the reverse inequalities hold, for example,

$$n \geqslant N \Rightarrow a_n \leqslant a_m + |a_n - a_m| < -\epsilon/2.$$

Definition 4.4

A metric space is **complete** when every Cauchy sequence in it converges.

In a complete metric space, there are no "missing" points and any divergent sequence is "truly" divergent—there is no bigger metric space which makes it convergent.

It follows that the space of rational numbers $\mathbb{Q}$ (with the standard metric) is not complete, a fact that allegedly deeply troubled Pythagoras and his followers. They shouldn't have worried because there *is* a way of creating the missing numbers (but skip the proof if it worries you on a first reading!):

Theorem 4.5

The real number space $\mathbb{R}$ is complete.

Proof (i) For this to be a theorem, we need to be clear about what constitutes $\mathbb{R}$. The usual definition is that it is a set with an addition $+$ and multiplication $\cdot$ which satisfy the axioms of a field (see p. 10), and with a linear order relation $\leqslant$ that is compatible with these operations:

$$x \leqslant y \Rightarrow x + z \leqslant y + z, \qquad x, y \geqslant 0 \Rightarrow xy \geqslant 0,$$

and in addition satisfies the *completeness axiom*:

Every non-empty subset A of $\mathbb{R}$ with an upper bound has a least upper bound.

Assuming all these axioms, let $(a_n)_{n \in \mathbb{N}}$ be a Cauchy sequence in $\mathbb{R}$, that is, for any $\epsilon > 0$, there is an N beyond which $|a_n - a_m| < \epsilon$. Let

$$B := \{ x \in \mathbb{R} : \exists M, \ n \geqslant M \Rightarrow x \leqslant a_n \}.$$

Its elements might be called eventual lower bounds of $\{ a_n : n \in \mathbb{N} \}$. The fact that Cauchy sequences are bounded implies that $\{ a_n : n \in \mathbb{N} \}$ has a lower bound and so $B \neq \varnothing$, while any upper bound of $\{ a_n : n \in \mathbb{N} \}$ is also one of B. Hence, by the completeness axiom, B has a least upper bound α. Two facts follow:

(a) $\alpha + \epsilon$ is not an element of B, so there must be an infinite number of terms $a_{n_i} < \alpha + \epsilon$;

(b) $\alpha - \epsilon$ is not an upper bound of B, so there must exist an $x \in B$ and an M such that $n \geqslant M \Rightarrow \alpha - \epsilon < x \leqslant a_n$.

These facts together imply that for $n_i \geqslant M$ we have $\alpha - \epsilon \leqslant a_{n_i} \leqslant \alpha + \epsilon$. Then, given any $n \geqslant N$, choose any $n_i \geqslant \max(M, N)$, so that

$$n \geqslant N \implies |a_n - \alpha| \leqslant |a_n - a_{n_i}| + |a_{n_i} - \alpha| < 2\epsilon$$

as required to show $a_n \to \alpha$.

This proof is open to the criticism that we have not proved whether, in fact, there exists such a set with all these properties. We need to fill this logical gap by giving a construction of $\mathbb{R}$ that satisfies these axioms.

(ii) The whole idea is to treat the Cauchy sequences of rational numbers themselves as the missing numbers! How can a *sequence* be a *number*? Actually, this is not really that novel—the familiar decimal representation of a real number is a particular Cauchy sequence: $e := 2.71828\ldots$ is just short for $(2, 2.7, 2.71, 2.718, \ldots)$. But there are several other Cauchy sequences that converge to e. For example, there is nothing special about the decimal system—the binary expansion $(2, 2\frac{1}{2}, 2 + \frac{1}{2} + \frac{1}{8}, \ldots)$ also converges to e. We should be grouping these asymptotic Cauchy sequences together, and treat each class as one real number. For example, the asymptotic sequences $0.32999\ldots$ and $0.33000\ldots$ represent the same real number.

Accordingly, $\mathbb{R}$ is defined as the set of equivalence classes of asymptotic Cauchy sequences of rational numbers; each real number is here written as $x = [a_n]$ (instead of the cumbersome $[(a_n)]$). We now develop the structure of $\mathbb{R}$: addition and multiplication, its order and distance function. Define

$$x + y = [a_n] + [b_n] := [a_n + b_n], \quad xy = [a_n][b_n] := [a_n b_n].$$

That addition is well-defined follows from an application of the triangle inequality in $\mathbb{Q}$; that it has the associative and commutative properties follows from the analogous properties for addition of rational numbers. The new real zero is $[0, 0, \ldots]$, and the negatives are $-x = -[a_n] = [-a_n]$. Similarly, multiplication is well defined and has all the properties that make $\mathbb{R}$ a field.

It is less straightforward to define an inequality relation on $\mathbb{R}$. Let $(a_n) > 0$ mean that the Cauchy sequence $(a_n)_{n \in \mathbb{N}}$ is eventually strictly positive (Example 4.3(7)),

$$\exists \epsilon \in \mathbb{Q}^+, \; \exists N, \quad n \geqslant N \implies a_n \geqslant \epsilon > 0.$$

Any other asymptotic Cauchy sequence must also eventually be strictly positive. Correspondingly, let $x < y$ mean that $y - x > 0$, or equivalently,

$$[a_n] < [b_n] \iff \exists \epsilon \in \mathbb{Q}^+, \; \exists N, \; \forall n \geqslant N, \; a_n + \epsilon \leqslant b_n.$$

This immediately shows that $x < y \iff x + z < y + z$. We make a few more observations about this relation:

1. If $a_n \geqslant 0$ for all n, then $[a_n] \geqslant 0$,
2. If $0 < x$ and $0 < y$ then $0 < xy$ and $0 < x + y$ (gives transitivity of $\leqslant$),

3. $x > 0$ OR $x = 0$ OR $x < 0$ (Example 4.3(7)).
4. If $x < 0$ then $-x > 0$.

Anti-symmetry of $\leqslant$ follows from the fact that $(b_n - a_n)_{n \in \mathbb{N}}$ cannot eventually be both strictly positive and strictly negative. This makes $\mathbb{R}$ a linearly ordered field.

Given a real number $x = [a_n] = [b_n]$, let $|x| := [|a_n|]$, which makes sense since

$$\big||a_n| - |a_m|\big| \leqslant |a_n - a_m| \to 0 \text{ as } n, m \to \infty,$$

$$\big||a_n| - |b_n|\big| \leqslant |a_n - b_n| \to 0 \text{ as } n \to \infty.$$

In fact $|x| = x$ when $x > 0$ and $|x| = -x$ when $x < 0$, so it satisfies the properties $|x| \geqslant 0$, $|x| = 0 \Leftrightarrow x = 0$, $|-x| = |x|$, and $|x + y| \leqslant |x| + |y|$. Thus $d(x, y) := |x - y|$ is a distance, as in Example 2.2(1).

$\mathbb{Q}$ *is dense in* $\mathbb{R}$: Note that a rational number a can be represented in $\mathbb{R}$ by the constant sequence $[a, a, \ldots]$. The Archimedean property holds since $[a_n] > 0$ implies that eventually $a_n \geqslant p > 0$, for some $p \in \mathbb{Q}$, so $[a_n] \geqslant [p/2] > 0$. Also, if $x = [a_n]$ then $a_n \to x$ in $\mathbb{R}$, since for any $\epsilon > 0$, let $p \in \mathbb{Q}, 0 < p < \epsilon$, so

$$\exists N, \quad n, m \geqslant N \ \Rightarrow \ |a_n - a_m| < p$$
$$\Rightarrow \ d(a_n, x) = d([a_n, a_n, \ldots], [a_1, a_2, \ldots])$$
$$= [|a_n - a_1|, |a_n - a_2|, \ldots] < \epsilon.$$

The completeness axiom is satisfied: Let A be any non-empty subset of $\mathbb{R}$ that is bounded above. Split $\mathbb{R}$ into the set B of upper bounds of A, and its complement B^c, both of which are non-empty, say $a_0 \in B^c$, $b_0 \in B$; these can even be taken to be rational, by the Archimedean property.

Divide $[a_0, b_0]$ in two using the midpoint $c := (a_0 + b_0)/2$; if $c \in B$ then select $[a_1, b_1] = [a_0, c]$, otherwise take $[a_1, b_1] = [c, b_0]$. Continue dividing and selecting sub-intervals like this, to get two asymptotic Cauchy sequences $(a_n)_{n \in \mathbb{N}}$, $(b_n)_{n \in \mathbb{N}}$, with $b_n \in B$, $a_n \in B^c$ (Example 4.3(3)). Let $\alpha := [a_n]$, so $a_n \to \alpha$, $b_n \to \alpha$, and (Exercise 3.6(4))

$(\forall a \in A, \ a \leqslant b_n) \ \Rightarrow \ (\forall a \in A, \ a \leqslant \alpha)$,   "$\alpha$ is an upper bound of A",

$(\forall b \in B, \ a_n \leqslant b) \ \Rightarrow \ (\forall b \in B, \ \alpha \leqslant b)$,   "$\alpha$ is the least upper bound".

A dual argument shows that every non-empty set with a lower bound has a greatest lower bound, denoted $\inf A$.

$\mathbb{R}$ *is complete*: This now follows from part (i), but we can see this directly in this context. Start with any Cauchy sequence of real numbers (in decimal form, say) and replace each number by a rational number to an increasing number of significant places, for example:

$$
\begin{array}{ll}
x_n \in \mathbb{R} & \mapsto a_n \in \mathbb{Q} \\
2.6280\ldots & 2 \\
2.7087\ldots & 2.7 \\
2.7173\ldots & 2.71 \\
2.7181\ldots & 2.718 \\
\ldots &
\end{array}
$$

The crucial point is that the two sequences are asymptotic by construction. Since the first one is Cauchy, so must be the second one. But a Cauchy sequence of rational numbers is, by definition, a real number x. Moreover, $a_n \to x$ implies $x_n \to x$.   □

This "completion" process generalizes readily to any metric space.

Theorem 4.6

> **Every metric space X can be completed, that is, there is a complete metric space $\widetilde{X}$, containing a dense copy of X and extending its distance function.**

Any such complete metric space $\widetilde{X}$ is called the *completion* of X.

Proof *Construction of* $\widetilde{X}$: Let C be the set of Cauchy sequences of X. For any two Cauchy sequences $a = (x_n)_{n\in\mathbb{N}}$, $b = (y_n)_{n\in\mathbb{N}}$, the real sequence $d(x_n, y_n)$ is also Cauchy (Exercise 4.11(6)), and since $\mathbb{R}$ is complete, it converges to a real number $D(a, b) := \lim_{n\to\infty} d(x_n, y_n)$. Symmetry and the triangle inequality of D follow from that of d, by taking the limit $n \to \infty$ in the following:

$$
\left.
\begin{array}{l}
d(y_n, x_n) = d(x_n, y_n) \\
d(x_n, y_n) \leqslant d(x_n, z_n) + d(z_n, y_n)
\end{array}
\right\}
\Rightarrow
\left\{
\begin{array}{l}
D(b, a) = D(a, b) \\
D(a, b) \leqslant D(a, c) + D(c, b).
\end{array}
\right.
$$

The only problem is that $D(a, b) = 0$, meaning $d(x_n, y_n) \to 0$, is perfectly possible without $a = b$. It happens when the Cauchy sequences $(x_n)_{n\in\mathbb{N}}$, $(y_n)_{n\in\mathbb{N}}$ are asymptotic. We have already seen that this is an equivalence relation, so C partitions into equivalence classes. Write $\tilde{d}([a], [b]) := D(a, b)$; it is well-defined since for any other representative sequences $a' \in [a]$ and $b' \in [b]$, we have

$$
D(a', b') \leqslant D(a', a) + D(a, b) + D(b, b') = D(a, b);
$$

similarly $D(a, b) \leqslant D(a', b')$; so $D(a, b) = D(a', b')$. Let $\widetilde{X}$ be the space of equivalence classes of Cauchy sequences, with the metric $\tilde{d}$.

There is a dense copy of X in $\widetilde{X}$: For any $x \in X$, there corresponds the constant sequence $x := (x, x, \ldots)$ in C. Since

$$\tilde{d}([x], [y]) = D((x), (y)) = \lim_{n \to \infty} d(x, y) = d(x, y),$$

this set of constant sequences is a true copy of X, preserving distances between points. To show that this copy is dense in $\widetilde{X}$, we need to show that any representative Cauchy sequence $a = (x_n)_{n \in \mathbb{N}}$ in C has constant sequences arbitrarily close to it. By the definition of Cauchy sequences, for any $\epsilon > 0$, there is an $N \in \mathbb{N}$ with $d(x_n, x_N) < \epsilon$ for $n \geqslant N$. Let x be the constant sequence (x_N). Then $D(a, x) = \lim_{n \to \infty} d(x_n, x_N) \leqslant \epsilon < 2\epsilon$ proves that $[x]$ is within 2ϵ of $[a]$.

$\widetilde{X}$ *is complete:* Let $([a_n])$ be a Cauchy sequence in $\widetilde{X}$; this means $\tilde{d}([a_n], [a_m]) = D(a_n, a_m) \to 0$, as $n, m \to \infty$. For each n, we can find a constant sequence x_n which is as close to a_n as needed, i.e., $D(x_n, a_n) < \epsilon_n$; by choosing $\epsilon_n \to 0$, we can select (x_n) to be asymptotic to (a_n). As (a_n) is Cauchy, so is (x_n). In fact, $x_n \to x := (x_n)$ since

$$\lim_{n \to \infty} D(x_n, x) = \lim_{m,n \to \infty} d(x_n, x_m) = 0,$$

so that the asymptotic sequence a_n also converges to x, and $[a_n]$ to $[x]$. □

Proving that a given metric space is complete is normally quite hard. Even showing that a particular Cauchy sequence converges may not be an easy matter because one has to identify which point it converges to, let alone doing this for arbitrary Cauchy sequences. But once a space is shown to be complete, one need not go through the same proof process to show that a subspace or a product is complete:

Proposition 4.7

> **Let X, Y be complete metric spaces. Then,**
>
> (i) **A subspace $F \subseteq X$ is complete $\Leftrightarrow$ F is closed in X,**
> (ii) **$X \times Y$ is complete.**

Proof (i) Let $F \subseteq X$ be complete, i.e., any Cauchy sequence in F converges to a limit in F. Let $x \in \bar{F}$, with a sequence $x_n \to x$, $x_n \in F$ (Proposition 3.4). Since convergent sequences are Cauchy and F is complete, x must be in F. Thus $F = \bar{F}$ is closed. The completeness of X has not been used, so in fact a complete subspace of *any* metric space is closed.

Conversely, let F be a closed set in X and let $(x_n)_{n \in \mathbb{N}}$ be a Cauchy sequence in F. Then (x_n) is a Cauchy sequence in X, which is complete. Therefore $x_n \to x$ for some $x \in X$; in fact $x \in \bar{F} = F$. Thus any Cauchy sequence of F converges in F.

(ii) Let $\binom{x_n}{y_n}$ be a Cauchy sequence in $X \times Y$. Recall that

$$d\left(\binom{x_n}{y_n}, \binom{x_m}{y_m} \right) := d_X(x_n, x_m) + d_Y(y_n, y_m) \geqslant d_X(x_n, x_m).$$

Since the left-hand term converges to 0 as $n, m \to \infty$, we get $d_X(x_n, x_m) \to 0$, so that the sequence $(x_n)_{n \in \mathbb{N}}$ is Cauchy in the complete space X. It therefore converges: $x_n \to x \in X$. By similar reasoning, $y_n \to y \in Y$. Consequently,

$$d\left(\binom{x_n}{y_n}, \binom{x}{y} \right) = d_X(x_n, x) + d_Y(y_n, y) \to 0 \text{ as } n \to \infty,$$

which is equivalent to $\binom{x_n}{y_n} \to \binom{x}{y}$ in $X \times Y$.   □

Examples 4.8

1. The completion of a subset A in a complete metric space X is $\bar{A}$.
   Proof: The completion Y of A must satisfy two criteria: Y must be complete, and A must be dense in Y. Now, $\bar{A}$ is closed in X, so is complete, and A is dense in $\bar{A}$ (by definition).

2. Two metric spaces may be homeomorphic yet one space may be complete and the other not. For example, $\mathbb{R}$ is homeomorphic to $]0, 1[$ (Exercise 3.13(18)), but the latter is not closed in $\mathbb{R}$.

3. Let $f : X \to Y$ be a continuous function. *If* it can be extended to the completions as a continuous function $\tilde{f} : \tilde{X} \to \tilde{Y}$, then this extension is unique.
   Proof: Any $x \in \tilde{X}$ has a sequence $(a_n)_{n \in \mathbb{N}}$ in X converging to it (Proposition 3.4). As $\tilde{f}$ is continuous, we find that $\tilde{f}(x)$ is uniquely determined by

$$\tilde{f}(x) = \lim_{n \to \infty} \tilde{f}(a_n) = \lim_{n \to \infty} f(a_n).$$

4. But not every continuous function $f : X \to Y$ can be extended continuously to the completions $\tilde{f} : \tilde{X} \to \tilde{Y}$. For example, the continuous function $f(t) := 1/t$ on $]0, \infty[$ cannot be extended continuously to $[0, \infty[$.

5. (Cantor) The completion of $\mathbb{Q}$ to $\mathbb{R}$ has come at a price: $\mathbb{R}$ is not countable. Prove this by taking the binary expansion of a list of real numbers in $[0, 1]$, arranged in an infinite array, and creating a new number from the diagonal that is different from all of them. The corollary of the next theorem is a strong generalization of this statement.

René-Louis Baire (1874–1932) After graduating in Paris around 1894, Baire tackled the problem of convergence and limits of functions, namely that no space of functions then known was "closed" under pointwise convergence. His Ph.D. dissertation, under the supervision of C.E. Picard, introduced the concept of 'nowhere dense' and proved his famous theorem. Progress on this issue was made by his colleague Borel in the direction of measurable sets.

Theorem 4.9 (Baire's Category Theorem)

In a complete metric space, a countable intersection of open dense subsets is again dense.

Proof Let $Y := \bigcap_{n=1}^{\infty} U_n$, where U_n are open dense subsets of the complete metric space X, and let $B_r(x)$ be any ball in X. To find a $y \in Y \cap B_r(x)$, we are going to create a nested sequence of balls of diminishing radius whose centers therefore form a Cauchy sequence. To start with, U_1 intersects the ball $B_{r_1}(x_1) := B_r(x)$, since it is dense; so the open set $U_1 \cap B_{r_1}(x_1)$ contains a point x_2 and some neighborhood $B_{r_2}(x_2)$. Now U_2 is dense, so the open set $U_2 \cap B_{r_2}(x_2)$ is non-empty and there is a ball $B_{r_3}(x_3) \subseteq U_2 \cap B_{r_2}(x_2)$.

Continuing like this, we can find a sequence of points (using the Axiom of Choice)

$$x_{n+1} \in B_{r_{n+1}}(x_{n+1}) \subseteq U_n \cap B_{r_n}(x_n).$$

Moreover at each stage, r_n can be chosen small enough that

$$r_n \to 0 \quad (\text{e.g., } r_n \leqslant 1/n),$$

$$\overline{B_{r_{n+1}}(x_{n+1})} \subseteq B_{r_n}(x_n) \quad (\text{e.g., } r_{n+1} < r_n - d(x_n, x_{n+1})).$$

Thus $(x_n)_{n \in \mathbb{N}}$ is a Cauchy sequence (Example 4.3(4)) which converges $x_n \to y$ since X is complete. For all $m > n$ we have $x_m \in B_{r_{n+1}}(x_{n+1})$ and taking the limit $x_m \to y$ we find $y \in \overline{B_{r_{n+1}}(x_{n+1})} \subseteq B_{r_n}(x_n)$. Since this holds for any n we obtain

$$y \in \bigcap_n B_{r_n}(x_n) \subseteq B_{r_1}(x_1) \cap \bigcap_n U_n = B_r(x) \cap Y.$$

Since Y intersects all balls, Y is dense in X. $\qquad \square$

Corollary 4.10

> **A nonempty complete metric space cannot be covered by a countable number of nowhere-dense subsets.**

Proof Suppose that the metric space $X = \bigcup_{n=1}^{\infty} A_n$ $(= \bigcup_{n=1}^{\infty} \bar{A}_n)$, where A_n are nowhere dense. Then, $\bigcap_n \bar{A}_n^c = \varnothing$, with $\bar{A}_n^c$ being open dense subsets (Exercise 2.23(9)). This clearly contradicts the theorem. $\qquad\square$

Exercises 4.11

1. Any sequence in $\mathbb{Q}$ of the type $(3.1, 3.14, 3.141, 3.1415, \ldots)$ is Cauchy.
2. The sequences $(1, 2, 3, \ldots)$ and $(1, -1, 1, -1, \ldots)$ are not Cauchy.
3. ⋆ Try to prove that the sequence defined by $a_0 := 1, a_{n+1} := \frac{a_n}{2} + \frac{1}{a_n}$ is Cauchy.
   (Hint: Use the principle of induction to show that $|a_{n+1} - a_n| \leqslant (\frac{1}{2})^{n+1}$.)
4. If a sequence $(x_n)_{n \in \mathbb{N}}$, chosen from a finite set of points, e.g., $(x, y, x, x, y, \ldots)$, is Cauchy then it must eventually become constant $(x_0, \ldots, x_N, x_N, \ldots)$.
5. The following give sufficient conditions for Cauchy sequences:

   (a) $d(x_{n+1}, x_n) \leqslant c\, d(x_n, x_{n-1})$ with $c < 1$,
   (b) $d(x_{n+1}, x_n) \leqslant c\, d(x_n, x_{n-1})^2$ with $c\, d(x_1, x_0) < 1$.

   But a sequence which decreases at the rate $d(x_{n+1}, x_n) \leqslant 1/n$ need not be Cauchy.
6. If $(x_n)_{n \in \mathbb{N}}$, $(y_n)_{n \in \mathbb{N}}$ are Cauchy sequences in X, then so is $d_n := d(x_n, y_n)$ in $\mathbb{R}$.
7. ▶ A continuous function need not map Cauchy sequences to Cauchy sequences.
8. If $x_n \to x$ and $y_n \to x$, then $(x_n)_{n \in \mathbb{N}}$, $(y_n)_{n \in \mathbb{N}}$ are asymptotic.
9. $\sqrt{n}$ and $\sqrt{n+1}$ are asymptotic divergent sequences in $\mathbb{R}$.
10. ▶ A subsequence of a Cauchy sequence is itself Cauchy, and if it converges so does its parent sequence.
11. If $(x_n)_{n \in \mathbb{N}}$ is a Cauchy sequence, and the set of values $\{x_n : n \in \mathbb{N}\}$ has a limit point x, then $x_n \to x$.
12. The completion of $]0, 1[$ and of $[0, 1[$ is $[0, 1]$. Any Cauchy sequence in the Cantor set C must converge in C. However a Cauchy sequence of rational numbers need not converge to a rational number because $\mathbb{Q}$ is not closed in $\mathbb{R}$.
13. ▶ $\mathbb{R}^n := \mathbb{R} \times \cdots \times \mathbb{R}$ and $\mathbb{C}$ are complete.
14. Is $\mathbb{N}$ complete? Any discrete metric space is complete.
15. (Cantor) We have already seen that the centers of a nested sequence of balls with $r_n \to 0$ form a Cauchy sequence (Example 4.3(4)). Show, furthermore, that in a complete metric space, $\bigcap_n \overline{B_{r_n}(x_n)} = \{\lim_{n \to \infty} x_n\}$.
16. The only functions $f : \mathbb{Q} \to \mathbb{Q}$ satisfying $f(x + y) = f(x) + f(y)$ are $f : x \mapsto \lambda x$. Deduce that the only continuous functions $\tilde{f} : \mathbb{R} \to \mathbb{R}$ with this property are of the same type.

17. ⋆ The *completion* of X is essentially unique, in the sense that any two such completions (such as the one defined in the theorem) are homeomorphic to each other.
18. The Cantor set is complete and nowhere dense in $\mathbb{R}$; why doesn't this contradict Baire's theorem (corollary)?

4.2 Uniformly Continuous Maps

We have seen that a continuous function need not preserve completeness, or even Cauchy sequences. If one analyzes the root of the problem, one finds that its resolution lies in the following strengthening of continuity:

Definition 4.12

A function $f : X \to Y$ is said to be **uniformly continuous** when

$$\forall \epsilon > 0, \ \exists \delta > 0, \ \forall x \in X, \quad f[B_\delta(x)] \subseteq B_\epsilon(f(x)).$$

The difference from continuity is that, here, δ is independent of x.

Easy Consequences
1. Uniformly continuous functions are continuous.
2. But not every continuous map is uniformly so; an example is $f(t) := 1/t$ on $]0, \infty[$.
3. ▶ The composition of uniformly continuous maps is again uniformly continuous.
   Proof: $\forall \epsilon > 0, \ \exists \delta, \delta' > 0, \ \forall x, \ g[f[B_\delta(x)]] \subseteq g[B_{\delta'}(f(x))] \subseteq B_\epsilon(g(f(x)))$.

The key properties of uniformly continuous maps are the following two propositions:

Proposition 4.13

A uniformly continuous function maps any Cauchy sequence to a Cauchy sequence.

Proof By definition $f : X \to Y$ is uniformly continuous when

$$\forall \epsilon > 0, \ \exists \delta > 0, \ \forall x, x', \quad d_X(x, x') < \delta \Rightarrow d_Y(f(x), f(x')) < \epsilon.$$

In particular, for a Cauchy sequence $(x_n)_{n\in\mathbb{N}}$ in X, with this δ,

$$\exists N, \quad n, m > N \implies d_X(x_n, x_m) < \delta$$
$$\implies d_Y(f(x_n), f(x_m)) < \epsilon,$$

proving that $(f(x_n))_{n\in\mathbb{N}}$ is a Cauchy sequence in Y. $\square$

More generally, practically the same proof shows that a uniformly continuous function $f : X \to Y$ maps any asymptotic sequences $(a_n)_{n\in\mathbb{N}}$, $(b_n)_{n\in\mathbb{N}}$ in X to asymptotic sequences $(f(a_n))_{n\in\mathbb{N}}$, $(f(b_n))_{n\in\mathbb{N}}$ in Y.

Theorem 4.14

> **Every uniformly continuous function $f : X \to Y$ has a unique uniformly continuous extension to the completions $\tilde{f} : \widetilde{X} \to \widetilde{Y}$.**

Proof In order not to complicate matters unnecessarily, let us suppose that X and Y are dense *subsets* of $\widetilde{X}$ and $\widetilde{Y}$ respectively, instead of being embedded in them. Nothing is lost this way, except quite a few extra symbols!

Let $x_n \to x \in \widetilde{X}$, with $x_n \in X$. The sequence $f(x_n)_{n\in\mathbb{N}}$ is Cauchy in Y by the previous proposition, so must converge to some element $y \in \widetilde{Y}$. Furthermore, if $a_n \to x$ as well ($a_n \in X$), then $(x_n)_{n\in\mathbb{N}}$ and $(a_n)_{n\in\mathbb{N}}$ are asymptotic (Exercise 4.11(8)) forcing $f(x_n)_{n\in\mathbb{N}}$ and $f(a_n)_{n\in\mathbb{N}}$ to be asymptotic in Y, hence $f(a_n) \to y$. This allows us to define $\tilde{f}(x) := y$ without ambiguity. Moreover, this choice is imperative and $\tilde{f}$ is unique, if it is to be continuous.

The uniform continuity of $\tilde{f}$ follows from that of f. For any $\epsilon > 0$, there is a $\delta > 0$ for which

$$\forall a, b \in X, \quad d(a, b) < \delta \implies d(f(a), f(b)) < \epsilon.$$

Let $x, x' \in \widetilde{X}$ with $d(x, x') < \delta$, let $a_n \to x$, $b_n \to x'$ with $a_n, b_n \in X$ and, by the above, $f(a_n) \to \tilde{f}(x)$, $f(b_n) \to \tilde{f}(x')$. Among these terms, we can find a close to x and b close to x' to within $r := (\delta - d(x, x'))/2 < \delta$, while also $f(a)$ is close to $\tilde{f}(x)$ and $f(b)$ is close to $\tilde{f}(x')$ to within ϵ. Then

$$d(a, b) \leqslant d(a, x) + d(x, x') + d(x', b) < 2r + d(x, x') = \delta$$
$$\implies d(\tilde{f}(x), \tilde{f}(x')) \leqslant d(\tilde{f}(x), f(a)) + d(f(a), f(b)) + d(f(b), \tilde{f}(x')) < 3\epsilon.$$

$\square$

The following are easily shown to be uniformly continuous functions:

Definition 4.15

A function $f : X \to Y$ is called a **Lipschitz** map when

$$\exists c > 0, \ \forall x, x' \in X, \quad d_Y(f(x), f(x')) \leqslant c \, d_X(x, x').$$

Furthermore, it is called

- an **equivalence** (or *bi-Lipschitz*) when f is bijective and both f and f^{-1} are Lipschitz;
- a **contraction** when it is Lipschitz with constant $c < 1$;
- an **isometry**, and X, Y are said to be **isometric**, when f preserves distances, i.e.,

$$\forall x, x' \in X, \quad d_Y(f(x), f(x')) = d_X(x, x').$$

Examples 4.16

1. Any $f : [a, b] \to \mathbb{R}$ with continuous derivative is Lipschitz.
   Proof: As f' is continuous, it is bounded on $[a, b]$, say $|f'(x)| \leqslant c$. The result then follows from the mean value theorem,

   $$f(x) - f(x') = f'(\xi)(x - x'), \qquad \exists \xi \in \,]a, b[.$$

2. To show $f : \mathbb{R}^2 \to \mathbb{R}^2$ is Lipschitz, where $f = (f_1, f_2)$, it is enough to show that

   $$|f_i(x_1, y_1) - f_i(x_2, y_2)| \leqslant c(|x_1 - x_2| + |y_1 - y_2|), \quad i = 1, 2,$$

   for then (using $(a + b)^2 \leqslant 2(a^2 + b^2)$ for $a, b \in \mathbb{R}$)

   $$\left\| \begin{pmatrix} f_1(x_1, y_1) \\ f_2(x_1, y_1) \end{pmatrix} - \begin{pmatrix} f_1(x_2, y_2) \\ f_2(x_2, y_2) \end{pmatrix} \right\| \leqslant |f_1(x_1, y_1) - f_1(x_2, y_2)| + |f_2(x_1, y_1) - f_2(x_2, y_2)|$$

   $$\leqslant 2c(|x_1 - x_2| + |y_1 - y_2|)$$

   $$\leqslant 2c\sqrt{2} \left\| \begin{pmatrix} x_1 \\ y_1 \end{pmatrix} - \begin{pmatrix} x_2 \\ y_2 \end{pmatrix} \right\|.$$

3. ▶ Lipschitz maps are uniformly continuous, since for any $\epsilon > 0$, we can let $\delta := \epsilon/2c$ independent of x to obtain $d(x, x') < \delta \Rightarrow d(f(x), f(x')) \leqslant c\delta < \epsilon$.

4. But not every uniformly continuous function is Lipschitz. For example, $\sqrt{x}$ on [0, 1] is uniformly continuous (show!); were it also Lipschitz, it would satisfy $|\sqrt{x} - \sqrt{0}| \leqslant c|x - 0|$ which leads to $\sqrt{x} \geqslant 1/c$.

The next theorem is one of the important unifying principles of mathematics. It has applications in such disparate fields as differential equations, numerical analysis, and fractals.

Theorem 4.17 (The Banach Fixed Point Theorem)

Let X be a nonempty complete metric space. Then every contraction map $f : X \to X$ has a unique *fixed point* $x = f(x)$, and the iteration

$$x_{n+1} := f(x_n)$$

converges to it for any x_0.

The rate of convergence is given at least by $d(x, x_n) \leqslant \frac{c^n}{1-c} d(x_1, x_0)$.

Proof Consider the iteration $x_{n+1} := f(x_n)$ starting with any x_0 in X. Note that

$$d(x_{n+1}, x_n) = d(f(x_n), f(x_{n-1})) \leqslant c\, d(x_n, x_{n-1}).$$

Hence, by induction on n,

$$d(x_{n+1}, x_n) \leqslant c^n d(x_1, x_0),$$

so $(x_n)_{n \in \mathbb{N}}$ is Cauchy since $c < 1$ (Example 4.3(5)). As X is complete, x_n converges to, say, x, and by continuity of f,

$$f(x) = f(\lim_{n \to \infty} x_n) = \lim_{n \to \infty} f(x_n) = \lim_{n \to \infty} x_{n+1} = x.$$

Suppose there are two fixed points $x = f(x)$ and $y = f(y)$; then

$$d(x, y) = d(f(x), f(y)) \leqslant c\, d(x, y)$$

implying $d(x, y) = 0$ since $c < 1$.

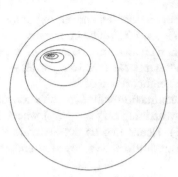

□

Exercises 4.18

1. Show that

   (a) $f : [a, b] \to \mathbb{R}$, $f(t) := t + 1/t$, is a contraction when $a > 2^{-\frac{1}{2}}$;

   (b) $f : [0, 1]^2 \to \mathbb{R}^2$, $f(x, y) := \binom{2-xy}{x^2-y}$ is Lipschitz.

2. The composition of two Lipschitz maps is Lipschitz.
3. ▶ A Lipschitz map (with constant c) sends the ball $B_r(a)$ into the ball $B_{cr}(f(a))$.
4. Isometries are necessarily 1–1. Surjective isometric maps are equivalences, and the latter are homeomorphisms.
5. ▶ Two metric spaces are said to be *equivalent* when there is an equivalence map between them. Equivalent metric spaces must be both complete or both incomplete.
6. ▶ If a space has two distances, the inequality $d_1(x, y) \leqslant c\, d_2(x, y)$, where $c > 0$, states that the identity map is Lipschitz. In the same vein, two distances are equivalent when there are $c, c' > 0$ such that

$$c'\, d_2(x, y) \leqslant d_1(x, y) \leqslant c\, d_2(x, y).$$

   Show that two equivalent distances have exactly the same Cauchy sequences.
7. The unit circle has two natural distance functions, (i) the arc length θ and (ii) the induced Euclidean distance $2\sin(\theta/2)$, where θ is the angle (at the center) between two points ($\leqslant \pi$). Prove that the two are equivalent by first showing

$$2\theta/\pi \leqslant \sin\theta \leqslant \theta, \quad \text{for } 0 \leqslant \theta \leqslant \pi/2.$$

8. The distances D_1 and D_∞ for $X \times Y$ (Example 2.2(6)) are equivalent.
9. The fixed point theorem can be generalized to the case when $f : \overline{B_r(x_0)} \to X$ is a contraction map, as long as the starting point satisfies $d(x_0, x_1) < (1-c)r$. Use the triangle inequality to show that x_n remain in $\overline{B_r(x_0)}$.

10. The classic example of an iteration converging to a fixed point is that provided by the map $t_{n+1} := (1 + t_n)^{-1}$, which converges to the golden ratio. Show that the map is a contraction on an appropriate closed interval.

11. Any continuously differentiable function $f : \mathbb{R} \to \mathbb{R}$ with $|f'(t)| < 1$ is a contraction map in a neighborhood of t.

12. If $f : \mathbb{R} \to \mathbb{R}$ is a contraction with Lipschitz constant $c < 1$, then $f(t) = t$ can also be solved by iterating $t_{n+1} := F(t_n)$ where $F(t) := t - \alpha(t - f(t))$, $0 < \alpha < 2/(c + 1)$. Hence find an approximate solution of $t = \sin t + 1$; experiment by choosing different values of α and compare with the iteration $t_{n+1} := f(t_n)$.

4.3 Separable Spaces

Completeness is a "nice" property that a metric can have. A different type of property of a metric space is whether it is, in a sense, "computable" or "constructive". Starting from the simplest, and speaking non-technically, we find:

Finite metric spaces	There are a finite number of possible distances to compute.
Countable metric spaces	With an infinite number of points, an algorithm may still calculate distances precisely, but it may take longer and longer to do so.
Separable metric spaces	Points can be approximated by one of a countable number of points; in principle, any distance can be evaluated, not precisely, but to any accuracy.
Non-separable metric spaces	There may be no algorithm that finds the distance between two generic points, even approximately.

Non-separable metric spaces are, in a sense, too large, while countable metric spaces leave out most spaces of interest.

Definition 4.19

A metric space is **separable** when it contains a countable dense subset,

$$\exists A \subseteq X, \ A \text{ is countable AND } \bar{A} = X.$$

Examples 4.20

1. Countable metric spaces, such as $\mathbb{N}, \mathbb{Z}, \mathbb{Q}$, are obviously separable.

2. ▶ $\mathbb{R}$ is separable because the countable subset $\mathbb{Q}$ is dense in it. By the next proposition, $\mathbb{C}$ and $\mathbb{R}^n$ are also separable.[1]

Proposition 4.21

> **Any subset of a separable metric space is separable.**
> **The product of two separable spaces is separable.**
> **The image of a separable space under a continuous map is separable.**

Proof (i) Let $Y \subseteq X$ and $\bar{A} = X$, with $A = \{a_n : n \in \mathbb{N}\}$ countable. For each a_n, let $Y_{n,m} := \{y \in Y : d(a_n, y) < 1/m\}$, and pick a representative point from each, $y_{n,m} \in Y_{n,m}$, whenever the set is non-empty. This array of points is certainly countable, and we now show that it is dense in Y.

Fix $0 < \epsilon < \frac{1}{2}$; any $y \in Y$ can be approximated by some $a_n \in A$ with $d(a_n, y) < \epsilon$. Pick the smallest integer m such that $m > 1/2\epsilon$; then $m - 1 \leqslant 1/2\epsilon$, so $m \leqslant 1/\epsilon$; therefore $\epsilon \leqslant 1/m < 2\epsilon$. Then $y \in Y_{n,m} \neq \varnothing$, so that there must be a representative $y_{n,m}$ with $d(a_n, y_{n,m}) < 1/m < 2\epsilon$. Combining the two inequalities, we get

$$d(y_{n,m}, y) \leqslant d(y_{n,m}, a_n) + d(a_n, y) < 3\epsilon.$$

(ii) Let $\{a_1, a_2, \ldots\}$ be dense in X, and $\{b_1, b_2, \ldots\}$ dense in Y. Then for any $\epsilon > 0$ and any pair $\binom{x}{y} \in X \times Y$, x can be approximated by some a_n such that $d_X(a_n, x) < \epsilon/2$, and y by some b_m with $d_Y(b_m, y) < \epsilon/2$; then

$$d\left(\begin{pmatrix} a_n \\ b_m \end{pmatrix}, \begin{pmatrix} x \\ y \end{pmatrix}\right) = d_X(a_n, x) + d_Y(b_m, y) < \epsilon$$

shows that the countable set of points $\binom{a_n}{b_m}$ $(n, m \in \mathbb{N})$ is dense in $X \times Y$.

(iii) Let $f : X \to Y$ be continuous and let A be countable and dense in X. Then fA is countable because the number of elements of a set cannot increase by a mapping. Moreover, as f is continuous, fA is dense in fX (Example 3.9(3)), and fX is separable. □

Exercises 4.22

1. A metric space X is separable when there is a countable number of points a_n such that the set of balls $B_\epsilon(a_n)$ covers X for any ϵ.
2. ⋆ In a separable space, we can do with a countable number of balls (with say rational radii), in the sense that every open set is a countable union of some of

---

[1] There is a catch here: The metric used in the proposition is not the Euclidean one. But the inequalities used there remain valid for the Euclidean metric, $\sqrt{d_X(a_n, x)^2 + d_Y(b_m, y)^2} < \sqrt{\epsilon^2/4 + \epsilon^2/4} < \epsilon$.

these. It then follows that every cover of the space using open sets has a countable subcover.

3. The union of a (countable) list of separable subsets is separable.
4. ▶ If there are an uncountable number of disjoint balls, then the space is non-separable, e.g., an uncountable set with the discrete metric is non-separable. We shall meet some non-trivial examples of non-separable metric spaces later on (Theorem 9.1).

Remarks 4.23

1. Note that $d(f(x), f(y)) < d(x, y)$ does not necessarily give a contraction map. For example, $f(t) := 2/(\sqrt{t^2 + 4} - t)$. In this case, the iteration $x_{n+1} := f(x_n)$ may satisfy $d(x_{n+1}, x_n) \to 0$ but need not be a Cauchy sequence.
2. The reader has most probably seen images of fractals; many of these are the fixed 'point', or *attractor*, of a contraction on the space of shapes (Example 2.2(4)) (see [19]).

3. The Banach fixed point theorem is also valid when $f^n := f \circ \cdots \circ f$, rather than f, is a contraction map; in this case the convergence is "cyclic".

Chapter 5
Connectedness

5.1 Connected Sets

We have an intuitive notion of what it means for a shape to be in one piece. The following definition makes this idea precise:

Definition 5.1

A metric space X is **disconnected** when it has a non-trivial partition of non-empty open subsets. Otherwise it is called **connected**.

Note that if $X = \bigcup_i A_i$ with A_i open, non-empty, and pairwise disjoint, then $X = A \cup B$, where $A = A_1$ and $B = \bigcup_{i \neq 1} A_i$, an open partition of two open sets. So a space X is connected when it cannot be split into *two* (or more) disjoint non-empty open subsets. To make the definition more useful we need to adapt it to the case of *subsets* of a metric space, since that is where we need it most:

Proposition 5.2

A subspace C of a metric space X is disconnected when it is the union of (at least) two disjoint non-empty subsets $C = A \cup B$ such that each subset is covered exclusively by an open set, that is,

(continued)

© The Author(s), under exclusive license to Springer Nature Switzerland AG 2024
J. Muscat, *Functional Analysis*, https://doi.org/10.1007/978-3-031-27537-1_5

$$A \subseteq U, \quad B \cap U = \varnothing, \quad U \text{ open in } X,$$
$$B \subseteq V, \quad A \cap V = \varnothing, \quad V \text{ open in } X.$$

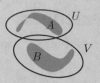

Proof By definition, C is disconnected when $C = A \cup B$ with A, B disjoint and open *relative to* C. This in turn means that $A = C \cap U$, $B = C \cap V$ for some open subsets U, V of X (Proposition 2.12). The disjointness of A, B means that $\varnothing = A \cap B = C \cap U \cap V$, which then implies $A \cap V = \varnothing = B \cap U$.

Conversely, given the conditions in the proposition, note that

$$C \cap U = (A \cup B) \cap U = (A \cap U) \cup (B \cap U) = A$$

and similarly $C \cap V = B$. Hence A, B are open in C, yet disjoint, thus forming an open partition of C.   □

Note carefully that it is *not* required that U be disjoint from V; only that they are disjoint *in* C, that is, $C \cap U \cap V = \varnothing$.

Examples 5.3

1. Single points are always connected because they cannot be split into two non-empty sets. Similarly the empty set is connected.
2. ▶ Any subset of $\mathbb{Z}$ (or any discrete metric space) is disconnected except the single points and the empty set. Metric spaces with this property are called *totally disconnected*.
   Proof: Let C contain more than one point, say a and b. Take $A = U := \{a\}$ and $B = V := C \smallsetminus \{a\} \neq \varnothing$. Then U and V are open (any subset is open) and respectively contain A and B exclusively.
3. ▶ A set A is connected when every continuous function $f : A \to \{0, 1\} \subset \mathbb{Z}$ is constant. Otherwise the open sets $f^{-1}\{0\}$ and $f^{-1}\{1\}$ cover and disconnect A.

Proposition 5.4

A subset C is connected $\Leftrightarrow$ every non-trivial subset of C has a non-empty boundary in C, that is,

$$\varnothing \neq A \subset C \Rightarrow \partial_C A \neq \varnothing.$$

Recall the definition of the relative boundary following Proposition 2.12, $\partial_C A = \{x \in C : \forall \epsilon > 0, \exists a \in A, \exists b \in C \smallsetminus A, \ a, b \in B_\epsilon(x)\}$.

Kazimierz Kuratowski (1896–1980) The Polish mathematician Kuratowski started his engineering studies in 1913 at the University of Glasgow but returned to the University of Warsaw because of World War I, changing his degree to mathematics. He rewrote much of Hausdorff's theory in 1921, introducing his closure axioms and expanding on topological connectedness. Similarly Aleksandrov and Urysohn, and later Tikhonov, in Moscow, built upon Hausdorff's work with compactness.

Proof Let $\varnothing \neq A \subset C$ be without a boundary in C. Then all the points of C are either interior points or exterior points of A; thus A and $B := C \setminus A$ are open in C. But then there are open sets U, V in X, with $A = U \cap C$ and $B = V \cap C$, and

$$U \cap B = U \cap (C \setminus A) = U \cap C \cap A^{\mathbf{c}} = A \cap A^{\mathbf{c}} = \varnothing,$$
$$V \cap A = V \cap C \cap U = B \cap U = \varnothing,$$

so $C = A \cup (C \setminus A) = A \cup B$ is disconnected.

Conversely, if C is disconnected, then $C = A \cup B$, with $A \subseteq U$, $B \subseteq V$, both non-empty, and U, V open sets in X with $A \cap V = \varnothing = B \cap U$. For any point $a \in A$, $a \in B_r(a) \subseteq U$; hence

$$a \in \{ x \in C : d(x, a) < r \} = B_r(a) \cap C \subseteq U \cap C = A$$

shows that A is open *in* C. Similarly $B = C \setminus A$ is open *in* C, thus leaving A without a boundary in C. $\qquad\square$

Theorem 5.5

The connected subsets of $\mathbb{R}$ are precisely the intervals.

Proof *Every non-trivial subset of an interval* $I \subseteq \mathbb{R}$ *has a boundary point:* Let A be a non-trivial subset of I; that A is non-trivial means that there exist $a_0 \in A$ and $b_0 \in I \setminus A$. We can assume $a_0 < b_0$, otherwise switch the roles of A and $I \setminus A$ in what follows.

Divide the interval $[a_0, b_0]$ into halves, $[a_0, c]$ and $[c, b_0]$, where $c := (a_0 + b_0)/2$ is the midpoint. If $c \in A$ let $[a_1, b_1] := [c, b_0]$, otherwise if $c \in A^{\mathbf{c}}$ let $[a_1, b_1] := [a_0, c]$. Continue taking midpoints to get a nested sequence of intervals $[a_n, b_n]$ in I, with $a_n \in A$, $b_n \in I \setminus A$.

$$A$$

$$
\begin{array}{ll}
a_0 \circ\!\!\!-\!\!\!\circ\ b_0 \\
a_1 \circ\!\!\!-\!\!\!-\!\!\!-\!\!\!-\!\!\!-\!\!\!-\!\!\!-\!\!\!-\!\!\!-\!\!\!\circ\ b_1 \\
\quad a_2 \circ\!\!\!-\!\!\!-\!\!\!-\!\!\!\circ\ b_2 \\
\quad\quad a_3 \circ\!\!-\!\!\!\circ\ b_3
\end{array}
$$

By the bisection property (Example 4.3(3)), the sequences $(a_n)_{n\in\mathbb{N}}$ and $(b_n)_{n\in\mathbb{N}}$ are Cauchy and asymptotic, and since $\mathbb{R}$ is complete, they converge: $a_n \to a$ and $b_n \to a$. The consequence is that, inside any open neighborhood $B_\epsilon(a)$, there are points $a_n \in A$ and $b_n \in I \setminus A$, making a a boundary point of A. By the preceding proposition, this translates as "every interval is connected".

Every connected subset C of $\mathbb{R}$ has the interval property $a, b \in C \Rightarrow [a, b] \subseteq C$: Let C be a connected set, and let $a, b \in C$ (say, $a < b$). Any $x \in [a, b]$ which is not in C would disconnect C using the disjoint open sets $]-\infty, x[$ and $]x, \infty[$.

Every subset of $\mathbb{R}$ with the interval property is an interval: Let A have the interval property. If $A \neq \varnothing$, say $x \in A$, and has an upper bound, then it has a least upper bound b. The interval $[x, b[$ is a subset of A because there are points of A arbitrarily close to b. Similarly if a is the greatest lower bound then $]a, x] \subseteq A$. Going through all the possibilities of whether A has upper bounds or lower bounds or none, and whether these belong to A or not, results in all the possible cases of intervals. For example, if it contains its least upper bound b but has no lower bound, then $[x, b] \subseteq A$ for any $x < b$, so that $A =]-\infty, b]$. □

By contrast, the connected sets in other metric spaces may be very difficult to describe and imagine. Even in $\mathbb{R}^2$, there are infinite connected sets such that when a single point is removed, the remaining set is totally disconnected! (For further information search for "Cantor's teepee".) Connectedness is an important intrinsic property that a set may have: it is preserved by any continuous function. Even though the codomain space may be very different from the domain, a connected set remains in 'one piece'.

Proposition 5.6

Continuous functions map connected sets to connected sets,

$f : X \to Y$ **is continuous** AND $C \subseteq X$ **is connected** $\Rightarrow$ fC **is connected.**

Proof Let C be a subset of X, and suppose fC is disconnected into the non-empty disjoint sets A and B, covered exclusively by the open sets U and V, that is,

$$fC = A \cup B \subseteq U \cup V, \qquad U \cap B = \varnothing = V \cap A.$$

Then,

$$C = f^{-1}A \cup f^{-1}B \subseteq f^{-1}U \cup f^{-1}V, \qquad f^{-1}U \cap f^{-1}B = \varnothing = f^{-1}V \cap f^{-1}A.$$

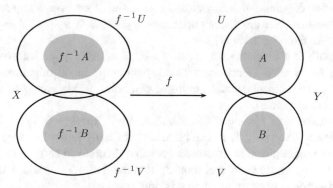

Moreover $f^{-1}A$ and $f^{-1}B$ are non-empty and disjoint, and $f^{-1}U$ and $f^{-1}V$ are open sets (Theorem 3.8). Hence fC disconnected implies C is disconnected.

□

Almost surprisingly, this simple proposition is the generalization of the classical "Intermediate Value Theorem" of Bolzano and Weierstraß. In effect, IVT has been dissected into this abstract, but transparent, statement and the previous one that intervals are connected. It embodies why abstraction is pursued in mathematics— the power of being applicable to very general spaces, with a proof that makes no use of irrelevant properties of some concrete space such as $\mathbb{R}$.

Corollary 5.7 (Intermediate Value Theorem)

> Let X be a connected space, and $f : X \to \mathbb{R}$ a continuous function. For any c with $f(a) < c < f(b)$ there exists an $x \in X$ such that $f(x) = c$.

Proof fX is connected in $\mathbb{R}$ and so must be an interval. By the interval property, $f(a), f(b) \in fX \Rightarrow c \in fX$, so $c = f(x)$ for some $x \in X$.   □

Exercises 5.8

1. Any two distinct points of a metric space are disconnected. More generally, (i) any set of n points ($n \geqslant 2$), (ii) the union of two disjoint closed sets, are disconnected.
2. The space of rational numbers $\mathbb{Q}$ is disconnected, e.g., using the open sets $]-\infty, \sqrt{2}[\cap \mathbb{Q}$ and $]\sqrt{2}, \infty[\cap \mathbb{Q}$. In fact $\mathbb{Q}$ is totally disconnected.
3. Suppose that there is an $x \in X$ and an $r > 0$ such that $d(x, y) \neq r$ for all $y \in X$, but there are points y with $d(x, y) > r$. Show that X is disconnected.

4. ▶ An open set (such as the whole metric space) is disconnected precisely when it consists of (at least) two disjoint open subsets. Find a connected set whose interior is disconnected.

5. ⋆ Any two disjoint non-empty closed sets A and B are completely separated in the sense that there are disjoint open sets $A \subseteq U$, $B \subseteq V$, $U \cap V = \varnothing$. (Hint: use Exercise 3.13(17).)

6. ▶ A **path** is a continuous function $I \to X$ where I is an interval in $\mathbb{R}$. Its image is connected. Hence show that the parametric curves of geometry, such as straight line segments, circles, ellipses, parabolas, and branches of hyperbolas in $\mathbb{R}^2$, are connected.

7. (a) The function $f(t) := t^n$ is continuous on $\mathbb{R}$, for $n = 0, 1, \dots$. Show that, for any fixed $n \geqslant 1$, t^n can be made arbitrarily large. Let x be a positive real number; use the intermediate value theorem to show that $\sqrt[n]{x}$ exists. More generally every real monic polynomial $t^n + \cdots + a_1 t + a_0$ ($n \geqslant 1$), where a_0 is negative or when n is odd, has a root.

   (b) Every continuous function $f : [0, 1] \to [0, 1]$ has a fixed point. (Hint: consider $f(t) - t$.)

8. If $f : [0, 1]^2 \to \mathbb{R}$ is continuous and $f(\boldsymbol{a}) < c < f(\boldsymbol{b})$ then there is an $\boldsymbol{x} \in [0, 1]^2$ such that $c = f(\boldsymbol{x})$. (Assume $[0, 1]^2$ is connected.)

9. Suppose X is connected and $f : X \to \mathbb{R}$ is continuous and locally constant, that is, every $x \in X$ has a neighborhood taking the value $f(x)$. Then f is constant on X. (Hint: Show $f^{-1}\{f(a)\}$ is closed and open in X.)

10. $\mathbb{Q}$ has non-interval subsets with the interval property (e.g. $[0, \sqrt{2}[\cap \mathbb{Q})$.

11. Use the intermediate value theorem to show that an injective continuous function on $[a, b]$ must be increasing or decreasing

$$s \leqslant t \Rightarrow f(s) \leqslant f(t) \quad \text{OR} \quad s \leqslant t \Rightarrow f(s) \geqslant f(t).$$

5.2   Components

It seems intuitively clear that every space is the disjoint union of connected subsets. To make this rigorous, let us present some more propositions that go some way in helping us show whether a set is connected, especially the principle that *whenever connected sets intersect, their union is connected*. This allows us to build connected sets from smaller ones.

Proposition 5.9

> **If C is connected then so is C with some boundary points (such as $\bar{C}$).**

Proof Let D be C with the addition of some boundary points. Suppose it separates as $D = A \cup B$ each covered exclusively by open sets U and V. Then C would also split up in the same way, unless $C \subseteq U$ say. This cannot be the case, for if $x \in B$ is a boundary point covered by V, then there is a ball $B_r(x) \subseteq V$ containing points of C, a contradiction. Thus D disconnected implies C is disconnected.   □

Theorem 5.10

> **If A_i, B are connected sets and $\forall i \; A_i \cap B \neq \varnothing$ then $B \cup \bigcup_i A_i$ is connected. If A_n are connected for $n = 1, 2, \ldots$, and $A_n \cap A_{n+1} \neq \varnothing$ then $\bigcup_n A_n$ is connected.**

Proof (i) Suppose the union $B \cup \bigcup_i A_i$ is disconnected and splits up into two parts covered exclusively by open sets U and V. Then B would split up into the two parts $B \cap U$ and $B \cap V$ were these to contain elements. But as B is known to be connected, one of these must be empty, say $B \cap U = \varnothing$. For any other $A := A_i$ that *is* partly covered by U (and there must be at least one) we get $A \cap V = \varnothing$ and $A \subseteq U$, for the same reason. But then $A \cap B \subseteq U \cap B = \varnothing$, contradicting the assumptions.
In particular, note that if A, B are connected and $A \cap B \neq \varnothing$, then $A \cup B$ is also connected. But the statement is true even for an uncountable number of A_i.

(ii) If $C_N := \bigcup_{n=1}^{N} A_n$ is connected, then $C_{N+1} = C_N \cup A_{N+1}$ is also connected by the first part of the theorem, since $C_N \cap A_{N+1} \neq \varnothing$. By induction, starting from the connected set $C_1 = A_1$, C_N is connected for all N. As $A_1 \subseteq C_N$ for all N, it follows from (i) that $\bigcup_{N=1}^{\infty} C_N = \bigcup_{n=1}^{\infty} A_n$ is also connected.   □

The converses of both these statements are false, but the following holds:

Proposition 5.11

Given non-empty connected sets A, B,

$A \cup B$ **is connected** $\Leftrightarrow \exists x \in A \cup B$, $\{x\} \cup A$ **and** $\{x\} \cup B$ **are connected.**

Proof Suppose no point $x \in A$ makes $\{x\} \cup B$ connected. That is, for each $x \in A$ there are two open sets which separate $\{x\} \cup B$. Call the set which contains x, U_x, and the other one V_x. They would also separate B unless $B \subseteq V_x$, and $U_x \cap B = \emptyset$. So $\bigcup_x U_x$ is an open set containing A but disjoint from B. If the same were to hold for points in B, then there would be an open set containing B but disjoint from A, making $A \cup B$ disconnected. The converse is a special case of the previous proposition.   □

Theorem 5.12

A metric space partitions into disjoint closed maximal connected subsets, called *components*. Any connected subset is contained in a component.

By a *maximal* connected set is meant a connected set C such that any $A \supsetneq C$ is disconnected.

Proof The relation $x \sim y$, defined by $\{x, y\} \subseteq C$ for some connected set C, is trivially symmetric; it is reflexive since $\{x, x\} = \{x\}$ is connected, and it is transitive because if $x, y \in C_1$ and $y, z \in C_2$ then $x, z \in C_1 \cup C_2$, which is connected by Theorem 5.10 as $y \in C_1 \cap C_2$. Moreover, another way of writing the relation $x \sim y$ is as

$$y \in \bigcup \{ C \subseteq X : x \in C, \ C \text{ connected} \},$$

so that the equivalence class $[x]$ (called the *component*) of x is the union of all the connected sets containing x. What this implies is that any connected set C that contains x must be part of the component of x. In addition, the component is connected by Theorem 5.10 and it is maximally so, as no strictly larger connected set containing x can exist. In particular, since $\overline{[x]}$ is connected (Proposition 5.9), it must be the case that $\overline{[x]} = [x]$ and $[x]$ is closed (Proposition 2.16).   □

Exercises 5.13

1. Show that $\mathbb{R}^2$ is connected by considering the radial lines all intersecting the origin.
2. ▶ More generally, if there exists a path between any two points, then the metric space is connected. (It is enough to find a path between any point and a single fixed point; why?) Such a space is said to be *path-connected*.
3. The square $[0, 1]^2$ and the half-plane $]a, \infty[\times \mathbb{R}$ are connected.
4. Intervals in $\mathbb{R}$, disks in $\mathbb{R}^2$, and balls in $\mathbb{R}^3$ are path-connected. Do balls in a general metric space have to be connected?
5. ▶ If X, Y are connected spaces then so is $X \times Y$.
6. The set $\mathbb{R}^2 \setminus \{x\}$ is connected. But $\mathbb{R} \setminus \{x\}$ is disconnected. Deduce that $\mathbb{R}$ and $\mathbb{R}^2$ are not homeomorphic.
   Using the same idea, show that $[a, b]$, $[a, b[$ and $]a, b[$ are not homeomorphic to each other, and neither is a circle to a parabola.
7. A connected metric space, such as $\mathbb{R}$, has one component, itself. At the other extreme, in *totally disconnected* spaces, the components are the single points $\{a\}$, e.g., $\mathbb{Q}$ and $\mathbb{Z}$.
8. If a subset of X has no boundary (so is closed and open) then it is the union of components of X.
9. Components need not be open sets.
10. A metric space X in which $B_r(x)$ is connected for any x and any r sufficiently small is said to be *locally connected*. Show that for a locally connected space,

    (a) the components are open in X,
    (b) any convergent sequence converges inside some component,
    (c) if X is also separable, then the components are countable in number.

Chapter 6
Compactness

In this final chapter of Part I, we encounter the second major descriptive concept available in metric spaces, after connectedness, namely the idea of *boundedness* of a subset, which is what is normally meant when one refers to "finiteness" in a geometric sense. This is not meant literally, that is, when one says "a circle is finite", one does not mean that it has a finite number of points, but rather that it does not reach out to infinity. Although this notion will be made rigorous in the first section, it is not even preserved by homeomorphisms, and therefore is not a proper metric characteristic. The concept needs to be strengthened somewhat to arrive at a property, called *compactness*, that is preserved by continuous maps.

6.1 Bounded Sets

Definition 6.1

A set B is **bounded** when the distance between any two points in the set has an upper bound,

$$\exists r > 0, \quad \forall x, y \in B, \quad d(x, y) \leqslant r.$$

The least such upper bound is called the **diameter** of the set:

$$\operatorname{diam} B := \sup_{x,y \in B} d(x, y).$$

$\operatorname{diam} B$

© The Author(s), under exclusive license to Springer Nature Switzerland AG 2024
J. Muscat, *Functional Analysis*, https://doi.org/10.1007/978-3-031-27537-1_6

The characteristic properties of bounded sets are:

Proposition 6.2

> **Any subset of a bounded set is bounded.**
> **The union of a finite number of bounded sets is bounded.**

Proof (i) Let B be a bounded set with $d(x, y) \leqslant r$ for any $x, y \in B$. In particular this holds for x, y in any subset $A \subseteq B$, so A is bounded.

(ii) Given a finite number of bounded sets $B_1,\ldots,B_N$, with diameters $r_1,\ldots,r_N$, respectively, let $r := \max(r_1, \ldots, r_N)$. Pick a representative point from each set, $a_n \in B_n$, and take the maximum distance between any two, $\widetilde{r} := \max_{m,n} d(a_m, a_n)$; it certainly exists as there are only a finite number of such pairs. Now, for any two points $x, y \in \bigcup_n B_n$, that is, $x \in B_i$, $y \in B_j$, for some i, j, and using the triangle inequality twice,

$$d(x, y) \leqslant d(x, a_i) + d(a_i, a_j) + d(a_j, y)$$
$$\leqslant r_i + \widetilde{r} + r_j$$
$$\leqslant 2r + \widetilde{r},$$

which furnishes an upper bound for the distances between points in $\bigcup_{n=1}^{N} B_n$.

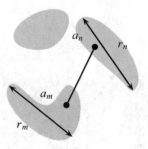

□

Examples 6.3

1. In any metric space, finite subsets are bounded. In $\mathbb{N}$, only the finite subsets are bounded. $\mathbb{N}$, $\mathbb{Q}$, $\mathbb{R}$, and $\mathbb{C}$ are all unbounded.
2. In a discrete metric space, every subset is bounded. A metric space may be non-separable ("large") yet be bounded.
3. ▶ A set B is bounded $\Leftrightarrow$ it is a subset of a ball,

$$\exists r > 0, \ \exists a \in X, \quad B \subseteq B_r(a).$$

Proof: Balls (and their subsets) are obviously bounded,

$$\forall x, y \in B_r(a), \quad d(x, y) \leqslant d(x, a) + d(y, a) < 2r.$$

Conversely, if a non-empty set is bounded by $R > 0$, fix any points $a \in X$ and $b \in B$ to conclude $B \subseteq B_r(a)$:

$$\forall x \in B, \quad d(x, a) \leqslant d(x, b) + d(b, a) < R + 1 + d(b, a) =: r.$$

4. The set $[0, 1[\cup]2, 3[\subset \mathbb{R}$ is bounded because it can be covered by the ball $B_3(0)$, or because it is the union of two bounded sets.
5. ▶ Boundedness is not necessarily preserved by continuous functions: If B is bounded and f is a continuous function, then fB need not be bounded. Worse, a set may be bounded in one metric space X, but unbounded in a homeomorphic copy Y.
   For example, $\mathbb{N}$ with the standard metric is unbounded, but its homeomorphic copy, $\mathbb{N}$ with the discrete metric, is bounded.

Exercises 6.4

1. The set $[-1, 1[$ is bounded in $\mathbb{R}$, with diameter 2; in fact, diam $[a, b[= b - a$.
2. Show that if $A \cap B \neq \varnothing$ then $\mathrm{diam}(A \cup B) \leqslant \mathrm{diam}(A) + \mathrm{diam}(B)$.
3. Any closed ball $\overline{B_r(a)} \subseteq \{ x : d(x, a) \leqslant r \}$ is bounded; hence the closure of a bounded set is bounded.
4. ▶ Cauchy sequences are bounded (Example 4.3(6)). So unbounded sequences cannot possibly converge.
5. ▶ Prove that Lipschitz functions map bounded sets to bounded sets (Exercise 4.18(3)). So equivalent metric spaces have corresponding bounded subsets.

6.2   Totally Bounded Sets

We have seen that boundedness is not an intrinsic property of a set, as it is not necessarily preserved by continuous functions. Let us try to capture the "finiteness" of a set with another definition:

Definition 6.5

A subset $B \subseteq X$ is **totally bounded** when it can be covered by a finite number of ϵ-balls, however small their radii ϵ,

$$\forall \epsilon > 0, \ \exists N \in \mathbb{N}, \ \exists a_1, \ldots, a_N \in X, \quad B \subseteq \bigcup_{n=1}^{N} B_\epsilon(a_n).$$

(continued)

Easy Consequences

1. Any subset of a totally bounded set is totally bounded (the same ϵ-cover of the parent covers the subset).
2. A finite union of totally bounded sets is totally bounded (the finite collection of ϵ-covers remains finite).
3. A totally bounded set is bounded (it is a subset of a finite union of bounded balls).

Examples 6.6

1. The interval $[0, 1]$ is totally bounded in $\mathbb{R}$ because it can be covered by the balls $B_\epsilon(n\epsilon)$ for $n = 0, \ldots, N$, where $\frac{1}{\epsilon} - 1 < N \leqslant \frac{1}{\epsilon}$.
2. Not all bounded sets are totally bounded. For example, in a discrete metric space, any subset is bounded but only finite subsets are totally bounded (take $\epsilon < 1$).
3. ▶ A totally bounded space X is separable.
   Proof: For each $n = 1, 2, \ldots$, consider finite covers of X by balls $B_{1/n}(a_{i,n})$ and let $A_n := \{a_{i,n}\}$ be the finite set of the centers, so $A := \bigcup_{n=1}^{\infty} A_n$ is countable. For any $\epsilon > 0$ and any point $x \in X$, let $\frac{1}{n} < \epsilon$, then x is covered by some ball $B_{1/n}(a_{i,n})$, i.e., $d(x, a_{i,n}) < \epsilon$, thus $\bar{A} = X$.
4. The center points a_n of the definition may, without loss of generality, be assumed to lie in B. Otherwise cover B with balls $B_{\epsilon/2}(x_n)$, and take representative points $a_n \in B \cap B_{\epsilon/2}(x_n)$ whenever non-empty; then

$$y \in B \Rightarrow d(y, x_n) < \frac{\epsilon}{2} \Rightarrow d(y, a_n) \leqslant d(y, x_n) + d(x_n, a_n) < \epsilon,$$

so $B \subseteq \bigcup_n B_\epsilon(a_n)$.

Proposition 6.7

A uniformly continuous function maps totally bounded sets to totally bounded sets.

Proof Let $f : X \to Y$ be a uniformly continuous function,

$$\forall \epsilon > 0, \ \exists \delta > 0, \ \forall x \in X, \quad f[B_\delta(x)] \subseteq B_\epsilon(f(x)).$$

Let A be a totally bounded subset of X, covered by a finite number of balls of radius δ, $A \subseteq \bigcup_{n=1}^{N} B_{\delta}(x_n)$. Then

$$fA \subseteq \bigcup_{n=1}^{N} f[B_{\delta}(x_n)] \subseteq \bigcup_{n=1}^{N} B_{\epsilon}(f(x_n)).$$

$\square$

A totally bounded set has more stringent properties than a bounded one: an infinite sequence of points in a totally bounded set is caged in, so to speak, with nowhere to escape to:

Theorem 6.8

A set B is totally bounded $\Leftrightarrow$ every sequence in B has a Cauchy subsequence.

Proof Let the totally bounded set B be covered by a finite number of balls of radius 1, and let $\{x_1, x_2, \dots\}$ be an infinite subset of B. (If B is finite, a selected sequence must take some value x_i infinitely often and so has a constant subsequence.) A finite number of balls cannot cover an infinite set of points, unless at least one of the balls, $B_1(a_1)$, has an infinite number of these points, say $\{x_{1,1}, x_{2,1}, \dots\}$.

Now cover B with a finite number of $\frac{1}{2}$-balls. For the same reason as above, at least one of these balls, $B_{1/2}(a_2)$ covers an infinite number of points of $\{x_{n,1}\}$, say the new subset $\{x_{1,2}, x_{2,2}, \dots\}$. Continue this process forming covers of $\frac{1}{m}$-balls and infinite subsets $\{x_{n,m}\}$ of $B_{1/m}(a_m)$. The sequence $(x_{n,n})$ is Cauchy, since for $m \leqslant n$, both $x_{m,m}$ and $x_{n,n}$ are elements of the set $\{x_{1,m}, x_{2,m}, \dots\}$, and so $d(x_{n,n}, x_{m,m}) < \frac{2}{m} \to 0$ as $n, m \to \infty$.

For the converse, start with any $a_1 \in B$. If $B_{\epsilon}(a_1)$ covers B then there is a single-element ϵ-ball cover. If not, pick a_2 in B but not in $B_{\epsilon}(a_1)$. Continue like this to get a sequence of distinct points $a_n \in B$ with $a_n \notin \bigcup_{i=1}^{n-1} B_{\epsilon}(a_i)$, all of which are at least ϵ distant from each other. This process cannot continue indefinitely else we get a sequence $(a_n)_{n \in \mathbb{N}}$ whose points are not close to each other, and so has no Cauchy subsequence. So after some N steps we must have $B \subseteq \bigcup_{i=1}^{N} B_{\epsilon}(a_i)$. $\square$

Exercises 6.9

1. ▶ If X and Y are totally bounded metric spaces, then so is $X \times Y$.
   (Hint: If $B_{\epsilon}(x_n)$ $(n = 1, \dots, N)$ cover X and $B_{\epsilon}(y_m)$ $(m = 1, \dots, M)$ cover Y, show that every point $(x, y) \in X \times Y$ lies in $B_{2\epsilon}(x_i, y_j)$ for some $i \leqslant N$, $j \leqslant M$.)
2. ▶ In $\mathbb{R}^n$ (and $\mathbb{C}^n$), a set is bounded $\Leftrightarrow$ it is totally bounded.
   (Hint: Show that if B is a bounded set in $\mathbb{R}^n$, with a bound $R > 0$, then B is a subset of $[-R, R]^n$, which is totally bounded by the previous exercise.)
3. The set of values of a Cauchy sequence is totally bounded.
4. The closure of a totally bounded set is totally bounded.

5. Let $B \subseteq Y \subseteq X$, then B is totally bounded in $Y \Leftrightarrow$ it is totally bounded in X (See Proposition 2.12).
6. Any bounded sequence in $\mathbb{R}^n$ (or $\mathbb{C}^n$) contains a convergent subsequence.
7. A continuous function $f : X \to Y$, with X, Y complete metric spaces, maps totally bounded subsets of X to totally bounded subsets of Y. (Hint: Consider a sequence in fB for a totally bounded set $B \subseteq X$.)

6.3 Compact Sets

In the presence of completeness, continuous functions preserve totally bounded sets. Alternatively, we can strengthen the definition of boundedness even further to a property that is preserved by continuous functions; such a property is *compactness*, but it will emerge that compact sets are precisely the complete and totally bounded subsets.

Definition 6.10

A set K is said to be **compact** when given any cover of balls (of possibly unequal radii), there is a finite sub-collection of them that still cover the set (a *subcover*),

$$K \subseteq \bigcup_i B_{\epsilon_i}(a_i) \quad \Rightarrow \quad \exists N, \exists i_1, \ldots, i_N, \quad K \subseteq \bigcup_{n=1}^{N} B_{\epsilon_{i_n}}(a_{i_n}).$$

Examples 6.11

1. Any finite set, including $\varnothing$, is compact.
2. The subset $[0, 1[\subset \mathbb{R}$ is totally bounded but *not* compact. For example, the cover using balls $B_{1-1/n}(0)$ for $n = 2, \ldots$ has no finite subcover. On the other hand, we will soon see that the closed intervals $[a, b]$ are compact.
3. ▶ Compact metric spaces are totally bounded, and so bounded and separable (consider the cover by all ϵ-balls). Thus, $\mathbb{R}$ and $\mathbb{N}$ are not compact.

An equivalent formulation of compactness is the following.

Proposition 6.12

A set is compact $\Leftrightarrow$ any open cover of it has a finite subcover.

By an *open cover* is meant a cover consisting of open sets, $K \subseteq \bigcup_i A_i$ (A_i open subsets of X).

Proof Let open sets A_i cover a compact set, $K \subseteq \bigcup_j A_j$. Each open set A_j consists of a union of balls. It follows that K is included in a union of balls. By the definition of compactness, there is a finite number of these balls $B_{\epsilon_1}(a_1), \ldots, B_{\epsilon_N}(a_N)$ that still cover the set K. Each of these balls is inside one of the open sets, say $B_{\epsilon_i}(a_i) \subseteq A_{j_i}$, and

$$K \subseteq \bigcup_{i=1}^{N} B_{\epsilon_i}(a_i) \subseteq \bigcup_{i=1}^{N} A_{j_i}$$

as claimed.

Conversely, suppose K is such that any open cover of it has a finite subcover. This holds in particular for a cover of (open) balls, so K is compact. $\square$

We will soon strengthen the following proposition to show that compact sets are complete, but the following proof is instructive, and remains valid in more general topological spaces:

Proposition 6.13

Compact sets are closed.

Proof Let K be compact and $x \in X \setminus K$. To show x is exterior to K, we need to surround it by a ball outside K. We know that x can be separated from any $y \in K$ by disjoint open balls $B_{r_y}(x)$ and $B_{r_y}(y)$ (Proposition 2.5). Since $y \in B_{r_y}(y)$, these latter balls cover K. But K is compact, so there is a finite sub-collection of these balls that still cover K,

$$K \subseteq B_{r_1}(y_1) \cup \cdots \cup B_{r_N}(y_N).$$

Now let $r := \min\{r_1, \ldots, r_N\}$; then $B_r(x) \cap K = \varnothing$ since

$$z \in B_r(x) \;\Rightarrow\; z \in B_{r_i}(x) \text{ for } i = 1, \ldots, N$$

$$\Rightarrow\; z \notin B_{r_1}(y_1) \cup \cdots \cup B_{r_N}(y_N) \supseteq K.$$

Therefore, $x \in B_r(x) \subseteq X \setminus K$. $\square$

Proposition 6.14

A closed subset of a compact set is compact.
A finite union of compact sets is compact.

Proof (i) Let F be a closed subset of a compact set K, and let the open sets A_i cover F; then

$$K \subseteq F \cup (X \smallsetminus F) \subseteq \bigcup_i A_i \cup (X \smallsetminus F).$$

The right-hand side is the union of open sets since $X \smallsetminus F$ is open when F is closed. But K is compact and therefore a finite number of these open sets are enough to cover it,

$$K \subseteq \bigcup_{i=1}^{N} A_i \cup (X \smallsetminus F), \qquad \text{so} \qquad F \subseteq \bigcup_{i=1}^{N} A_i.$$

(ii) Let the open sets A_i cover the finite union of compact sets $K_1 \cup \cdots \cup K_N$. Then they cover each individual K_n, and a finite number will then suffice in each case, $K_n \subseteq \bigcup_{k=1}^{R_n} A_{i_k}$. For $n = 1, \ldots, N$, the collection of chosen A_{i_k} remains finite, and together cover all the K_n.   □

Compactness is robust enough a concept that it is preserved by continuous functions; it is thus a truly intrinsic property of a set, as any homeomorphic copy of a compact set must also be compact.

Proposition 6.15

Continuous functions map compact sets to compact sets,

$$f : K \subseteq X \to Y \text{ continuous } \text{AND } K \text{ compact} \Rightarrow fK \text{ compact}.$$

Proof Let the sets A_i be an open cover for fK,

$$fK \subseteq \bigcup_i A_i.$$

From this can be deduced

$$K \subseteq f^{-1}[\bigcup_i A_i] = \bigcup_i f^{-1} A_i.$$

But $f^{-1} A_i$ are open sets since f is continuous (Theorem 3.8). Therefore the right-hand side is an open cover of K. As K is compact, a finite number of these open sets will do to cover it,

$$K \subseteq \bigcup_{k=1}^{N} f^{-1} A_{i_k}.$$

It follows that there is a finite subcover, $fK \subseteq \bigcup_{k=1}^{N} A_{i_k}$, as required to show fK compact. □

To summarize some previous results,

> **Continuous functions preserve compactness,**
> **Uniformly continuous functions preserve total boundedness,**
> **Lipschitz continuous functions preserve boundedness.**

An immediate corollary is this statement from classical real analysis:

Corollary 6.16

> Let $f : K \to \mathbb{R}$ be a continuous function on a compact space **K**. Then its image fK is bounded, and the function attains its bounds,
>
> $$\exists x_0, x_1 \in K, \ \forall x \in K, \quad f(x_0) \leqslant f(x) \leqslant f(x_1).$$

Proof The image fK is compact, and so bounded, $fK \subseteq B_R(0)$, i.e., $|f(x)| \leqslant R$ for all $x \in K$. Moreover compact sets are closed and so contain their boundary points. In particular fK contains $\inf fK$ and $\sup fK$ (Example 2.8(3)), i.e., $\inf fK = f(x_0)$, $\sup fK = f(x_1)$ for some $x_0, x_1 \in K$. □

A property that holds locally, i.e., in a ball around any point, will often also hold in a compact set by using a finite number of these balls. As an example of this, consider a continuous function with compact domain. By the definition of continuity, any x in the domain is surrounded by a small ball $B_{\delta_x}(x)$ on which the function varies by at most a small fixed amount ϵ; on a compact domain, a finite number of these balls and radii suffice to cover the set, so a single δ can be chosen irrespective of x. More formally:

Proposition 6.17

> Any continuous function, $f : K \to Y$, from a compact space to a metric space is uniformly continuous.
> If, moreover, f is bijective, then f is a homeomorphism.

Proof (i) By continuity of f, every $x \in K$ has a δ_x for which $f B_{\delta_x}(x) \subseteq B_\epsilon(f(x))$ (Theorem 3.8). As a preliminary step, the balls $B_{\delta_x/2}(x)$ cover K as x varies over K, so it has a finite subcover, from which can be chosen the smallest value of δ.

Now let $a, b \in K$ be any points with $d(a, b) < \delta/2$. The point a is covered by a ball $B_{\delta_x/2}(x)$ from the finite list. Indeed, $B_{\delta_x}(x)$ covers b as well since

$$d(x, b) \leqslant d(x, a) + d(a, b) < \delta_x/2 + \delta/2 \leqslant \delta_x.$$

As both a and b belong to $B_{\delta_x}(x)$, their images under f satisfy $f(a), f(b) \in B_\epsilon(f(x))$, so that

$$d(f(a), f(b)) \leqslant d(f(a), f(x)) + d(f(x), f(b)) < 2\epsilon.$$

This inequality was achieved with one δ independently of a and b, so f is uniformly continuous.

(ii) If f is continuous and onto, then $Y = fK$ is compact. But when in addition it is also 1–1, it preserves open sets: if A is open in K, then $K \smallsetminus A$ is closed, hence compact, in K; this is mapped 1–1 to the closed compact set $f[K \smallsetminus A] = Y \smallsetminus fA$, implying that fA is open in Y. This is precisely what is needed for f^{-1} to be continuous, and thus for f to be a homeomorphism. $\square$

We are now ready for some concrete examples, starting with that of $\mathbb{R}$, the simplest non-trivial complete space.

Proposition 6.18 (Heine-Borel's Theorem)

The closed interval $[a, b]$ is compact in $\mathbb{R}$.

Proof Let $\bigcup_i A_i \supseteq [a, b]$ be an open cover of the closed interval. We seek to obtain a contradiction by supposing there is no finite subcover. One of the two subintervals $[a, (a+b)/2]$ and $[(a+b)/2, b]$ (and possibly both) does not admit a finite subcover: call it $[a_1, b_1]$. Repeat this process of dividing, each time choosing a nested interval $[a_n, b_n]$ of length $(b - a)/2^n$ which does not admit a finite subcover.

Now $(a_n)_{n\in\mathbb{N}}$ and $(b_n)_{n\in\mathbb{N}}$ are asymptotic Cauchy sequences, which must therefore converge to the same limit, say, $a_n \to x$ and $b_n \to x$ (Example 4.3(3), Proposition 4.2 and Theorem 4.5). This limit x is in the set $[a, b]$ (Proposition 3.4) and is therefore covered by some open set A_{i_0}. As an interior point of it, x can be surrounded by an ϵ-ball (in this case, an interval)

$$x \in B_\epsilon(x) \subseteq A_{i_0}.$$

But $a_n \to x$ and $b_n \to x$ imply that there is an N such that $a_N, b_N \in B_\epsilon(x)$, and so $[a_N, b_N] \subseteq B_\epsilon(x) \subseteq A_{i_0}$. This contradicts how $[a_N, b_N]$ was chosen not to be covered by a finite number of A_i's, so there must have been a finite subcover to start with. □

Combined with Proposition 6.15, Theorem 5.5, Proposition 5.6, and Proposition 6.13, this proposition implies that any continuous real function maps intervals of type $[a, b]$ to intervals of the same type. The Heine-Borel theorem generalizes readily to arbitrary metric spaces.

Theorem 6.19

A set K is compact $\Leftrightarrow$ K is complete and totally bounded.

Proof *Compact sets are totally bounded*: Let K be a compact set. For any $\epsilon > 0$, cover K with the balls $B_\epsilon(x)$ for all $x \in K$. This open cover has a finite sub-cover. *Compact sets are complete*: Let $(x_n)_{n\in\mathbb{N}}$ be a Cauchy sequence which has no limit in K, so that for each $x \in K$,

$$\exists \epsilon > 0, \ \forall N, \ \exists n \geqslant N, \quad d(x_n, x) \geqslant \epsilon.$$

For this ϵ (which may depend on x),

$$\exists M, \quad n, m \geqslant M \ \Rightarrow \ d(x_n, x_m) < \epsilon/2,$$

$$\therefore \quad \epsilon \leqslant d(x_n, x) \leqslant d(x_n, x_m) + d(x_m, x) < \epsilon/2 + d(x_m, x),$$

$$\therefore \quad m \geqslant M \ \Rightarrow \ d(x_m, x) \geqslant \epsilon/2.$$

For $m < M$, the distances $d(x_m, x)$ take only a finite number of values. Hence, for each $x \in K$, there is a small enough ball $B_{r(x)}(x)$ which contains no points x_n unless $x_n = x$. This gives an open cover of K, which must have a finite sub-cover. But this implies that the sequence takes a finite set of values and so must eventually repeat and converge (Exercise 4.11(4)). In any case, there must be a limit in K. *Complete and totally bounded sets are compact*: Let K be a complete and totally bounded set. Suppose it to be covered by open sets V_i, but that no finite number of these open sets is enough to cover K. Since K is totally bounded,

$$K \subseteq \bigcup_{i=1}^{N} B_1(y_i)$$

for some $y_i \in K$ (Example 6.6(4)). If each of these balls were covered by a finite number of the open sets V_i, then so would K. So at least one of these balls needs an infinite number of V_i's to cover it; let us call this ball $B_1(x_1)$.

Now consider $B_1(x_1) \cap K$, also totally bounded. Once again, it can be covered by a finite number of balls of radius $1/2$, one of which does not have a finite subcover, say $B_{1/2}(x_2)$. Repeat this process to get a nested sequence of balls $B_{1/2^n}(x_n)$, with $x_n \in K$, none of which has a finite subcover. The sequence $(x_n)_{n \in \mathbb{N}}$ is Cauchy since $d(x_n, x_m) < 1/2^n$ (for $m > n$), and K is complete, hence $x_n \to x$ in K.

But x is covered by some open set V_{i_0}. Therefore there is an $\epsilon > 0$ such that

$$x \in B_\epsilon(x) \subseteq V_{i_0}.$$

Moreover since $1/2^n \to 0$ and $x_n \to x$, an N can be found such that $1/2^N < \epsilon/2$ and $d(x_N, x) < \epsilon/2$, so that for $d(y, x_N) < 1/2^N$,

$$d(y, x) \leqslant d(y, x_N) + d(x_N, x) < \epsilon$$

i.e., $$B_{1/2^N}(x_N) \subseteq B_\epsilon(x) \subseteq V_{i_0},$$

which contradicts the way that the balls $B_{1/2^n}(x_n)$ were chosen.   □

Corollary 6.20

> In a complete metric space, a subset K is compact $\Leftrightarrow$ K is closed and totally bounded.
> In $\mathbb{R}^n$, K is compact $\Leftrightarrow$ K is closed and bounded.

Proof In a complete metric space, a subset is complete if, and only if, it is closed (Proposition 4.7).

In the complete space $\mathbb{R}^n$, a set is totally bounded if, and only if, it is bounded (Exercise 6.9(2)). Note carefully that this remains true whether the distance is Euclidean, D_1, or D_∞ (Example 2.2(6)).   □

Theorem 6.21 (Bolzano-Weierstraß Property)

> In a metric space, a subset K is compact
>
> $\Leftrightarrow$ every sequence in K has a subsequence that converges in K
> $\Leftrightarrow$ every infinite subset of K has a limit point in K.

Proof We prove the logical equivalences in a cyclic manner.

(i) A compact set is totally bounded, and so every sequence in it has a Cauchy subsequence (Theorem 6.8). But compact metric spaces are also complete, implying convergence of this subsequence in K.

(ii) Let A be an infinite subset of K, and select a sequence of distinct terms $a_1, a_2, \ldots$ in A. Assuming that every sequence in K has a convergent subsequence, then $a_{n_i} \to a \in K$, as $i \to \infty$. For any ball $B_\epsilon(a)$, there are an infinite number of points $a_{n_i} \in B_\epsilon(a)$, making a a limit point of A (a can be equal to at most one of these distinct points). Thus K satisfies the Bolzano-Weierstraß property that every infinite subset has a limit point in K.

(iii) Let K have the Bolzano-Weierstraß property, let $(x_n)_{n \in \mathbb{N}}$ be any sequence in K and let A be the set of its values $\{x_0, x_1, x_2, \dots\}$. If A is infinite, then it has a limit point $x \in K$ and so there is a convergent subsequence $x_n \to x$ with $x_n \in A$ (Proposition 3.4). Otherwise, if A is finite, one can pick a constant subsequence. In either case there is a (Cauchy) convergent subsequence in K.

This shows, firstly, that K is totally bounded, and secondly, that every Cauchy sequence in K converges in K (Exercise 4.11(10)), that is, K is complete. Complete and totally bounded subsets are compact. □

Exercises 6.22

1. A compact set that consists of isolated points is finite.
2. In $\mathbb{Z}$, and any discrete metric space, the compact subsets are finite.
3. Show that $[0, 1] \cap \mathbb{Q}$ is closed and totally bounded in $\mathbb{Q}$ but not compact. (Hint: First show that $[0, r[\cap \mathbb{Q}$ is not compact when r is irrational.)
4. Every bounded real sequence has a convergent subsequence. Show this in two ways: (i) by bisecting intervals and choosing one that has an infinite number of values, (ii) using the Bolzano-Weierstraß property.
5. (Cantor) Let K_n be a decreasing nested sequence of non-empty compact sets. If $\bigcap_n K_n = \varnothing$ then $X \setminus K_n$ ($n = 2, 3, \dots$) form an open cover of K_1. Deduce that $\bigcap_n K_n$ is compact and non-empty. Moreover, if diam $K_n \to 0$ then $\bigcap_n K_n$ consists of a single point.
6. The Cantor set is compact, totally disconnected, and has no isolated points (Exercise 2.20(7)). (In fact, it is the only non-empty space with these properties, up to homeomorphism.)
7. The least distance between a compact set and a disjoint closed subset of a metric space is strictly positive.
8. Suppose K is a compact subset of $\mathbb{R}^2$ which lies in the half-plane $\{(x, y) : x > 0\}$. Show that the open disks with centers $(x + x^{-1}, 0)$ and radii $x > 1$ cover the half-plane, and deduce that K is enclosed by a circle that does *not* meet the y-axis.
9. The circle $\mathbb{S}^1$ is compact; more generally, any continuous path $[0, 1] \to X$ has a compact image.
10. Show that there can be no continuous map (i) $\mathbb{S}^1 \to [0, 2\pi[$ which is onto, or (ii) $\mathbb{S}^1 \to \mathbb{R}$ which is 1–1.
11. A continuous function $f : \mathbb{R}^2 \to \mathbb{R}$ takes a maximum, and a minimum, value on a continuous path $\gamma : [0, 1] \to \mathbb{R}^2$. For example, there is a maximum and a minimum distance between points on the path and the origin. Give an example to show that this is false if $[0, 1]$ is replaced by $]0, 1]$.
12. If $f : X \to K$ is bijective and continuous, and K is compact, it does not follow that X is compact. Show that the mapping $f(\theta) := (\cos\theta, \sin\theta)$ for $0 \leqslant \theta < 2\pi$, is a counter-example.

13. Generalize the Heine-Borel theorem to closed rectangles $[a, b] \times [c, d]$ in $\mathbb{R}^2$, by repeatedly dividing it into four sub-rectangles and adapting the same argument of the proof. Can you extend this further to $\mathbb{R}^n$?

14. ▶ The spheres and the closed balls in $\mathbb{R}^n$ are compact.

15. Verify that $[a, b] \cap \mathbb{Q}$ is *not* compact by finding an infinite set of rational numbers in $[a, b]$ that does not have a rational limit point.

16. Let $f : \mathbb{R}^N \to \mathbb{R}^N$ be a continuous function; consider the following iteration $x_{n+1} := f(x_n)/|f(x_n)|$ of mapping by f and normalizing. Show that there is a convergent subsequence (one for each limit point), assuming $f(x_n) \neq 0$.

17. ▶ If X, Y are compact metric spaces then so is $X \times Y$.

18. It is instructive to find an alternative proof that a continuous function maps a compact set to a compact set, using the BW property.

6.4   The Space $C(X, Y)$

The last section of this chapter is, in a sense, the culmination of Part I as it brings many strands together to tackle problems related to convergence of functions. To appreciate the difficulty involved, note that if we were to define $f_n \to f$ to mean *pointwise convergence*, that is, $f_n(x) \to f(x)$ for all $x \in X$, then no metric is involved and we could get an incomplete space: Even if we restrict to functions $[0, 1] \to [0, 1]$, the polynomials t^n converge pointwise to a discontinuous function as $n \to \infty$.

But there is a way to turn the set of continuous functions $f : [0, 1] \to \mathbb{C}$ into a complete metric space $C[0, 1]$, thereby giving one precise meaning to $f_n \to f$. In fact, we consider the more general case of bounded functions from any set to a metric space. A *bounded function* is one such that im f is bounded in the codomain Y, that is,

$$\exists r > 0, \ \forall a, b \in X, \quad d_Y(f(a), f(b)) \leqslant r.$$

Theorem 6.23

> **The space of bounded functions from a set X to a metric space Y is itself a metric space, with distance defined by**
>
> $$d(f, g) := \sup_{x \in X} d_Y(f(x), g(x)),$$
>
> **which is complete when Y is.**
>
> **It contains the closed subspace $C_b(X, Y)$ of bounded continuous functions, when X is a metric space.**

Proof *Distance*: The distance is well-defined because if im f and im g are bounded, then so is their union, and $d_Y(f(x), g(x)) \leqslant \text{diam}(\text{im } f \cup \text{im } g)$ for all $x \in X$.

That d satisfies the distance axioms follows from the same properties for d_Y;

$$d(f, g) = 0 \Leftrightarrow \forall x \in X, \quad d_Y(f(x), g(x)) = 0$$
$$\Leftrightarrow \forall x \in X, \quad f(x) = g(x)$$
$$\Leftrightarrow f = g,$$

$$d(f, g) = \sup_{x \in X} d_Y(f(x), g(x))$$
$$\leqslant \sup_{x \in X} \big(d_Y(f(x), h(x)) + d_Y(h(x), g(x))\big)$$
$$\leqslant \sup_{x \in X} d_Y(f(x), h(x)) + \sup_{x \in X} d_Y(h(x), g(x)) \quad \text{(Exercise 3.6(7b))}$$
$$= d(f, h) + d(h, g).$$

The axiom of symmetry $d(g, f) = d(f, g)$ is easily verified.

Completeness: Let $f_n : X \to Y$ be a Cauchy sequence of bounded functions, then for every $x \in X$,

$$d_Y(f_n(x), f_m(x)) \leqslant d(f_n, f_m) \to 0, \quad \text{as } n, m \to \infty$$

so $(f_n(x))$ is a Cauchy sequence in Y. When Y is complete, $f_n(x)$ converges to, say, $f(x)$.

Normally, this convergence would be expected to depend on x, being slower for some points than others. In this case however, the convergence is *uniform*, as the generic distance $d(f_n, f_m) := \sup_x d_Y(f_n(x), f_m(x))$ converges to 0. So given any $\epsilon > 0$ there is an N, such that $d_Y(f_n(x), f_m(x)) < \epsilon/2$ for any $n, m \geqslant N$ and *any* $x \in X$. For each x, we can choose $m \geqslant N$, dependent on x and large enough so that $d_Y(f_m(x), f(x)) < \epsilon/2$, and this implies

$$\forall x \in X, \quad d_Y(f_n(x), f(x)) \leqslant d_Y(f_n(x), f_m(x)) + d_Y(f_m(x), f(x)) < \epsilon \quad (6.1)$$

for any $n \geqslant N$. Since this N is independent of x, it follows that $d(f_n, f) \to 0$.

The function f is bounded because for any $x, y \in X$, using (6.1),

$$d_Y(f(x), f(y)) \leqslant d_Y(f(x), f_N(x)) + d_Y(f_N(x), f_N(y)) + d_Y(f_N(y), f(y))$$
$$< \epsilon + R_{f_N} + \epsilon \quad (6.2)$$

with N independent of x and y, and where R_{f_N} is the diameter of im f_N.

$C_b(X, Y)$ is closed: If X is a metric space and f_n are continuous, then this same inequality (6.2) shows that f is also continuous: if δ_N is small enough, then

$$d_X(x, y) < \delta_N \ \Rightarrow \ d_Y(f_N(x), f_N(y)) < \epsilon$$
$$\Rightarrow \ d_Y(f(x), f(y)) < 3\epsilon,$$

so that $f_n \to f \in C_b(X, Y)$. □

Any continuous function on a compact space is automatically bounded, so $C_b(K, Y) = C(K, Y)$, when K is compact. Moreover, we often write $C(K)$ for the complete metric space $C_b(K, \mathbb{C})$.

The convergence $f_n \to f$ in $C_b(X, Y)$ is called *uniform convergence*. It is much stronger than pointwise convergence $\forall x \in X, f_n(x) \to f(x)$; since $d(f_n, f) = \sup_x d(f_n(x), f(x))$ is decreasing to 0, f_n approximates f for large n at all values of x uniformly.

Having created these metric spaces of continuous functions, we can explore what properties they may have: connectedness, compactness, etc. Let us start with separability. Recall that continuous functions on a compact domain are uniformly continuous (Proposition 6.17). Thus any ball of a fixed radius δ is mapped by a real-valued continuous function f into a ball of radius ϵ. So, if $[a, b] \subset \mathbb{R}$ is partitioned into intervals $[x_i, x_i + \delta[$, then f maps each into an interval of length at most ϵ. Letting f take a constant value $f(x_i)$ on each interval gives a uniform approximation by a "step" function. Of course, step functions are usually discontinuous. We can improve the approximation by constructing a function consisting of straight-line segments from one end-point $(x_i, f(x_i))$ to the next $(x_i + \delta, f(x_i + \delta))$. In fact, extending this idea further, one can find quadratic or cubic polynomial fits, called "splines" that are widely used to approximate real continuous functions. Such a line of argument does give a valid proof that $C[a, b]$ is separable; in fact one can even generalize it to show that $C(K)$ is separable whenever K is a compact metric space. Stone's theorem goes further than splines and shows that the complex-valued functions on any compact subset K of $\mathbb{C}$ can be approximated by polynomials on K.

Karl Weierstraß (1815–1897) After belatedly becoming a secondary school mathematics teacher at 26 years, Weierstraß privately studied Abel's exposition of integrals and elliptic functions, until in 1854 he wrote a paper on his work and was given an honorary degree by the University of Königsberg. He then became famous with his programme of "arithmetization" of analysis: the construction of the real numbers, the rigorous derivation of calculus and calculus of variations, including a precise definition of uniform continuity; and his example of a function that is continuous but nowhere differentiable.

Theorem 6.24 (Stone–Weierstraß)

> **The polynomials (in z and $\bar{z}$) are dense in $C(K)$, when $K \subset \mathbb{C}$ is compact.**

Proof The proof is in five steps. The first two steps show that if a real-valued function $f \in C(K)$ can be approximated by a polynomial p, then another polynomial can be found that approximates $|f|$. Since the maximum of two functions $\max(f, g)$ can be written in terms of $|f - g|$, it can also be approximated by polynomials if f and g can. The fourth step, which is the main one, shows how a piecewise-linear approximation of $f \in C(K)$ can be written in terms of max and min. Together these steps prove that the polynomials $\mathbb{R}[x, y]$ are dense in the space of real continuous functions on K. The final step extends this to complex-valued continuous functions.

(i) *There are real polynomials that approximate $|t|$ on $-1 \leqslant t \leqslant 1$:* For example, let $q_1(t) := t^2$, $q_2(t) := 2t^2 - t^4, \ldots$, defined iteratively by

$$q_{n+1}(t) := q_n(t) + (t^2 - q_n(t)^2), \quad \text{starting from } q_0(t) := 0.$$

Let $y_n := q_n(t)$ for brevity, where $0 < t < 1$. Notice that

$$y_{n+1} - t = y_n - t - (y_n - t)(y_n + t)$$
$$= (y_n - t)(1 - t - y_n).$$

When $|y_n - t| \leqslant |y_1 - t| = t - t^2$, we get

$$0 < t^2 \leqslant y_n \leqslant 2t - t^2 < 1$$
$$\Rightarrow \quad -t < 1 - t - y_n < 1 - t$$
$$\Rightarrow \quad |y_{n+1} - t| \leqslant c|y_n - t|$$

where $c := \max(t, 1 - t) < 1$.

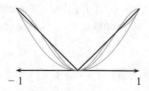

By induction, it follows that as $n \to \infty$,

$$|y_{n+1} - t| \leqslant c^n |y_1 - t| \to 0.$$

The special cases $t = 0$ and $t = 1$ converge immediately to 0 and 1 respectively, while $q_n(t) \to |t|$ when $t \in [-1, 0[$ by the symmetry of the expression in the definition of q_n.

Moreover, the convergence is uniform in t (certainly for $0 \leqslant t < \epsilon$ and $1 - \epsilon < t \leqslant 1$, but for the other positive values of t it takes at most $-2 \log \epsilon / \epsilon$ iterates for $|y_n - t| \leqslant c^n t < \epsilon$).

(ii) Let $f \in C(K, \mathbb{R})$ ($f \neq 0$) with $c := \max_{x \in K} |f(x)| + 1 > 0$ (Corollary 6.16). By the above, let $\tilde{q}(u) := cq(u/c)$, then for $u \in [-c, c]$,

$$\big| |u| - \tilde{q}(u) \big| = \big| c|u/c| - cq(u/c) \big| < c\epsilon. \tag{6.3}$$

Note that $\tilde{q}$ is uniformly continuous on $[-1, 1]$,

$$\forall \epsilon > 0, \ \exists \delta > 0, \ \forall u_1, u_2 \in [-c, c]$$

$$|u_1 - u_2| < \delta \Rightarrow |\tilde{q}(u_1) - \tilde{q}(u_2)| < \epsilon. \tag{6.4}$$

If the polynomial p approximates f to within δ, it can be expected that $\tilde{q} \circ p$ approximates $|f|$ on $C(K)$. This indeed holds since, for any $x \in K$,

$$\big| |f(x)| - \tilde{q} \circ p(x) \big| \leqslant \big| |f(x)| - \tilde{q} \circ f(x) \big| + |\tilde{q} \circ f(x) - \tilde{q} \circ p(x)| \tag{6.5}$$

$$\leqslant c\epsilon + \epsilon \qquad \text{(by (6.3) and (6.4)),} \tag{6.6}$$

$$\therefore \quad d(f, \tilde{q} \circ p) \leqslant (c + 1)\epsilon$$

(iii) For real functions, define $\max(f, g)(x) := \max(f(x), g(x))$ as well as $\min(f, g)(x) := \min(f(x), g(x))$; a short exercise shows that

$$\max(f, g) = (f + g + |f - g|)/2, \quad \min(f, g) = (f + g - |f - g|)/2.$$

If f and g can be approximated by polynomials, then so can their sum and difference, and by (ii), also $|f - g|$, and hence $\max(f, g)$ and $\min(f, g)$.

(iv) *The real polynomials are dense among the real continuous functions* $C(K, \mathbb{R})$: Let $f \in C(K, \mathbb{R})$; for any $z \neq w$ in K, there is a linear function (a polynomial) $p_{z,w}$ which agrees with f at the points z, w, i.e., $p_{z,w}(z) = f(z)$, $p_{z,w}(w) = f(w)$. For a fixed z, let

$$U_{z,w} := \{a \in K : p_{z,w}(a) < f(a) + \epsilon\} = (f - p_{z,w})^{-1}]{-\epsilon}, \infty[$$

a non-empty open set (since $f - p_{z,w}$ is continuous and $U_{z,w}$ contains z). As $w \in U_{z,w}$, we have $K \subseteq \bigcup_{z \neq w} U_{z,w}$; but K is compact so it can be covered by a finite number of subsets of this open cover,

$$K = U_{z,w_1} \cup \cdots \cup U_{z,w_M}.$$

Let $g_z := \min(p_{z, w_1}, \ldots, p_{z, w_M}) < f + \epsilon$; it is continuous and can be approximated by polynomials, from (iii). Now let

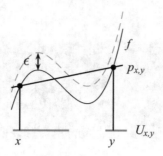

$$V_z := \{ a \in K : g_z(a) > f(a) - \epsilon \} = (f - g_z)^{-1}]-\infty, \epsilon[$$

a non-empty open set ($f - g_z$ is continuous, and $z \in V_z$). Once again, $K \subseteq \bigcup_z V_z$, and so $K = V_{z_1} \cup \cdots \cup V_{z_N}$. Let $h := \max(g_{z_1}, \ldots, g_{z_N})$, a continuous function which can be approximated by polynomials, since g_{z_i} can. Furthermore $f - \epsilon < h < f + \epsilon$; and as this holds uniformly in z, we have $d(f, h) < \epsilon$.

(v) *The set of polynomials in z and $\bar{z}$ is dense in $C(K)$*: If $f \in C(K)$ is complex-valued, then it can be written as $f = u + iv$ with u, v real-valued and continuous, that can be approximated by real polynomials p, q, say. Then,

$$\forall z \in K, \quad |(p(z) + iq(z)) - (u(z) + iv(z))| \leqslant |p(z) - u(z)| + |q(z) - v(z)|$$

$$\Rightarrow d(p + iq, u + iv) \leqslant d(p, u) + d(q, v)$$

shows that $p + iq$ approximates f. But is, say, $x^2 y + i(x^3 - xy^2)$ a polynomial in z? Not necessarily: for example, take the polynomial x itself and suppose $\text{Re}(z) = x = a_m z^m + \cdots + a_n z^n$ with $a_m \neq 0$ being the first non-zero coefficient; then $a_m = \lim_{z \to 0} \frac{x}{z^m}$, but $\text{Re}(z)/z^m$ can be made real or imaginary, so $a_m = 0$, a contradiction. Nevertheless, writing $x = (z + \bar{z})/2$ and $y = (z - \bar{z})/2i$ shows that every polynomial $p(x, y) + iq(x, y)$ is a polynomial in z and $\bar{z}$. $\qquad\square$

An immediate corollary is that $C(K)$ is separable, since the polynomials with rational coefficients are dense in the subspace of polynomials.

When is a subset of functions compact? The last theorem in this section characterizes the totally bounded sets of the space $C(K, Y)$ of continuous functions on a compact space K. Returning to the example of the polynomials t^n not converging uniformly to 0 on $[0, 1]$, it must be the case that the set $\{t^n : n \in \mathbb{N}\}$ is not totally bounded in $C[0, 1]$, otherwise there would be a convergent subsequence. The problem appears to arise because of the large slopes that they have near to $t = 1$;

t^n is uniformly continuous in t but not in n. The next definition remediates this with a property of families of functions:

Definition 6.25

A subset $F \subseteq C(X, Y)$ of continuous functions on metric spaces is said to be **equicontinuous** when

$$\forall \epsilon > 0, \ \exists \delta > 0, \ \forall f \in F, \ \forall x, x' \in X, \quad d(x, x') < \delta \Rightarrow d(f(x), f(x')) < \epsilon.$$

Theorem 6.26 (Arzelà-Ascoli)

Let K and Y be metric spaces, with K compact. Then

$F \subseteq C(K, Y)$ is totally bounded $\Leftrightarrow$ FK is totally bounded in Y and F is equicontinuous.

FK denotes the set $\{ f(x) : f \in F, x \in K \}$.

Proof (i) Let F be a totally bounded subset of $C(K, Y)$. This means that for any $\epsilon > 0$, there are a finite number of continuous functions $f_1, \ldots, f_n \in F$ that are close to within ϵ of every other function in F.

FK is totally bounded: Let $\epsilon > 0$ be arbitrary. Each $f_i K$ is compact (Proposition 6.15), so $\bigcup_{i=1}^n f_i K$ is totally bounded (Proposition 6.14 and Theorem 6.19), and covered by a finite number of balls $B_\epsilon(y_j)$, $j = 1, \ldots, m$. This means that for every $x \in K$ and $i = 1, \ldots, n$, $f_i(x)$ is close to some $y_j \in Y$. Combining this with the fact that any function $f \in F$ is close to some f_i, gives

$$d(f(x), y_j) \leqslant d(f(x), f_i(x)) + d(f_i(x), y_j) < 2\epsilon.$$

Thus each $f(x)$, where $f \in F$ and $x \in K$, is close to some y_j (j depends on x and f), in other words the finite number of balls $B_{2\epsilon}(y_j)$ cover FK.

F is equicontinuous: We have seen previously that functions $f \in C(K)$, in particular f_i, are uniformly continuous (Proposition 6.17): each $\epsilon > 0$ gives parameters δ_i. But we can say more. Since there are only a finite number of the functions f_i, the minimum $\delta := \min_i \delta_i$ can be chosen such that

$$\forall \epsilon > 0, \ \exists \delta > 0, \ \forall i, \ \forall x, x' \in K, \quad d(x, x') < \delta \Rightarrow d(f_i(x), f_i(x')) < \epsilon.$$

Marshall Stone (1903–1989) Stone studied at Harvard under Birkhoff (1926), with a thesis on ordinary differential equations and orthogonal expansions (Hermite, etc.). He then worked on spectral theory in Hilbert spaces, obtaining his big breakthrough in 1937 when he generalized the Weierstrass approximation theorem, which led him to the Stone-Čech compactification theory.

But indeed this works for any $f \in F$:

$$\forall \epsilon > 0, \ \exists \delta > 0, \ \forall f \in F, \ \forall x, x' \in K, \quad d(x, x') < \delta \Rightarrow$$

$$d(f(x), f(x')) \leqslant d(f(x), f_i(x)) + d(f_i(x), f_i(x')) + d(f_i(x'), f(x')) < 3\epsilon.$$

The *equi* in equicontinuous refers to the fact that δ is independent of $f \in F$.

(ii) Let FK be totally bounded and F be equicontinuous. Then FK can be covered by a finite number of balls $B_\epsilon(y_j)$, $j = 1, \ldots, m$, i.e., any value $f(x)$ for $f \in F$ and $x \in K$ is close to some y_j to within ϵ. 'F is equicontinuous' means that for any $\epsilon > 0$, the distance $d(f(x), f(x')) < \epsilon$ for any $f \in F$, whenever x and x' are sufficiently close together to within some $\delta > 0$ that does not depend on x, x', or f. We also require that K is totally bounded, so that it can be covered by a finite number of balls of diameter δ. By removing any overlaps between the balls, we can replace them by a finite partition of subsets B_i, $i = 1, \ldots, n$, each of diameter at most δ.

For any $f \in F$ and $x \in B_i$, $f(x)$ is close to some y_j, $d(f(x), y_j) < \epsilon$. Indeed, for any other $x' \in B_i$, we have

$$d(f(x'), y_j) \leqslant d(f(x'), f(x)) + d(f(x), y_j) < 2\epsilon,$$

because $d(x, x') \leqslant \delta$ and F is equicontinuous. In other words, the function f maps each B_i into a ball $B_{2\epsilon}(y_j)$ (j depending on i), and the whole partitioned space K into some of these balls. That is, we know f to within the approximation 2ϵ, if we know precisely how it maps each B_i to which ball $B_{2\epsilon}(y_j)$; this is equivalent to an "encoding" $i \mapsto j$ from $i = 1, \ldots, n$ to $j = 1, \ldots, m$. There are at most m^n such maps, although not all need be represented by the functions in F. For those combinations that are in fact represented by functions in F, select one from each and denote it by g_k, $k = 1, \ldots, N$.

Going back to $f \in F$, with an encoding $i \mapsto j$, pick g_k with the same encoding. Then for any $x \in K$, pick y_j close to $f(x)$ (and $g_k(x)$),

$$d(f(x), g_k(x)) \leqslant d(f(x), y_j) + d(y_j, g_k(x)) < 4\epsilon$$

and taking the supremum over x, we have $d(f, g_k) \leqslant 4\epsilon$. To summarize, the finite number of functions g_k are close to within 4ϵ to any function $f \in F$, so that F is totally bounded.   □

Examples 6.27

1. If $f_n : [0, 1] \to \mathbb{R}$ are continuous, uniformly bounded ($|f_n(x)| \leqslant c$ for all $x \in [0, 1], n \in \mathbb{N}$) and equicontinuous, then there is a subsequence that converges uniformly to a continuous function.
   Proof: The sequence $(f_n)_{n \in \mathbb{N}}$ belongs to $C[0, 1]$, a complete metric space with the supremum metric. The set $F := \{f_n : n \in \mathbb{N}\}$ is bounded by c, so $F[0, 1] \subseteq [-c, c]$, a totally bounded set in $\mathbb{R}$. Since F is also equicontinuous, it follows by the Arzelà-Ascoli theorem, that F is totally bounded. Therefore it contains a Cauchy subsequence, which converges (uniformly) in $C[0, 1]$.
2. Suppose $f_n : K \to \mathbb{R}^n$ are continuous functions on a compact space K, converging pointwise to f. If f_n are equicontinuous and uniformly bounded, then f is also continuous.
   Proof: As in the example above, f_n has a uniformly convergent subsequence $f_{n_i} \to g$ in $C(K, \mathbb{R}^n)$. But $\forall x \in K$, $f_{n_i}(x) \to f(x)$, so $f = g$, which is continuous.

Exercises 6.28

1. For the space $C[0, 1]$, (i) describe the ball $B_r(f)$, and (ii) show it is connected. (Hint: Consider $(1 - t)f + tg$.)
2. Show that $C_b(X, \mathbb{C})$ contains the closed subset $C_b(X, \mathbb{R})$.
3. Plot the functions $f(nt)$, where (i) $f(t) := \max(0, t(1 - t))$ on $[0, 1]$, and (ii) $t \mapsto 1/(1 + nt)$ on $]0, \infty[$; then show they converge pointwise to 0 as $n \to \infty$, but not uniformly.
4. $\int f_n \to \int f$ and $f_n'(t) \to f'(t)$ need not hold if f_n converges to f pointwise. Show that $t \mapsto nt^n$ and $\frac{\sin nt}{n}$ are counterexamples in $C(0, 1)$.
5. ⋆ (Dini) If K is compact and $f_n \in C(K)$ is an increasing sequence of real-valued functions, converging pointwise to $f \in C(K)$, then $f_n \to f$ in $C(K)$. (Hint: Cover K by balls $B_\delta(x)$ inside which $f - \epsilon < f_n < f$ for $n > N_x$.)
6. ⋆ The space $C[a, b]$ is separable (using piecewise linear functions with kinks at rational numbers), but $C_b(\mathbb{R}^+)$ is not.
7. The subspace of polynomials in $C[a, b]$ is not closed (and so is incomplete): construct a sequence of polynomials that converges to a non-polynomial continuous function in $C[0, 1]$.
8. Let $y_{n+1}(t) := 1 + \int_0^t y_n$. Show that this iteration converges in $C[0, 1 - \epsilon]$ to e^t, by using the fixed point theorem.
9. If $J : X \to \tilde{X}$ and $L : Y \to \tilde{Y}$ are homeomorphisms then $f \mapsto L \circ f \circ J^{-1}$ is a homeomorphism between $C(X, Y)$ and $C(\tilde{X}, \tilde{Y})$.
10. Follow the proof of the Stone-Weierstrass theorem to find a quadratic approximation to the function $f(t) := \begin{cases} t, & 0 \leqslant t \leqslant 1 \\ 0, & -1 \leqslant t < 0 \end{cases}$.

11. A set of Lipschitz functions $f : [a, b] \to \mathbb{R}$ (Definition 4.15) with the same Lipschitz constant c, $|f(s) - f(t)| \leqslant c|s - t|$, form a totally bounded set in $C[a, b]$. The fact that one c works for all, implies that they are equicontinuous; and their collective image in $\mathbb{R}$ is bounded ($|s - t| \leqslant |b - a|$), hence totally bounded.

12. Show that the sets of functions $\{\sin t, \sin 2t, \ldots\}$ and $\{t, t^2, t^3, \ldots\}$ on $[0, 1]$ are not equicontinuous.

Part II
Banach and Hilbert Spaces

Chapter 7
Normed Spaces

7.1 Vector Spaces

It is assumed that the reader has already encountered vectors and matrices before but a brief summary of their theory is provided here for reference purposes.

Definition 7.1

A **vector space** V over a field $\mathbb{F}$ is a set on which are defined an operation of *vector addition*, $+ : V^2 \to V$, satisfying associativity, commutativity, zero and inverse axioms, and an operation of *scalar multiplication*, $\mathbb{F} \times V \to V$, that satisfies the respective distributive laws: For every $x, y, z \in V$ and $\lambda, \mu \in \mathbb{F}$,

$$
\begin{aligned}
x + (y + z) &= (x + y) + z, & \lambda(x + y) &= \lambda x + \lambda y, \\
x + y &= y + x, & (\lambda + \mu)x &= \lambda x + \mu x, \\
0 + x &= x, & (\lambda \mu)x &= \lambda(\mu x), \\
x + (-x) &= 0, & 1x &= x.
\end{aligned}
$$

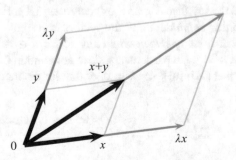

Review 7.2

1. $(-1)x = -x$, $-(-x) = x$, $0x = 0$, $\lambda 0 = 0$. There is little danger that confusing the zero scalar with the zero vector causes errors, so no attempt is made to distinguish them.
2. The field $\mathbb{F}$ is itself a vector space with scalar multiplication being plain multiplication. The smallest vector space is $\{0\}$, often written as 0.
3. The product of vector spaces (over the same field), $V \times W$, is a vector space with addition and scalar multiplication defined by

$$\begin{pmatrix} x_1 \\ y_1 \end{pmatrix} + \begin{pmatrix} x_2 \\ y_2 \end{pmatrix} := \begin{pmatrix} x_1 + x_2 \\ y_1 + y_2 \end{pmatrix}, \qquad \lambda \begin{pmatrix} x \\ y \end{pmatrix} := \begin{pmatrix} \lambda x \\ \lambda y \end{pmatrix}.$$

The zero in this case is $\begin{pmatrix} 0 \\ 0 \end{pmatrix}$ and the negatives are $-\begin{pmatrix} x \\ y \end{pmatrix} = \begin{pmatrix} -x \\ -y \end{pmatrix}$. By extension, $\mathbb{F}^n := \mathbb{F} \times \cdots \times \mathbb{F}$ is a vector space.

4. If V is a vector space, then so is the set of functions $V^A := \{ f : A \to V \}$ (for any set A) with

$$(f + g)(x) := f(x) + g(x), \qquad (\lambda f)(x) := \lambda f(x).$$

The zero of V^A is $0(x) := 0$, and the negatives are $(-f)(x) := -f(x)$.

5. A subset of a vector space V which is itself a vector space with respect to the inherited vector addition and scalar multiplication is called a *linear subspace*. Since associativity and commutativity are inherited properties, one need only check that the non-empty subset is "closed" under vector addition and scalar multiplication (then the zero $0 = 0x$ and inverses $-x = (-1)x$ are automatically in the set). Equivalently, one needs to verify that for $x, y \in W$, $\lambda \in \mathbb{F}$, then $\lambda x + y \in W$. There are always the trivial linear subspaces $\{0\}$ and V.

6. The intersection of linear subspaces is itself a linear subspace.

7. An important example of a linear subspace is that generated by a set of vectors

$$[\![A]\!] := \{ \lambda_1 v_1 + \cdots + \lambda_n v_n : v_i \in A, \ \lambda_i \in \mathbb{F}, \ n \in \mathbb{N} \},$$

with the convention that $[\![\varnothing]\!] := \{0\}$. It is the smallest linear subspace that includes A, and we say that A *spans*, or *generates*, $[\![A]\!]$. Each element of $[\![A]\!]$ is said to be a *linear combination* of the vectors in A.

8. The set A is *linearly independent* when any vector $v \in A$ is not generated by the rest, $v \notin [\![A \smallsetminus \{v\}]\!]$. (In particular A does not contain 0.) This is equivalent to saying that $\lambda v \in [\![A \smallsetminus \{v\}]\!] \Leftrightarrow \lambda = 0$, or that for distinct $v_i \in A$,

$$\sum_{i=1}^{n} \lambda_i v_i = 0 \Leftrightarrow \lambda_i = 0, \ i = 1, \ldots, n.$$

A vector generated by a linearly independent set A has unique coefficients λ_i,

$$x = \sum_{i=1}^{n} \lambda_i v_i = \sum_{i=1}^{n} \mu_i v_i \Leftrightarrow \lambda_i = \mu_i, \ i = 1, \ldots, n.$$

9. A *basis* is a minimal set of generating vectors; it must be linearly independent. Conversely, every generating set of linearly independent vectors is a basis. $\mathbb{F}^n$ has the *standard* basis $\{ e_1, \ldots, e_n \}$, where $e_i := (0, \ldots, 0, 1, 0, \ldots, 0)$, with the 1 occurring in the ith position.

10. A vector space is said to be *finite-dimensional* when it is generated by a finite number of vectors, $V = [\![v_1, \ldots, v_n]\!]$ $(:= [\![\{ v_1, \ldots, v_n \}]\!])$. The smallest such number of generating vectors is called the *dimension* of the vector space, denoted $\dim V$, and is equal to the number of vectors in a basis.

    For example, $\mathbb{F}$ has dimension 1, because it is generated by any non-zero element, while $\dim\{0\} = 0$. The linear subspace generated by two linearly independent vectors $[\![x, y]\!]$ is 2-dimensional and is called a *plane* (passing through the origin).

11. The space of $m \times n$ matrices is a finite-dimensional vector space, generated by the mn matrices E_{ij} consisting of 0s everywhere with the exception of a 1 at row i and column j.

12. We write $A + B := \{ a + b \in V : a \in A \ \text{AND} \ b \in B \}$ and $\lambda A := \{ \lambda a \in V : a \in A \}$ for any subsets $A, B \subseteq V$, e.g., $\mathbb{Q} + \mathbb{Q} = \mathbb{Q}$, $\mathbb{C} = \mathbb{R} + i\mathbb{R}$. Thus $\lambda(A \cup B) = \lambda A \cup \lambda B$, and $\lambda(A \cap B) = \lambda A \cap \lambda B$ (for $\lambda \neq 0$); a non-empty set A is a linear subspace when $\lambda A + \mu A \subseteq A$ for all $\lambda, \mu \in \mathbb{F}$. For brevity, $x + A$ is written instead of $\{x\} + A$; it is a *translation* of the set A by the vector x. Care must be taken in interpreting these symbols: $A - A = \{ a - b : a, b \in A \}$ is not usually $\{0\}$.

13. For non-empty subsets of $\mathbb{R}$, and $\lambda \geq 0$,

$$\sup(A + B) \leq \sup A + \sup B, \qquad \sup(\lambda A) = \lambda \sup A$$

Proof: Let $a + b \in A + B$, then $a \leq \sup A$ and $b \leq \sup B$, so $\sup A + \sup B$ is an upper bound of $A + B$, and hence greater than its least upper bound. Similarly, for all $a \in A$, $a \leq \sup A \Rightarrow \lambda a \leq \lambda \sup A$, so $\sup(\lambda A) \leq \lambda \sup A$. Hence, $\sup A = \sup(\frac{1}{\lambda} \lambda A) \leq \frac{1}{\lambda} \sup(\lambda A)$ and equality holds.

14. If V is finite-dimensional, then so is any linear subspace W, and $\dim W \leq \dim V$ (strictly less if it is a proper subspace).

    If V, W are finite-dimensional, then so is $V + W$, and $\dim(V + W) \leq \dim V + \dim W$.

15. In general, $[\![A \cup B]\!] = [\![A]\!] + [\![B]\!]$; so for a finite-dimensional space generated by $v_1, \ldots, v_n$, $X = [\![v_1]\!] + \cdots + [\![v_n]\!]$.

    The space V is said to decompose as a *direct sum* of its subspaces M and N, written $V = M \oplus N$, when $V = M + N$ and $M \cap N = 0$. For example, $\mathbb{R}^2 = [\![\binom{1}{1}]\!] \oplus [\![\binom{2}{1}]\!]$.

16. ► For vector spaces over $\mathbb{R}$ or $\mathbb{C}$, a subset C is said to be *convex* when it contains the line segment between any two of its points,

$$\forall x, y \in C, \quad 0 \leqslant t \leqslant 1 \Rightarrow tx + (1-t)y \in C,$$

equivalent to $sC + tC = (s+t)C$ for $s, t \geqslant 0$. This generalizes easily to $t_1 x_1 + \cdots + t_n x_n \in C$ when $t_1 + \cdots + t_n = 1$, $t_i \geqslant 0$, and $x_i \in C$. Clearly, linear subspaces are convex.

17. The intersection of convex sets is convex. There is a smallest convex set containing a subset A of a vector space, called its *convex hull*, defined as the intersection of all convex sets containing A, $\text{Conv}(A) := \bigcap_{A \subseteq C \text{ convex}} C$, which equals

$$\{ t_1 x_1 + \cdots + t_n x_n : x_i \in A, \, t_i \geqslant 0, \, t_1 + \cdots + t_n = 1, \, n \in \mathbb{N} \}.$$

If A, B are convex sets, then λA and $A + B$ are convex.

Hausdorff's Maximality Principle

The Hausdorff Maximality Principle is a statement that can be used to possibly extend arguments that work in the finite or countable case to sets of arbitrary size. There are a few proofs in this book that make use of this principle; it is only needed to extend results to "uncountably infinite" dimensions. As such, it is mainly of theoretical value, and this section can be skipped if the main interest is in applications.

Consider a collection $\mathcal{M}$ of subsets $M \subseteq X$ that satisfy a certain property $\mathcal{P}$. A *chain* $\mathcal{C} = \{M_\alpha\}_{\alpha \in I}$ of such sets is a nested sub-collection, meaning that for any two sets $M_\alpha, M_\beta \in \mathcal{C}$, either $M_\alpha \subseteq M_\beta$ or $M_\beta \subseteq M_\alpha$. A chain can contain any number of nested subsets, even uncountable. A chain is called *maximal* when it cannot be added to by the insertion of any subset in $\mathcal{M}$. Hausdorff's Maximality Principle states that

Every chain in $\mathcal{M}$ is contained in some maximal chain in $\mathcal{M}$.

Hausdorff's Maximality Principle is often used to show there is a maximal set E that satisfies some property $\mathcal{P}$ as follows: The empty chain can be extended to a maximal chain of sets M_α; *if* it can be shown that the union of this chain $E := \bigcup_\alpha M_\alpha$ also satisfies $\mathcal{P}$, then there are no sets properly containing E which satisfy $\mathcal{P}$, by the maximality of the chain M_α, i.e., E is a maximal set in $\mathcal{M}$.

At the end of this chapter, it is shown that Hausdorff's Maximality Principle implies the Axiom of Choice. In fact, it can be proved (using the other standard set axioms) that it is logically equivalent to the Axiom of Choice, as well as to a number of other formulations such as Zorn's lemma and the Well-Ordering principle. These statements are not constructive in the sense that they give no explicit way of finding the choice function or the maximal chain, but simply assert their existence.

The purpose in introducing Hausdorff's Maximality Principle here is to prove:

Every vector space has a basis.

Proof Consider the collection of all linearly independent sets of vectors in V. By Hausdorff's maximality principle, there is a maximal chain $\mathcal{M}$ of nested linearly independent sets A_α. We show that $E := \bigcup_\alpha A_\alpha$ is linearly independent and spans V, hence forms a basis. If $\sum_{i=1}^n \lambda_i v_i = 0$ for $v_i \in E$, then each of the vectors v_i ($i = 1, \ldots, n$) belongs to some A_{α_i}, and hence they all belong to some single A_α because these sets are nested in each other; but as A_α is linearly independent, $\lambda_i = 0$ for $i = 1, \ldots, n$. Thus E is linearly independent. Suppose E does not span V, meaning there is a vector $v \notin [\![E]\!]$, so that $E \cup \{v\}$ is linearly independent. As it properly contains E and every A_α, it contradicts the maximality of the chain $\mathcal{M}$.

$\square$

7.2 Norms

With the intention of extending the operations of $\mathbb{R}^n$ to infinite dimensional spaces, we would like to consider vector spaces having a metric space structure. Any set can be given a metric, so this is quite possible, but it is more interesting to have a metric that is related to vector addition and scalar multiplication in a natural way. Taking cue from Euclid's ideas of congruence and similarity, the properties that we have in mind are:

(a) *translation invariance*: distances between vectors should remain the same when they are translated by the same amount,

$$d(x + a, y + a) = d(x, y),$$

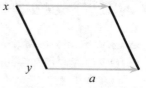

(b) *scaling homogeneity*: distances should scale in proportion when vectors are scaled,

$$d(\lambda x, \lambda y) = |\lambda| d(x, y).$$

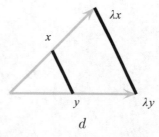

These properties are valid only for special types of metric. When d is translation invariant, then $d(x, y) = d(x - y, y - y) = d(x - y, 0)$ and d becomes essentially a function of one variable, namely the *norm* function $\|x\| := d(x, 0)$ with $d(x, y) = \|x - y\|$. Conversely, any such d defined this way is translation invariant because $d(x + a, y + a) = \|x + a - y - a\| = d(x, y)$. This function is then scaling-homogeneous precisely when

$$\|\lambda x\| = d(\lambda x, 0) = |\lambda| d(x, 0) = |\lambda| \|x\|.$$

What properties does a norm need to have, for d to be a distance? It is easy to see that

$$\left.\begin{array}{r} d(x, z) \leqslant d(x, y) + d(y, z) \\ d(y, x) = d(x, y) \\ d(x, y) \geqslant 0 \\ d(x, y) = 0 \Leftrightarrow x = y \end{array}\right\} \Leftrightarrow \left\{\begin{array}{l} \|a + b\| \leqslant \|a\| + \|b\| \\ \|-a\| = \|a\| \\ \|a\| \geqslant 0 \\ \|a\| = 0 \Leftrightarrow a = 0 \end{array}\right.$$

where $a = x - y$, $b = y - z$. Of these, the symmetry property follows from scaling-homogeneity, while positivity follows from $0 = \|x - x\| \leqslant \|x\| + \|-x\| = 2\|x\|$.

Definition 7.3

A **normed space** X is a vector space over $\mathbb{F} = \mathbb{R}$ or $\mathbb{C}$ with a function called the **norm** $\| \cdot \| : X \to \mathbb{R}^+$ such that for any $x, y \in X$, $\lambda \in \mathbb{F}$,

$$\|x + y\| \leqslant \|x\| + \|y\|, \qquad \|\lambda x\| = |\lambda| \|x\|, \qquad \|x\| = 0 \Leftrightarrow x = 0.$$

If necessary, norms on different spaces are distinguished by a subscript such as $\| \cdot \|_X$. A non-negative function that satisfies the first two axioms is termed a *semi-norm*.

Easy Consequences
1. $\|x - y\| \geqslant \big| \|x\| - \|y\| \big|$.
2. $\|x_1 + \cdots + x_n\| \leqslant \|x_1\| + \cdots + \|x_n\|$ (by induction).

Examples 7.4

1. The absolute value functions, $|\cdot|$, for $\mathbb{R}$ and $\mathbb{C}$ are themselves norms, making these the simplest normed spaces.
2. ▶ The spaces $\mathbb{R}^n$ and $\mathbb{C}^n$ of geometric vectors have a *Euclidean* norm defined by

$$\|x\|_2 = \|(a_1, \ldots, a_n)\|_2 := \left(\sum_{i=1}^{n} |a_i|^2 \right)^{1/2}.$$

There are other possibilities, e.g., $\|x\|_1 := \sum_{i=1}^{n} |a_i|$, or $\|x\|_\infty := \max_i |a_i|$. Thus

$$\left\|\binom{3}{-4}\right\|_1 = 3+4 = 7, \quad \left\|\binom{3}{-4}\right\|_2 = \sqrt{9+16} = 5, \quad \left\|\binom{3}{-4}\right\|_\infty = \max(3,4) = 4.$$

The different norms give the different distances already defined in Example 2.2(6).

3. ▶ A sequence of vectors $x_n = (a_{1,n}, \ldots, a_{N,n})$ in $\mathbb{F}^N$ converges, $x_n \to x$ (in any of these norms), precisely when each coefficient converges in $\mathbb{F}$, $a_{i,n} \to a_i$ for $i = 1, \ldots, N$.
   Proof: Using the 2-norm, for any fixed i,

$$|a_{i,n} - a_i|^2 \leqslant |a_{1,n} - a_1|^2 + \cdots + |a_{N,n} - a_N|^2 = \|x_n - x\|_2^2$$

so when the latter diminishes to 0, so does the left-hand side.
Conversely, if $a_{i,n} \to a_i$ for $i = 1, \ldots, N$, then

$$\|x_n - x\|_2 = \sqrt{|a_{1,n} - a_1|^2 + \cdots + |a_{N,n} - a_N|^2} \to 0,$$

by continuity of the various constituent functions.
With minor changes, the same proof works for the other norms as well.

4. More generally, we can define the p-norm on $\mathbb{F}^n$, $\|x\|_p := \sqrt[p]{\sum_{i=1}^{n} |a_i|^p}$ for $p \geqslant 1$. Shortly, we will see that all these norms are equivalent in finite dimensions, so we usually take the most convenient ones, such as $p = 1, 2, \infty$.

5. ▶ **Sequences**: sequences can be added and multiplied by scalars, and form a vector space.

$$(a_0, a_1, \ldots) + (b_0, b_1, \ldots) := (a_0 + b_0, a_1 + b_1, \ldots),$$

$$\lambda(a_0, a_1, \ldots) := (\lambda a_0, \lambda a_1, \ldots).$$

The zero sequence is $(0, 0, \ldots)$ and $-(a_0, a_1, \ldots) = (-a_0, -a_1, \ldots)$.

The different norms introduced above generalize to sequences; the three most important normed sequence spaces are:

(a) $\ell^1 := \{ (a_n)_{n \in \mathbb{N}} : \sum_{n=0}^{\infty} |a_n| < \infty \}$ with norm defined by

$$\|(a_n)\|_{\ell^1} := \sum_{n=0}^{\infty} |a_n|.$$

(b) $\ell^2 := \{ (a_n)_{n \in \mathbb{N}} : \sum_{n=0}^{\infty} |a_n|^2 < \infty \}$ with norm defined by

$$\|(a_n)\|_{\ell^2} := \Big(\sum_{n=0}^{\infty} |a_n|^2 \Big)^{1/2}.$$

(c) $\ell^\infty := \{ (a_n)_{n \in \mathbb{N}} : \exists c, \ |a_n| \leqslant c \}$ with norm defined by

$$\|(a_n)\|_{\ell^\infty} := \sup_{n \in \mathbb{N}} |a_n|.$$

For example, for the sequence $(1/n) = (1, \frac{1}{2}, \frac{1}{3}, \ldots)$,

$$\|(1/n)\|_{\ell^1} = \infty, \quad \|(1/n)\|_{\ell^2} = \pi/\sqrt{6}, \quad \|(1/n)\|_{\ell^\infty} = 1.$$

In each case there are two versions of the spaces, depending on whether $a_n \in \mathbb{R}$ or $\mathbb{C}$; the scalar field is then, correspondingly, real or complex. By default, we take the complex spaces as standard, unless specified otherwise.

   Note carefully that an implicit assumption is being made here that adding two sequences in a space gives another sequence in the same space. This follows from the triangle inequality for the respective norm; it is left as an exercise for ℓ^1 and ℓ^∞, but is proved for ℓ^2 in the next proposition. See Proposition 9.12 for ℓ^p.

6. These spaces are different from each other. Not only do they contain different sequences, but convergence is different in each. For example, the sequences

$$x_1 := (1, 0, 0, \ldots)$$
$$x_2 := (\tfrac{1}{2}, \tfrac{1}{2}, 0, \ldots)$$
$$x_3 := (\tfrac{1}{3}, \tfrac{1}{3}, \tfrac{1}{3}, \ldots)$$

$$\ldots$$

are all in ℓ^1, ℓ^2, and ℓ^∞. They converge $x_n \to 0$ in ℓ^∞ and ℓ^2 as $n \to \infty$,

$$\|x_n\|_{\ell^\infty} = \sup\{ \tfrac{1}{n}, 0 \} = \frac{1}{n} \to 0, \qquad \|x_n\|_{\ell^2} = \Big(\sum_{i=1}^{n} \frac{1}{n^2} \Big)^{1/2} = \frac{1}{\sqrt{n}} \to 0.$$

But they do *not* converge in ℓ^1,

$$\|x_n\|_{\ell^1} = \frac{1}{n} + \cdots + \frac{1}{n} = 1 \nrightarrow 0.$$

Thus, convergence of each coefficient is necessary, but not sufficient, for the convergence of x_n.

7. ▶ **Functions** $A \to \mathbb{F}$, where A is an interval in $\mathbb{R}$, say, also form a vector space, with

$$(f + g)(t) := f(t) + g(t), \qquad (\lambda f)(t) := \lambda f(t),$$

and different norms can be defined for them as well (once again, there are two versions of each space, depending on whether the functions are real- or complex-valued):

(a) The space $L^1(A) := \{\, f \in \mathbb{C}^A : \int_A |f(t)|\, dt < \infty \,\}$ with norm defined by

$$\|f\|_{L^1} := \int_A |f(t)|\, dt.$$

Or rather, this would be a norm, except that $\|f\|_{L^1} = \int_A |f(t)|\, dt = 0$ not when $f = 0$ but when $f = 0$ a.e. (Sect. 9.2). The failure of this axiom is not drastic, and those functions that are equal almost everywhere can be identified into equivalence classes to create a proper normed space, called *Lebesgue space* (Remark 2.24(1)). But to adopt a special notation for them, such as $[f]$, would be too pedantic to be useful; the symbol f, when used in the context of Lebesgue spaces, represents any function in its equivalence class. (The same comment holds for the next two spaces.)

(b) The space $L^2(A) := \{\, f \in \mathbb{C}^A : \int_A |f(t)|^2\, dt < \infty \,\}$, with norm defined by $\|f\|_{L^2} := \left(\int_A |f(t)|^2\, dt \right)^{\frac{1}{2}}$. More generally there are the $L^p(A)$ spaces for $p \geqslant 1$.

(c) The space

$$L^\infty(A) := \{\, f \in \mathbb{C}^A : f \text{ is measurable AND } \exists c\ |f(t)| \leqslant c \text{ a.e.} t \,\},$$

with norm defined by $\|f\|_{L^\infty} := \sup_{t \text{ a.e.}} |f(t)|$ (i.e., the smallest c such that $|f(t)| \leqslant c$ a.e.t). The term 'measurable' is explained in Sect. 9.2.

(d) The space $C_b(X, Y)$ of bounded continuous functions, defined previously (Theorem 6.23), is a normed space when Y is, with

$$\|f\|_C := \sup_{x \in X} \|f(x)\|_Y.$$

(Check that d as defined on $C_b(X, Y)$ is translation-invariant and scaling-homogeneous.) $C_b(X)$ is a linear subspace of $L^\infty(X)$, with the same norm. Note that $C_b(\mathbb{N}) = \ell^\infty$. $C_b(\mathbb{R})$ contains the closed subspace

$$C_0(\mathbb{R}) := \{\, f \in C(\mathbb{R}) : \lim_{t \to \pm\infty} f(t) = 0 \,\}.$$

For example, on $A := [0, 2\pi]$, $\|\sin\|_{L^1} = 4$, $\|\sin\|_{L^2} = \sqrt{\pi}$, and $\|\sin\|_{L^\infty} = 1$. More details and proofs for the first three spaces can be found in Sect. 9.2.

8. ▶ When X, Y are normed spaces over the same field, $X \times Y$ is also a normed space, with

$$\begin{pmatrix} x_1 \\ y_1 \end{pmatrix} + \begin{pmatrix} x_2 \\ y_2 \end{pmatrix} := \begin{pmatrix} x_1 + x_2 \\ y_1 + y_2 \end{pmatrix}, \quad \lambda \begin{pmatrix} x \\ y \end{pmatrix} := \begin{pmatrix} \lambda x \\ \lambda y \end{pmatrix}, \quad \left\| \begin{pmatrix} x \\ y \end{pmatrix} \right\| := \|x\|_X + \|y\|_Y.$$

The induced metric is D_1, defined previously for $X \times Y$ as metric spaces (Example 2.2(6)).

9. ▶ Suppose a vector space has two norms $\|\cdot\|$ and $\|\!|\cdot|\!\|$. Convergence with respect to one norm is the same as convergence with respect to the other norm when they are *equivalent* in the sense of metrics (Exercise 4.18(6)), i.e., there are positive constants $c, d > 0$,

$$c\|x\| \leqslant \|\!|x|\!\| \leqslant d\|x\|.$$

Proof: Suppose the inequalities hold and $\|x_n - x\| \to 0$, then

$$\|\!|x_n - x|\!\| \leqslant d\|x_n - x\| \to 0$$

as well; similarly if $\|\!|x_n - x|\!\| \to 0$ then $\|x_n - x\| \leqslant c^{-1}\|\!|x_n - x|\!\| \to 0$.
Conversely, suppose the ratios $\|\!|x|\!\|/\|x\|$ approach 0 as x varies in X. By rescaling, a sequence of vectors x_n can be found such that $\|x_n\| = 1$ but $\|\!|x_n|\!\| \leqslant 1/n$, i.e., $x_n \to 0$ with respect to $\|\!|\cdot|\!\|$ but not with respect to $\|\cdot\|$. For this not to happen, $\|\!|x|\!\|/\|x\| \geqslant c > 0$, and similarly, $\|x\|/\|\!|x|\!\| \geqslant 1/d > 0$.

Let us justify the claim that ℓ^2 is a normed space, by showing that the standard norm $\|(a_n)\|_{\ell^2} := \sqrt{\sum_n |a_n|^2}$ satisfies the triangle inequality, even in infinite dimensions.

Proposition 7.5 (Cauchy's Inequality)

For $a_n, b_n \in \mathbb{C}$,

$$\left| \sum_{n=0}^{\infty} a_n b_n \right| \leqslant \sqrt{\sum_{n=0}^{\infty} |a_n|^2} \sqrt{\sum_{n=0}^{\infty} |b_n|^2},$$

$$\sqrt{\sum_{n=0}^{\infty} |a_n + b_n|^2} \leqslant \sqrt{\sum_{n=0}^{\infty} |a_n|^2} + \sqrt{\sum_{n=0}^{\infty} |b_n|^2}.$$

Proof (i) Let $x = (a_n)$ and $y = (b_n)$ be sequences in ℓ^2, and let $u_n := a_n/\|x\|_{\ell^2}$, $v_n := b_n/\|y\|_{\ell^2}$. Trivially, $\sum_n |u_n|^2 = 1 = \sum_n |v_n|^2$. It is easy to show from $(a - b)^2 \geqslant 0$ that $ab \leqslant (a^2 + b^2)/2$ for any real numbers a, b. Hence,

$$\left| \sum_n u_n v_n \right| \leqslant \sum_n |u_n||v_n| \leqslant \sum_n \frac{|u_n|^2 + |v_n|^2}{2} = \frac{1}{2} + \frac{1}{2} = 1$$

Substituting back u_n, v_n, gives the required result $|\sum_n a_n b_n| \leqslant \|x\|_{\ell^2} \|y\|_{\ell^2}$.

(ii) $\sum_n |a_n + b_n|^2 \leqslant \sum_n \left(|a_n|^2 + |b_n|^2 + 2|a_n b_n| \right)$ $\qquad\qquad\qquad$ □

$$\leqslant \sum_n |a_n|^2 + \sum_n |b_n|^2 + 2\sqrt{\sum_n |a_n|^2 \sum_n |b_n|^2}$$

$$= \left(\sqrt{\sum_n |a_n|^2} + \sqrt{\sum_n |b_n|^2} \right)^2.$$

Thus for any two real sequences $x = (a_n)_{n \in \mathbb{N}}$, $y = (b_n)_{n \in \mathbb{N}}$ in ℓ^2, one can define their 'dot product'

$$x \cdot y := \sum_{n=0}^{\infty} a_n b_n$$

whose convergence is assured by Cauchy's inequality. The identity $\|x\|^2 = x \cdot x$, familiar for Euclidean spaces, remains valid for ℓ^2. Note that the two inequalities above can be written as $|x \cdot y| \leqslant \|x\| \|y\|$ and $\|x + y\| \leqslant \|x\| + \|y\|$, and that $x \cdot y$ or $x + y$ need not be finite unless both x and y are in ℓ^2.

A good strategy to adopt when tackling a question about normed spaces, is to try to answer it first for concrete examples such as $\mathbb{R}$ or $\mathbb{C}$, then $\mathbb{R}^n$, then for a sequence space such as ℓ^∞ or ℓ^2, and finally for a function space $C[0, 1]$, $L^\infty[0, 1]$, or $L^1(\mathbb{R})$. Theoretically, sequence spaces are useful as model spaces that are rich enough to exhibit most generic properties of normed spaces. But they are also indispensable in practice: a real-life function $f(t)$ is *discretized*, or digitized, into a sequence of numbers $a_i = f(t_i)$, before it can be manipulated by algorithms.

Since the metric of a normed space is translation invariant, it is not surprising that balls do not change their shape when translated.

Proposition 7.6

All balls in a normed space have the same convex shape:

$$B_r(x) = x + rB_1(0),$$
$$B_r(x) + B_s(y) = B_{r+s}(x + y), \quad \lambda B_r(x) = B_{|\lambda|r}(\lambda x).$$

Proof The norm axioms can be recast as axioms for the shape of balls. The translation-invariance and scaling-homogeneity of the distance are equivalent to

$$B_r(x + a) = \{ y : d(y, x + a) < r \} = \{ y : d(y - a, x) < r \}$$
$$= \{ a + z : d(z, x) < r \} = B_r(x) + a,$$

$$\lambda B_1(0) = \{ \lambda y : \|y\| < 1 \} = \{ z : \|z\| < |\lambda| \} = B_{|\lambda|}(0), \quad (\lambda \neq 0).$$

Combining the two gives $B_r(a) = a + rB_1(0)$, showing that all balls have the same shape as the ball of radius 1 centered at the origin.

The third norm axiom is equivalent to $\bigcap_{r>0} B_r(0) = \{0\}$, while the triangle inequality becomes $B_r(0) + B_s(0) = B_{r+s}(0)$ since

$$\|x\| < r \text{ AND } \|y\| < s \implies \|x + y\| < r + s,$$

$$\|x\| < r + s \implies x = \frac{r}{r+s}x + \frac{s}{r+s}x \in B_r(0) + B_s(0).$$

Recasting this equation as $(r + s)B_1(0) = rB_1(0) + sB_1(0)$ for $r, s \geqslant 0$ shows that $B_1(0)$, and hence all other balls, are convex: for $x, y \in B_1(0)$ and $0 \leqslant t \leqslant 1$,

$$(1 - t)x + ty \in (1 - t)B_1(0) + tB_1(0) = B_1(0).$$

In particular,

$$B_r(x) + B_s(y) = x + rB_1(0) + y + sB_1(0) = B_{r+s}(x + y),$$

$$\lambda B_r(x) = \lambda x + \lambda r B_1(0) = B_{|\lambda|r}(\lambda x).$$

$\square$

The unit ball is often denoted by $B_X := B_1(0)$ and takes a central role as representative of all other balls; its shape contains all the information about the norm of X.

Examples 7.7

1. The boundary of a ball $B_r(x)$ is the sphere $S_r(x) := \{ y \in X : d(x, y) = r \}$. Any point on the sphere has nearby points inside and outside the ball (for example, $(1 - \epsilon)y$ and $(1 + \epsilon)y$). Thus $\overline{B_r(x)} = \{ y \in X : d(x, y) \leqslant r \}$.

2. ⋆ Balls can have quite counter-intuitive properties. For example, consider the path of functions $f_t(x) := 2|x - t| - 1$ in $C[0, 1]$, starting from the function $f_0(x) = 2x - 1$ and ending at the function $f_1 = -f_0$. It lies on the unit sphere of $C[0, 1]$, but has a total length equal to the distance between f_0 and f_1,

$$\text{length} = \int_0^1 \| \frac{d}{dt} f_t \| \, dt = \int_0^1 2 \, dt = 2,$$

$$\text{distance} = \| f_1 - f_0 \|_{C[0,1]} = 2 \| f_0 \|_{C[0,1]} = 2.$$

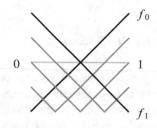

Exercises 7.8

1. For any vectors x, y, either $\|x + y\| \geqslant \|y\|$ or $\|x - y\| \geqslant \|y\|$.

2. (a) Prove that $\| \cdot \|_1$ and $\| \cdot \|_\infty$ are norms on $\mathbb{R}^n$. Which norm axiom does $\| \cdot \|_p$ fail when $p < 1$?

   (b) Show $\|\| f \|\| := \| f \|_{C[-1,0]} + \| f \|_{C[0,1]}$ is a norm on $C[-1, 1]$, equivalent to the standard supremum norm.

3. What do the unit balls of $\mathbb{R}^2$ in each norm of Example 7.4(2) look like?

4. Show that $\| \binom{a}{b} \| := |a + b| + 2|a - b|$ is a norm on $\mathbb{R}^2$. What is its unit ball?

5. The sequence $(1, 1, \ldots, 1, 0, 0, \ldots)$ is not a good approximation to the constant sequence $(1, 1, \ldots)$ in ℓ^∞; but $(1 - \epsilon, 1 - \epsilon, \ldots)$ is.

6. The norm axioms for ℓ^1 and ℓ^∞ are, when interpreted correctly,

$$\sum_n |a_n + b_n| \leqslant \sum_n |a_n| + \sum_n |b_n|, \quad \sup_n |a_n + b_n| \leqslant \sup_n |a_n| + \sup_n |b_n|,$$
$$\sum_n |\lambda a_n| = |\lambda| \sum_n |a_n|, \quad \sup_n |\lambda a_n| = |\lambda| \sup_n |a_n|,$$
$$\sum_n |a_n| = 0 \Leftrightarrow \forall n, a_n = 0, \quad \sup_n |a_n| = 0 \Leftrightarrow \forall n, a_n = 0.$$

Prove these, assuming any results about series (Sect. 7.5). Write these axioms for ℓ^2 and prove them.

7. A subset A is bounded when there is a $c > 0$ such that $\forall x \in A, \|x\| \leqslant c$ (Sect. 6.1). A non-zero normed space is not bounded.
8. For any subset A, and $r > 0$, $A + B_r(0)$ is an open set containing A.
9. ▶ The 1-, 2-, and ∞-norms are all equivalent on $\mathbb{R}^n$ since (prove!)

$$\|x\|_\infty \leqslant \|x\|_2 \leqslant \|x\|_1 \leqslant n\|x\|_\infty.$$

But they are not equivalent for sequences or functions! Find sequences of functions that converge in $L^1[0, 1]$ but not in $L^\infty[0, 1]$, or vice-versa. Can a sequence converge in ℓ^1 but not in ℓ^∞?

10. ⋆ *Minkowski semi-norm*: Let C be a convex set which is *balanced*, $e^{i\theta}C = C$ ($\forall \theta \in \mathbb{R}$), and such that $\bigcup_{r>0} rC = X$. Then

$$\|x\| := \inf\{r > 0 : x \in rC\}$$

is a semi-norm on X.

7.3   Metric and Vector Properties

By construction, normed spaces are metric spaces, as well as vector spaces. We can apply ideas related to both, in particular open/closed sets, convergence, completeness, continuity, connectedness, and compactness, as well as linear subspaces, linear independence and spanning sets, convexity, linear transformations, etc. Many of these notions have better characterizations in normed spaces, as the following propositions attest.

Proposition 7.9

Vector addition,	$(x, y) \mapsto x + y,$	$X^2 \to X,$
scalar multiplication,	$(\lambda, x) \mapsto \lambda x,$	$\mathbb{F} \times X \to X,$
and the norm	$x \mapsto \|x\|,$	$X \to \mathbb{R},$
are continuous.		

Proof Vector addition and the norm are in fact Lipschitz maps,

$$\|(x_1 + y_1) - (x_2 + y_2)\| \leqslant \|x_1 - x_2\| + \|y_1 - y_2\| = \|(x_1, y_1) - (x_2, y_2)\|_{X^2},$$

$$\big| \|x\| - \|y\| \big| \leqslant \|x - y\|.$$

Scalar multiplication is continuous: for any $\epsilon > 0$, take $|\lambda - \mu|$ to be smaller than $\epsilon/3(1 + \|x\|)$ and $\|x - y\| < \min(\epsilon/3(1 + |\lambda|), 1)$, to get

$$
\begin{aligned}
\|\lambda x - \mu y\| &\leqslant \|\lambda x - \mu x\| + \|\mu x - \mu y\| \\
&= |\lambda - \mu| \|x\| + |\mu| \|x - y\| \\
&\leqslant |\lambda - \mu| \|x\| + |\lambda| \|x - y\| + |\lambda - \mu| \|x - y\| \\
&< \epsilon.
\end{aligned}
$$

$\square$

Corollary 7.10

When $(x_n)_{n \in \mathbb{N}}$ and $(y_n)_{n \in \mathbb{N}}$ converge,

$$
\lim_{n \to \infty} (x_n + y_n) = \lim_{n \to \infty} x_n + \lim_{n \to \infty} y_n,
$$

$$
\lim_{n \to \infty} \lambda x_n = \lambda \lim_{n \to \infty} x_n,
$$

$$
\lim_{n \to \infty} \|x_n\| = \| \lim_{n \to \infty} x_n \|.
$$

Of particular importance are closed linear subspaces, because they are "closed" not only with respect to the algebraic operations of addition $+$ and scalar multiplication $\lambda \cdot$, but also with respect to convergence $\to$.

Proposition 7.11

If M is a linear subspace of X, then so is $\overline{M}$.

$\llbracket A \rrbracket$ **is the smallest closed linear space containing A.**

Proof (i) Let $x, y \in \overline{M}$, with sequences $x_n \in M$, $y_n \in M$, converging to them, $x_n \to x$ and $y_n \to y$ (Proposition 3.4). As $x_n + y_n$ and λx_n both belong to M, then

$$
x + y = \lim_{n \to \infty} x_n + \lim_{n \to \infty} y_n = \lim_{n \to \infty} (x_n + y_n) \in \overline{M},
$$

$$
\lambda x = \lambda \lim_{n \to \infty} x_n = \lim_{n \to \infty} (\lambda x_n) \in \overline{M}.
$$

Thus $\overline{M}$ is closed under vector addition and scalar multiplication. In particular this holds when M is generated by A.

(ii) $[\![A]\!]$ is the smallest linear subspace containing A, and $\overline{[\![A]\!]}$ is the smallest closed set containing $[\![A]\!]$. So any closed linear subspace containing A must also contain $[\![A]\!]$, and its closure $\overline{[\![A]\!]}$.   □

Examples 7.12

1. The following sets are closed linear subspaces of their respective spaces:

   (a) $A := \{\, (a_i)_{i \in \mathbb{N}} \in \ell^1 : \sum_{i=0}^{\infty} a_i = 0 \,\}$,
   (b) $B := \{\, f \in C[a, b] : f(a) = f(b) \,\}$.

   The proofs for closure (linearity is left as an exercise) depend on the following inequalities that hold when $x_n \to x$ in ℓ^1, $x_n = (a_{i,n})_{i \in \mathbb{N}} \in A$, and $f_n \to f$ in $C[a, b]$, $f_n \in B$,

   $$\left| \sum_{i=0}^{\infty} a_i \right| = \left| \sum_{i=0}^{\infty} a_{i,n} + \sum_{i=0}^{\infty} (a_i - a_{i,n}) \right| \leqslant \sum_{i=0}^{\infty} |a_i - a_{i,n}| = \|x - x_n\|_{\ell^1} \to 0$$

   $$f(a) = \lim_{n \to \infty} f_n(a) = \lim_{n \to \infty} f_n(b) = f(b)$$

2. ⋆ If M and N are closed subsets of a normed space, $M + N$ need not be closed (see also Exercise 7.15(8)).

   (a) Let $f : X \to Y$ be a continuous function between normed spaces; let $M := \{\, (x, f(x)) : x \in X \,\}$, $N := \{\, (x, 0) : x \in X \,\}$; they are closed subsets of $X \times Y$ (prove!). But $M + N = \{\, (\tilde{x}, f(x)) : x, \tilde{x} \in X \,\}$ is closed if, and only if, im f is closed, which need not be the case. To take a specific example,

   $$\{ (x, 0) : x \in \mathbb{R} \} + \{ (x, e^x) : x \in \mathbb{R} \} = \mathbb{R} \times \,]0, \infty[.$$

   (b) This is true even if M, N are linear subspaces. Let M be the set of ℓ^2-sequences $(a_1, 0, a_2, 0, \ldots)$ whose even terms vanish, and let N consist of ℓ^2-sequences of the type $(a_1, a_1/1, a_2, a_2/2, a_3, a_3/3, \ldots)$. They are both closed subspaces of ℓ^2 (check!). Now consider

   $$x_n := (1, 1, 1, \tfrac{1}{2}, 1, \tfrac{1}{3}, 1, \tfrac{1}{4}, \ldots, 1, \tfrac{1}{n}, 0, 0, \ldots) \in N$$

   $$y_n := (1, 0, 1, 0, 1, 0, 1, 0, \ldots, 1, 0, 0, 0, \ldots) \in M$$

   $$x_n - y_n = (0, 1, 0, \tfrac{1}{2}, 0, \tfrac{1}{3}, 0, \tfrac{1}{4}, \ldots, 0, \tfrac{1}{n}, 0, 0, \ldots) \in M + N$$

   $x_n - y_n$ converges to the sequence $(0, 1, 0, \tfrac{1}{2}, \ldots) \in \ell^2$ which cannot be expressed as a vector in $M + N$.

Connected and Compact Subsets

Recall that connected sets may be complicated objects in general metric spaces. This is still true in normed spaces, but at least for open subsets, connectedness reduces to path-connectedness, which is more intuitive and usually easier to prove.

Proposition 7.13

An open connected set in a normed space is path-connected.

Proof Let C be a non-empty open connected set in X. Recall that "path-connected" means that any two points in C can be joined by a continuous path $r : [0, 1] \to C$ starting at one point and ending at the other. Fix any $x \in C$, and let P be the subset of C consisting of those points that are path-connected to x. We wish to show that $P = C$.

P has no boundary in C: Given any boundary point z of P in C, there is a ball $B_\epsilon(z) \subseteq C$ since C is open, and thus a point $y \in P$ in the ball. This means that there is a path r from x to y. In normed spaces, it is obvious that balls, like all convex sets, are path-connected (by straight paths). So we can extend the path r to one that starts from x and ends at any other $w \in B_\epsilon(z)$, simply by adjoining the straight line at the end. More rigorously, the function $\tilde{r} : [0, 1] \to C$ defined by

$$\tilde{r}(t) := \begin{cases} r(2t), & t \in [0, \tfrac{1}{2}] \\ y + (2t - 1)(w - y), & t \in]\tfrac{1}{2}, 1] \end{cases}$$

is continuous. So z is surrounded by points of P, a contradiction.

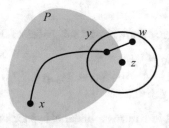

But a connected set such as C cannot contain a subset, such as P, without a boundary in C (Proposition 5.4), unless $P = \varnothing$ (which is not the case here) or $P = C$. $\qquad\square$

There is quite a bit to say about bounded and totally bounded sets. As we will see later on, they are the same in finite dimensional normed spaces, but in infinite dimensional ones, no open set can be totally bounded, although balls are bounded sets. For now, let us show that translations and scalings of bounded and totally bounded sets remain so.

Proposition 7.14

> **If A, B are both bounded, totally bounded, or compact sets, then so are, respectively, λA and $A + B$.**

Proof Proposition 7.6 is used throughout the following.

Boundedness: If $A \subseteq B_r(x)$ and $B \subseteq B_s(y)$, then

$$\lambda A \subseteq \lambda B_r(x) = B_{|\lambda|r}(\lambda x),$$

$$A + B \subseteq B_r(x) + B_s(y) = B_{r+s}(x + y).$$

Total boundedness:

$$\lambda A \subseteq \lambda \bigcup_{i=1}^{n} B_{\epsilon/|\lambda|}(x_i) = \bigcup_{i=1}^{n} B_\epsilon(\lambda x_i),$$

$$A + B \subseteq \bigcup_{i=1}^{n} B_{\epsilon/2}(x_i) + \bigcup_{j=1}^{m} B_{\epsilon/2}(y_j) = \bigcup_{i,j} B_\epsilon(x_i + y_j).$$

Compactness: If A is compact, then scalar multiplication, being continuous, sends it to the compact set λA (Proposition 6.15). If B is also compact, then $A \times B$ is compact (Exercise 6.22(17)), and vector addition, being a continuous function $X \times X \to X$, maps it to the compact set $A + B$.   □

Exercises 7.15

1. Show that the following sets are closed subspaces of their respective spaces:

    (a) $\{ (a_n)_{n \in \mathbb{N}} \in \ell^\infty : a_0 = 0 \}$,
    (d) $\{ (a_n)_{n \in \mathbb{N}} \in \ell^2 : a_1 = a_3 \text{ AND } a_0 = \sum_{n=1}^{\infty} a_n/n \}$,
    (c) $\{ f \in C[0, 1] : \int_0^1 f = 0 \}$.

2. The set of polynomials in t forms a linear subspace of $C[0, 1]$. Its dimension is infinite because the elements $1, t, t^2, \ldots$ are linearly independent. Is it closed, or if not, what could be the closure of the polynomials in this space?

3. Why is the example in 7.12(2)(b) not valid for the space ℓ^∞? Let M and N be the spaces of bounded sequences of type $(a_1, 0, a_2, 0, \ldots)$ and $(a_1, a_1, \ldots, a_n, a_n/n^2, 0, \ldots)$, respectively. Modify x_n and y_n to show that $M + N$ is not closed in ℓ^∞.

4. Show that $\{ e_n : n \in \mathbb{N} \}$ is bounded but not totally bounded in ℓ^2; and the set $\{ \frac{1}{n} e_n : n \in \mathbb{N} \}$ is totally bounded in ℓ^2.

5. The convex hull of a closed set need not be closed; a counterexample is given by $(\mathbb{R} \times \{0\}) \cup \{(0, 1)\}$. But the closure of a convex set remains convex.
6. (Mazur) The convex hull of a bounded subset is again bounded, and of a totally bounded subset is again totally bounded.
   (Hint: Cover B with balls $B_\epsilon(x_i)$; and the finite number of line segments x_i-x_j with ϵ-balls.)
7. Line segments are path-connected; so linear subspaces and convex subsets (such as balls) are connected.
8. The continuity of $+$ and $\lambda \cdot$ imply that $\overline{\lambda A} = \lambda \bar{A}$ and $\bar{A} + \bar{B} \subseteq \overline{A + B}$. Find an example to show that equality need not necessarily hold.

7.4 Complete and Separable Normed Vector Spaces

Definition 7.16

When the induced metric $d(x, y) := \|x - y\|$ is complete, the normed space is called a **Banach space**.

Examples 7.17

1. ▶ $\mathbb{R}^n$ and $\mathbb{C}^n$ are separable Banach spaces. It is later shown that the sequence spaces ℓ^p and the Lebesgue function spaces $L^p[0, 1]$, $1 \leqslant p < \infty$, are also separable Banach spaces, but ℓ^∞ is a non-separable Banach space. (See Propositions 9.14, 9.25, Exercise 9.32(8), Theorem 9.1)
2. (a) A closed linear subspace of a Banach space is itself a Banach space.
   (b) When X, Y are Banach spaces over the same field, so is $X \times Y$.
      (Proposition 4.7)

Stefan Banach (1892–1945) After WW1, at 24 years, a chance event led Banach to meet Steinhaus, who had studied under Hilbert in 1911, and was then at Krakow university. His 1920 thesis on abstract normed real vector spaces earned him a post at the University of Lwow; working mostly in the "Scottish café", he continued research on "linear operations", where he introduced weak convergence and proved various theorems such as the Hahn-Banach, Banach-Steinhaus, Banach-Alaoglu, his fixed-point theorem, and the Banach-Tarski paradox.

3. $C_b(X, Y)$ is a Banach space whenever Y is (Theorem 6.23).
4. Not every normed space is complete (when infinite dimensional).

(i) The set c_{00} of finite sequences $(a_0, \ldots, a_n, 0, 0, \ldots)$, $n \in \mathbb{N}$, is an incomplete linear subspace of ℓ^∞. For example, the vectors $(1, 0, 0, \ldots)$, $(1, \frac{1}{2}, 0, 0, \ldots)$, $\ldots$, $(1, \frac{1}{2}, \ldots, \frac{1}{n}, 0, 0, \ldots)$, $\ldots$, form a Cauchy sequence which does not converge in c_{00}.

(ii) Take the vector space of continuous functions $C[-1, 1]$ with the 1-norm $\|f\| := \int_{-1}^{1} |f(t)| \, dt$. This is indeed a norm but it is not complete on that space. For consider the sequence of continuous functions defined by

$$f_n(t) := \begin{cases} 0, & -1 \leqslant t < 0 \\ nt & 0 \leqslant t \leqslant 1/n \\ 1 & 1/n < t \leqslant 1 \end{cases}.$$

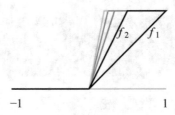

It is Cauchy:

$$\|f_n - f_m\| = \int_{-1}^{1} |f_n - f_m| = \frac{1}{2} \left| \frac{1}{n} - \frac{1}{m} \right| \to 0, \text{ as } n, m \to \infty$$

but were it to converge to some $f \in C[-1, 1]$, i.e., $\int_{-1}^{1} |f_n(t) - f(t)| \, dt \to 0$, then

$$\int_{-1}^{0} |f(t)| \, dt = 0 = \int_{1/n}^{1} |1 - f(t)| \, dt,$$

so that $f(t) = 0$ on $[-1, 0[$ and $f(t) = 1$ on $]0, 1]$, implying it is discontinuous. Similarly, the set $C[a, b]$ is not closed as a linear subspace of $L^2[a, b]$.

Proposition 7.18

Every normed space can be completed to a Banach space.

Proof Let $\widetilde{X}$ be the completion of the normed space X (Theorem 4.6). We need to prove that vector addition, scalar multiplication and the norm on X can be extended to $\widetilde{X}$. Using the notation of Theorem 4.6, let $x = [x_n]$, $y = [y_n]$ be elements of $\widetilde{X}$, with $(x_n)_{n\in\mathbb{N}}$, $(y_n)_{n\in\mathbb{N}}$ being Cauchy sequences in X. Since

$$\|x_n + y_n - x_m - y_m\| \leqslant \|x_n - x_m\| + \|y_n - y_m\| \to 0$$

$$\|\lambda x_n - \lambda x_m\| = |\lambda| \|x_n - x_m\| \to 0$$

$$\big| \|x_n\| - \|x_m\| \big| \leqslant \|x_n - x_m\| \to 0,$$

as $n, m \to \infty$, we find that $(x_n + y_n)_{n\in\mathbb{N}}$, $(\lambda x_n)_{n\in\mathbb{N}}$ and $(\|x_n\|)_{n\in\mathbb{N}}$ are all Cauchy sequences. For the same reasons, if $(x'_n)_{n\in\mathbb{N}}$ is asymptotic to $(x_n)_{n\in\mathbb{N}}$, and $(y'_n)_{n\in\mathbb{N}}$ to $(y_n)_{n\in\mathbb{N}}$, then the pairs $(x'_n + y'_n)_{n\in\mathbb{N}}$ and $(x_n + y_n)_{n\in\mathbb{N}}$, $(\lambda x'_n)_{n\in\mathbb{N}}$ and $(\lambda x_n)_{n\in\mathbb{N}}$, and $\|x'_n\|$ and $\|x_n\|$, are asymptotic to each other, respectively. So we can define

$$x + y := [x_n + y_n], \quad \lambda x := [\lambda x_n], \quad \|x\| := \lim_{n\to\infty} \|x_n\|.$$

Note that $\tilde{d}(x, y) = \|x - y\|$. It is easy to check that they give a legitimate vector addition, scalar multiplication and a norm; the required axioms follow from the same properties in X and the continuity of these operations, e.g.,

$$\|x + y\| = \lim_{n\to\infty} \|x_n + y_n\| \leqslant \lim_{n\to\infty} (\|x_n\| + \|y_n\|) = \|x\| + \|y\|,$$

$$\|x\| = 0 \Rightarrow \|x_n\| \to 0 \Rightarrow x = [x_n] = [0] = \mathbf{0}.$$

Note that the zero can be represented by the Cauchy sequence (0), and $-x$ by $(-x_n)_{n\in\mathbb{N}}$. Furthermore, recall that there is a copy of X in $\widetilde{X}$ (as constant sequences); the operations just defined on $\widetilde{X}$ reduce to the given operations on X, when restricted to it. $\qquad\square$

Proposition 7.19

A normed space X is separable if, and only if, there is a countable subset A such that $X = \overline{[\![A]\!]}$.

Proof If $X = \bar{A}$, such as when X is separable, then $X = \bar{A} \subseteq \overline{[\![A]\!]} \subseteq X$.

Conversely, suppose $X = \overline{[\![A]\!]}$ with A countable; this means that for any vector x, there is a linear combination of $a_i \in A$ $(a_i \neq 0)$, such that

$$\|\lambda_1 a_1 + \cdots + \lambda_n a_n - x\| < \epsilon \qquad \lambda_i \in \mathbb{R} \text{ or } \mathbb{C}. \tag{7.1}$$

$[\![A]\!]$ is not countable (unless $A \subseteq \{0\}$), but the set of (finite) linear combinations of vectors in A using coefficients from $\mathbb{Q} + i\mathbb{Q}$ is countable (Why? Hint: $\bigcup_n (\mathbb{Q}^2)^n$ is countable). Choosing $r_i = p_i + iq_i \in \mathbb{Q} + i\mathbb{Q}$, such that $|r_i - \lambda_i| < \frac{\epsilon}{n\|a_i\|}$, and combining with (7.1), we get

$$\|r_1 a_1 + \cdots + r_n a_n - x\| \leqslant \|(r_1 - \lambda_1)a_1 + \cdots + (r_n - \lambda_n)a_n\| + \|\lambda_1 a_1 + \cdots + \lambda_n a_n - x\|$$

$$\leqslant |r_1 - \lambda_1|\|a_1\| + \cdots + |r_n - \lambda_n|\|a_n\| + \epsilon < 2\epsilon.$$

This shows that X is separable.   □

7.5   Series

Sequences and convergence play a big role in metric spaces. Normed spaces allow sequences to be combined with summation, thereby obtaining series $x_1 + \cdots + x_n$.

Definition 7.20

A **series** $\sum_n x_n$ is a sequence of vectors in a normed space obtained by addition, $(x_0, x_0 + x_1, x_0 + x_1 + x_2, \ldots)$; the general term of the sequence is denoted by $\sum_{k=0}^n x_k$. Therefore, a series converges when $\|x - \sum_{k=0}^n x_k\| \to 0$ for some $x \in X$, as $n \to \infty$; in this case the limit x is called its **sum**,

$$x_0 + x_1 + x_2 + \cdots = \sum_{n \in \mathbb{N}} x_n = \sum_{n=0}^\infty x_n := \lim_{N \to \infty} \sum_{n=0}^N x_n = x.$$

A series is said to converge **absolutely** when $\sum_n \|x_n\|$ converges in $\mathbb{R}$.

Examples 7.21

1. Results about convergence of sequences can be converted to series:

   (a) $\sum_{n \in \mathbb{N}} (x_n + y_n) = \sum_{n \in \mathbb{N}} x_n + \sum_{n \in \mathbb{N}} y_n$ when the latter converge.
       For example, $\sum_{n \in \mathbb{N}} x_n = \sum_{n \in \mathbb{N}} x_{2n} + \sum_{n \in \mathbb{N}} x_{2n+1}$ if the latter converge.
   (b) $\sum_{n \in \mathbb{N}} \lambda x_n = \lambda \sum_{n \in \mathbb{N}} x_n$.
   (c) A series is Cauchy when $x_n + \cdots + x_m \to 0$ as $n, m \to \infty$.

2. If a series converges both normally and absolutely, then $\left\| \sum_{n=0}^\infty x_n \right\| \leqslant \sum_{n=0}^\infty \|x_n\|$.

   Proof: Take the limit of $\|x_0 + \cdots + x_n\| \leqslant \|x_0\| + \cdots + \|x_n\|$ as $n \to \infty$.
3. There are series that converge but not absolutely. As an example, take any decreasing sequence of positive real numbers $a_n \to 0$, then $\sum_n (-1)^n a_n$ converges in $\mathbb{R}$ (Leibniz); yet $\sum_n a_n$ may diverge.

Indeed, when $\sum_n a_n = \infty$ and $0 \leqslant a_n \to 0$, the series $\sum_n \pm a_n$ can converge to any $a \in \mathbb{R}$ by a judicious choice of signs. Take enough terms a_n to just exceed a, then reverse sign to lower the sum to just less than a, then reverse sign again and continue.

4. A rearrangement of a series need not converge; even if it does, it need not have the same sum. For example,

$$1 - \tfrac{1}{2} + \tfrac{1}{3} - \tfrac{1}{4} + \tfrac{1}{5} - \tfrac{1}{6} + \tfrac{1}{7} - \tfrac{1}{8} + \cdots \to \log 2,$$
$$1 - \tfrac{1}{2} + \tfrac{1}{3} + \tfrac{1}{5} + \tfrac{1}{7} + \tfrac{1}{9} - \tfrac{1}{4} + \tfrac{1}{11} + \cdots \to \infty,$$
$$1 - \tfrac{1}{2} - \tfrac{1}{4} + \tfrac{1}{3} - \tfrac{1}{6} - \tfrac{1}{8} + \tfrac{1}{5} - \tfrac{1}{10} + \cdots \to \tfrac{1}{2}\log 2,$$
$$1 - \tfrac{1}{2} + \tfrac{1}{3} + \tfrac{1}{5} - \tfrac{1}{4} + \tfrac{1}{7} + \tfrac{1}{9} - \tfrac{1}{6} + \cdots \to 1.$$

5. The sum of a 'sequence' $(x_n)_{n \in \mathbb{Z}}$ can also be given a meaning:

$$\sum_{n \in \mathbb{Z}} x_n = \sum_{n=-\infty}^{\infty} x_n := \sum_{n=1}^{\infty} x_{-n} + \sum_{n=0}^{\infty} x_n,$$

when the latter two series converge.

In general, absolute convergence is logically independent of convergence of $\sum_n x_n$. But for Banach spaces, absolute convergence implies convergence. This can be very useful, as sums of real numbers are sometimes more amenable.

Proposition 7.22

A normed space X is complete if, and only if, any absolutely convergent series in X converges.

Proof Let X be a Banach space, and suppose that $\sum_n \|x_n\|$ converges. Let $y_N := \sum_{n=0}^{N} x_n$, so that for $M > N$,

$$\|y_M - y_N\| = \Big\| \sum_{n=N+1}^{M} x_n \Big\| \leqslant \sum_{n=N+1}^{M} \|x_n\| \to 0 \quad \text{as } N, M \to \infty.$$

Hence (y_N) is a Cauchy sequence in the complete space X, and so converges.

Conversely, let X be a normed space for which every absolutely convergent series converges. Let $(x_n)_{n \in \mathbb{N}}$ be a Cauchy sequence in X, so that for $n, m \geqslant N_\epsilon$ large enough, $\|x_n - x_m\| < \epsilon$. Letting $\epsilon := 1/2^r$, $r = 1, 2, \ldots$, we can find ever larger numbers n_r such that $\|x_{n_r} - x_{n_{r+1}}\| < 1/2^r$. Thus,

$$\sum_{r=1}^{\infty} \|x_{n_{r+1}} - x_{n_r}\| \leqslant \sum_{r=1}^{\infty} \frac{1}{2^r} = 1.$$

By assumption, since its absolute series converges, so does $\sum_r (x_{n_r} - x_{n_{r+1}})$, i.e.,

$$x_{n_1} - x_{n_r} = (x_{n_1} - x_{n_2}) + (x_{n_2} - x_{n_3}) + \cdots + (x_{n_{r-1}} - x_{n_r})$$

converges as $r \to \infty$. This forces the subsequence x_{n_r} to converge, and so must the parent Cauchy sequence $(x_n)_{n \in \mathbb{N}}$ (Proposition 4.2).   □

Series can be used to extend the idea of a basis as follows: a fixed list of unit vectors e_n is called a (Schauder) **basis** when for any $x \in X$ there are unique coefficients $\alpha_n \in \mathbb{F}$ such that

$$x = \sum_{n=1}^{\infty} \alpha_n e_n.$$

The set $E := \{ e_1, e_2, \dots \}$ has to be linearly independent and dense $X = \overline{[\![E]\!]}$; by necessity X must be separable (though not every separable space has a Schauder basis [31]). Note that a Schauder basis need *not* be a linearly independent set of spanning vectors; the latter is called a *Hamel basis* for distinction.

Since a vector $x = \sum_{n \in \mathbb{N}} \alpha_n e_n$ is identified by its sequence of coefficients $(\alpha_n)_{n \in \mathbb{N}}$ with respect to a Schauder basis, the space X is essentially a sequence space with norm $\| (\alpha_n) \| := \| \sum_{n \in \mathbb{N}} \alpha_n e_n \|_X$. Ideally, shuffling a basis should not make a difference, but not every basis has this property; if it does, the basis is termed *unconditional*. There are examples of Banach spaces which have no unconditional bases.

Convergence Tests

Real series are easier to handle than series of vectors, and a number of tests for absolute convergence have been devised:

Comparison Test If $\| x_n \| \leqslant a_n$ then $\sum_{n=0}^{N} \| x_n \| \leqslant \sum_{n=0}^{N} a_n$. If the latter converges to $\sum_{n=0}^{\infty} a_n$, then $\sum_n \| x_n \|$ is increasing and bounded above, so converges.

An important special case is comparison with the geometric series, $\| x_n \| \leqslant r^n$ with $r < 1$, because $1 + r + r^2 + \cdots = 1/(1 - r)$. This leads to:

Juliusz Schauder (1899–1943) Schauder, after fighting in WWI, graduated at 24 years from the University of Lwow under Steinhaus with a dissertation on statistics. He continued researching in the Banach/Steinhaus school, giving the theory of compact operators its modern shape; he proved that the adjoint of a compact operator is compact, the Schauder fixed point theorem, and generalized aspects of orthonormal bases to Banach spaces; later he specialized to partial differential equations. Along with many other Polish academics, he was killed by the Nazis during WWII.

Root Test Let $r := \limsup_n \|x_n\|^{1/n}$;

(a) if $r < 1$ then the series $\sum_n x_n$ is absolutely convergent,
(b) if $r = 1$ then the series may or may not converge,
(c) if $r > 1$ then the series diverges.

Proof: (a) $\|x_n\| \leqslant (r + \epsilon)^n$ except for finitely many terms. Since the right-hand side is a convergent geometric series when $r < 1$ and ϵ is taken small enough, the left-hand side series also converges by comparison.

(b) The series $\sum_{n=1}^{\infty} \frac{1}{n} = \infty$ and $\sum_{n=1}^{\infty} \frac{1}{n^2} < 2$ both have $r = 1$.

(c) When $r > 1$, $\|x_n\| \geqslant (1 + \epsilon)^n > 1$ for infinitely many terms, so the series $\sum_n \|x_n\|$ cannot possibly converge.

Ratio Test (D'Alembert's) If the ratios $\|x_{n+1}\|/\|x_n\| \to r$ then $\|x_n\|^{1/n} \to r$ and the root test applies; it is often easier to find the first limit, if it exists, than the second.

Proof: The idea is that for large n, $\|x_n\| \approx r\|x_{n-1}\| \approx r^n\|x_0\|$, so $\|x_n\|^{1/n} \approx r$. More precisely, for $n \geqslant N$ large enough,

$$r - \epsilon < \|x_n\|/\|x_{n-1}\| < r + \epsilon,$$

$$\therefore \quad (r - \epsilon)^{n-N}\|x_N\| < \|x_n\| < (r + \epsilon)^{n-N}\|x_N\|,$$

$$\therefore \quad r - 2\epsilon < \|x_n\|^{1/n} < r + 2\epsilon,$$

since $(r \pm \epsilon)^{-N/n}\|x_N\|^{1/n} \to 1$.

Cauchy's Test If $\|x_n\|$ is decreasing, then $\sum_n \|x_n\|$ converges $\Leftrightarrow$ $\sum_n 2^n\|x_{2^n}\|$ converges.

Proof: Let $r_n := \|x_n\|$; the test follows from two comparisons,

$$r_1 + r_2 + \cdots + r_{2^{n+1}-1} = r_1 + (r_2 + r_3) + \cdots + (r_{2^n} + \cdots + r_{2^{n+1}-1})$$

$$\leqslant r_1 + 2r_2 + \cdots + 2^n r_{2^n}.$$

$$r_1 + 2r_2 + 4r_4 + \cdots + 2^n r_{2^n} \leqslant r_1 + 2r_2 + 2(r_3 + r_4) + \cdots$$

$$+ 2(r_{2^{n-1}+1} + \cdots + r_{2^n})$$

$$\leqslant 2(r_1 + r_2 + \cdots + r_{2^n}).$$

Kummer's Test Let $\sum_n \frac{1}{r_n}$ be a divergent series of positive terms and

$$\frac{r_{n+1}}{r_n} \frac{\|x_{n+1}\|}{\|x_n\|} = 1 - \frac{\alpha}{r_n} + o(1/r_n).$$

If $\alpha > 0$, then the series $\sum_n x_n$ converges absolutely, otherwise when $\alpha < 0$ the series diverges. For example, $r_n := 1$ gives the ratio test, $r_n := n$ is Gauss's or Raabe's test, and $r_n := n \log n$ is Bertrand's test.

Proof: When $\alpha > 0$, we are given that $c\|x_n\| \leqslant r_n \|x_n\| - r_{n+1}\|x_{n+1}\|$ for $n \geqslant N$ large enough, and some $0 < c < \alpha$. Summing up these inequalities results in

$$c(\|x_N\| + \cdots + \|x_m\|) \leqslant r_N\|x_N\| - r_{m+1}\|x_{m+1}\| \leqslant r_N\|x_N\|$$

so the series converges as it is increasing but bounded above.

When $\alpha < 0$, we have $r_n\|x_n\| < r_{n+1}\|x_{n+1}\|$ for $n \geqslant N$ large enough. Hence

$$\|x_n\| > \frac{r_N\|x_N\|}{r_n}$$

and the series diverges by comparison with the series $\sum_n \frac{1}{r_n}$.

There are yet other tests, for example, Cauchy's inequality shows that $\sum_n a_n b_n$ converges when $\sum_n a_n^2$ and $\sum_n b_n^2$ do.

Exercises 7.23

1. If a series $\sum_n x_n$ converges, then $x_n \to 0$ as $n \to \infty$. The converse is false:

$$1 + \frac{1}{2} + \frac{1}{3} + \frac{1}{4} + \cdots + \frac{1}{n} \to \infty.$$

   More generally, for any fixed k, $x_m + x_{m+1} + \cdots + x_{m+k} \to 0$ and $\sum_{n=m}^{\infty} x_n \to 0$, as $m \to \infty$.

2. From the geometric series, it follows that $1 - a + a^2 - a^3 + \cdots$ and $\sum_n a^{r_n}$ ($r_n \geqslant n$) converge for $|a| < 1$ in $\mathbb{R}$.

3. The series $\sum_n \frac{1}{n!}$, $\sum_n \frac{n}{2^n}$, and $\sum_n \frac{2^n}{n!}$ converge by comparison with a geometric series (or using the ratio test).

4. $1 + \frac{1}{2^2} + \frac{1}{3^2} + \cdots = \frac{\pi^2}{6}$. This series was too hard to sum before Euler; show at least that it converges, using the comparison $\frac{1}{n^2} \leqslant \frac{1}{n(n-1)} = \frac{1}{n-1} - \frac{1}{n}$. Generalize this to the case $\frac{1}{(n-1)^{p-1}} - \frac{1}{n^{p-1}} \geqslant \frac{p-1}{n^p}$ to show that $\sum_n \frac{1}{n^p}$ converges for $p > 1$. Deduce that $\sum_n \frac{\sqrt{n}+1}{n^2-1}$ converges, by comparison.

5. These last series are examples that converge slower than the geometric series; in fact they are not decided by the root and ratio tests. Are there series that converge even slower?

6. The Cauchy or Raabe tests can also be used to show that $1 + \frac{1}{2^p} + \frac{1}{3^p} + \cdots$ converges when $p > 1$. Show further that $\sum_n \frac{1}{n}$, $\sum_n \frac{1}{n \log n}$, $\sum_n \frac{1}{n \log n \log \log n}$, ..., diverge.

7. Determine whether the following series converge, converge absolutely, or diverge in ℓ^1, ℓ^2, and c_0: (i) $\sum_n e_n$, (ii) $\sum_n \frac{(-1)^n}{n} e_n$, (iii) $\sum_n \frac{1}{n^2} e_n$, (iv) $\sum_n \frac{2^n}{n!} e_n$.

8. *The Weierstraß M-test* (comparison test for L^∞): If $\|f_n\|_{L^\infty} \leqslant M_n$ where $\sum_n M_n$ converges, then $\sum_n f_n$ converges in $L^\infty(A)$ (i.e., uniformly). Use it to show that the function $\sum_n \frac{t^n}{n^2}$ converges uniformly on $[-1, 1]$.

9. Let $f_n(t) := e^{-nt}/n$, then $\|f_n\|_{L^1[0,1]} \leqslant 1/n^2$, and so $\sum_n f_n$ converges in $L^1[0, 1]$.

10. If $\sum_{n\in\mathbb{N}} \|x_{n,m} - x_n\| \to 0$ as $m \to \infty$ and $\sum_n x_n$ converges,

$$\lim_{m\to\infty} \sum_{n\in\mathbb{N}} x_{n,m} = \sum_{n\in\mathbb{N}} \lim_{m\to\infty} x_{n,m} = \sum_{n\in\mathbb{N}} x_n$$

11. What is wrong with this argument: When $\|x_n\|^{1/n} \to 1$, then $\|x_n\| > (1 - \epsilon)^n$ for infinitely many terms; the right-hand side sums to $1/\epsilon$, which is arbitrarily large; hence the series cannot converge absolutely.

12. A rearrangement of an absolutely convergent series also converges, to the same sum. (Hint: Eventually, the rearranged series will contain the first n terms.)

13. Suppose a series $x_1 + x_2 + \cdots$ is split up into two subseries, say $x_1 + x_4 + \cdots$ and $x_2 + x_3 + \cdots$, denoted by $\sum_i x_{n_i}$ and $\sum_j x_{n'_j}$. If they both converge, to x and y respectively, then the original series $\sum_n x_n$ also converges, to $x + y$. If one converges, and the other diverges, then the series $\sum_n x_n$ diverges. But it is possible for two subseries to diverge, yet the original series to converge; for example, $1 - \frac{1}{2} + \frac{1}{3} - \frac{1}{4} + \cdots \to \log 2$.

14. (a) The sequences e_n form an unconditional (Schauder) basis for ℓ^1 and c_0.
    (b) The polynomials t^n, $n \in \mathbb{N}$, do not form a basis for $C[a, b]$. (Hint: For $C[-1, 1]$, take $f(t) := \begin{cases} 0, & t < 0 \\ t, & t \geqslant 0 \end{cases}$.

15. *Cesáro limit*: A sequence $(x_n)_{n\in\mathbb{N}}$ is said to converge in the sense of Cesáro when $\frac{x_1 + \cdots + x_n}{n}$ converges. Show that if $a = \lim_{n\to\infty} x_n$ exists then the Cesáro limit is also a. Show that the divergent sequence $(-1)^n$ is Cesáro convergent to 0.

Remarks 7.24

1. Weighted spaces are defined similarly to ℓ^p and L^p but with a different measure or *weight*. For example, an ℓ^1_w space with weights $w_n > 0$ consists of sequences with bounded norms $\|x\|_{\ell^1_w} := \sum_n |a_n| w_n$. Similarly, $L^2_w(A)$ has norm $(\int |f(t)|^2 w(t)\, dt)^{\frac{1}{2}}$. In fact, weighted spaces are isomorphic to the unweighted spaces; for example $\ell^1_w \cong \ell^1$ via the map $(a_n)_{n\in\mathbb{N}} \mapsto (w_n a_n)_{n\in\mathbb{N}}$.

2. The second norm axiom requires that the field be normed. A famous theorem by Frobenius states that the only normed fields over the reals are $\mathbb{R}$ and $\mathbb{C}$.

3. Cauchy's inequality was known to Lagrange in the form

$$\sum_{n=1}^{N} a_n^2 \sum_{m=1}^{N} b_m^2 - \left(\sum_{n=1}^{N} a_n b_n \right)^2 = \sum_{n=1}^{N} \sum_{m>n} (a_m b_n - a_n b_m)^2.$$

4. Hausdorff's Maximality Principle $\Rightarrow$ Axiom of Choice.

Proof: Let $\mathcal{A} = \{ A_\alpha \subseteq X : \alpha \in I \}$ be a collection of non-empty subsets of a set X. Consider choice functions $g : J \to X$, i.e., $g(\alpha) \in A_\alpha$ for all $\alpha \in J \subseteq I$. To prove the axiom of choice we need to show that there is a choice function f with domain I.

Let these choice functions be ordered by extension, that is, $g_1 \leqslant g_2$ when g_2 extends g_1. By Hausdorff's maximality principle, there exists a maximal chain of choice functions g_i. Let J be the union of all their domains. For each $\alpha \in J$, α must be in the domain of some choice function g_i; so define $f(\alpha) := g_i(\alpha)$. This function is well-defined since the choice functions extend one another; and it is a choice function itself since $f(\alpha) = g_i(\alpha) \in A_\alpha$.

Finally, if there is some set A_β which is missed by f, i.e., $\beta \notin J$, let $x_\beta \in A_\beta$. Then f can be extended further by defining $f(\beta) := x_\beta$, contradicting the maximality of the chain g_i. Hence f is defined on all $\alpha \in I$.

Chapter 8
Continuous Linear Maps

8.1 Operators

In every branch of mathematics which concerns itself with sets having some particular structure, the functions which preserve that structure, called *morphisms*, feature prominently. Such maps allow us to transfer equations from one space to another, to compare spaces with each other and state when two of them are essentially the same, or if not, whether one can be embedded in the other, etc. Even in applications, it is often the case that certain aspects of a process are *conserved*. For example, a rotation of geometric space yields essentially the same space, and rotating the axes might simplify a problem. The morphisms on normed spaces are formalized by the following definition.

Definition 8.1

An **operator**[1] is a continuous linear transformation $T : X \to Y$ between normed spaces (over the same field), that is, it preserves vector addition, scalar multiplication, and convergence,

$$T(x + y) = Tx + Ty, \qquad T(\lambda x) = \lambda Tx, \qquad T(\lim_{n \to \infty} x_n) = \lim_{n \to \infty} Tx_n.$$

A **functional** is a continuous linear map $\phi : X \to \mathbb{F}$ from a normed space to its field. The set of operators from X to Y is denoted by $B(X, Y)$, and the set of functionals, denoted by X^*, is called the **dual space** of X.

---

[1] The use of the term operator is not standardized: it may simply mean a linear transformation, or even just a function, especially outside Functional Analysis. But it is standard to write Tx instead of $T(x)$.

© The Author(s), under exclusive license to Springer Nature Switzerland AG 2024
J. Muscat, *Functional Analysis*, https://doi.org/10.1007/978-3-031-27537-1_8

Easy Consequences
1. $T0 = 0$.
2. Linearity is equivalent to showing $T(\lambda x + y) = \lambda Tx + Ty$.
3. $T(\sum_{n=0}^{\infty} \lambda_n x_n) = \sum_{n=0}^{\infty} \lambda_n T x_n$.
4. A linear map is determined by the values it takes on the unit sphere.

A simple test for continuity of a linear transformation is the following Lipschitz property, due to Banach.

Proposition 8.2

A linear transformation $T : X \to Y$ is continuous if, and only if, T is a Lipschitz map,

$$\exists c > 0, \ \forall x \in X, \quad \|Tx\|_Y \leqslant c\|x\|_X.$$

Proof The definition of a Lipschitz map reads, when applied for normed spaces, $\|f(x) - f(y)\| \leqslant c\|x - y\|$ for some $c > 0$. When f is in fact a linear map T, it becomes $\|T(x - y)\| \leqslant c\|x - y\|$, or equivalently, $\|Tv\| \leqslant c\|v\|$ for all $v \in X$. That Lipschitz maps are (uniformly) continuous is true in every metric space (Example 4.16(3)), but can easily be seen in this context. If $x_n \to x$, then $Tx_n \to Tx$, since

$$\|Tx_n - Tx\| = \|T(x_n - x)\| \leqslant c\|x_n - x\| \to 0.$$

Conversely, suppose the ratios $\|Tx\|/\|x\|$ are unbounded. Since scaling x does not affect this ratio (because T is linear), there must be vectors x_n such that $\|Tx_n\| = 1$ but $\|x_n\| \leqslant 1/n$. So $x_n \to 0$ yet $Tx_n \not\to 0$, and T is not continuous. □

Equivalently, T sends bounded sets in X to bounded sets in Y, since it maps the ball $B_r(x)$ into the ball $B_{cr}(Tx)$ (Exercise 4.18(3)). Because of this, continuous operators are widely referred to as being "bounded", but, except for the zero operator, their image is certainly not bounded! This usage of the word "bounded" is avoided in this text, in favor of the equivalent term "continuous".

Examples 8.3

1. An operator T maps the linear subspace $[\![A]\!]$ to $[\![TA]\!]$ because

$$x = \sum_{i=1}^{n} \alpha_i v_i \ \Rightarrow \ Tx = \sum_{i=1}^{n} \alpha_i T v_i.$$

In particular it maps a straight line to another straight line (or to the origin), hence the name "linear" applied to operators.

If T is 1–1, then $E \subseteq X$ is linearly independent iff TE is linearly independent.

2. ▶ A linear transformation from $\mathbb{F}^n$ to $\mathbb{F}^m$ takes the form of a *matrix*. Letting $\mathbb{F}^n = [\![e_1, \dots, e_n]\!]$, $\mathbb{F}^m = [\![e'_1, \dots, e'_m]\!]$, $x = \sum_{j=1}^n a_j e_j = (a_1, \dots, a_n)$, and $Te_j = \sum_{i=1}^m T_{ij} e'_i$ (for some $T_{ij} \in \mathbb{F}$), then

$$Tx = \sum_{j=1}^n a_j Te_j = \sum_{i=1}^m \left(\sum_{j=1}^n T_{ij} a_j \right) e'_i = \begin{pmatrix} T_{11} & \cdots\cdots & T_{1n} \\ \vdots & \ddots & \vdots \\ T_{m1} & \cdots\cdots & T_{mn} \end{pmatrix} \begin{pmatrix} a_1 \\ \vdots \\ a_n \end{pmatrix}.$$

Notice that the column vectors of T are Te_j.

Every matrix is continuous,

$$\|Tx\|_2 \leqslant \sum_{i=1}^m \sum_{j=1}^n |T_{ij} a_j| \leqslant \left(\sum_{i,j=1}^{m,n} |T_{ij}| \right) \|x\|_2$$

3. A functional from $\mathbb{F}^n$ to $\mathbb{F}$ is then a $1 \times n$ matrix, otherwise known as a *row vector*,

$$\phi \begin{pmatrix} a_1 \\ \vdots \\ a_n \end{pmatrix} = \phi \left(\sum_{i=1}^n a_i e_i \right) = \sum_{i=1}^n \phi(e_i) a_i = \sum_{i=1}^n b_i a_i = \begin{pmatrix} b_1 & \dots & b_n \end{pmatrix} \begin{pmatrix} a_1 \\ \vdots \\ a_n \end{pmatrix}.$$

4. Generalizing this to functionals on complex sequences, let $y^\top(x) = y \cdot x := \sum_n b_n a_n$, when the series exists, where $x = (a_n)_{n \in \mathbb{N}}$ and $y = (b_n)_{n \in \mathbb{N}}$. Then $y^\top$ is linear,

$$y \cdot (x + x') = \sum_n b_n (a_n + a'_n) = \sum_n b_n a_n + \sum_n b_n a'_n = y \cdot x + y \cdot x',$$

$$y \cdot (\lambda x) = \sum_n b_n \lambda a_n = \lambda \sum_n b_n a_n = \lambda y \cdot x,$$

but may or may not be continuous, depending on y and the normed spaces involved. For example, to show that $\phi(a_n) := \sum_{n=1}^\infty \frac{(-1)^n a_n}{n^2}$ defined on ℓ^∞ is continuous, note

$$|\phi x| = \left| \sum_{n=1}^\infty \frac{(-1)^n}{n^2} a_n \right| \leqslant \sum_{n=1}^\infty \frac{1}{n^2} \sup_n |a_n| \leqslant 2\|x\|_{\ell^\infty}.$$

5. When X has a Schauder basis $(e_n)_{n \in \mathbb{N}}$, a functional must have the above form:

$$\phi x = \phi \Big(\sum_{n \in \mathbb{N}} a_n e_n \Big) = \sum_{n \in \mathbb{N}} a_n \phi e_n = \sum_{n \in \mathbb{N}} b_n a_n, \qquad (b_n := \phi e_n, \ a_n \in \mathbb{F}).$$

6. The *identity operator* $I : X \to X, \ x \mapsto x$, is trivially linear and continuous. Similarly for scalar multiplication, $\lambda : x \mapsto \lambda x$.

7. ▶ The *left-shift operator* $L : \ell^1 \to \ell^1$ defined by $(a_n)_{n \in \mathbb{N}} \mapsto (a_{n+1})_{n \in \mathbb{N}}$, i.e.,

$$L(a_0, a_1, a_2, \ldots) := (a_1, a_2, a_3, \ldots),$$

is onto, linear, continuous, and satisfies $\|Lx\| \leqslant \|x\|$; it is not 1–1.
Proof: That L is onto is obvious; linearity and continuity follow from

$$L(a_n + b_n) = (a_1 + b_1, a_2 + b_2, \ldots) = (a_1, a_2, \ldots) + (b_1, b_2, \ldots)$$
$$= L(a_n) + L(b_n),$$
$$L(\lambda a_n) = (\lambda a_1, \lambda a_2, \ldots) = \lambda(a_1, a_2, \ldots) = \lambda L(a_n),$$
$$\|Lx\|_{\ell^1} = \sum_{n=1}^{\infty} |a_n| \leqslant \sum_{n=0}^{\infty} |a_n| = \|x\|_{\ell^1}.$$

Any two sequences, which differ in the first coefficient only, map to the same sequence.

8. ▶ In general, the multiplication of sequences $x \mapsto yx$, defined by $(b_n)(a_n) := (b_n a_n)_{n \in \mathbb{N}}$, is linear on the vector space of sequences. When $|b_n| \leqslant c$, it is continuous as a map $\ell^p \to \ell^p$ $(p \geqslant 1)$; e.g., for $p = 1$,

$$\|yx\|_{\ell^1} = \sum_{n=0}^{\infty} |b_n a_n| \leqslant c \sum_{n=0}^{\infty} |a_n| = c\|x\|_{\ell^1}.$$

In finite dimensions, this is equivalent to multiplying x by a diagonal matrix.

9. Integration, $f \mapsto \int_A f$, is a functional on $L^1(A)$.

10. The 'delta function' $\delta_{x_0}(f) := f(x_0)$, is a functional on $C_b(X)$, where $x_0 \in X$.
Proof: Linearity is immediate,

$$\delta_{x_0}(\lambda f + g) = (\lambda f + g)(x_0) = \lambda f(x_0) + g(x_0) = \lambda \delta_{x_0}(f) + \delta_{x_0}(g).$$

For continuity, $|\delta_{x_0} f| = |f(x_0)| \leqslant \sup_{x \in X} |f(x)| = \|f\|_{C(X)}$.

11. Differentiation of functions is linear (say on the vector space of differentiable functions) but it is not continuous in the ∞-norm, e.g.,

$$\|D \cos(nt)\|_{C_b(\mathbb{R})} = \|-n \sin(nt)\|_{C_b(\mathbb{R})} = n$$

whereas $\| \cos(nt) \|_{C_b(\mathbb{R})} = 1$. Similarly, $\| Dt^n \|_{C[0,1]} / \| t^n \|_{C[0,1]} \to \infty$ as $n \to \infty$. (Here, t^n and $\cos(nt)$ denote functions in t.)

12. Conjugation in $\mathbb{C}$, $z \mapsto \bar{z}$, is continuous but not linear. It is *conjugate*-linear, because $\overline{\lambda z} = \bar{\lambda} \bar{z} \neq \lambda \bar{z}$ in general.

Proposition 8.4

> **If $T : X \to Y$ is an operator,**
>
> (i) **the image of a linear subspace A of X is again a linear subspace of Y,**
> $$TA := \{ Tx \in Y : x \in A \},$$
> (ii) **the pre-image of a closed linear subspace B of Y is a closed linear subspace of X, $T^{-1}B := \{ x \in X : Tx \in B \}$.**
>
> **The image and pre-image of convex subsets are convex.**

In particular, its *image* $\operatorname{im} T := TX$ is a linear subspace; and its *kernel* $\ker T := T^{-1}0$ is a closed linear subspace. Their dimensions are called the *rank* and *nullity* of T, respectively. The kernel of a non-zero functional, $\ker \phi$, is called a *hyperplane*.

Proof (i) Let $Tx, Ty \in TA$, then $\lambda Tx + Ty = T(\lambda x + y) \in TA$.

(ii) Let $x, y, x_n \in T^{-1}B$, that is, $Tx, Ty, Tx_n \in B$, and let $\lambda \in \mathbb{F}$. Then

$$T(x + y) = Tx + Ty \in B, \qquad T(\lambda x) = \lambda Tx \in B,$$

$$x_n \to v \Rightarrow Tv = T(\lim_{n \to \infty} x_n) = \lim_{n \to \infty} Tx_n \in B,$$

show that $T^{-1}B$ is a closed linear subspace.

(iii) Let A be a convex subset of X and let $Tx, Ty \in TA$, where $x, y \in A$. Then for any $0 \leqslant t \leqslant 1$, $z := tx + (1 - t)y$ is in A, so

$$t Tx + (1 - t)Ty = T(tx + (1 - t)y) = Tz \in TA$$

shows TA is also convex. Now let $B \subseteq Y$ be convex, and let $x, y \in T^{-1}B$, i.e., Tx, Ty are both in B. Then, by convexity of B,

$$T(tx + (1 - t)y) = tTx + (1 - t)Ty \in B$$

and $tx + (1 - t)y \in T^{-1}B$ as required. $\qquad\qquad\qquad\qquad \square$

Solving linear equations $Tx = y$, where T and y are given, is probably the single most useful application in the whole of mathematics. The key to finding the *general* solution of this equation is to know $\operatorname{im} T$ and $\ker T$.

- To say that $Tx = y$ has a solution is the same as saying $y \in \operatorname{im} T$. Hence $Tx = y$ always has a solution precisely when T is surjective, $\operatorname{im} T = Y$.
- $\ker T$ is the set of solutions of the *homogeneous* equation $Tv = 0$. If $v \in \ker T$ then $T(x + v) = Tx = y$, so both x and $x + v$ are solutions. Hence $Tx = y$ has unique solutions (if any), and T is injective, precisely when $\ker T = 0$.
- If x_0 is any individual or *particular* solution, $Tx_0 = y$, then for any other solution of $Tx = y$, we get $T(x - x_0) = 0$ so $x - x_0 \in \ker T$, called a *complementary* solution. Thus we have proved:

The complete set of solutions of $Tx = y$ is $x_0 + \ker T$.

Examples 8.5

1. For a matrix, the image of T is often called its *column space*, since $\operatorname{im} T = [\![Te_j]\!]_{j=1}^n$. Gaussian column operations can be performed on the columns to simplify it to a basis, for example,

$$\operatorname{im} T = \operatorname{ColSpace} \begin{pmatrix} 1 & 3 & 7 & -17 \\ -4 & -4 & -20 & 36 \\ 1 & 1 & 5 & -9 \end{pmatrix} = \begin{bmatrix} 1 & 0 & 0 & 0 \\ -4 & 8 & 8 & -32 \\ 1 & -2 & -2 & 8 \end{bmatrix} = \begin{bmatrix} 1 & 0 \\ 0 & 4 \\ 0 & -1 \end{bmatrix}.$$

   (Column vector parentheses are suppressed for clarity.)
   Similarly, the rows can be simplified by row operations,

$$\operatorname{RowSpace} \begin{pmatrix} 1 & 3 & 7 & -17 \\ -4 & -4 & -20 & 36 \\ 1 & 1 & 5 & -9 \end{pmatrix} = \operatorname{RowSpace} \begin{pmatrix} 1 & 0 & 4 & -5 \\ 0 & 1 & 1 & -4 \end{pmatrix}$$

   so a vector belongs to $\ker T$ iff it is annihilated by the row-space; in this case they are the vectors $\begin{pmatrix} 5t - 4s \\ 4t - s \\ s \\ t \end{pmatrix}$, so $\ker T = \begin{bmatrix} 5 & -4 \\ 4 & -1 \\ 0 & 1 \\ 1 & 0 \end{bmatrix}$.

2. The kernel of the left-shift operator is spanned by e_0, since if $x \in \ker L$, $(a_1, a_2, \ldots) = Lx = 0$, so $a_n = 0$ for all $n \neq 0$, i.e., $x = a_0 e_0$; in fact $Le_0 = 0$.
3. An open linear mapping must be surjective.
   Proof: As TB_X is open in Y, it contains a neighborhood $B_\epsilon(0) = \epsilon B_Y$. But $X = \bigcup_n B_n(0)$, so $TX = \bigcup_n nTB_X \supseteq \bigcup_n n\epsilon B_Y = Y$.
4. Let V be a linear subspace of $\operatorname{im} T \subseteq Y$, with a basis E; for each element $e \in E$, choose a pre-image $u \in X$, i.e., $Tu = e$, and form the set E_0 of such vectors, one for each e; these are the particular solutions mentioned above. Then $T^{-1}V = [\![E_0]\!] + \ker T$ and $\dim(T^{-1}V) \leqslant \dim V + \dim \ker T$.

Proof: Let E_1 be a basis for $\ker T$; then $E_0 \cup E_1$ is a basis for $T^{-1}V$ since the solutions of $Tx = v \in V$ are obtained as $x = x_0 + w \in [\![E_0]\!] + [\![E_1]\!] = [\![E_0 \cup E_1]\!]$.

In particular, taking $V = \operatorname{im} T$, the *rank-nullity formula* holds:

$$\operatorname{rank}(T) + \operatorname{nullity}(T) = \dim X$$

Note that $\operatorname{rank}(T) \leqslant \dim X$. (These formulae hold even for infinite dimensions, in the sense of cardinal numbers.)

5. A *finite-rank* operator is one whose image is finite-dimensional. If S, T are finite-rank then so are $S + T$ and ST (when defined), with

$$\operatorname{rank}(S + T) \leqslant \operatorname{rank}(S) + \operatorname{rank}(T),$$

$$\operatorname{rank}(ST) \leqslant \min(\operatorname{rank}(S), \operatorname{rank}(T)).$$

Proof: The domain of S in the composition ST can be taken to be $\operatorname{im} T$, so $\dim(\operatorname{im}(ST)) \leqslant \dim(\operatorname{im}(T))$ by the rank-nullity formula; the rest follow from $\operatorname{im}(ST) \subseteq \operatorname{im}(S)$ and $\operatorname{im}(S + T) \subseteq \operatorname{im} S + \operatorname{im} T$.

6. *Sylvester's inequality*: If S and T both have finite nullity, then so does ST with

$$\operatorname{nullity}(ST) \leqslant \operatorname{nullity}(S) + \operatorname{nullity}(T).$$

Proof: $\ker(ST) = T^{-1}(\ker S)$, so $\operatorname{nullity}(ST) \leqslant \dim \ker S + \dim \ker T$.

Exercises 8.6

1. Show that the following are continuous functionals,

   (a) $\phi(a_n) := \sum_{n=1}^{\infty} \frac{1}{n} a_n$ on ℓ^2;
   (b) $\phi(a_n) := \sum_{n=0}^{\infty} e^{in\omega} a_n$ on ℓ^1, ($\omega \in \mathbb{R}$);
   (c) $\delta_1(a_n) := a_1$ on $\ell^1, \ell^2, \ell^\infty$.

   Their best Lipschitz constants are $\frac{\pi}{\sqrt{6}}$, 1, and 1, respectively.
2. If $(e_n)_{n \in \mathbb{N}}$ is a Schauder basis, with $x = \sum_n a_n(x) e_n$ for each x, show that the map $x \mapsto a_k(x)$ is linear. (That it is also continuous is true in a Banach space, but not obviously.)
3. ▶ The *right-shift operator* is defined by $R(a_n) := (0, a_0, a_1, \ldots)$. Show that it is 1–1, isometric, and has a closed image. Is the left-shift operator its inverse?
4. Other examples of operators (on ℓ^1 or ℓ^∞) are

   $$S(a_n) := (a_1, a_0, a_3, a_2, \ldots), \quad T(a_n) := (a_{n+4} - a_n)_{n \in \mathbb{N}}.$$

5. $T\overline{[\![A]\!]} \subseteq \overline{[\![TA]\!]}$ for a continuous linear operator T and a subset $A \subseteq X$.
   In particular, $M := \overline{[\![x, Tx, T^2x, \ldots]\!]}$ is a T-invariant subspace of X: $TM \subseteq M$.

6. Solve for the functional equation $f(t + 1) = f(t) + t$ as follows: (i) the map $T : f(t) \mapsto f(t + 1) - f(t)$ is linear in f; find (ii) $\ker T$; (iii) a particular solution in the space of polynomials; (iv) the general solution.
7. If a linear map is continuous at one point, say 0, then it is continuous everywhere.
8. When Y is a normed space and $T : X \to Y$ is 1–1 and linear, then the map $x \mapsto \|Tx\|$ is a norm on X.
9. Typical examples of functionals acting on functions are of the form $f \mapsto \int k(t) f(t) \, dt$, where k has to satisfy some conditions for the functional to be continuous. For example, $\phi f := \int_0^\infty e^{-t} f(t) \, dt$ is a functional on $L^\infty[0, \infty[$.
10. If $S, T \in B(X)$ commute, $ST = TS$, then S maps $\ker T$ and $\operatorname{im} T$ into themselves.

8.2 Operator Norms

Proposition 8.2 states that a linear transformation T is continuous when it satisfies an inequality $\|Tx\|_Y \leqslant c\|x\|_X$. The smallest such constant c is denoted by $\|T\|$, because it turns out to be a norm on operators. The sharp inequality

$$\|Tx\| \leqslant \|T\| \|x\|$$

is used extensively in the rest of the text.

Theorem 8.7

$B(X, Y)$ **is a vector space with a norm defined by**

$$\|T\| := \sup_{x \neq 0} \frac{\|Tx\|_Y}{\|x\|_X} = \sup_{\|x\|=1} \|Tx\|_Y.$$

$B(X, Y)$ **is complete when Y is complete. In particular, X^* is a Banach space, with norm**

$$\|\phi\| = \sup_{x \neq 0} \frac{|\phi x|}{\|x\|}.$$

Proof The norm is well-defined in the sense that if T is an operator, then the ratios $\|Tx\|/\|x\|$ are bounded above, and so have a supremum $\|T\|$. In fact, a linear map belongs to $B(X, Y)$ if, and only if, $\|T\| < \infty$.

Addition and scalar multiplication of operators is defined by

$$(S + T)x := Sx + Tx, \qquad (\lambda T)x := \lambda Tx.$$

That $B(X, Y)$ with these operations is a vector space is a straightforward calculation, using the linearity and continuity of these operations in X and Y (Proposition 7.9).

$$(\lambda T)(\alpha x + y) = \lambda T(\alpha x + y) = \lambda \alpha T x + \lambda T y = \alpha (\lambda T) x + (\lambda T) y.$$

More crucially,

$$\|S + T\| = \sup_{\|x\|=1} \|Sx + Tx\| \leqslant \sup_{\|x\|=1} (\|Sx\| + \|Tx\|)$$

$$\leqslant \sup_{\|x\|=1} \|Sx\| + \sup_{\|x\|=1} \|Tx\|$$

$$= \|S\| + \|T\|$$

$$\|\lambda T\| = \sup_{\|x\|=1} \|\lambda Tx\| = \sup_{\|x\|=1} |\lambda| \|Tx\| = |\lambda| \|T\|$$

$$\|T\| = 0 \Leftrightarrow \forall x \ \|Tx\| = 0 \Leftrightarrow T = 0.$$

$B(X, Y)$ *is complete if* Y *is*: Let T_n be a Cauchy sequence of operators in $B(X, Y)$, that is, $\|T_n - T_m\| \to 0$ as $n, m \to \infty$. Then, for each $x \in X$,

$$\|T_n x - T_m x\| \leqslant \|T_n - T_m\| \|x\| \to 0$$

implies that $(T_n x)_{n \in \mathbb{N}}$ is a Cauchy sequence in Y, so that $T_n x$ converges to some vector which can be denoted by $T(x)$, if Y is complete. We now show that T is linear:

$$\begin{array}{cccc}
T_n(x + y) = & T_n x + T_n y, & T_n(\lambda x) = & \lambda T_n x, \\
\downarrow & \downarrow & \downarrow & \downarrow \qquad \text{as } n \to \infty, \\
T(x + y) & T(x) + T(y), & T(\lambda x) & \lambda T(x),
\end{array}$$

by continuity of addition and scalar multiplication.

Finally, for any $\epsilon > 0$ and any $x \in X$,

$$\|(T_n - T)x\| \leqslant \|T_n - T_m\| \|x\| + \|T_m x - Tx\| < \epsilon \|x\| + \epsilon \|x\|,$$

where m is chosen large enough, depending on x, to make $\|T_m x - Tx\| < \epsilon \|x\|$, and $n, m \geqslant N$ large enough to make $\|T_n - T_m\| < \epsilon$. Hence $\|T_n - T\| < 2\epsilon$ for $n \geqslant N$. This shows that $T_n - T$, and so T, are continuous, and furthermore that $T_n \to T$. $\qquad \square$

Proposition 8.8

> **If $T : X \to Y$ and $S : Y \to Z$ are operators, then so is their composition ST, with $\|ST\| \leqslant \|S\|\|T\|$.**
>
> $B(X) := B(X, X)$ **is closed under multiplication.**

Proof That ST is linear is obvious: $ST(x + y) = S(Tx + Ty) = STx + STy$ and $ST(\lambda x) = S(\lambda Tx) = \lambda STx$. Also,

$$\|STx\| = \|S(Tx)\| \leqslant \|S\|\|Tx\| \leqslant \|S\|\|T\|\|x\|,$$

and the result follows by taking the supremum for unit vectors x.   □

Examples 8.9

1. $\|0\| = 0$, $\|I\| = 1$; more generally, $\|\lambda I\| = |\lambda|$.
2. The norm of the functional $y^\top$ is $\|y\|_{\ell^\infty}$ when considered as a map $\ell^1 \to \mathbb{F}$.
   Proof: Taking $x = (a_n)_{n\in\mathbb{N}}$, $y = (b_n)_{n\in\mathbb{N}}$,

$$|y \cdot x| \leqslant \sum_{n\in\mathbb{N}} |b_n||a_n| \leqslant (\sup_{n\in\mathbb{N}} |b_n|) \sum_{n\in\mathbb{N}} |a_n| = \|y\|_{\ell^\infty}\|x\|_{\ell^1},$$

gives $\|y^\top\| \leqslant \|y\|_{\ell^\infty}$. Since the supremum $\|y\|_{\ell^\infty}$ is a boundary point of the set $\{ |b_n| : n \in \mathbb{N} \}$, there is a subsequence $|b_{n_i}| \to \|y\|_{\ell^\infty}$, so that $\|y^\top\| \geqslant \|y\|_{\ell^\infty}$,

$$\|y^\top\| \geqslant |y \cdot e_{n_i}| = |b_{n_i}| \to \|y\|_{\ell^\infty} \qquad (\|e_{n_i}\| = 1).$$

3. $\|T\| \leqslant \|S\| \nRightarrow \|Tx\| \leqslant \|Sx\|$, for example, $T = I$, $S = \begin{pmatrix} 2 & 0 \\ 0 & 0 \end{pmatrix}$, $x = \begin{pmatrix} 0 \\ 1 \end{pmatrix}$.
4. If S extends T, with domains $X \supseteq Y$, then $\|T\| \leqslant \|S\|$, since

$$\|T\| = \sup_{x\in B_Y} \|Tx\| \leqslant \sup_{x\in B_X} \|Tx\| = \|S\|, \qquad (B_Y \subseteq B_X).$$

5. ▶ Any linear continuous operator on normed spaces, $T : X \to Y$, is Lipschitz, hence uniformly continuous. By Theorem 4.14, it can be extended uniquely to an operator on their (Banach) completion spaces, $\widetilde{T} : \widetilde{X} \to \widetilde{Y}$. This extension remains linear and continuous, and retains the same norm, $\|\widetilde{T}\| = \|T\|$.
   Proof: For any vector $x \in \widetilde{X}$, there exist vectors $x_n \in X$ such that $x_n \to x$; let $\widetilde{T}x := \lim_{n\to\infty} Tx_n \in \widetilde{Y}$. Then, for any vector $v \in \widetilde{X}$, with $v_n \to v$, $v_n \in X$,

$$\widetilde{T}(\lambda x + v) = \lim_{n\to\infty} T(\lambda x_n + v_n) = \lim_{n\to\infty} (\lambda Tx_n + Tv_n) = \lambda\widetilde{T}(x) + \widetilde{T}(v),$$

$$\|\widetilde{T}x\| = \lim_{n\to\infty} \|Tx_n\| \leqslant \|T\| \lim_{n\to\infty} \|x_n\| = \|T\|\|x\|.$$

So $\|\widetilde{T}\| \leqslant \|T\|$, but, as the domain of $\widetilde{T}$ includes that of T, equality holds.

6. Let $\phi \in X^*$ and $y \in Y$; then the map $y\phi : x \mapsto (\phi x)y$ is continuous and linear, with $\|y\phi\| = \|y\|\|\phi\|$.

   Proof: $\|y\phi\| = \sup_{\|x\|=1} \|y\phi x\| = (\sup_{\|x\|=1} |\phi x|)\|y\| = \|\phi\|\|y\|$.

7. An 'affine' map $f(x) := v + Tx$ with $T \in B(X)$ is a contraction mapping when $\|T\| < 1$. The iteration $x_{n+1} := v + Tx_n$, starting from any x_0, converges to its fixed point $y = v + Ty$ (Theorem 4.17). Try it out as a plot with an affine map such as $\frac{2}{5}\begin{pmatrix} 2 & -1 \\ 1 & 2 \end{pmatrix}x + \begin{pmatrix} 1 \\ 0 \end{pmatrix}$.

Matrix Norms

Every matrix $T : \mathbb{F}^n \to \mathbb{F}^m$ is continuous, hence has a finite norm $\|T\|$. But this operator norm needs to be disabused of some notions. It is not a number that depends only on T; it also depends on which norms are being used for $\mathbb{F}^n$ and $\mathbb{F}^m$ and therefore it is customary to denote it by $\|T\|_{p,q}$ when the p- and q-norms are used in $\mathbb{F}^n$ and $\mathbb{F}^m$, respectively, unless it is obvious from the context. Moreover, it may be hard to compute a norm in general, so any estimate for it is most welcome. Finally, there exist other more convenient norms that are based on specific formulas, foremost of which is the *Frobenius norm* of a matrix defined by

$$\|T\|_F := \left(\sum_{i,j=1}^{m,n} |T_{ij}|^2 \right)^{1/2}.$$

It is just the Euclidean norm of the matrix thought of as a vector with mn components.

Proposition 8.10

Let a matrix T have coefficients T_{ij}, then

$$\|T\|_{1,\infty} \leqslant \max_{i,j} |T_{ij}|, \qquad \|T\|_{1,1} = \max_j \sum_i |T_{ij}| =: c,$$

$$\|T\|_{\infty,1} \leqslant \sum_{i,j} |T_{ij}|, \qquad \|T\|_{\infty,\infty} = \max_i \sum_j |T_{ij}| =: r,$$

$$\|T\|_{2,2} \leqslant \|T\|_F, \qquad \|T\|_{2,2} \leqslant \sqrt{rc}.$$

The numbers c and r measure the matrix's "largest" column and row, respectively. Then the second inequality for $\|T\|_{2,2}$, known as *Schur's test* and sometimes an improvement on the first inequality, states that it is at most their geometric mean.

Proof Let $x = (a_j)_{j=1}^n$ and $Tx = (\sum_{j=1}^n T_{ij} a_j)_{i=1}^m$.

(i) $\|Tx\|_\infty = \max_i \left| \sum_j T_{ij} a_j \right| \leqslant (\max_i \max_j |T_{ij}|) \|x\|_1$.

(ii) $\|Tx\|_1 = \sum_i \left| \sum_j T_{ij} a_j \right| \leqslant \left(\sum_i \sum_j |T_{ij}| \right) \|x\|_\infty$.

(iii) By Cauchy's inequality, $\left| \sum_j T_{ij} a_j \right|^2 \leqslant \sum_j |T_{ij}|^2 \sum_j |a_j|^2$ for each i, so

$$\|Tx\|_2^2 = \sum_i \left| \sum_j T_{ij} a_j \right|^2 \leqslant \sum_{ij} |T_{ij}|^2 \|x\|_2^2.$$

(iv) $\|Tx\|_1 = \sum_i \left| \sum_j T_{ij} a_j \right| \leqslant \sum_j \sum_i |T_{ij}||a_j| \leqslant \sum_j c|a_j| = c\|x\|_1$. If the largest column is the kth one, then $\|Te_k\|_1 = c = c\|e_k\|_1$, so $\|T\|_{1,1} = \sup \frac{\|Tx\|_1}{\|x\|_1} = c$.

(v) $\|Tx\|_\infty = \max_i \left| \sum_j T_{ij} a_j \right| \leqslant \left(\max_i \sum_j |T_{ij}| \right) \|x\|_\infty$. If the largest row is the kth one, consider the unit vector $x := (|T_{kj}|/T_{kj})_{j=1}^n$ (take 1 if $T_{kj} = 0$); then $\|Tx\|_\infty = \sum_j |T_{kj}| = r$.

(vi) Let $y := (b_i)_{i=1}^m$; then again by Cauchy's inequality, over $\mathbb{F}^{nm}$,

$$|y \cdot Tx| \leqslant \sum_{i,j} |T_{ij}||a_j||b_i| = \sum_{i,j} \left(|T_{ij}|^{\frac{1}{2}} |a_j| \right) \left(|T_{ij}|^{\frac{1}{2}} |b_i| \right)$$

$$\leqslant \sqrt{\sum_{i,j} |T_{ij}||a_j|^2} \sqrt{\sum_{i,j} |T_{ij}||b_i|^2},$$

$$\leqslant \sqrt{c \sum_j |a_j|^2} \sqrt{r \sum_i |b_i|^2} = \sqrt{rc}\,\|x\|_2 \|y\|_2$$

In particular, putting $y = Tx$ gives $\|Tx\|_2^2 \leqslant \sqrt{rc}\,\|x\|_2 \|Tx\|_2$. $\square$

Proposition 8.11

The (p, q)-norm of a matrix T can only increase if

  (i) **a row or column is added; or**

 (ii) **the coefficients T_{ij} are replaced by $|T_{ij}|$ or larger.**

Proof (i) Adding a row increases $\|Tx\|_q$ without affecting $\|x\|_p$. Adding a column enlarges the domain, since $\mathbb{F}^n$ is embedded in $\mathbb{F}^{n+1}$. Since all the original vectors are still present with a zero at the position of the new column, the supremum of $\|Tx\|$ among all the unit vectors can only increase.

(ii) For any vector $x = (a_j)_{j=1}^n$, let $x_+ := (|a_j|)_{j=1}^n$, both vectors having the same norm. Let S be a matrix with coefficients satisfying $S_{ij} \geqslant |T_{ij}|$. Then

$$\|Tx\|_q \leqslant \|(\sum_j T_{ij}a_j)_{i=1}^m\|_q \leqslant \|(\sum_j |T_{ij}||a_j|)_{i=1}^m\|_q \leqslant \|Sx_+\|_q$$

$$\leqslant \|S\|_{p,q}\|x\|_p.$$

$\square$

Examples 8.12

1. By deleting rows and columns, it follows that the norm of a matrix is larger than the norm of any submatrix, including that of any row or column, or individual components.

2. The $(1,1)$- and (∞, ∞)-norms are easy to calculate and are achieved by vectors:

$$\begin{pmatrix} 7 & -7 & 5 \\ -2 & 9 & 5 \end{pmatrix} \begin{pmatrix} 0 \\ 1 \\ 0 \end{pmatrix} = \begin{pmatrix} -7 \\ 9 \end{pmatrix}, \qquad \begin{pmatrix} 7 & -7 & 5 \\ -2 & 9 & 5 \end{pmatrix} \begin{pmatrix} 1 \\ -1 \\ 1 \end{pmatrix} = \begin{pmatrix} 19 \\ -6 \end{pmatrix}.$$

But for the $(2,2)$-norm, the above propositions only tell us that

$$11.4 \approx \sqrt{130} \leqslant \left\| \begin{pmatrix} 7 & -7 & 5 \\ -2 & 9 & 5 \end{pmatrix} \right\|_{2,2} \leqslant \sqrt{233} \approx 15.3,$$

using columns, rows, and the Frobenius norm.

3. If an operator $T : X \to X$ has an eigenvector, $Tx = \lambda x$, then $\|T\| \geqslant \frac{\|Tx\|}{\|x\|} = |\lambda|$. For example, the (p,p)-norm of a square matrix is at least equal to its largest eigenvalue.

4. Any norm $\|T\|_{p,q}$ depends continuously on its coefficients: changing them slightly by at most ϵ does not change T drastically, e.g.,

$$\|S - T\|_{\infty,1} \leqslant mn \max_{i,j} |S_{ij} - T_{ij}| \leqslant mn\epsilon.$$

5. Finite matrices have a whole set of attributes that do not generalize to operators, so it is important to 'unlearn' them for infinite dimensions, so to speak. The following is a list of properties that generally hold *only* in finite dimensions:

   (a) A matrix is injective iff surjective.
   (b) Every matrix has finite rank and nullity.
   (c) Every matrix is continuous.
   (d) The image of any matrix is closed.
   (e) Every square matrix satisfies some polynomial; that monic polynomial of smallest degree is called its *minimal polynomial*.
   (f) Every square matrix has a determinant and a trace.

Exercises 8.13

1. The mapping $T : \ell^1 \to \ell^1$, defined by $T(a_n) := (a_0, a_1/2, a_2/3, \ldots)$, is linear and continuous, with norm 1. It is 1–1, and its image, denoted $\ell_1^1 := \operatorname{im} T \subset \ell^1$, is not closed in ℓ^1. (Hint: Consider $(1, 1/2, \ldots, 1/n, 0, 0, 0, \ldots)$.)
2. The mapping $D : \ell_1^1 \to \ell^1$, defined by $D(a_n) := (na_n)_{n \in \mathbb{N}}$, is linear and invertible, but not continuous. (Hint: $D(e_n/n) = e_n$.)
3. The right-shift operator satisfies $\|Rx\| = \|x\|$ as $\ell^p \to \ell^p$, so $\|R\|_{p,p} = 1$. Show further that $\|R\|_{1,\infty} = 1$ and $\|L\|_{p,p} = 1$ (where L is the left-shift operator).
4. Some examples of continuous linear maps on $C_b(\mathbb{R})$ are:

   (a) $Tf(t) := (f(t) + f(-t))/2$,
   (b) Translations $T_a f(t) := f(t - a)$; they are isometries and form a group with $T_a T_b = T_{a+b}$, $I = T_0$, $T_a^{-1} = T_{-a}$,
   (c) Warping of the domain: $T_g f(t) := f \circ g(t)$, where g is invertible;
   (d) Multipliers $M_g f(t) := g(t) f(t)$, where $g \in C_b(\mathbb{R})$.

   What are their kernels and image subspaces? and their norms?
5. It is not so easy to calculate $\|T\|$ in general, even when T is a matrix. Show that, with the Euclidean norms,

   (a) $\left\| \begin{pmatrix} \lambda & 0 \\ 0 & \mu \end{pmatrix} \right\| = \max(|\lambda|, |\mu|)$.
   (b) $\left\| \begin{pmatrix} 0 & 1 \\ 0 & 0 \end{pmatrix} \right\| = 1 = \left\| \begin{pmatrix} 0 & 1 \\ -1 & 0 \end{pmatrix} \right\|$.
   (c) If you feel up to it, show that for real 2×2 matrices,

   $$\left\| \begin{pmatrix} a & b \\ c & d \end{pmatrix} \right\| = \sqrt{\frac{a^2+b^2+c^2+d^2+\sqrt{((a-d)^2+(b+c)^2)((a+d)^2+(b-c)^2)}}{2}}$$

   (Hint: Use Lagrange multipliers to find the maximum of $(ax + by)^2 + (cx + dy)^2$ subject to $x^2 + y^2 = 1$. See also Exercise 15.21(8).)

6. Prove that if a matrix decomposes as $T = \begin{pmatrix} A & 0 \\ 0 & B \end{pmatrix}$ then $\|T\|_{2,2} \leqslant \max(\|A\|_{2,2}, \|B\|_{2,2})$. Does this generalize to $\|T\|_{p,p}$ or $\|T\|_{2,1}$?
7. If $T_n x_n \to 0$ for any choice of unit vectors x_n, then $T_n \to 0$.
8. * A real matrix T has two norms, in principle: when considered as a matrix mapping $\mathbb{R}^n \to \mathbb{R}^m$, and as $\mathbb{C}^n \to \mathbb{C}^m$. The 'complex' norm is always larger than the 'real' norm, but need not be equal. For example, taking $T := \begin{pmatrix} 1 & -1 \\ 1 & 1 \end{pmatrix}$ and $x := \begin{pmatrix} 1+i \\ 1-i \end{pmatrix}$ gives $\|Tx\|_1 = 2\sqrt{2}\|x\|_\infty$, yet $\|T\|_{\infty,1} = 2$ over the reals. Show, however, that the two norms are equal when the 2-norms are used in both domain and codomain. (Hint: $\|x + iy\|^2 = \|x\|^2 + \|y\|^2$.)

8.3   Isomorphisms and Projections

We sometimes need to show that two normed spaces are essentially the same, meaning that any process involving addition, scalar multiplication, or convergence, in one space is mirrored in precise fashion in the other space, and vice-versa. This is the idea of an isomorphism.

Definition 8.14

An **isomorphism** between normed vector spaces is a bijective map $T : X \to Y$ such that both T and T^{-1} are linear and continuous. The spaces are then said to be *isomorphic* to each other, $X \cong Y$.

An *isometric isomorphism* is one that preserves distance, $\|Tx\|_Y = \|x\|_X$ for all $x \in X$, and isometrically isomorphic spaces are denoted by $X \equiv Y$.

We say that X is *embedded* in Y, denoted $X \subsetneq Y$ when $X \cong Z \subseteq Y$ for some subspace Z, and the isomorphism $X \to Z$ is called an *embedding*.

Thus, isomorphic normed spaces are isomorphic as vector spaces and homeomorphic (in fact equivalent) as metric spaces. Intuitively speaking, if X is embedded in Y, then one can treat it as if it were a subspace of Y even though its elements are not in Y.

Proposition 8.15

If $T : X \to Y$ is a bijective linear map, then T^{-1} is linear, and is continuous when $c\|x\|_X \leqslant \|Tx\|_Y$ for some $c > 0$.

When T is an isomorphism, $\|T^{-1}\| \geqslant \|T\|^{-1}$.

Proof Let T be a bijective linear map, let $x, y \in Y$, and let $u := T^{-1}x$, $v := T^{-1}y$; then $T(u + v) = Tu + Tv = x + y$, so that $u + v = T^{-1}(x + y)$. Similarly $T(\lambda u) = \lambda Tu = \lambda x$ gives $T^{-1}(\lambda x) = \lambda u = \lambda T^{-1}(x)$. This shows T^{-1} is linear.

The inverse is continuous when $\|T^{-1}y\| \leqslant c\|y\|$ for all $y \in Y$, in particular for $y = Tx$: $\|x\| \leqslant c\|Tx\|$ for all $x \in X$. Since T is surjective, the two inequalities are logically equivalent.

By Proposition 8.8, $1 = \|I\| = \|TT^{-1}\| \leqslant \|T\|\|T^{-1}\|$. □

Isomorphisms are also important in practical applications of functional analysis, where linear equations of the type $Tx = y$, with y given, are very common. Three requirements are prescribed for such an equation to be *well-posed*:

(i) a solution exists; in operator terminology, this means that T is onto;
(ii) the solution is unique, that is, T is 1–1;

(iii) the solution is stable; small variations in y do not lead to sudden large changes in x, in other words, x depends continuously on y, that is, T^{-1} is continuous.

Collectively, these three conditions entail that T has a continuous inverse. They not only have theoretical implications but practical ones as well. Existence and uniqueness of a solution are of obvious practical importance; stability implies that an algorithm *can* give a meaningful approximate solution, in the sense that small numerical errors in the initial conditions or algorithmic steps do not render the output completely wrong.

To measure how well-posed an equation is, we can consider the maximum relative error in x given a relative error in y. That is, if an error δy in y gives a corresponding fluctuation δx in the solution x, $T(x + \delta x) = y + \delta y$, then $T(\delta x) = \delta y$. Thus combining $\|\delta x\| \leqslant \|T^{-1}\|\|\delta y\|$ with $\|y\| \leqslant \|T\|\|x\|$, gives

$$\frac{\|\delta x\|}{\|x\|} \leqslant \|T^{-1}\|\|T\|\frac{\|\delta y\|}{\|y\|}.$$

The number $\|T^{-1}\|\|T\|$ is called the *condition number* of T. If it is relatively large, then the equation is said to be *ill-conditioned* because the relative error of the solution could be larger than that of the data.

Examples 8.16

1. ▶ Suppose a vector space X is normed in two ways, giving two normed spaces $X_{\|\cdot\|}$ and $X_{\|\cdot\|}$. The two norms are equivalent if, and only if, the identity map $I : X_{\|\cdot\|} \to X_{\|\cdot\|}$ is an isomorphism (Example 7.4(9)); equivalently, there are constants $c, d > 0$,

$$\forall x, \qquad c\|x\| \leqslant \|x\| \leqslant d\|x\|.$$

   For example, $\mathbb{R}^n$ with the 1-norm is equivalent to $\mathbb{R}^n$ with the ∞-norm.
2. ℓ^1 is not isomorphic to ℓ^∞. It is not enough to exhibit a sequence, such as $(1, 1, \ldots)$, which belongs to ℓ^∞ but not to ℓ^1, because such a sequence may, in principle, correspond to some other sequence in ℓ^1. One must demonstrate a property that ℓ^1 satisfies but ℓ^∞ doesn't; e.g., we will show later on that the former, but not the latter, is separable.
3. ▶ The inequality $c\|x\| \leqslant \|Tx\|$ $(c > 0)$, valid for all x in a Banach space X, implies that $\operatorname{im} T$ is closed and T is 1–1.
   Proof: If $Tx = Ty$, then $c\|x - y\| \leqslant \|Tx - Ty\| = 0$ and $x = y$. Suppose $Tx_n \to y$ in Y; then $c\|x_n - x_m\| \leqslant \|Tx_n - Tx_m\| \to 0$ as $n, m \to \infty$, so $(x_n)_{n \in \mathbb{N}}$ is Cauchy and converges to, say, $x \in X$. By continuity of T, $Tx_n \to Tx = y$, hence $y \in \operatorname{im} T$ and $\operatorname{im} T$ is closed.
4. Suppose we wish to find the solution of $Tx = y$ $(T \in B(X, Y))$, but it is time-consuming or impossible to calculate T^{-1}. If $S \in B(X, Y)$ is easily inverted and close to T, i.e., $T = S + R$ and $\|R\| < \|S^{-1}\|^{-1}$, then $\|S^{-1}R\| < 1$, and the iteration

$$x_{n+1} := x_n + S^{-1}(y - Tx_n) = S^{-1}(y - Rx_n)$$

converges to the solution of the equation by the Banach fixed point theorem.

Projections

Our next aim is to show firstly that all n-dimensional spaces are isomorphic to each other (for each n), and secondly to seek an analogue of the first isomorphism theorem of vector spaces, namely $V/\ker T \cong \operatorname{im} T$. Accordingly we need to introduce an important type of operator called a projection, and then construct quotient spaces.

Definition 8.17

A **projection** is a continuous linear map $P : X \to X$ such that $P^2 = P$.

For example, shadows are the projection of objects in $\mathbb{R}^3$ to shapes in a two-dimensional plane; a flat object on the ground is its own shadow. Playing around with the definition gives a number of consequences:

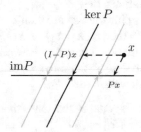

Examples 8.18

1. $(I - P)^2 = I - 2P + P^2 = I - P$ is also a projection.
2. $(I - P)P = 0$, so $x \in \operatorname{im} P \Leftrightarrow x - Px = 0$, and $\operatorname{im} P = \ker(I - P)$ is a closed subspace. Similarly $\operatorname{im}(I - P) = \ker(I - I + P) = \ker P$.
3. Any $x \in X$ can be written as $x = Px + (I - P)x \in \operatorname{im} P + \ker P$. If $x \in \operatorname{im} P \cap \ker P = \ker(I - P) \cap \ker P$, then $x = Px + (I - P)x = 0$, so that $X = \operatorname{im} P \oplus \ker P$.
4. Any linear map on a Banach space, which satisfies $P^2 = P$, is automatically continuous when $\operatorname{im} P$ and $\ker P$ are closed subspaces, but more powerful results are needed to show this (Proposition 11.5).
5. To create a projection onto the space M along the space K, find bases for these spaces and form a matrix out of their column vectors, $T := [M K]$. The projection needs to satisfy $PM = M$ and $PK = 0$, hence $P = [M 0]T^{-1}$. For example to map onto $\begin{pmatrix} 1 \\ -1 \end{pmatrix}$ along $\begin{pmatrix} 1 \\ 1 \end{pmatrix}$, $P = \begin{pmatrix} 1 & 0 \\ -1 & 0 \end{pmatrix}\begin{pmatrix} 1 & 1 \\ -1 & 1 \end{pmatrix}^{-1} = \frac{1}{2}\begin{pmatrix} 1 & -1 \\ -1 & 1 \end{pmatrix}$.

Exercises 8.19

1. (a) The map $\begin{pmatrix} a_1 \\ a_2 \end{pmatrix} \mapsto (0, a_1, a_2, 0, 0, \ldots)$ embeds $\mathbb{R}^2$ in the real space ℓ^1.
   (b) The map $J : (a_n)_{n \in \mathbb{N}} \mapsto (a_n/2^n)_{n \in \mathbb{N}}$, $\ell^\infty \to \ell^1$, is injective, linear, and continuous, but is not an embedding ($\|x\|_{\ell^\infty} \not\leqslant c\|Jx\|_{\ell^1}$).

2. An infinite-dimensional space may be properly embedded in itself: for example, the right-shift operator $R : \ell^\infty \to \text{im } R \subset \ell^\infty$ is an embedding. This cannot happen in finite dimensions.

3. Separate each sequence $x = (a_n)_{n\in\mathbb{N}}$ into two parts $x_e := (a_0, a_2, \ldots)$ and $x_o := (a_1, a_3, \ldots)$. Then the map $x \mapsto (x_e, x_o)$ is an isometric isomorphism $\ell^1 \equiv \ell^1 \times \ell^1$.

4. The space $\ell^1(\mathbb{Z})$ consists of 'sequences' $\ldots, a_{-2}, a_{-1}, a_0, a_1, a_2, \ldots$ such that $\sum_{n=-\infty}^{\infty} |a_n| < \infty$. It contains ℓ^1 as a proper subspace, even if $\ell^1 \equiv \ell^1(\mathbb{Z})$.

5. Show that if $T : X \to Y$ is an operator and P, Q are isometric isomorphisms on X, Y respectively, then $\|QTP\| = \|T\|$.

6. ⋆ Let $T : \ell^\infty \to \ell^\infty$ be an operator with matrix coefficients T_{ij}, i.e., it maps a sequence $(a_j)_{j\in\mathbb{N}} \in \ell^\infty$ to $(\sum_{j=0}^{\infty} T_{ij}a_j)_{i\in\mathbb{N}} \in \ell^\infty$. Suppose also that the matrix is dominated by its diagonal, meaning that for some $c > 0$,

$$|T_{ii}| - \sum_{j\neq i} |T_{ij}| \geqslant c.$$

Then $\|Tx\| \geqslant c\|x\|$. (Hint: use $|a + b| \geqslant |a| - |b|$.)

7. ⋆ If X_1 and X_2 are isomorphic then so are their completions $\widetilde{X}_1 \cong \widetilde{X}_2$.

8. ⋆ If $X_1 \cong X_2$ and $Y_1 \cong Y_2$ then $B(X_1, Y_1) \cong B(X_2, Y_2)$.

9. Let $Ax = b$ be a matrix equation, where A is a square matrix. Use Example 8.16(4) above to describe iterative algorithms for finding the solution of the equation in the following cases:

   (a) (*Jacobi*) A is almost diagonal in the sense that $A = D + R$, with D being the diagonal of A, and $\|R\| < \|D^{-1}\|^{-1}$.

   (b) (*Gauss-Seidel*) A is almost a lower triangular matrix, in the sense that $A = L + U$ where L is lower triangular and $\|U\| < \|L^{-1}\|^{-1}$. The inverse of a triangular matrix is fairly easy to compute.

10. *Perturbation theory*: When the solution of an invertible linear equation $Sx_0 = y$ is known, one can also find the solutions of 'nearby' equations $(S+\epsilon E)x = y$, where ϵE is a 'perturbation'. Writing $E = -ST$, the new solution satisfies $(I - \epsilon T)x = x_0$. We might try an expansion of the type $x = x_0 + \epsilon x_1 + \epsilon^2 x_2 + \cdots$; show that $x_{n+1} = Tx_n$, and the series converges if $\|E\| < \|S^{-1}\|^{-1}$ and $\epsilon < 1$.

11. Show that the following are projections:

   (a) $\begin{pmatrix} 1 & 0 \\ 1 & 0 \end{pmatrix}$ and $\frac{1}{2}\begin{pmatrix} 1 & 1 \\ 1 & 1 \end{pmatrix}$; they have the same image, but different kernels, and their norms are $\sqrt{2}$ and 1 respectively.

   (b) $P := \begin{pmatrix} 0 & 0 \\ 0 & 1 \end{pmatrix}$ and $Q := \begin{pmatrix} 1 & 1 \\ 0 & 0 \end{pmatrix}$; $\ker P = \text{im } Q$, so $PQ = 0$ is a projection but QP is not.

   (c) RL, where R and L are the shift-operators.

   (d) $x\phi \in B(X)$, where $\phi \in X^*$ and $x \in X$ such that $\phi x = 1$; in this case, $X = [\![x]\!] \oplus \ker \phi$.

12. If P and Q are commutative projections, then PQ projects onto $\operatorname{im} P \cap \operatorname{im} Q$, and $P + Q - PQ$ projects onto $\operatorname{im} P + \operatorname{im} Q$.
13. By induction, if $I = P_1 + \cdots + P_n$, with the projections P_i satisfying $P_i P_j = 0$ for $i \neq j$, then $X = \operatorname{im} P_1 \oplus \cdots \oplus \operatorname{im} P_n$.
14. $\star$ Given a closed linear subspace, is there always a projection that maps onto it?

8.4   Quotient Spaces

A linear subspace M of a vector space can be translated to form *cosets* $x + M$. For example, a straight line $L \subset \mathbb{R}^2$ passing through the origin, gives the parallel copies $x + L$. Except that with some translations, the resulting line is indistinguishable from L; it is easy to see that $x + L = L \Leftrightarrow x \in L$. More generally, $x + L = y + L \Leftrightarrow x - y \in L$. This latter is an equivalence relation (check!), so the space $\mathbb{R}^2$ 'foliates' into a stack of parallel lines, each a coset $x + L$. It is obvious that when a line L is translated by x, and then by y, the result is the line $(x + y) + L$; in fact, since translation in the direction of $v \in L$ is irrelevant to the coset, one can even talk about the addition of lines, $(x + L) + (y + L)$ as meaning $x + (y + L)$. Similarly lines can be stretched, $\lambda(x + L) = \lambda x + L$ (unless $\lambda = 0$), and the distance between lines is defined in elementary geometry as the minimum distance between them. This space of parallel lines is a good candidate for a normed space.

Turning to the general case, a vector space partitions into the cosets of M to form a vector space X/M, which is normed when M is closed, and complete when X is complete:

Proposition 8.20

If X is a normed space and M is a closed linear subspace, then the space of cosets

$$X/M := \{ x + M : x \in X \}$$

is a normed space with addition, scalar multiplication, and norm defined by

$$(x + M) + (y + M) := (x + y) + M,$$

$$\lambda(x + M) := \lambda x + M,$$

$$\| x + M \| = d(x, M) := \inf_{v \in M} \| x - v \|.$$

If M is complete, then X/M is complete $\Leftrightarrow$ X is complete.

Proof That the relation $x - y \in M$ is an equivalence relation with equivalence classes $x + M$, and that the defined addition and scalar multiplication of these classes satisfy the axioms of a vector space should be clear; the zero coset is M and the negative of $x + M$ is $-x + M$. Let us show that we do indeed get a norm:

$$\|(x + M) + (y + M)\| = \|x + y + M\| = \inf_{w \in M} \|x + y - w\|$$

$$= \inf_{u,v \in M} \|x + y - u - v\|$$

$$\leqslant \inf_{u,v \in M} (\|x - u\| + \|y - v\|)$$

$$= \inf_{u \in M} \|x - u\| + \inf_{v \in M} \|y - v\|$$

$$= \|x + M\| + \|y + M\|$$

$$\|\lambda(x + M)\| = \|\lambda x + M\| = \inf_{v \in M} \|\lambda x - v\|$$

$$= \inf_{u \in M} \|\lambda x - \lambda u\| \qquad \text{(for } \lambda \neq 0)$$

$$= \inf_{u \in M} |\lambda| \|x - u\|$$

$$= |\lambda| \|x + M\|$$

$$\|0(x + M)\| = \|M\| = d(0, M) = 0 = 0\|x + M\|$$

$$\|x + M\| = \inf_{v \in M} \|x - v\| \geqslant 0.$$

$$\|x + M\| = 0 \iff d(x, M) = 0 \iff x \in \overline{M} = M$$

$$\iff x + M = 0 + M \text{ (Exercise 2.20(9)).}$$

Completeness Let $x_n + M$ be an absolutely convergent series in X/M, i.e., $\sum_n \|x_n + M\|$ converges. Now, for each n, there is a $v_n \in M$ such that

$$\|x_n - v_n\| \leqslant \|x_n + M\| + 1/2^n.$$

The left-hand side can be summed by comparison with the right, so $\sum_n (x_n - v_n)$ converges to some x, since X is complete (Proposition 7.22). Thus

$$\left\| \sum_{n=1}^{N} (x_n + M) - (x + M) \right\| = \left\| \sum_{n=1}^{N} x_n - x + M \right\| \leqslant \left\| \sum_{n=1}^{N} (x_n - v_n) - x \right\| \to 0$$

since in general $\|a + M\| \leqslant \|a + v\|$ for any $v \in M$. Hence $\sum_n (x_n + M)$ converges, along with every other absolutely summable series, and X/M is complete.

Conversely, let $(x_n)_{n \in \mathbb{N}}$ be a Cauchy sequence in X; then

$$\| (x_n + M) - (x_m + M) \| = \| x_n - x_m + M \| \leqslant \| x_n - x_m \|$$

implies that $(x_n + M)$ is Cauchy in X/M, so converges to, say, $x + M$. This means there are $v_n \in M$ such that $x_n - (x + v_n) \to 0$; but then,

$$\| v_n - v_m \| \leqslant \| x_n - x_m - v_n + v_m \| + \| x_n - x_m \| \to 0$$

shows $(v_n)_{n \in \mathbb{N}}$ is Cauchy in M and converges to, say, $v \in M$. Thus $x_n \to x + v$. $\square$

If M is a linear subspace of X such that X/M is finite dimensional, then its *codimension* is defined by codim $M := \dim(X/M)$.

Examples 8.21

1. The cosets of the closed subspace $M := [\![\binom{1}{1}]\!] \subset \mathbb{R}^2$ are the lines parallel to M, and $\mathbb{R}^2/M \cong \mathbb{R}$.
   Proof: A vector x belongs to $x_0 + M$ when $x = \binom{a_0}{b_0} + t \binom{1}{1}$ for some $t \in \mathbb{R}$, which is the equation of a line parallel to $\binom{1}{1}$. The map $a \mapsto \binom{a}{0} + M$, $\mathbb{R} \to \mathbb{R}^2/M$ is linear and continuous. It is bijective since $\binom{a}{b} + M = \binom{a-b}{0} + M$ and

$$\binom{a_1}{0} - \binom{a_2}{0} \in M \Leftrightarrow \exists \lambda, \ \binom{a_1 - a_2}{0} = \lambda \binom{1}{1} \Leftrightarrow a_1 = a_2.$$

   The inverse map is continuous as the distance $\| \binom{a}{0} + M \|$ equals $|a|/\sqrt{2}$.
2. If X is finite-dimensional, then so is X/M, with

$$\text{codim } M = \dim X/M = \dim X - \dim M.$$

   Proof: Let $e_1, \ldots, e_m$ be a basis for M, extended by $e_{m+1}, \ldots, e_n$ to a basis for X. Then, for any vector $x = \sum_{i=1}^{n} \lambda_i e_i$, its coset,

$$x + M = \sum_{i=1}^{n} \lambda_i e_i + M = \sum_{i=m+1}^{n} \lambda_i (e_i + M),$$

   is generated by $e_{m+1} + M, \ldots, e_n + M$. Moreover, these are linearly independent, since

$$\sum_{i=m+1}^{n} \lambda_i e_i + M = 0 + M \Leftrightarrow \sum_{i=m+1}^{n} \lambda_i e_i = \sum_{i=1}^{m} \alpha_i e_i \in M$$

$$\Leftrightarrow \lambda_i = 0, \ i = m + 1, \ldots, n.$$

   Hence $\dim X/M = n - m$.

3. If $\phi \in X^*$ then $\|x + \ker \phi\| = \frac{|\phi x|}{\|\phi\|}$.

   Proof: When $\phi x \neq 0$, then $X = [\![x]\!] \oplus \ker \phi$, and

$$\|\phi\| = \sup_{y \neq 0} \frac{|\phi y|}{\|y\|} = \sup_{v \in \ker \phi} \frac{|\lambda||\phi x|}{\|\lambda x + v\|} = \frac{|\phi x|}{\inf_{v \in \ker \phi} \|x + v\|} = \frac{|\phi x|}{\|x + \ker \phi\|}$$

The following proposition states, in effect, that when one translates a closed linear subspace to any distance $c < 1$ from the origin, the resulting coset intersects the unit sphere:

Proposition 8.22 (Riesz's lemma)

> **For any non-trivial closed linear subspace M, and $0 \leqslant c < 1$, there is a unit vector x such that $\|x + M\| = c$.**

Proof Let $y \notin M$ so that $\|y + M\| > 0$; by re-scaling y if necessary, one can assume $\|y + M\| = c$. The map $f : M \to \mathbb{R}$, defined by $f(v) := \|y + v\|$, takes values close to c, as well as arbitrarily large values ($\|y + \lambda v\| \geqslant |\lambda| \|v\| - \|y\| \to \infty$ as $\lambda \to \infty$, for $M \neq 0$). Since M is connected, and f is continuous, its image must include $]c, \infty[$ by the intermediate value theorem (Corollary 5.7). In particular there is a $v \in M$ such that $\|y + v\| = 1$, so letting $x := y + v$ gives $\|x + M\| = \|y + M\| = c$. □

Exercises 8.23

1. ▶ The mapping $x \mapsto x + M$, $X \to X/M$, is linear and continuous.
2. Let $M := \{ f \in C[0, 1] : f(0) = 0 \}$, then $2 + M = \{ f \in C[0, 1] : f(0) = 2 \}$, and $C[0, 1]/M \cong \mathbb{C}$.
3. (a) $X/X \equiv 0$, $X/0 \equiv X$.

   (b) If X, Y are normed spaces, then $\dfrac{X \times Y}{X \times 0} \equiv Y$.
4. Let X be a finite-dimensional space generated by a set of unit vectors $E := \{ e_i : i = 1, \ldots, n \}$, and let $M_i := [\![E \smallsetminus \{e_i\}]\!]$. Then the coefficient $|\alpha_i|$ in $x = \sum_{i=1}^{n} \alpha_i e_i$ is at most $\|x\|/\|e_i + M_i\|$. Thus, in finding a basis for X, it is best to select unit vectors that are as 'far' from each other as possible.
5. Let M be a closed subspace of X. If both M and X/M are separable, then so is X.

8.5 $\mathbb{R}^n$ and Totally Bounded Sets

That finite-dimensional normed spaces ought to be better behaved than infinite-dimensional ones is to be expected. What is slightly surprising is the following result that they allow only a unique way of defining convergence: Any norm on $\mathbb{C}^n$

is equivalent to the complete Euclidean norm. This is an example of a mathematical "small is beautiful" principle, in the same league of results as "finite integral domains are fields".

Theorem 8.24

Every n-dimensional normed space over $\mathbb{C}$ is isomorphic to $\mathbb{C}^n$, and so is complete.

The theorem is also true for real finite-dimensional normed spaces: they are isomorphic to $\mathbb{R}^n$.

Proof Let X be an n-dimensional normed space, with a basis of unit vectors $v_1, \ldots, v_n$, and let $\mathbb{C}^n$ be given the complete 1-norm (Example 7.17(2)). There is a map between them, $J : \mathbb{C}^n \to X$, defined by

$$x = \begin{pmatrix} \alpha_1 \\ \vdots \\ \alpha_n \end{pmatrix} \mapsto \alpha_1 v_1 + \cdots + \alpha_n v_n.$$

J is linear: This follows from the distributive laws of vectors; that it is 1–1 and onto follow from the linear independence and spanning of $\{v_i\}_{i=1}^n$ respectively.

J is continuous:
$$\|Jx\|_X = \|\alpha_1 v_1 + \cdots + \alpha_n v_n\|_X$$
$$\leqslant |\alpha_1| + \cdots + |\alpha_n|$$
$$= \|x\|_1$$

J^{-1} *is continuous*: Let $f(x) := \|Jx\|_X$, which is a composition of two continuous functions: the norm and J. The unit sphere $S := \{u \in \mathbb{C}^n : \|u\|_1 = 1\}$ is a compact set (since it is closed and bounded in $\mathbb{C}^n = \mathbb{R}^{2n}$ (Corollary 6.20)), so fS is also compact (thus closed in $\mathbb{R}$). One point that is outside fS is 0,

$$f(x) = 0 \Leftrightarrow \|Jx\| = 0 \Leftrightarrow Jx = 0 \Leftrightarrow x = 0.$$

Zero is therefore an exterior point contained in an open interval $]-c, c[$ outside fS. This means that $c \leqslant \|Ju\|$ for any unit vector u. Applying this to $u = x/\|x\|_1$ for any (non-zero) vector $x \in \mathbb{C}^n$, we find $c\|x\|_1 \leqslant \|Jx\|$ as required (Proposition 8.15).

Clearly, the proof does not depend critically on the use of complex rather than real scalars. $\square$

Proposition 8.25 (Riesz's theorem)

A subset K of a normed space X is totally bounded $\Leftrightarrow$ K is bounded and lies arbitrarily close to finite-dimensional subspaces, meaning

$\forall \epsilon > 0, \ \exists Y$ finite-dimensional subspace of $X, \ \forall x \in K, \quad \|x + Y\| < \epsilon.$

Balls are totally bounded only in finite-dimensional normed spaces.

Proof (i) Let $K \subseteq \bigcup_{i=1}^{n} B_\epsilon(x_i)$ be a totally bounded set in the normed space X, and let $Y := [\![x_1, \ldots, x_n]\!]$. Any point $x \in K$ is covered by some ball $B_\epsilon(x_i)$, i.e., $\|x - x_i\| < \epsilon$, so that $\|x + Y\| = \inf_{y \in Y} \|x - y\| < \epsilon$. Since ϵ can be chosen arbitrarily small, this proves one implication in the first statement.

In a finite-dimensional normed space, bounded sets are totally bounded: This is true for $\mathbb{C}^n$ because balls (and their subsets) are totally bounded (Exercise 6.9(2)). Any finite-dimensional space Y has an isomorphism $J : \mathbb{C}^n \to Y$ by the previous theorem. If A is a bounded subset of Y, $J^{-1}A$ is a bounded set in $\mathbb{C}^n$ (Exercise 4.18(3)), hence totally bounded; mapping back to Y, $A = JJ^{-1}A$ is totally bounded (Proposition 6.7).

For the converse of the proposition, suppose K is bounded by r, and lies within ϵ of an n-dimensional subspace Y. This means that if $x \in K$ then $\|x\| \leqslant r$, and there is a $y \in Y$ such that $\|x - y\| < \epsilon$, so

$$\|y\| \leqslant \|x\| + \|y - x\| < r + \epsilon.$$

But we have just seen that the ball $B_{r+\epsilon}(0) \cap Y$ is totally bounded in Y, and can be covered by a finite number of ϵ-balls, $B_\epsilon(y_i)$, $i = 1, \ldots, m$. In particular, there is some y_i for which $\|y - y_i\| < \epsilon$, and so

$$\|x - y_i\| \leqslant \|x - y\| + \|y - y_i\| < 2\epsilon,$$

$$\Rightarrow \quad K \subseteq \bigcup_{i=1}^{m} B_{2\epsilon}(y_i).$$

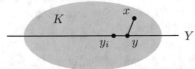

(ii) Suppose X has a totally bounded ball, which by re-scaling and translation can be taken to be the unit ball B_X (Proposition 7.6). It must be within $\epsilon < \frac{1}{2}$ of a finite-dimensional closed subspace Y. In fact $X = Y$, otherwise we can use Riesz's lemma to find a vector $y \in B_X$ with $d(y, Y) = \|y + Y\| \geqslant \frac{1}{2} > \epsilon$.   □

Examples 8.26

1. All norms on $\mathbb{C}^n$ are equivalent.
2. Given a point $x \in X$ and a finite-dimensional subspace M, there is always a best approximation $v \in M$ to x. We need only look in the compact ball $B :=$ $\overline{B_{\|x\|}(0)} \cap M$, and since the function $v \mapsto \|v - x\|$ on it is continuous, it achieves the minimum (Corollary 6.16).

    For example, there is always a polynomial of degree at most n that best approximates a function with respect to any given norm.
3. Every proper finite-dimensional subspace is nowhere dense since it is closed yet cannot contain any ball. Hence a countable union of finite-dimensional spaces cannot be complete, by the Baire category theorem. For example, (i) $c_{00} := [\![e_0, e_1, \dots]\!]$, (ii) the space of polynomials, cannot have a complete norm, since they are such unions. (Note that the two spaces are isomorphic as vector spaces.)
4. If M, N are subspaces of a normed space, with M complete and N finite-dimensional, then $M + N$ is complete (see Example 7.12(2)).
   Proof: It is enough to show that $M + [\![e]\!]$ is complete when $e \notin M$; the result then follows by induction. For any $x \in M, \alpha \in \mathbb{C}$,

$$|\alpha| \|e + M\| = \|\alpha e + M\| \leqslant \|\alpha e + x\|,$$

$$\|x\| \leqslant \|x + \alpha e\| + |\alpha| \|e\| \leqslant c \|\alpha e + x\|.$$

So if $(x_n + \alpha_n e)$ is a Cauchy sequence in $M + [\![e]\!]$, then so are $(\alpha_n)_{n \in \mathbb{N}}$ and $(x_n)_{n \in \mathbb{N}}$, in $\mathbb{C}$ and M respectively. Hence, $x_n + \alpha_n e \to x + \alpha e \in M + [\![e]\!]$.

Exercises 8.27

1. Totally bounded sets cannot be open (or have a proper interior) in an infinite dimensional normed space.
2. The set of polynomials of degree at most n forms a closed linear subspace of $L^1[a, b]$ with dimension $n + 1$; a basis for this space is $1, t, \dots, t^n$.
3. As an illustration of Riesz's theorem, the unit ball in the infinite-dimensional space ℓ^∞ (or ℓ^1) is not totally bounded. (Hint: Show $(e_n)_{n \in \mathbb{N}}$ has no Cauchy subsequence.)
4. Among normed spaces, only in finite dimensions are closed and bounded subsets compact.
5. Totally bounded sets need not lie in a finite-dimensional subspace, just arbitrarily close to them. Can you think of an infinite-dimensional totally bounded set?

Remarks 8.28

1. By analogy with matrices, it is customary to write Tx instead of $T(x)$. This is a slight abuse of notation; a linear map on the vector space of *matrices* need not act on the left, e.g., $A \mapsto AB$, $A \mapsto AB + BA$, $A \mapsto A^\top$, and $A \mapsto B^{-1}AB$ are all linear.
2. For the initiated, the idea of continuous linear maps can be extended to continuous multi-linear maps (tensors); they also form a Banach space with norm

$$\|T\| := \sup |T(x_1, \ldots, \phi_1, \ldots)| / \|x_1\| \ldots \|\phi_1\| \ldots.$$

3. $B(X, Y)$ forms part of the larger space of Lipschitz functions $X \to Y$. For such functions, $\|f\| := \sup_{x_1 \neq x_2 \in X} \|f(x_1) - f(x_2)\| / \|x_1 - x_2\|$ satisfies the norm axioms, except that $\|f\| = 0 \Leftrightarrow f$ is constant.

Chapter 9
The Classical Spaces

Having fleshed out a substantial amount of abstract theory, we turn to the concrete examples of normed spaces and identify which are complete and separable. Unavoidably, the proofs become more technical once we leave the familiarity of finite dimensions and enter the realm of infinite-dimensional spaces, having to deal as it were with sequences of sequences and limits of functions in different norms. However, a careful study of this chapter will be rewarded by having an armory of spaces, so to speak, ready to serve as examples to confirm or refute conjectured statements. We can barely scratch the surface of all the properties that these spaces possess, concentrating mostly on completeness, separability, and duality.

9.1 Sequence Spaces

The Space ℓ^∞

A sequence in ℓ^∞ is a sequence of sequences, $x_n = (a_{n,i})_{i\in\mathbb{N}}$. Convergence in ℓ^∞ means uniform convergence of the components, that is,

$$x_n \to 0 \;\Leftrightarrow\; \sup_{i\in\mathbb{N}} |a_{n,i}| \to 0 \text{ as } n \to \infty$$

$$\Leftrightarrow |a_{n,i}| \to 0 \text{ as } n \to \infty, \text{ uniformly for all components } i,$$

$$\Leftrightarrow \forall \epsilon > 0,\ \exists N,\ \forall n \geqslant N, \forall i \in \mathbb{N},\quad |a_{n,i}| < \epsilon.$$

J. Muscat, *Functional Analysis*, https://doi.org/10.1007/978-3-031-27537-1_9

For example, of the following three sequences of sequences, only the first converges to $\mathbf{0}$, even though each component converges to 0.

$$(1, 1, 1, 1, \ldots) \qquad (1, 1, 1, 1, 1, \ldots) \qquad (1, 0, 0, 0, \ldots)$$
$$(\tfrac{1}{2}, \tfrac{1}{2}, \tfrac{1}{2}, \tfrac{1}{2}, \ldots) \qquad (0, 0, 1, 1, 1, \ldots) \qquad (0, 1, 0, 0, \ldots)$$
$$(\tfrac{1}{3}, \tfrac{1}{3}, \tfrac{1}{3}, \tfrac{1}{3}, \ldots) \qquad (0, 0, 0, 0, 1, \ldots) \qquad (0, 0, 1, 0, \ldots)$$
$$\vdots \qquad\qquad\qquad \vdots \qquad\qquad\qquad \vdots$$
$$\downarrow \qquad\qquad\qquad \not\downarrow \qquad\qquad\qquad \not\downarrow$$
$$(0, 0, 0, 0, \ldots) \qquad (0, 0, 0, 0, \ldots) \qquad (0, 0, 0, 0, \ldots)$$

Theorem 9.1

ℓ^∞ **is complete but not separable.**

Proof (i) Let $(x_n)_{n \in \mathbb{N}}$ be a Cauchy sequence in ℓ^∞, i.e., $\|x_n - x_m\|_{\ell^\infty} \to 0$ as $n, m \to \infty$. Note that $\|x_n\|_{\ell^\infty} \leqslant c$ since Cauchy sequences are bounded (Example 4.3(6)).

$$
\begin{array}{c|llll}
x_0 & a_{00}\ a_{01}\ a_{02}\ \ldots & \leqslant \|x_0\|_{\ell^\infty} \\
x_1 & a_{10}\ a_{11}\ a_{12}\ \ldots & \leqslant \|x_1\|_{\ell^\infty} \\
\vdots & \vdots\ \ \vdots\ \ \vdots & \vdots \\
\downarrow & \downarrow\ \ \downarrow\ \ \downarrow \\
x & a_0\ \ a_1\ \ a_2\ \ldots & \leqslant c
\end{array}
$$

(The absolute signs of $a_{n,i}$ are omitted in the horizontal rows.)

For each column i, $|a_{n,i} - a_{m,i}| \leqslant \|x_n - x_m\|_{\ell^\infty} \to 0$, so $(a_{n,i})_{n \in \mathbb{N}}$ is a Cauchy sequence in $\mathbb{C}$, which converges to, say, $a_i := \lim_{n \to \infty} a_{n,i}$.

That $x := (a_i)_{i \in \mathbb{N}}$ is in ℓ^∞ follows from taking the limit $n \to \infty$ of

$$|a_{n,i}| \leqslant \|x_n\|_{\ell^\infty} \leqslant c.$$

More crucially, $x_n \to x$ in ℓ^∞ since, for any $\epsilon > 0$, $\|x_m - x_n\|_{\ell^\infty} < \epsilon$ for $m, n \geqslant N$, large enough; and for any column i, one can choose an $m \geqslant N$ large enough that $|a_{m,i} - a_i| < \epsilon$, so that

$$|a_i - a_{n,i}| \leqslant |a_i - a_{m,i}| + |a_{m,i} - a_{n,i}| < \epsilon + \|x_m - x_n\|_{\ell^\infty} < 2\epsilon,$$

implying $a_{n,i} \to a_i$, independently of i.

(ii) To show ℓ^∞ is not separable we display an uncountable number of disjoint balls (Exercise 4.22(4)). Consider the sequences that consist of 1s and 0s. The distance between any two of them is exactly 1, so that the balls centered on them with radius

1/2 are disjoint. Moreover, these sequences are uncountable for the same reason that the real numbers are uncountable: If one were able to list them as

$$x_0 = (a_{00}, \ a_{01}, \ a_{02}, \ldots)$$
$$x_1 = (a_{10}, \ a_{11}, \ a_{12}, \ldots)$$
$$x_2 = (a_{20}, \ a_{21}, \ a_{22}, \ldots)$$
$$\vdots$$

one could take the diagonal sequence $(a_{00}, a_{11}, a_{22} \ldots)$, and swap its 1s and 0s, giving a sequence $(1 - a_{n,n})_{n \in \mathbb{N}}$ that cannot be in the list because it disagrees with any x_n in the nth position, as $1 - a_{n,n} \neq a_{n,n}$. $\qquad \square$

To appreciate how large ℓ^∞ is, consider that even if given an immense number of terms of a sequence $(a_n)_{n \in \mathbb{N}} \in \ell^\infty$, one cannot tell how large are the remaining terms, and they cannot be ignored. Contrast this with ℓ^1, where any sequence can be approximated by a finite set of values and the rest replaced by zero. Crucially, $(a_n)_{n \leqslant N} \to (a_n)_{n \in \mathbb{N}}$, as $N \to \infty$, in ℓ^1 but not necessarily in ℓ^∞. However, ℓ^∞ does have separable complete subspaces:

Proposition 9.2

> **The space of convergent complex sequences, and of those sequences that converge to 0,**
>
> $$c := \{\, (a_n)_{n \in \mathbb{N}} : \exists a \in \mathbb{C}, \ \lim_{n \to \infty} a_n = a \,\},$$
>
> $$c_0 := \{\, (a_n)_{n \in \mathbb{N}} : \lim_{n \to \infty} a_n = 0 \,\},$$
>
> **are complete separable subspaces of ℓ^∞.**

Proof The spaces are nested in each other as $c_0 \subset c \subset \ell^\infty$ since convergent sequences are bounded. They are easily shown to be linear subspaces: $\lambda a_n + b_n \to \lambda a + b$ when $a_n \to a$ and $b_n \to b$ as $n \to \infty$.

c_0 *is closed in* ℓ^∞: Let $x_n \to x$ in ℓ^∞, with $x_n \in c_0$; their components converge uniformly $a_{n,i} \to a_i$ as $n \to \infty$.

$$
\begin{array}{c|cccc}
x_0 & a_{00} & a_{01} & a_{02} & \ldots \ \to 0 \\
x_1 & a_{10} & a_{11} & a_{12} & \ldots \ \to 0 \\
\vdots & \vdots & & \vdots & \vdots \\
& \downarrow & \downarrow & \downarrow & \downarrow \\
x & a_0 & a_1 & a_2 & \ldots \ \overset{?}{\to} 0
\end{array}
$$

Now, for any $\epsilon > 0$, there is an x_n in c_0 such that $\|x_n - x\|_{\ell^\infty} < \epsilon$, and for this sequence, there is an integer N, such that

$$i \geqslant N \implies |a_{n,i}| < \epsilon.$$

It follows that for $i \geqslant N$,

$$|a_i| \leqslant |a_{n,i}| + |a_i - a_{n,i}| \leqslant |a_{n,i}| + \|x - x_n\|_{\ell^\infty} < 2\epsilon$$

so $\lim_{i \to \infty} a_i = 0$ and $x \in c_0$.

c_0 *is separable :* The vectors $e_n := (\delta_{n,i}) = (0, \ldots, 0, 1, 0, \ldots)$, with the 1 occurring at the nth position, form a Schauder basis for c_0: for any $x = (a_n)_{n \in \mathbb{N}} \in c_0$,

$$\left\| x - \sum_{n=0}^{N} a_n e_n \right\|_{\ell^\infty} = \sup_{n > N} |a_n| \to 0, \text{ as } N \to \infty.$$

If $\sum_n a_n e_n = \sum_n b_n e_n$, then $(a_0 - b_0, a_1 - b_1, \ldots) = \mathbf{0}$ hence $a_n = b_n$ and the coefficients are unique.

The spaces c and c_0 are isomorphic: Let $J : c \to c_0 \subset \ell^\infty$ be defined by

$$J(a_0, a_1, a_2, \ldots) := (-a, a_0 - a, a_1 - a, \ldots), \qquad \text{where } a := \lim_{n \to \infty} a_n.$$

J is 1–1 since

$$J(a_n) = J(b_n) \implies a = b \text{ and } \forall n \in \mathbb{N}, a_n - a = b_n - b$$

$$\implies (a_n) = (b_n).$$

J is onto c_0 for, given any $y = (b_n)_{n \in \mathbb{N}} \in c_0$, it is clear that $x := (b_1 - b_0, b_2 - b_0, \ldots)$ is in c and maps to y. In fact, writing $\mathbf{1} := (1, 1, \ldots)$,

$$Jx = Rx - a\mathbf{1}, \qquad J^{-1}y = Ly - b_0\mathbf{1},$$

where R and L are the shift operators. This observation shows that both J and J^{-1} are continuous and linear since $(a_n)_{n \in \mathbb{N}} \mapsto a\mathbf{1}$, as well as $(b_n)_{n \in \mathbb{N}} \mapsto b_0\mathbf{1}$, are operators

$$\|a\mathbf{1}\|_{\ell^\infty} = |a|\|\mathbf{1}\|_{\ell^\infty} = \lim_{n \to \infty} |a_n| \leqslant \sup_n |a_n| = \|(a_n)\|_{\ell^\infty}$$

$$\|b_0\mathbf{1}\|_{\ell^\infty} = |b_0|\|\mathbf{1}\|_{\ell^\infty} \leqslant \sup_n |b_n| = \|(b_n)\|_{\ell^\infty}.$$

It follows that $c \cong c_0$ and has the same properties of completeness and separability that c_0 enjoys.

$\square$

Theorem 9.3

> **Every functional on c_0 is of the type** $(a_n)_{n \in \mathbb{N}} \mapsto \sum_n b_n a_n$ **where** $(b_n)_{n \in \mathbb{N}} \in \ell^1$, **and**
>
> $$c_0^* \equiv \ell^1.$$

Proof Given $y = (b_n)_{n \in \mathbb{N}} \in \ell^1$ and $x = (a_n)_{n \in \mathbb{N}} \in c_0$, the inequality

$$|y \cdot x| = \left| \sum_{n=0}^{\infty} b_n a_n \right| \leqslant \sum_{n=0}^{\infty} |b_n||a_n| \leqslant \sup_n |a_n| \sum_{n=0}^{\infty} |b_n| = \|x\|_{\ell^\infty} \|y\|_{\ell^1}$$

shows that the linear map $y^\top : x \mapsto y \cdot x := \sum_{n=0}^{\infty} b_n a_n$ (Example 8.3(4)) is well-defined and continuous on ℓ^∞ (including c_0), with $\|y^\top\| \leqslant \|y\|_{\ell^1}$.

Every functional on c_0 is of this type: By the linearity and continuity of any $\phi \in c_0^*$,

$$\phi x = \phi \left(\sum_{n=0}^{\infty} a_n e_n \right) = \sum_{n=0}^{\infty} a_n b_n = y \cdot x, \qquad \text{where } b_n := \phi e_n, \; y := (b_n)_{n \in \mathbb{N}}.$$

Also, writing $b_n = |b_n| e^{i\theta_n}$ in polar form,

$$\sum_{n=0}^{\infty} |b_n| = \sum_{n=0}^{\infty} e^{-i\theta_n} \phi e_n = \lim_{N \to \infty} \phi \left(\sum_{n=0}^{N} e^{-i\theta_n} e_n \right) \leqslant \|\phi\| \|(e^{-i\theta_n})\|_{\ell^\infty} = \|\phi\|,$$

hence $y \in \ell^1$, with $\|y\|_{\ell^1} \leqslant \|\phi\| = \|y^\top\|$. Combined with the inequality above, we get $\|y\|_{\ell^1} = \|y^\top\|$.

Isometric isomorphism: Let $J : \ell^1 \to c_0^*$ be the map $y \mapsto y^\top$. The above conclusions can be summarized as stating that J is a surjective isometry. That J is linear is easily seen from the following statement that holds for every $x \in c_0$, $u, v, y \in \ell^1$,

$$(u + v) \cdot x = \sum_{n=0}^{\infty} (u_n + v_n) a_n = \sum_{n=0}^{\infty} u_n a_n + \sum_{n=0}^{\infty} v_n a_n = u \cdot x + v \cdot x,$$

$$(\lambda y) \cdot x = \sum_{n=0}^{\infty} (\lambda b_n) a_n = \lambda \sum_{n=0}^{\infty} b_n a_n = \lambda (y \cdot x),$$

so $(u + v)^\top = u^\top + v^\top$ and $(\lambda y)^\top = \lambda y^\top$.

$\square$

We often make remarks like "the dual space of c_0 is ℓ^1"—this is not literally true because a functional on c_0 is not a sequence, but the application of one, i.e., it is $y^\top$ not y. But the two are mathematically the same object in different clothing, and functionals on c_0 do behave like the sequences in ℓ^1.

Exercises 9.4

1. The kernel of the functional $\mathrm{Lim} : (a_n)_{n\in\mathbb{N}} \mapsto \lim_{n\to\infty} a_n$ on c, is c_0.
2. Any convergent complex sequence $a_n \to a$ can be written as

$$(a_n)_{n\in\mathbb{N}} = \sum_n (a_n - a)e_n + a\mathbf{1},$$

   where $\mathbf{1} := (1, 1, \ldots)$. Deduce that the vectors e_n together with $\mathbf{1}$ form a Schauder basis for c; what is its dual space c^*?
3. ▶ One can *multiply* bounded sequences together as $(a_n)(b_n) := (a_n b_n)_{n\in\mathbb{N}}$, to get another bounded sequence, $\|xy\|_{\ell^\infty} \leqslant \|x\|_{\ell^\infty}\|y\|_{\ell^\infty}$. This multiplication is commutative and associative, and has unity $\mathbf{1}$. Only those sequences which are bounded away from 0 (i.e., $|a_n| \geqslant c > 0$) have an inverse, $(a_n)_{n\in\mathbb{N}}^{-1} = (a_n^{-1})_{n\in\mathbb{N}}$.
4. ⋆ The inequality $\|xy\|_{\ell^1} \leqslant \|x\|_{\ell^\infty}\|y\|_{\ell^1}$ is also true, so the map $x \mapsto M_x$, where $M_x y := xy$, embeds ℓ^∞ in $B(\ell^1)$.
5. The closure of c_{00} in the ℓ^∞-norm is $\overline{c_{00}} = c_0$.
6. ℓ^∞ contains the space of sequences $\ell_s^\infty := \{ (a_n)_{n\in\mathbb{N}} : \exists c, \forall n \geqslant 1, |a_n| \leqslant c/n^s \}$ $(s > 0)$. What is its closure? Can you think of a sequence which is in c_0 but not in any ℓ_s^∞?
7. The distance between a sequence $(a_n)_{n\in\mathbb{N}} \in \ell^\infty$ and c_0 is $\limsup_n |a_n|$.
8. ⋆ $C[0, 1]$ can be embedded in ℓ^∞, since $f \in C[0, 1]$ is determined by its values on the dense subset $\mathbb{Q} \cap [0, 1]$ which can be listed as a sequence $(q_n)_{n\in\mathbb{N}}$. Check that the mapping $f \mapsto (f(q_n))_{n\in\mathbb{N}}$ is linear and isometric.

The Space ℓ^1

Convergence in ℓ^1 is more stringent than in ℓ^∞. This can be seen by the inequality

$$\forall x = (a_i)_{i\in\mathbb{N}} \in \ell^1, \qquad \|x\|_{\ell^\infty} = \sup_{i\in\mathbb{N}} |a_i| = \max_{i\in\mathbb{N}} |a_i| \leqslant \sum_{i=0}^\infty |a_i| = \|x\|_{\ell^1}$$

so $x_n \to \mathbf{0}$ in ℓ^∞ does not guarantee $x_n \to \mathbf{0}$ in ℓ^1. For the latter to occur, not only must the components approximate 0 together, but their sum must also diminish. Fewer sequences manage to do this, and this is reflected in the fact that ℓ^1 is separable.

Theorem 9.5

ℓ^1 **is complete and separable.**

Proof (i) Since $\ell^1 \equiv c_0^*$, one can argue that ℓ^1 is complete, as are all dual spaces (Theorem 8.7).

Alternatively, the following direct proof shows that every absolutely summable series in ℓ^1 converges using Proposition 7.22 (Note: as ℓ^1 is defined in terms of sums, it is more straight-forward to use series instead of Cauchy sequences). Suppose $x_0 + x_1 + x_2 + \cdots$ is a series such that $\sum_{n \in \mathbb{N}} \|x_n\|_{\ell^1} = s$. In the following diagram, we will show convergence of the various vertical sums.

$$
\begin{array}{c|cccc|c}
x_0 & a_{00} & + \; a_{01} & + \; a_{02} & + \cdots & \|x_1\|_{\ell^1} \\
+ & + & + & + & & + \\
x_1 & a_{10} & + \; a_{11} & + \; a_{12} & + \cdots & \|x_2\|_{\ell^1} \\
+ & + & + & + & & + \\
\vdots & & \vdots & & & \vdots \\
\downarrow & \downarrow & \downarrow & \downarrow & & \downarrow \\
x & a_0 & a_1 & a_2 & \cdots & s
\end{array}
$$

(Note that the absolute signs of $a_{n,i}$ are omitted in the horizontal sums.)

The main point of the proof is that any rectangular sum of terms in this array is less than the corresponding sum on the right-hand column:

$$
\left| \sum_{i=I}^{J} \sum_{n=N}^{M} a_{n,i} \right| \leqslant \sum_{i=I}^{J} \sum_{n=N}^{M} |a_{n,i}| \leqslant \sum_{n=N}^{M} \|x_n\|_{\ell^1}.
$$

In particular, taking the ith column, $|\sum_n a_{n,i}| \leqslant \sum_n |a_{n,i}| \leqslant s$ shows that its sum converges in $\mathbb{C}$ to, say, $a_i := \sum_{n=0}^{\infty} a_{n,i}$. In fact, the whole array sum is bounded, $\sum_i |a_i| = \sum_i |\sum_{n=0}^{\infty} a_{n,i}| \leqslant s$, so that $x := (a_i)_{i \in \mathbb{N}}$ belongs to ℓ^1.

Finally, note that any rectangular sum goes to 0 as it moves downward, because $\sum_{n=N}^{\infty} \|x_n\|_{\ell^1} \to 0$ as $N \to \infty$. Hence

$$
\left\| x - \sum_{n=1}^{N} x_n \right\|_{\ell^1} = \sum_{i=0}^{\infty} \left| a_i - \sum_{n=1}^{N} a_{n,i} \right| = \sum_{i=0}^{\infty} \left| \sum_{n=N+1}^{\infty} a_{n,i} \right| \to 0
$$

giving $x = \sum_{n=0}^{\infty} x_n$.

(ii) The sequences $e_n := (0, \ldots, 0, 1, 0, \ldots)$, with the 1 occurring at the nth position, is a Schauder basis because, firstly, for any vector $x = (a_n)_{n \in \mathbb{N}} \in \ell^1$,

$$\Big\| x - \sum_{n=0}^{N} a_n e_n \Big\|_{\ell^1} = \| (a_0, a_1, \ldots) - (a_0, \ldots, a_N, 0, 0, \ldots) \|_{\ell^1}$$

$$= \| (0, \ldots, 0, a_{N+1}, \ldots) \|_{\ell^1}$$

$$= \sum_{n=N+1}^{\infty} |a_n| \to 0 \quad \text{as } N \to \infty$$

since $\sum_n |a_n|$ converges. Secondly, if $x = \sum_{n=0}^{\infty} b_n e_n$, then $b_m = e_m \cdot x = a_m$ for each $m \in \mathbb{N}$, so e_n form a Schauder basis. $\square$

Proposition 9.6

Every functional on ℓ^1 is of the type $(a_n)_{n \in \mathbb{N}} \mapsto \sum_n b_n a_n$ where $(b_n)_{n \in \mathbb{N}} \in \ell^\infty$, and

$$\ell^{1*} \equiv \ell^\infty.$$

Proof The proof is practically identical to the one for $c_0^* \equiv \ell^1$, except that now $y = (b_n)_{n \in \mathbb{N}} \in \ell^\infty$ and $x = (a_n)_{n \in \mathbb{N}} \in \ell^1$. The inequality

$$|y \cdot x| \leqslant \sum_n |b_n| |a_n| \leqslant \sup_n |b_n| \sum_n |a_n| = \| y \|_{\ell^\infty} \| x \|_{\ell^1}$$

shows that the linear mapping $y^\top : \ell^1 \to \mathbb{C}$ is well-defined and continuous with $\| y^\top \| \leqslant \| y \|_{\ell^\infty}$.

Every functional on ℓ^1 is of this type: Let $\phi \in \ell^{1*}$, then by linearity and continuity of ϕ,

$$\phi x = \phi \Big(\sum_{n=0}^{\infty} a_n e_n \Big) = \sum_{n=0}^{\infty} a_n b_n = y \cdot x, \quad \text{where } b_n := \phi e_n, \, y := (b_n)_{n \in \mathbb{N}}.$$

Moreover $|b_n| = |\phi e_n| \leqslant \| \phi \| \| e_n \|_{\ell^1} = \| \phi \|$ so that $y \in \ell^\infty$, with $\| y \|_{\ell^\infty} \leqslant \| \phi \|$. As $\phi = y^\top$, $\| y \|_{\ell^\infty} = \| y^\top \|$.

Isomorphism: The mapping $J : \ell^\infty \to \ell^{1*}$, $y \mapsto y^\top$, is linear and the above assertions state that J is a surjective isometry. $\square$

Exercises 9.7

1. Suppose each coefficient of $x_n = (a_{n,i})_{i\in\mathbb{N}} \in \ell^1$ converges, $a_{n,i} \to a_i$ as $n \to \infty$, and suppose $x := (a_i)_{i\in\mathbb{N}}$ is in ℓ^1; then it does not follow that $x_n \to x$ in ℓ^1, e.g., $e_n \not\to 0$. But if $|a_{n,i} - a_i|$ is decreasing with n (for each i), then $x_n \to x$ in ℓ^1.

2. ▶ ℓ^1 has a natural product, called **convolution**:

$$(a_n) * (b_n) := (a_0 b_0, a_1 b_0 + a_0 b_1, a_2 b_0 + a_1 b_1 + a_0 b_2, \ldots, \sum_{i=0}^{n} a_{n-i} b_i, \ldots).$$

   This is indeed in ℓ^1 because the sum to n terms (a triangle of terms $a_i b_j$) is less than $(|a_0| + \cdots + |a_n|)(|b_0| + \cdots + |b_n|)$ (a square of terms), so that

$$\|x * y\|_{\ell^1} \leqslant \|x\|_{\ell^1} \|y\|_{\ell^1}.$$

   Convolution is commutative and associative, and e_0 acts as the identity element $e_0 * x = x$. The inverse of $(1, a, 0, \ldots)$ is $(1, -a, a^2, -a^3, \ldots)$, which is in ℓ^1 only when $|a| < 1$.

3. If $x \in \ell^1$ and $y \in \ell^\infty$, then $x * y$ is a bounded sequence

$$\|x * y\|_{\ell^\infty} \leqslant \|x\|_{\ell^1} \|y\|_{\ell^\infty}.$$

4. The right-shift operator can be written as a convolution $Rx = e_1 * x$. In general, $R^n x = e_n * x$, since $e_n * e_m = e_{n+m}$. The "running average" of a "time-series" x is $\frac{1}{n}(\underbrace{1, \ldots, 1}_{n}, 0, \ldots) * x$.

5. ⋆ A subset K of ℓ^1 is totally bounded $\Leftrightarrow$ it is bounded and

$$\forall \epsilon > 0, \ \exists N \in \mathbb{N}, \ \forall (a_n)_{n\in\mathbb{N}} \in K, \quad \|(a_n)_{n\geqslant N}\|_{\ell^1} < \epsilon.$$

   (Recall that K lies arbitrarily close to finite-dimensional subspaces.)

6. ℓ^1 has the functional $\mathrm{Sum}(b_n)_{n\in\mathbb{N}} := \sum_{n=0}^{\infty} b_n$. It corresponds to the bounded sequence $\mathbf{1} = (1, 1, \ldots)$, i.e., $\mathrm{Sum}\, x = \mathbf{1} \cdot x$. Hence if $\sum_{n,i} |a_{n,i}| < \infty$ then

$$\sum_{i\in\mathbb{N}} \sum_{n\in\mathbb{N}} a_{n,i} = \sum_{n\in\mathbb{N}} \sum_{i\in\mathbb{N}} a_{n,i}.$$

7. The functionals $\delta_N(a_n)_{n\in\mathbb{N}} := a_N$ correspond to $e_N \in \ell^\infty$, i.e., $\delta_N x = e_N \cdot x$. Similarly, the sum $\mathrm{Sum}_N(a_n) := \sum_{n=0}^{N} a_n$ corresponds to $e_0 + \cdots + e_N = (1, \ldots, 1, 0, \ldots)$. Since $(1, \ldots, 1, 0, \ldots) \not\to \mathbf{1}$ in ℓ^∞, we also have $\mathrm{Sum}_N \not\to \mathrm{Sum}$ in ℓ^{1*}, yet $\mathrm{Sum}_N(x) \to \mathrm{Sum}(x)$ for any sequence $x \in \ell^1$. We'll discuss this apparent paradox in Sect. 11.5.

The Space ℓ^2

This normed space has properties that are, in many respects, midway between ℓ^1 and ℓ^∞. Yet it stands out, as it has a dot product $x \cdot y$ defined for any two of its sequences, and $\bar{x} \cdot x = \|x\|^2$; we will have much more to say about normed spaces with such dot products in the next chapter.

Theorem 9.8

ℓ^2 **is complete and separable.**

Proof (i) Let $x_n = (a_{n,i})_{i \in \mathbb{N}}$ be a Cauchy sequence in ℓ^2; the terms are uniformly bounded $\|x_n\| \leqslant c$. For each i,

$$|a_{n,i} - a_{m,i}|^2 \leqslant \sum_i |a_{n,i} - a_{m,i}|^2 = \|x_n - x_m\|^2 \to 0 \text{ as } n, m \to \infty,$$

so $(a_{n,i})_{n \in \mathbb{N}}$ is a Cauchy sequence in $\mathbb{F}$ which converges to, say, $a_i := \lim_{n \to \infty} a_{n,i}$.
The sequence $x := (a_i)_{i \in \mathbb{N}}$ belongs to ℓ^2 by taking the limit $N \to \infty$ of

$$\sum_{i=0}^N |a_i|^2 = \lim_{n \to \infty} \sum_{i=0}^N |a_{n,i}|^2 \leqslant \lim_{n \to \infty} \|x_n\|^2 \leqslant c^2.$$

As x_n is Cauchy, for each $\epsilon > 0$ there is a positive integer M such that

$$n, m \geqslant M \implies \|x_n - x_m\| < \epsilon.$$

Moreover, for each $i \in \mathbb{N}$, there exists an integer M_i such that

$$m \geqslant M_i \implies |a_{m,i} - a_i| < \frac{\epsilon}{2^i}.$$

Therefore, for any $N \in \mathbb{N}$, picking m larger than $M, M_0, M_1, \ldots, M_N$, gives

$$\sqrt{\sum_{i=0}^N |a_{n,i} - a_i|^2} \leqslant \sqrt{\sum_{i=0}^N |a_{n,i} - a_{m,i}|^2} + \sqrt{\sum_{i=0}^N |a_{m,i} - a_i|^2}$$

$$< \|x_n - x_m\| + \sqrt{\sum_{i=0}^N \frac{\epsilon^2}{4^i}} < 3\epsilon,$$

which implies $\|x_n - x\| < 3\epsilon$ for $n \geqslant M$.

(ii) For separability, ℓ^2 has the Schauder basis e_n, since for any $x = (a_n)_{n \in \mathbb{N}} \in \ell^2$,

$$\left\| x - \sum_{n=0}^{N} a_n e_n \right\|_{\ell^2} = \| (0, \ldots, 0, a_{N+1}, \ldots) \|_{\ell^2} = \left(\sum_{n=N+1}^{\infty} |a_n|^2 \right)^{1/2} \to 0.$$

Uniqueness of the coefficients follows as in the proof of Theorem 9.5. □

Proposition 9.9

Every functional on ℓ^2 is of the type $(a_n)_{n \in \mathbb{N}} \mapsto \sum_n b_n a_n$ where $(b_n)_{n \in \mathbb{N}} \in \ell^2$, and

$$\ell^{2*} \equiv \ell^2.$$

'*Proof*': The argument is so similar to the previous ones about c_0^* and ℓ^{1*} that it is left as an exercise (use Cauchy's inequality at one point).

Exercises 9.10

1. Show that $|x \cdot y| = \|x\| \|y\|$ if, and only if, y is a multiple of x (or $x = 0$).
2. The map $(a_1, \ldots, a_n) \mapsto (a_1, \ldots, a_n, 0, 0, \ldots)$ embeds $\mathbb{C}^n$ in ℓ^2.
3. ℓ^2 contains the interesting compact convex set $\{ (a_n)_{n \in \mathbb{N}} : |a_n| \leqslant 1/n \}$, called the *Hilbert cube*. It is totally bounded in ℓ^2, as it is close within any ϵ to a finite-dimensional space $\{ (a_n)_{n \in \mathbb{N}} : \forall n > N_\epsilon, a_n = 0 \}$, yet it is infinite-dimensional; it cannot enclose any ball (else the ball would be totally bounded).
4. ▶ The various sequence spaces are subsets of each other as follows:

$$c_{00} \subset \ell^1 \subset \ell^2 \subset c_0 \subset c \subset \ell^\infty, \quad \text{because} \quad \|x\|_{\ell^\infty} \leqslant \|x\|_{\ell^2} \leqslant \|x\|_{\ell^1},$$

but $\ell^1 \subset \ell^2 \subset c_0$ are *not* Banach space embeddings! Show further that c_{00} with the respective norms is dense in ℓ^1, ℓ^2, and c_0 (c_{00} cannot be complete in any norm, Example 8.26(3)).

The Space ℓ^p

The space $\ell^p := \{ (a_n)_{n \in \mathbb{N}} : a_n \in \mathbb{C}, \ \sum_n |a_n|^p < \infty \}$, $p \geqslant 1$, is endowed with addition and scalar multiplication like the other sequence spaces, and the norm

$$\|x\|_{\ell^p} := \left(\sum_{n=0}^{\infty} |a_n|^p \right)^{1/p}.$$

Our aim in this section is to prove the triangle inequality for this norm, otherwise known as Minkowski's inequality, and show ℓ^p is complete and separable.

As the reader is probably becoming aware, it is inequalities that are at the heart of most proofs about continuity, including isomorphisms. They can be thought of as a 'process' transforming numbers from one form to another, perhaps more useful, form, but losing some information on the way. Much like tools to be chosen with care, some are "sharper" than others. (See [8] for much more.) The following three inequalities are continually used in analysis. The first is a gem, simple yet rich:

$$a^\alpha b^\beta \leqslant \alpha a + \beta b, \qquad \text{for } \alpha, \beta, a, b \geqslant 0, \ \alpha + \beta = 1. \tag{9.1}$$

This AM-GM inequality, as it is known, states that any weighted geometric mean is less than or equal to the same-weighted arithmetic mean. The special case $\sqrt{ab} \leqslant (a + b)/2$ has already been encountered previously. Writing $a = e^x, b = e^y$ gives

$$e^{\alpha x + \beta y} \leqslant \alpha e^x + \beta e^y.$$

This is equivalent to the convexity of the exponential function, and can be taken as its proof (any real function with a positive second derivative is convex).

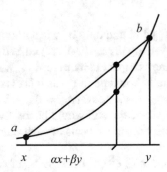

The same idea applied to the convexity of x^p, $p \geqslant 1$, gives

$$(\alpha a + \beta b)^p \leqslant \alpha a^p + \beta b^p, \qquad \text{for } \alpha, \beta, a, b \geqslant 0, \ \alpha + \beta = 1. \tag{9.2}$$

A third inequality of importance is

$$a^p + b^p \leqslant (a + b)^p, \qquad \text{for } p \geqslant 1, \ a, b \geqslant 0, \tag{9.3}$$

obtained by adding the inequalities $a^p \leqslant a(a + b)^{p-1}$ and $b^p \leqslant b(a + b)^{p-1}$.

Proposition 9.11

For $a, b, \alpha, \beta \geqslant 0, \alpha + \beta = 1, p \geqslant 1, q > 0$,

$$
\begin{aligned}
\min(a, b) &\leqslant \left(\alpha a^{-q} + \beta b^{-q}\right)^{-1/q} && \textbf{harmonic mean} \\
&\leqslant a^\alpha b^\beta && \textbf{geometric mean} \\
&\leqslant \left(\alpha a^{1/p} + \beta b^{1/p}\right)^p \\
&\leqslant \alpha a + \beta b && \textbf{arithmetic mean} \\
&\leqslant \sqrt[p]{\alpha a^p + \beta b^p}, && \textbf{root-mean-"square"} \\
&\leqslant \max(a, b).
\end{aligned}
$$

Proof (i) If $a \leqslant b$ (without loss of generality), then $a^q \leqslant b^q$, so

$$
\frac{\alpha}{a^q} + \frac{\beta}{b^q} \leqslant \frac{\alpha + \beta}{a^q} = \frac{1}{a^q}
$$

which is equivalent to the first inequality of the proposition.

(ii) The second inequality is equivalent to $a^{-\alpha q} b^{-\beta q} \leqslant \alpha a^{-q} + \beta b^{-q}$, which is (9.1) with a, b replaced by a^{-q}, b^{-q} respectively.

(iii) Similarly, the third inequality is essentially $a^{\alpha/p} b^{\beta/p} \leqslant \alpha a^{1/p} + \beta b^{1/p}$, which is (9.1) with a, b replaced by $a^{1/p}, b^{1/p}$ respectively.

(iv) If a, b in (9.2) are substituted by $a^{1/p}$ and $b^{1/p}$ one obtains $(\alpha a^{1/p} + \beta b^{1/p})^p \leqslant \alpha a + \beta b$.

(v) The fifth inequality is precisely (9.2), while the sixth one follows easily if we assume, say, $a \leqslant b$; for then, $a^p \leqslant b^p$, so $\alpha a^p + \beta b^p \leqslant (\alpha + \beta) b^p = b^p$. Substituting q/p for p in (9.2), when $p \leqslant q$, and a^p for a, b^p for b, yields

$$
(\alpha a^p + \beta b^p)^{1/p} \leqslant (\alpha a^q + \beta b^q)^{1/q} \qquad \text{for } 0 < p \leqslant q,
$$

which is implicitly implied in the scheme of inequalities above. $\qquad\square$

An induction proof generalizes all these inequalities to arbitrary sums or products,

$$
a_1^{\alpha_1} \cdots a_n^{\alpha_n} \leqslant \alpha_1 a_1 + \cdots + \alpha_n a_n \leqslant \sqrt[p]{\alpha_1 a_1^p + \cdots + \alpha_n a_n^p}, \tag{9.4}
$$

when $a_i, \alpha_i \geqslant 0, \alpha_1 + \cdots + \alpha_n = 1, p \geqslant 1$.

Hermann Minkowski (1864–1907) Minkowski studied under Lindemann (of π-transcendentality fame) at the University of Königsberg, together with Hilbert. At 19 years of age, 2 years before he graduated with a thesis on quadratic forms, he had already won the prestigious French Academy's Grand Prix. Starting 1889, he developed his "geometry of numbers" ideas on lattices, including his inequality. After teaching in Zurich (where Einstein was a student), he moved to Göttingen, became interested in physics and presented his version of special relativity as a unified space-time.

Finally, substituting q/p for p in (9.3), and a^q for a, b^q for b, gives

$$(a^q + b^q)^{1/q} \leqslant (a^p + b^p)^{1/p}, \qquad \text{for } 0 < p \leqslant q,$$

which generalizes by induction to

$$\sqrt[q]{a_1^q + \cdots + a_n^q} \leqslant \sqrt[p]{a_1^p + \cdots + a_n^p}, \qquad \text{for } 0 < p \leqslant q.$$

This last inequality remains valid for infinite sums, $\|x\|_{\ell^q} \leqslant \|x\|_{\ell^p}$ when $p \leqslant q$, implying $\ell^p \subseteq \ell^q$. Of course, $\|x\|_{\ell^\infty} \leqslant \|x\|_{\ell^p}$ is true since $|a_n| \leqslant \|x\|_{\ell^p}$ for all n. Thus a bounded sequence lies in a whole range of ℓ^p spaces, down to some infimum p.

Proposition 9.12 (Minkowski's Inequality)

$$\|x + y\|_{\ell^p} \leqslant \|x\|_{\ell^p} + \|y\|_{\ell^p}, \qquad \text{where } 1 \leqslant p \leqslant \infty.$$

Proof All norms in this proof are taken to be the ℓ^p-norm. Let $u = (a_n)_{n \in \mathbb{N}}$ and $v = (b_n)_{n \in \mathbb{N}}$ be two sequences in ℓ^p. Summing the arithmetic mean inequality $(\alpha|a| + \beta|b|)^p \leqslant \alpha|a|^p + \beta|b|^p$ ($\alpha + \beta = 1$, $\alpha, \beta \geqslant 0$) for a sequence of terms gives

$$\sum_{n \in \mathbb{N}} |\alpha a_n + \beta b_n|^p \leqslant \sum_{n \in \mathbb{N}} (\alpha|a_n| + \beta|b_n|)^p \leqslant \alpha \sum_{n \in \mathbb{N}} |a_n|^p + \beta \sum_{n \in \mathbb{N}} |b_n|^p,$$

that is, $\|\alpha u + \beta v\|^p \leqslant \alpha\|u\|^p + \beta\|v\|^p.$

Substituting $u = x/\|x\|$, $v = y/\|y\|$, $\alpha = \|x\|/(\|x\|+\|y\|)$, $\beta = \|y\|/(\|x\|+\|y\|)$, gives

$$\frac{\|x + y\|}{\|x\| + \|y\|} = \|\alpha u + \beta v\| \leqslant (\alpha + \beta)^{1/p} = 1.$$

The proof of the inequality for $p = \infty$ is Exercise 7.8(6). $\qquad\square$

Proposition 9.13 (Hölder's Inequality)

$$|x \cdot y| \leqslant \|x\|_{\ell^p} \|y\|_{\ell^{p'}}, \text{ where } \frac{1}{p} + \frac{1}{p'} = 1, \; p \geqslant 1.$$

Proof Substitute $a^{1/\alpha}$ and $b^{1/\beta}$ instead of a and b in $a^\alpha b^\beta \leqslant \alpha a + \beta b$, with $\alpha = 1/p$, $\beta = 1/p'$, to get

$$ab \leqslant \frac{a^p}{p} + \frac{b^{p'}}{p'}. \tag{9.5}$$

Summing this for a sequence of complex numbers leads to

$$|u \cdot v| \leqslant \sum_{n \in \mathbb{N}} |a_n b_n| \leqslant \sum_{n \in \mathbb{N}} \left(\frac{|a_n|^p}{p} + \frac{|b_n|^{p'}}{p'} \right) = \frac{1}{p} \|u\|_{\ell^p}^p + \frac{1}{p'} \|v\|_{\ell^{p'}}^{p'}.$$

In particular, for unit vectors $u = x/\|x\|_{\ell^p}$, $v = y/\|y\|_{\ell^{p'}}$, we obtain Hölder's inequality,

$$\frac{|x \cdot y|}{\|x\|_{\ell^p} \|y\|_{\ell^{p'}}} \leqslant \frac{1}{p} + \frac{1}{p'} = 1.$$

$\qquad\square$

Proposition 9.14

For $p \geqslant 1$, ℓ^p is a separable Banach space, with dual space $\ell^{p*} \equiv \ell^{p'}$, where $\frac{1}{p} + \frac{1}{p'} = 1$.

Proof Minkowski's inequality is the non-trivial part in showing that ℓ^p is indeed a normed space. It is separable with the Schauder basis e_n, since for any $x = (a_n)_{n\in\mathbb{N}} \in \ell^p$, the series $\sum_n |a_n|^p$ converges to $\|x\|_{\ell^p}^p$, so

$$\Big\|x - \sum_{n=0}^N a_n e_n\Big\|_{\ell^p}^p = \|(0,\dots,0,a_{N+1},\dots)\|_{\ell^p}^p = \sum_{n=N+1}^\infty |a_n|^p \to 0,$$

so $x = \sum_{n\in\mathbb{N}} a_n e_n$. The coefficients are unique since if $x = \sum_{n\in\mathbb{N}} b_n e_n = (b_0, b_1, \dots)$, then $b_n = a_n$.

Dual of ℓ^p: Any vector $y \in \ell^{p'}$ acts on ℓ^p via $y^\top : x \mapsto y \cdot x$, with the latter being finite by Hölder's inequality $|y \cdot x| \leqslant \|y\|_{\ell^{p'}} \|x\|_{\ell^p}$. By Exercise 2 below, there is an $x \in \ell^p$ which makes this an equality. Thus $\|y^\top\| = \|y\|_{\ell^{p'}}$.

Conversely, let ϕ be a functional on ℓ^p; then for $x = (a_n)_{n\in\mathbb{N}} = \sum_{n\in\mathbb{N}} a_n e_n$, $\phi x = \sum_{n=0}^\infty a_n b_n = y \cdot x$, where $b_n := \phi e_n$, $y := (b_n)_{n\in\mathbb{N}}$. Writing $b_n = |b_n| e^{i\theta_n}$ and noting $p(p'-1) = p'$,

$$\sum_{n=0}^N |b_n|^{p'} = \sum_{n=0}^N b_n e^{-i\theta_n} |b_n|^{p'-1} = |\phi(e^{-i\theta_n} |b_n|^{p'-1})_{n=0}^N| \leqslant \|\phi\| \left(\sum_{n=0}^N |b_n|^{p'}\right)^{1/p}.$$

Dividing the right-hand series gives $\left(\sum_{n=0}^N |b_n|^{p'}\right)^{1/p'} \leqslant \|\phi\|$; as N is arbitrary, $y \in \ell^{p'}$.

Completeness: In common with all dual spaces, $\ell^p \equiv \ell^{p'*}$ is complete (or from an argument similar to the one for ℓ^2).   □

We end this section with a couple of propositions about operators on sequence spaces, $T : \ell^p \to \ell^q$. They take the form of a matrix, albeit ones with an infinite number of rows and columns. Consider the output vector $y = (b_i)_{i\in\mathbb{N}} := Tx \in \ell^q$, where $x = (a_j)_{j\in\mathbb{N}} \in \ell^p$. The coefficients b_i can be obtained as follows, by linearity and continuity of T and $e_i^\top$,

$$b_i = e_i^\top y = e_i \cdot T\Big(\sum_{j\in\mathbb{N}} a_j e_j\Big) = \sum_{j\in\mathbb{N}} a_j (e_i^\top T e_j) = \sum_{j\in\mathbb{N}} t_{i,j} a_j,$$

where $t_{i,j} = e_i \cdot T e_j$. This can be thought of as a matrix equation with the matrix $[t_{i,j}]$ having a countable number of rows and columns:

$$\begin{pmatrix} b_1 \\ b_2 \\ \vdots \end{pmatrix} = \begin{pmatrix} t_{11} & t_{12} & \cdots \\ t_{21} & t_{22} & \cdots \\ \vdots & \vdots & \ddots \end{pmatrix} \begin{pmatrix} a_1 \\ a_2 \\ \vdots \end{pmatrix}.$$

Each column is $T e_j \in \ell^q$, while each row is a dual vector in $\ell^{p*} \equiv \ell^{p'}$, so the coefficients of such a matrix must eventually become small as we move down or to the right.

For practical purposes, to solve $T x = y$, one can truncate the matrix and vectors to yield a finite $N \times N$ matrix equation that can then be solved. This can be justified because the remainder terms of y and x, of the type $\sum_{n=N+1}^{\infty} \gamma_n e_n$, etc., converge to 0 as $N \to \infty$. Note carefully that the above does not hold for ℓ^{∞} since $x \neq \sum_n a_n e_n$ in general.

The next proposition makes the link between infinite and finite matrices, while the following one generalizes Proposition 8.10.

Proposition 9.15

> **Let $T : \ell^p \to \ell^q$ be an infinite matrix with upper-left $n \times n$ matrices A_n, then**
>
> $$\|T\| = \lim_{n \to \infty} \|A_n\|_{p,q}.$$

The following proof is valid even if the domain and/or codomain are c_0.

Proof Any submatrix S of T has a diminished norm since it has both a smaller domain and range. The unit vectors in the domain of S are included in those of T; furthermore, for such vectors, $\|S x\|_q \leqslant \|T x\|_q$ since $T x$ has the same components as $S x$ and more.

Let T be divided into sub-matrices A_n, B_n, C_n, with A_n having n rows and columns, and each having norm at most $\|T\|$, as follows:

$$\left(\begin{array}{c|c} A_n & B_n \\ \hline & C_n \end{array} \right)$$

Since A_n is a sub-matrix of A_{n+1}, $\|A_n\|$ is increasing and bounded above by $\|T\|$, so it converges to the least upper bound $s := \sup_n \|A_n\| \leqslant \|T\|$. To show $s = \|T\|$ we need to prove that $\|A_n\|$ approach $\|T\|$ arbitrarily closely. Given any unit vector $x \in \ell^p$, split both vectors $x = x_n + y_n$ and $T x = a_n + b_n \in \ell^q$ at the nth component. For n large enough, both y_n and b_n have small p- and q-norms, respectively,

$$\forall \epsilon > 0, \ \exists N, \quad n \geqslant N \Rightarrow \|y_n\|_p < \epsilon \text{ AND } \|b_n\|_q < \epsilon.$$

Therefore, for all unit vectors x,

$$\|Tx\|_q = \|a_n + b_n\|_q < \|a_n\|_q + \epsilon$$
$$= \|A_n x_n + B_n y_n\|_q + \epsilon$$
$$\leqslant \|A_n\|\|x_n\|_p + \|B_n\|\epsilon + \epsilon$$
$$\leqslant \|A_n\| + \|T\|\epsilon + \epsilon$$
$$\therefore \|T\| \leqslant s + \|T\|\epsilon + \epsilon.$$

Thus $s \leqslant \|T\| \leqslant \frac{s+\epsilon}{1-\epsilon}$ and, by squeezing, $\|T\| = s$.   □

Proposition 9.16

Let T be an infinite matrix with coefficients T_{ij}, then with $\frac{1}{p} + \frac{1}{p'} = 1$,
$c = \sup_j \sum_i |T_{ij}|, r = \sup_i \sum_j |T_{ij}|$,

$$\|T\|_{p,p'} \leqslant \left(\sum_{i,j} |T_{ij}|^{p'}\right)^{1/p'}, \qquad\qquad \|T\|_{p,p} \leqslant c^{1/p} r^{1/p'}.$$

Proof (i) For any vector $x = (a_n)_{n \in \mathbb{N}} \in \ell^p$ and $\alpha = \sum_{i,j} |T_{ij}|^{p'}$, by Hölder's inequality,

$$\|Tx\|_{p'}^{p'} = \sum_i \left|\sum_j T_{ij} a_j\right|^{p'} \leqslant \sum_i \left(\sum_j |T_{ij}|^{p'}\right)\left(\sum_j |a_j|^p\right)^{p'/p} = \alpha \|x\|_p^{p'}$$

so $\|T\| \leqslant \alpha^{1/p'}$, as required. (ii) In addition, for any vector $y = (b_i)_{i \in \mathbb{N}} \in \ell^{p'}$,

$$|y \cdot Tx| = \left|\sum_{i,j} b_i T_{ij} a_j\right| \leqslant \sum_{i,j} \left(|T_{ij}|^{1/p} |a_j|\right)\left(|T_{ij}|^{1/p'} |b_i|\right)$$

$$\leqslant \left(\sum_{i,j} |T_{ij}||a_j|^p\right)^{1/p}\left(\sum_{i,j} |T_{ij}||b_i|^{p'}\right)^{1/p'}$$

$$\leqslant c^{1/p} r^{1/p'} \|x\|_p \|y\|_{p'}$$

Choose a vector y such that $|y \cdot Tx| = \|y\|_{p'} \|Tx\|_p$ to get the required result.   □

Exercises 9.17

1. Given $1 \leqslant q < p$, find an example of a sequence which is in ℓ^p but not in ℓ^q.

2. For each $x \in \ell^p$, find a sequence $y \in \ell^{p'}$ which makes Hölder's inequality an equality.
3. If $p \leqslant q$ then ℓ^p is dense in ℓ^q. (Hint: Consider c_{00}.)
4. For an infinite matrix, $\|T\|_{p,q}$ is larger than the q-norm of any column and the p'-norm of any row.
5. Show that of the following matrices from Chap. 1,

$$
\begin{pmatrix} 1 & 1 & \cdots\cdots \\ 0 & 1 & \\ & & \ddots \\ & & & \ddots \end{pmatrix}, \quad
\begin{pmatrix} 1 & -1 & 0 & \cdots\cdots \\ 0 & 1 & -1 & \\ & & & \ddots \\ & & & & \ddots \end{pmatrix}, \quad
\begin{pmatrix} 0 & 1 & 0 & \cdots\cdots \\ & 0 & 2 & 0 \\ & & & \ddots \\ & & & & \ddots \end{pmatrix}
$$

only the middle one has a finite $(2, 2)$-norm.
6. The upper left $n \times n$ matrices A_n of an infinite matrix T need not converge to T. Show, for example, that $I_n \nrightarrow I$ as operators $\ell^1 \to \ell^1$.
7. *Generalized Hölder's inequalities*

   (a) $\|xy\|_{\ell^r} \leqslant \|x\|_{\ell^p} \|y\|_{\ell^q}$, where $\frac{1}{p} + \frac{1}{q} = \frac{1}{r}$,

   (b) $\left| \sum_n a_n b_n c_n \right| \leqslant \|(a_n)\|_{\ell^p} \|(b_n)\|_{\ell^q} \|(c_n)\|_{\ell^r}$, where $\frac{1}{p} + \frac{1}{q} + \frac{1}{r} = 1$.

   (Hint: Apply Hölder's inequality to the product $|a_n|^r |b_n|^r$.)
8. Any two p-norms on $\mathbb{R}^n$ are equivalent. Show, using Hölder's inequality, that for $p \leqslant q$ and any $x \in \mathbb{R}^n$,

$$
\|x\|_q \leqslant \|x\|_p \leqslant n^{\frac{1}{p} - \frac{1}{q}} \|x\|_q.
$$

9. (a) *Littlewood's inequality*: $\|x\|_{\ell^r} \leqslant \|x\|_{\ell^p}^{\alpha} \|x\|_{\ell^q}^{1-\alpha}$, where $\frac{1}{r} = \frac{\alpha}{p} + \frac{1-\alpha}{q}$.
   (Hint: Apply the generalized Hölder's inequality above to $|a_n|^{\alpha} |a_n|^{1-\alpha}$, using p/α and $q/(1-\alpha)$ instead of p and q.)
   (b) $\|x\|_{\ell^r} \to \|x\|_{\ell^{\infty}}$ as $r \to \infty$.
   (Hint: Use Littlewood's inequality with $q = \infty$, $\alpha = p/r$.)
10. ⋆ *Young's inequality*:

$$
\|x * y\|_{\ell^r} \leqslant \|x\|_{\ell^p} \|y\|_{\ell^q},
$$

where $\frac{1}{p} + \frac{1}{q} = 1 + \frac{1}{r}$, $p, q \geqslant 1$ (Exercises 9.7(2,3)).
   Justify the steps of the following proof. First note that $\frac{1}{p'} + \frac{1}{q'} + \frac{1}{r} = 1$ (where $\frac{1}{p'} = 1 - \frac{1}{p}$, etc.); then using the second generalized Hölder's inequality

above on the positive numbers a_n, b_n, c_n, and an exquisite juggling of indices, (where $k := n - m$)

$$\sum_{n=0}^{N}\sum_{m=0}^{n} a_{n-m}b_m c_n = \sum_{n=0}^{N}\sum_{m=0}^{n}(a_k)^{p/r}(b_m)^{q/r}(a_k)^{p/q'}(c_n)^{r'/q'}(b_m)^{q/p'}(c_n)^{r'/p'}$$

$$\leqslant \left(\sum_{n,k} a_k^p b_m^q\right)^{1/r}\left(\sum_{n,k} a_k^p c_n^{r'}\right)^{1/q'}\left(\sum_{n,m} b_m^q c_n^{r'}\right)^{1/p'}$$

$$= \left(\sum_{n=0}^{N} a_n^p\right)^{1/p}\left(\sum_{n=0}^{N} b_n^q\right)^{1/q}\left(\sum_{n=0}^{N} c_n^{r'}\right)^{1/r'}.$$

Hence if $(a_n) \in \ell^p$, and $(b_n) \in \ell^q$, and $(c_n) \in \ell^{r'}$, then $x * y \in (\ell^{r'})^* \equiv \ell^r$.

11. $\diamond$ Prove the reverse Minkowski inequality for $0 < p \leqslant 1$, and *positive* real sequences $x = (a_n)_{n\in\mathbb{N}}$, $y = (b_n)_{n\in\mathbb{N}}$, $a_n, b_n \geqslant 0$,

$$\|x\|_p + \|y\|_p \leqslant \|x + y\|_p.$$

(Hint: the reverse inequality has its roots in x^p being concave.)

9.2   Function Spaces

Even though it is function spaces that are at the heart of "functional analysis", they are technically more complicated to construct. The most familiar classes of functions, such as continuous functions or step functions, are lacking in one way or another when confronted with limits or integrals. Constructing a complete space of integrable functions proved to be a much harder task historically than was anticipated by mathematicians.

For example, the simplest convergence of functions is what's termed *pointwise*

$$\forall t, \quad f_n(t) \to f(t) \text{ as } n \to \infty.$$

It is not hard to find sequences of integrable functions which converge pointwise but whose integrals do not; for example, $nt^n \to 0$ on $[0, 1[$ yet $\int_0^1 nt^n \, dt \to 1$. Moreover, it is left as an exercise to show that pointwise convergence cannot be induced by any norm, so in practice little can be deduced from it.

Another type of convergence is *uniform*. The space of real- or complex-valued bounded continuous functions $C_b(A)$, where $A \subseteq \mathbb{R}$, is easily seen to be closed under addition and scalar multiplication, and was shown in Theorem 6.23 to be a complete metric space, with the metric induced from the ∞-norm; so $C_b(A)$ is a Banach space in which convergence means uniform convergence. Indeed, $C_b(\mathbb{N})$ is just the space ℓ^∞, since the functions $\mathbb{N} \to \mathbb{C}$ are sequences. However, once again, the usefulness of $C_b(A)$ is limited when it comes to integration. For example, the functions t^{-p} are in $C_b[1, \infty[$ and converge to $\frac{1}{t}$ as $p \to 1$ in this space (prove!), but $\int_1^\infty t^{-p} \, dt = \frac{1}{p-1} \to \infty$.

Is there a way to generalize the Banach space ℓ^1 to a space of functions, where summation $\sum_n a_n$ becomes integration $\int f(t) \, dt$? This is indeed possible and much from the section on sequences can be repeated for functions, at least in spirit. For example, the proof that ℓ^∞ is complete generalizes to the space $L^\infty(\mathbb{R})$, practically untouched. However, we do not prove all these generalizations here, as laying the groundwork for integration and measures would take us too far afield. Instead a review is provided, referring the reader to [1] for more details. On the other hand, we allow for vector-valued functions, $f : A \to X$, because it does not incur any extra difficulty. Note that when $f(x)$ is a vector, we write $|f|$ for the function $x \mapsto \|f(x)\|$, in order to avoid confusion with the scalar $\|f\|$.

Lebesgue Measure on $\mathbb{R}^n$

Review 9.18

1. A **measure** μ on $\mathbb{R}^N$ is an assignment of positive numbers or ∞ to certain subsets $E \subseteq \mathbb{R}^N$ with the properties that it be

   (i) additive, $\mu(E \cup F) = \mu(E) + \mu(F)$ for E, F, disjoint;
   (ii) continuous, $E_n \to E \Rightarrow \mu(E_n) \to \mu(E)$.
       It is enough for now to take $E_n \to E$ to mean that E_n is a decreasing sequence of sets of finite measure, with $\bigcap_n E_n = E$.
       One final property that we expect μ to satisfy, at least in $\mathbb{R}^N$, is that it be
   (iii) translation invariant, $\mu(E + x) = \mu(E)$.

Henri Lebesgue (1875–1941) Lebesgue graduated at the École Normale Supérieure of Paris at 27 years. His thesis built upon work of Baire, Borel and Jordan, to generalize lengths and areas, and so an integration powerful enough to tackle functions too discontinuous for Riemann's integration—the first complete space of integrable functions. After a century of attempts by other mathematicians, he finally proved that uniformly bounded series of integrable functions, such as the Fourier series, could be integrated term by term. Although his achievement was widely seen as abstract, in his words, "Reduced to general theories, mathematics would be a beautiful form without content. It would quickly die."

Examples of measures are the standard length, area, and volume of Euclidean geometry.

2. Taking $\mathbb{R}$ as our main example, and defining $\mu[0, 1[:= 1$, these properties completely determine the length of any interval, namely $\mu[a, b] = b - a = \mu[a, b[$. (Hint: divide $[0, 1[$ into equal intervals to show $\mu[0, m/n] = m/n$.)

3. As a first step in constructing μ on $\mathbb{R}$, therefore, the length of any interval is defined to be the difference of its endpoints, e.g., $m[a, b] := b - a$. This function can be extended in two ways to

   (a) the length of any countable union of disjoint intervals

$$m(\bigcup_n I_n) := \sum_n m(I_n),$$

   (b) the length of the set obtained by removing a countable union of disjoint subintervals from a bounded interval

$$m(I \smallsetminus \bigcup_n I_n) := m(I) - \sum_n m(I_n).$$

4. For general sets, define

$$m^*(A) := \inf\{ m(U) : A \subseteq U = \bigcup_n I_n \},$$

$$m_*(A) := \sup\{ m(K) : A \supseteq K = I \smallsetminus \bigcup_n I_n \}.$$

(Note that since we are taking the infimum and supremum, respectively, we might as well take I to be a closed and bounded interval and I_n to be open intervals, in which case U is an open set, and K a compact set.)

It is a fact that there exist sets for which these two values do not agree (see [1]). A "well-behaved" set, called **measurable**, satisfies $m^*(E) = m_*(E)$, which is then called its *Lebesgue measure* $\mu(E)$.

5. $m^*(\bigcup_n A_n) \leqslant \sum_n m^*(A_n)$ and $A \subseteq B \Rightarrow m^*(A) \leqslant m^*(B)$ (since open covers for each A_n provide an open cover for their union). Of course, these statements continue to hold for Lebesgue measure applied to measurable sets.

6. A useful equivalent criterion of measurability of E is:

   For any subset A, $m^*(E \cap A) + m^*(E^c \cap A) = m^*(A)$.

7. Using this criterion, it follows that, for E, F, and E_n measurable sets,

   (a) E^c, $E \cup F$, $E \cap F$, $E \smallsetminus F$, and $E \triangle F$ are measurable; when they are disjoint, $\mu(E \cup F) = \mu(E) + \mu(F)$.

(b) $\bigcup_{n=1}^{\infty} E_n$ and $\bigcap_{n=1}^{\infty} E_n$ are measurable, and when E_n are disjoint,

$$\mu\left(\bigcup_{n=1}^{\infty} E_n\right) = \sum_{n=1}^{\infty} \mu(E_n).$$

The sets that can be obtained by starting with the intervals and applying these constructions are called *Borel sets*; they include the open and closed sets.

8. Sets with (m^*-)measure 0 are obviously measurable and are called **null** sets. For example, any countable set is null; but most null sets are uncountable, e.g., the Cantor set. The countable union of null sets is null.

   Adding (or removing) a null set N from a measurable set E does not affect its measure,

$$\mu(E \cup N) = \mu(E) + \mu(N) = \mu(E).$$

   Because measures don't distinguish sets up to a null set, we say that two sets are equal *almost everywhere*, $E = F$ a.e., when they differ by a null set. More generally, we qualify a statement "$\mathcal{P}(t)$ a.e.t" when $\mathcal{P}(t)$ is true for all t except on a null set; for example, we say $f = g$ a.e. when $f(t) = g(t)$ for all t in their domain apart from a null set.

9. The distance between measurable sets is defined as $d(E, F) := \mu(E \triangle F)$. It is a metric, with the proviso that $d(E, F) = 0 \Leftrightarrow E = F$ a.e. The measure μ is continuous with respect to it, $E_n \to E \Rightarrow \mu(E_n) \to \mu(E)$.

10. A similar procedure gives the Lebesgue measure on $\mathbb{R}^n$, with the modification that cuboids are used instead of intervals to generate the measurable sets. Most subsets of $\mathbb{R}^n$ that the reader is likely to have encountered are measurable, including balls in $\mathbb{R}^3$.

Measurable Functions

Review 9.19

1. The *characteristic* function of a set is defined by

$$1_E(t) := \begin{cases} 1, & t \in E \\ 0, & t \notin E \end{cases}.$$

Linear combinations of characteristic functions $\sum_{n=1}^{k} 1_{E_n} x_n$, where E_n are bounded measurable subsets of $\mathbb{R}$ and $x_n \in \mathbb{C}$, are called **simple** functions. More generally, $\mathbb{R}$ can be replaced by a fixed measurable set A, and x_n can belong to a Banach space X. The simple functions form a vector space $\mathcal{S}$.

2. A function $f : A \to X$ is said to be **measurable** when it is almost everywhere
   the pointwise limit of simple functions, $s_n \to f$ a.e. For real-valued functions,
   this is equivalent to $f^{-1}[a, \infty[$ being measurable for all $a \in \mathbb{R}$.
   Note that simple functions supported in E (i.e., are zero outside E) can converge
   only to measurable functions supported in E (since $s_n 1_E \to f 1_E$ a.e.).
3. Measurable functions form a vector space: λf and $f + g$ are measurable when
   f, g are. It follows from $\big| |s_n| - |f| \big| \leqslant |s_n - f|$ that $|f| : A \to \mathbb{R}$ is measurable.
   For real-valued measurable functions, fg, $\max(f, g)$, and $\sup_n(f_n)$, are also
   measurable. Real-valued continuous functions are measurable.
4. ▶ In fact the space of measurable functions is in a sense complete: if f_n are
   measurable and $f_n \to f$ a.e., then f is measurable.
5. $L^\infty(A)$ is defined as the space of (equivalence classes of) bounded measurable
   functions $f : A \to \mathbb{C}$, over a measurable set A, with the supremum norm
   $\|f\|_{L^\infty} := \sup_{t \text{ a.e.}} |f(t)|$, that is, the smallest real number c such that $|f(t)| \leqslant$
   c a.e.t.
6. $L^\infty(\mathbb{R})$ contains the closed subspace of bounded continuous functions $C_b(\mathbb{R})$,
   which in turn contains $C_0(\mathbb{R}) := \{ f \in C(\mathbb{R}) : \lim_{t \to \pm\infty} f(t) = 0 \}$. The space
   $C[a, b]$ is embedded in $C_0(\mathbb{R})$. $C_b(\mathbb{R})$ is not separable for the same reason that
   ℓ^∞ is not (Theorem 9.1); replace a 0–1 sequence by a "0–1" tent function.
7. $L^\infty[a, b]$ is not separable: the uncountable number of characteristic functions
   $1_{[s,t]}$, $a < s < t < b$, are at unit distance from each other.

Proposition 9.20

$L^\infty(A)$ **is a Banach space.**

Proof If $|f(t)| \leqslant \|f\|_{L^\infty}$ except on the null set E_1, and $|g(t)| \leqslant \|g\|_{L^\infty}$ except on
the null set E_2, then for all $t \in A \setminus (E_1 \cup E_2)$,

$$|f(t) + g(t)| \leqslant |f(t)| + |g(t)|, \qquad |\lambda f(t)| = |\lambda| |f(t)|,$$
$$\text{so } \|f + g\|_{L^\infty} \leqslant \|f\|_{L^\infty} + \|g\|_{L^\infty}, \qquad \|\lambda f\|_{L^\infty} = |\lambda| \|f\|_{L^\infty}.$$

Clearly $\|f\|_{L^\infty} = 0$ only when $|f(t)| = 0$ a.e. It follows that $L^\infty(A)$ is a normed
space, as long as we identify ae-equal functions into equivalence classes.

Completeness: Let $f_n \in L^\infty(A)$ be a Cauchy sequence, where $|f_n(t)| \leqslant \|f_n\|_{L^\infty}$
for all $t \in A$ except in some null set E_n. Copying the proof of the completeness of
ℓ^∞ (Theorem 9.1),

$$|f_n(t) - f_m(t)| \leqslant \|f_n - f_m\|_{L^\infty} \to 0$$

for each $t \in A$, except possibly on the null set $\bigcup_n E_n$, so $f_n(t)$ is Cauchy and
converges $f_n(t) \to f(t)$ a.e.t. The function f is evidently measurable, and $f_n \to f$

uniformly away from this null set, since for any $\epsilon > 0$ and n large enough (but independent of t),

$$|f_n(t) - f(t)| \leqslant |f_n(t) - f_m(t)| + |f_m(t) - f(t)|$$

$$\leqslant \|f_n - f_m\|_{L^\infty} + |f_m(t) - f(t)| \quad \text{a.e.} t$$

$$< 2\epsilon$$

where $m \geqslant n$ is chosen, depending on t, to make $|f_m(t) - f(t)| < \epsilon$. This means that $f_n \to f$ in L^∞, and implies $\|f\|_{L^\infty} \leqslant \|f - f_n\|_{L^\infty} + \|f_n\|_{L^\infty} < \infty$, so $f \in L^\infty(A)$.

$\square$

Integrable Functions

Review 9.21

1. Given a set E of finite measure and its characteristic function, let $\int 1_E := \mu(E)$. For a simple function, define its **integral**

$$\int \sum_{n=1}^{N} 1_{E_n} x_n := \sum_{n=1}^{N} \mu(E_n) x_n.$$

It is well-defined, since a simple function has a unique representation in terms of disjoint E_n. It is straightforward to verify that $\int(s + r) = \int s + \int r$ and $\int \lambda s = \lambda \int s$ for $s, r \in \mathcal{S}$.

2. The function $\|s\| := \int |s| = \sum_n \mu(E_n)\|x_n\|$ is a norm on $\mathcal{S}$. Here, $|s|$ is the real-valued simple function $|s| = \sum_n 1_{E_n}\|x_n\| \geqslant 0$. In particular, for real-valued simple functions, $r \leqslant s \Rightarrow \int r \leqslant \int s$.
   Proof: (i) $\|\lambda s\| = \sum_n \mu(E_n)\|\lambda x_n\| = |\lambda|\|s\|$,
   (ii) $\|s + r\| = \sum_n \mu(E_n)\|x_n + y_n\| \leqslant \sum_n \mu(E_n)(\|x_n\| + \|y_n\|) = \|s\| + \|r\|$,
   (iii) $\int |s| = 0$ when $\sum_n \mu(E_n)\|x_n\| = 0$. This implies $\mu(E_n)\|x_n\| = 0$ for all n, i.e., $x_n = 0$ OR $\mu(E_n) = 0$, so $s = 0$ a.e.

3. The integral is a continuous functional on $\mathcal{S}$, $\|\int s\| \leqslant \int |s|$, since,

$$\left\| \int s \right\| = \left\| \sum_n \mu(E_n)x_n \right\| \leqslant \sum_n \mu(E_n)\|x_n\| = \int |s|.$$

4. The space of real (or complex) simple functions with this norm is separable (the simple functions with $x_n \in \mathbb{Q}$ and E_n equal to intervals with rational endpoints, are countable and dense), but not complete.

5. A Cauchy sequence of simple functions converges a.e. to a measurable function.

Proof: Let s_n be a Cauchy sequence in $\mathcal{S}$. Pick a subsequence such that $\|s_{k_{j+1}} - s_{k_j}\| \leqslant \frac{1}{2^j}$. For a fixed α, the subsets

$$E_{n,\alpha} := \{ t \in \mathbb{R} : \textstyle\sum_{j=1}^{n} |s_{k_{j+1}}(t) - s_{k_j}(t)| \geqslant \alpha \}$$

are increasing with n up to the set $E_\alpha = \bigcup_{n \in \mathbb{N}} E_{n,\alpha}$. The crucial observation is

$$\alpha\, \mu(E_{n,\alpha}) \leqslant \int_{E_{n,\alpha}} \sum_{j=1}^{n} |s_{k_{j+1}} - s_{k_j}| \leqslant \sum_{j=1}^{n} \|s_{k_{j+1}} - s_{k_j}\| \leqslant \sum_{j=1}^{\infty} \frac{1}{2^j} = 1.$$

As $\mu(E_{n,\alpha}) \leqslant \frac{1}{\alpha}$, it follows that $\mu(E_\alpha) \leqslant \frac{1}{\alpha}$. The subset E_α decreases as α increases, so $E := \bigcap_{\alpha > 0} E_\alpha$ is a null set. But this is precisely the set where $\sum_j |s_{k_{j+1}} - s_{k_j}|$ diverges. Thus, $\sum_j |s_{k_{j+1}} - s_{k_j}|$ converges a.e., as must do $\sum_j (s_{k_{j+1}} - s_{k_j})$ and $s_n = s_1 + \sum_{k=1}^{n-1}(s_{k+1} - s_k)$.

6. A function $f : \mathbb{R} \to X$ is said to be **integrable** when it is the ae-limit of a Cauchy sequence of simple functions $s_n \to f$ a.e. Its *integral* is given by the extension of the integral on $\mathcal{S}$,

$$\int f := \lim_{n \to \infty} \int s_n.$$

Note that $\int s_n$ is a Cauchy sequence in X ($\| \int s_n - \int s_m \|_X \leqslant \int |s_n - s_m| \to 0$). The space of (equivalence classes of) integrable functions $\mathbb{R} \to X$ is denoted by $L^1(\mathbb{R}, X)$; it is the completion of $\mathcal{S}$ (Theorem 4.6). By Proposition 7.18, the space $L^1(\mathbb{R}, X)$ is a normed vector space with

$$\|f\|_{L^1} := \lim_{n \to \infty} \|s_n\| = \lim_{n \to \infty} \int |s_n| = \int |f|,$$

so $f \in L^1(\mathbb{R}, X) \Leftrightarrow |f| \in L^1(\mathbb{R})$. It also follows that for real-valued integrable functions $f \leqslant g \Rightarrow \int f \leqslant \int g$.

7. ▶ The integral is a continuous functional on $L^1(\mathbb{R}, X)$ (Example 8.9(5)),

$$\int f + g = \int f + \int g, \quad \int \lambda f = \lambda \int f, \quad \left\| \int f \right\| \leqslant \int |f|.$$

Thus if $f_n \to f$ in $L^1(\mathbb{R}, X)$ then $\int f_n \to \int f$ in X.

8. (a) $f \in L^1(\mathbb{R}) \Rightarrow \int f(t)x \, dt = (\int f)x$,
   (b) $T \in B(X, Y) \Rightarrow \int Tf = T \int f$.
   Proof: (a) is a special case of (b) with $T : \mathbb{F} \to X$, $T(\lambda) := \lambda x$.

As an operator, $T : X \to Y$ acts linearly on simple functions $s = \sum_n 1_{E_n} x_n \in \mathcal{S}$,

$$Ts = \sum_{n=1}^{N} 1_{E_n} T x_n \Rightarrow \int Ts = \sum_{n=1}^{N} \mu(E_n) T x_n = T \int s.$$

If $s_n \to f$ in $L^1(\mathbb{R}, X)$ then $T s_n \to T f$ in $L^1(\mathbb{R}, Y)$, so $\int T f = T \int f$.

9. For a measurable set $A \subseteq \mathbb{R}$, define $L^1(A) := \{ f 1_A : f \in L^1(\mathbb{R}) \}$, and let $\int_A f := \int f 1_A$.

Note that $\int_A f = 0$ for any null set A. Hence if $f = g$ a.e., with $g \in L^1(\mathbb{R})$, and $E = F$ a.e., then $f \in L^1(\mathbb{R})$ as well and $\int_E f = \int_F g$.

10. For E, F disjoint measurable sets,

$$\int_{E \cup F} f = \int_E f + \int_F f$$

It follows that $E \subseteq F \Rightarrow \int_E |f| \leqslant \int_F |f|$.

11. A *signed* measure is defined to be a mapping from measurable subsets of $\mathbb{R}$ to real values (possibly negative), which satisfies the axioms of a measure. Similarly a *complex* measure is one which takes values in $\mathbb{C}$.

12. *Radon-Nikodym theorem*: If ν is a complex measure on $\mathbb{R}$ such that $\nu(A) = 0$ whenever A is a null set, then there is a complex-valued measurable function f such that

$$\nu(A) = \int_A f.$$

Refer to [1] for a proof.

Theorem 9.22

For $A \subseteq \mathbb{R}$, $L^1(A)$ is a separable Banach space.

Proof

Completeness: Let f_n be a Cauchy sequence in $L^1(A)$, i.e., $\| f_n - f_m \| \to 0$. Choose $s_n \in \mathcal{S}$ close to f_n, say $\| s_n - f_n \| < 1/n$. Then $(s_n)_{n \in \mathbb{N}}$ is a Cauchy sequence of simple functions, asymptotic to f_n. By Notes 5 and 7 above, s_n converges to an integrable function f in $L^1(A)$. Hence, so does the asymptotic sequence f_n.

Separability: By construction, the separable set $\mathcal{S}$ of simple functions is dense in $L^1(A)$: Any $f \in L^1(A)$ has a sequence of simple functions converging to it $(s_n)_{n \in \mathbb{N}} \to f$ a.e., so $\| f - s_n \|_{L^1} \to 0$ as $n \to \infty$.

$\square$

We can start reaping the immediate benefits of these Lebesgue spaces. They have excellent limit properties:

Proposition 9.23

If $f_n \to f$ in $L^\infty(\mathbb{R})$, that is, uniformly, and

   (i) f_n are continuous, then f is continuous,
  (ii) f_n are integrable, then f is integrable on $[a, b]$, and

$$\int_a^b f_n \to \int_a^b f,$$

 (iii) f_n' are continuous and converge uniformly, then $f_n' \to f'$.

Proof (i) The first assertion is a restatement of the fact that $C_b(\mathbb{R})$ is closed in $L^\infty(\mathbb{R})$ (Theorem 6.23).

(ii) The second follows from the completeness of $L^1[a, b]$ and the continuity of the integral

$$\left| \int_a^b f_n - \int_a^b f \right| \leqslant \int_a^b |f_n - f| \leqslant (b-a)\|f_n - f\|_{L^\infty[a,b]} \to 0.$$

(iii) If $f_n' \to g$ uniformly, then $f_n' \to g$ in $L^1[a, t]$ by (ii), and $\int_a^t f_n' \to \int_a^t g$. But, assuming the fundamental theorem of calculus (Theorem 12.8), $\int_a^t f_n' = f_n(t) - f_n(a)$, which converge to $f(t) - f(a)$ uniformly and in $L^1[a, t]$. So $\int_a^t g = f(t) - f(a)$, showing f is differentiable, with $f' = g$.   □

Much the same analysis can be made starting with the norm $\|s\|_p := \left(\int |s|^p \right)^{1/p}$, $1 \leqslant p < \infty$, on $\mathcal{S}(A)$. The completion of $\mathcal{S}$ in this norm is denoted by $L^p(A)$, which is thus complete and separable ($\mathcal{S}(A)$ dense in it).

The product $x \cdot y$ of sequences becomes $f \cdot g := \int fg$ for functions. Hölder's inequality is valid:

Proposition 9.24 (Generalized Hölder's Inequality)

For $A \subseteq \mathbb{R}$, if $f \in L^p(A)$, $g \in L^q(A)$, then $fg \in L^r(A)$ where $\frac{1}{r} = \frac{1}{p} + \frac{1}{q}$, and

$$\|fg\|_{L^r} \leqslant \|f\|_{L^p}\|g\|_{L^q}.$$

Proof The AM-GM inequality, for any complex numbers a, b, yields

$$|ab|^r = |a|^{p\frac{r}{p}}|b|^{q\frac{r}{q}} \leqslant \frac{r}{p}|a|^p + \frac{r}{q}|a|^q.$$

If a, b are now the values of functions, integrated over the set A,

$$\int_A |a(t)b(t)|^r \, dt \leqslant \frac{r}{p}\int_A |a(t)|^p \, dt + \frac{r}{q}\int_A |b(t)|^q \, dt.$$

Substituting $a(t) = f(t)/\|f\|_p$ and $b(t) = g(t)/\|g\|_q$,

$$\frac{\int |f(t)g(t)|^r \, dt}{\|f\|_p^r \|g\|_q^r} \leqslant \frac{r}{p} + \frac{r}{q} = 1.$$

$\square$

Theorem 9.25

> **For $1 < p < \infty$, $L^p[a, b]$ is a Banach space whose dual space is**
>
> $$L^p[a, b]^* \equiv L^{p'}[a, b],$$
>
> **where $\frac{1}{p} + \frac{1}{p'} = 1$.**

Proof That $L^p(A)$ is a complete normed vector space follows from its construction as the completion of the vector space of simple functions with the p-norm. The triangle inequality for this norm can be proved in an identical fashion to the proof of Minkowski's inequality for ℓ^p (Proposition 9.12).

Given any function $g \in L^{p'}(A)$ and $f \in L^p(A)$, let $\phi(f) := \int_A gf$ (clearly linear in f). Then by Hölder's inequality,

$$|\phi f| \leqslant \int_A |gf| \leqslant \|g\|_{L^{p'}}\|f\|_{L^p}$$

Equality can hold if we choose $f = |g|^{p'/p}e^{-i\theta}$ where $g = |g|e^{i\theta}$; note that $\|f\|_{L^p} = \|g\|_{L^{p'}}^{p'/p}$. Then

$$\phi f = \int |g|^{\frac{p'}{p}+1} = \int |g|^{p'} = \|g\|_{L^{p'}}^{p'} = \|g\|^{\frac{p'}{p}+1} = \|f\|_{L^p}\|g\|_{L^{p'}}$$

All this shows that every function $g \in L^{p'}(A)$ gives rise to a functional on $L^p(A)$, with norm $\|\phi\| = \|g\|_{L^{p'}}$.

Let A be a bounded interval and $\phi \in L^p(A)^*$. The map $E \mapsto \phi(1_E)$ can be seen to be a complex measure on A, which takes the value 0 on null sets since $1_E = 0$ a.e. when E is null. Hence by the Radon-Nikodym theorem, there exists a measurable function g such that $\phi(1_E) = \int g 1_E$. This extends, by linearity, to any simple function, $\phi(s) = \int gs$. Let s_n be an increasing sequence of non-negative simple functions converging pointwise a.e. to $|g| = ge^{-i\theta}$. Then

$$\phi(s_n^{p'/p} e^{-i\theta}) = \int |g| s_n^{p'/p} \geqslant \int s_n^{1+p'/p} = \int s_n^{p'}$$

Hence

$$\|s_n\|_{p'}^{p'} \leqslant \|\phi\| \|s_n^{p'/p}\|_p = \|\phi\| \|s_n\|_{p'}^{p'/p},$$

that is, $\|s_n\|_{p'} \leqslant \|\phi\|$. In the limit as $s_n \to |g|$, $\|g\|_{p'} \leqslant \|\phi\|$. Finally, since ϕ is Lipschitz, the identity $\phi(f) = \int_A gf$ for $f \in S$ must continue to hold for the completion space $\overline{S} = L^p(A)$ (Theorem 4.14).

$\square$

Note that $L^2(\mathbb{R})$ is its own dual.

Examples 9.26

1. Convergence in $L^1(\mathbb{R})$ is quite different from uniform convergence. For example, the sequence of functions $\frac{1}{n} 1_{[0,n]}$ converge uniformly to 0, but not in $L^1(\mathbb{R})$, whereas the sequence $1_{[0,\frac{1}{n}]}$ converges to 0 in $L^1(\mathbb{R})$ but not uniformly.

2. $\|f\|_{L^r} \leqslant \|f\|_{L^p}^{\alpha} \|f\|_{L^q}^{1-\alpha}$, where $\frac{1}{r} = \frac{\alpha}{p} + \frac{1-\alpha}{q}$; thus f lies in $L^p(A)$ for p in an interval of values.
   Proof: $\||f|^{\alpha}|f|^{1-\alpha}\|_r \leqslant \||f|^{\alpha}\|_{p/\alpha} \||f|^{1-\alpha}\|_{q/(1-\alpha)} = \|f\|_p^{\alpha} \|f\|_q^{1-\alpha}$, using Hölder's inequality.

3. ▶ When the domain of the functions is compact, the spaces are included in each other *as sets*, in the reverse order of the sequence spaces,

$$C[a,b] \subseteq L^{\infty}[a,b] \subseteq L^2[a,b] \subseteq L^1[a,b].$$

The identity maps $L^{\infty}[a,b] \to L^2[a,b] \to L^1[a,b]$ are continuous,

$$\|f\|_{L^1[a,b]} \leqslant (b-a)^{\frac{1}{2}} \|f\|_{L^2[a,b]} \leqslant (b-a)\|f\|_{L^{\infty}[a,b]}.$$

Proof: By Hölder's inequality, with $p < q$, $\|f\|_p \leqslant \|1\|_s \|f\|_q$ where $\frac{1}{p} = \frac{1}{s} + \frac{1}{q}$.

4. The notation $\int_{-\infty}^{\infty} f$ is capable of at least three interpretations, as (i) $\int_{\mathbb{R}} f$ when $f \in L^1(\mathbb{R})$, (ii) $\lim_{R,S\to\infty} \int_{-S}^{R} f$, (iii) $\lim_{R\to\infty} \int_{-R}^{R} f$. It should be clear that the finiteness of these integrals follow (i) $\Rightarrow$ (ii) $\Rightarrow$ (iii), but the examples $\int_{-R}^{R} t \, dt = 0$ and $\int_0^R \frac{\sin t}{t} \, dt \to \pi/2$ as $R \to \infty$ show that the converses are false.

Integral Operators

We now consider a broad class of operators that act on spaces of functions. An **integral operator** (or *transform*) is a mapping on functions

$$Tf(s) := \int_A k(s, t) f(t) \, dt,$$

where k is called the *kernel* of T (not to be confused with ker T). To motivate this definition, suppose T is a linear operator that inputs a function $f : A \subseteq \mathbb{R} \to \mathbb{C}$ and outputs a function $g : B \subseteq \mathbb{R} \to \mathbb{C}$. If A and B are partitioned into small subintervals, the functions f and g are discretized into vectors (f_j) and (g_i), and the linear operator T becomes approximately some matrix $[T_{ij}]$. As the partitions

$$Tf = g$$

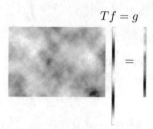

are refined, one might hope that T_{ij} would converge to some function $k(s, t)$ on $A \times B$, and the finite sums involved in the matrix multiplication $\sum_j T_{ij} f_j$ become integrals $\int_A k(s, t) f(t) \, dt$. (This is not necessarily the case, as the identity map attests.)

An integral functional on a function space is then of the form $\phi f := \int_A k(t) f(t) \, dt$.

Proposition 9.27

> **An integral operator $Tf(s) := \int_A k(s, t) f(t) \, dt$ is linear, and is continuous as $L^p(A) \to L^q(B)$ when:**
>
> $$\|T\|_{1,1} \leqslant \int_B \sup_{t \in A} |k(s, t)| \, ds, \qquad \|T\|_{1,\infty} \leqslant \sup_{t \in A, s \in B} |k(s, t)|,$$
>
> $$\|T\|_{\infty,1} \leqslant \int_B \int_A |k(s, t)| \, dt \, ds, \qquad \|T\|_{\infty,\infty} \leqslant \sup_{s \in B} \int_A |k(s, t)| \, dt,$$
>
> $$\|T\|_{2,2} \leqslant \left(\int_B \int_A |k(s, t)|^2 \, dt \, ds \right)^{1/2}.$$

Proof Linearity follows easily from

$$\int_A k(s,t)(\lambda f(t) + g(t))\,dt = \lambda \int_A k(s,t)f(t)\,dt + \int_A k(s,t)g(t)\,dt.$$

(i) $\|Tf\|_{L^1(B)} = \int_B \left| \int_A k(s,t)f(t)\,dt \right| ds \leq \int_B \sup_{t\in A} |k(s,t)|\,ds \int_A |f(t)|\,dt.$

(ii) $\|Tf\|_{L^\infty(B)} \leq \sup_{s\in B} \int_A |k(s,t)f(t)|\,dt \leq \sup_{s,t} |k(s,t)| \int_A |f(t)|\,dt.$

(iii) $\|Tf\|_{L^1(B)} \leq \int_B \int_A |k(s,t)f(t)|\,dt\,ds \leq \int_{A\times B} |k(s,t)|\,dt\,ds\,\|f\|_{L^\infty(A)}.$

(iv) $\|Tf\|_{L^\infty(B)} \leq \sup_{s\in B} \int_A |k(s,t)f(t)|\,dt \leq \sup_{s\in B} \int_A |k(s,t)|\,dt\,\|f\|_{L^\infty(A)}.$

(v) $\|Tf\|_{L^2(B)}^2 \leq \int_B \left| \int_A k(s,t)f(t)\,dt \right|^2 ds \leq \int_B \int_A |k(s,t)|^2\,dt\,ds \int_A |f(t)|^2\,dt,$

by Cauchy's inequality for functions.   □

Examples 9.28

1. The *Volterra operator* on $L^1[0,1]$ is $Vf(t) := \int_0^t f$. It is an integral operator
   with $k(s,t) := \begin{cases} 1, & s \leq t \\ 0, & t < s \end{cases}$.

2. For integral operators S, T, with kernels k_S, k_T respectively,

   (a) $S = T$ only when $k_S = k_T$ a.e., (since for all f, $(S-T)f(s) = \int (k_S(s,t) - k_T(s,t))f(t)\,dt = 0$);

   (b) $S+T$ has kernel $k_S + k_T$, and λT has kernel λk_T,

   (c) ST has kernel $k_{ST}(s,t) := \int k_S(s,u)k_T(u,t)\,du.$

   The kernel acts like a "matrix" with real-valued indices, $k_{s,t}$ in place of $A_{i,j}$. The properties listed here are analogous to those of the addition and multiplication of matrices.

3. Which integral operators on $L^1(\mathbb{R})$ are translation invariant, meaning $TT_a f = T_a Tf$, where $T_a f(t) = f(t-a)$? The requirement is, for all $f \in L^1(\mathbb{R})$,

   $$\int k(s,t)f(t-a)\,dt = \int k(s-a,t)f(t)\,dt.$$

   By changing the t-variable in the left-hand integral to $\tilde{t} = t - a$, we obtain $k(s,t) = k(s-a, t-a)$ a.e., as f is arbitrary. Equivalently, $k(s,t) = k(s-t, 0) =: k(s-t)$ a.e.(s,t) for some function $k \in L^1(\mathbb{R})$. That is,

   $$Tf = k * f := \int k(s-t)f(t)\,dt$$

called the **convolution** of k with f. Just like the same-named operation in ℓ^1, convolution is well-defined on $L^1(\mathbb{R})$ and is associative and commutative.

Approximation of Functions

The approximation of functions by polynomials is commonly used in many algorithms because they are much faster to compute than many analytical functions. This applies to the computing of many statistical functions and several engineering applications, including interpolation and curve fitting, which is the approximation of data points by functions such as polynomials and splines.

Proposition 9.29

> **The polynomials are dense in $L^1[a, b]$, $L^2[a, b]$, and $C[a, b]$.**

Proof By construction, the simple functions are dense in $L^1(\mathbb{R})$. Now, intuitively speaking, any real-valued step function s can be "nudged" into a continuous function g by replacing its discontinuities with steep slopes, and the distance $\|s - g\|_{L^1}$ can be made as small as needed by making the slopes steeper. More precisely and more generally, any bounded measurable set E in $\mathbb{R}$ lies between a compact set K and an open set U, such that $\mu(U \smallsetminus K) < \epsilon$ (Review 9.18(4)); also, there is a continuous function g_E taking values in $[0, 1]$ such that $g_E[K] = 1$, $g_E[U^c] = 0$ (Exercise 3.13(17)). So

$$\forall \epsilon > 0, \, \exists g_E \in C(\mathbb{R}), \quad \|g_E - 1_E\|_{L^1} = \int_{U \smallsetminus K} |g_E - 1_E| \leqslant \mu(U \smallsetminus K) < \epsilon.$$

Vito Volterra (1860–1940) Volterra studied hydrodynamics at Pisa under Betti (1883); this led him over the next 10 years to consider integral equations of the type $f(x) - \int_a^x k(x, y) f(y) \, dy = g(x)$, which he showed can be solved by iteration. He applied such "functionals" to the theory of optics and distortions, Hamilton-Jacobi dynamics, elasticity and electro-magnetism. He moved from one professorship in Turin to another in Rome, becoming a senator in 1905, and finding the time to write his Volterra equations about the numbers of predators and prey in mathematical biology, until in 1931 he preferred exile to the reign of Mussolini.

Consequently, taking any non-zero simple function $s = \sum_{n=1}^{N} 1_{E_n} x_n$ and replacing each 1_{E_n} with continuous functions g_n, where $\|g_n - 1_{E_n}\|_{L^1} < \epsilon / \sum_{n=1}^{N} \|x_n\|$, gives a continuous function $g := \sum_{n=1}^{N} g_n x_n$, which approximates s in L^1,

$$\|s - g\|_{L^1} \leqslant \sum_{n=1}^{N} \|1_{E_n} - g_n\|_{L^1} \|x_n\| < \epsilon.$$

Thus any function $f \in L^1(\mathbb{R})$ has a simple function approximation s, which in turn can be approximated by a continuous function g. Combining these two facts gives

$$\|f - g\|_{L^1} \leqslant \|f - s\|_{L^1} + \|s - g\|_{L^1} < 2\epsilon$$

showing that *the set of (integrable) continuous functions is dense in $L^1(\mathbb{R})$*. Note further that precisely the same arguments work for $L^2(\mathbb{R})$.

We have already seen, in the Stone-Weierstraß theorem (Theorem 6.24), that the set of polynomials $p(z, \bar{z})$ is dense in $C[a, b]$. But, in this case, $z = \bar{z} = t \in [a, b]$, so such polynomials are of the usual form $p \in \mathbb{C}[t]$. Combining this with the above result shows that $\mathbb{C}[t]$ is also dense in $L^1[a, b]$ and $L^2[a, b]$: for any $\epsilon > 0$, there is a polynomial $p \in \mathbb{C}[t]$ such that

$$\|f - p\|_{L^1[a,b]} \leqslant \|f - g\|_{L^1[a,b]} + \|g - p\|_{L^1[a,b]} < 3\epsilon$$

since $\|g - p\|_{L^1[a,b]} \leqslant (b - a)\|g - p\|_{C[a,b]}$ can be made arbitrarily small. □

More generally, the polynomial splines are dense in the real version of these spaces. A *spline* of degree N is a function $\sum_n 1_{E_n} p_n$, where E_n are disjoint intervals and p_n are polynomials of degree at most N such that the first $N - 1$ derivatives match at the endpoints of E_n. They are often used in numerical techniques and graphics computing.

There is another very useful way of approximating integrable functions by smooth functions using convolution.

Proposition 9.30 (Approximation to the Identity)

If $h_n \in L^1(\mathbb{R})$ are such that $h_n \geqslant 0$, $\int h_n = 1$, and $\int_{\mathbb{R} \setminus [-\delta, \delta]} h_n \to 0$ as $n \to \infty$, then $h_n * f \to f$ in $C(\mathbb{R})$ and $L^1[a, b]$.

Proof Let g be a continuous function, and let $t \in \mathbb{R}$; on the one hand,

$$\forall \epsilon > 0, \ \exists \delta > 0, \ |s| < \delta \Rightarrow |g(t + s) - g(t)| < \epsilon, \tag{9.6}$$

and on the other hand, for this δ,

$$\exists N,\ n \geqslant N \implies \int_{\mathbb{R} \setminus [-\delta,\delta]} |h_n| < \epsilon \tag{9.7}$$

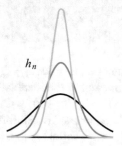

h_n

Therefore, for all t and $n \geqslant N$,

$$|h_n * g(t) - g(t)| = \left| \int h_n(s)\big(g(t-s) - g(t)\big)\, ds \right|$$

$$\leqslant \int |h_n(s)|\big|g(t-s) - g(t)\big|\, ds$$

$$\leqslant \int_{-\delta}^{\delta} h_n(s)\,\epsilon\, ds + 2\|g\|_C \int_{\mathbb{R} \setminus [-\delta,\delta]} h_n(s)\, ds \qquad \text{by (9.6)}$$

$$\leqslant \epsilon(1 + 2\|g\|_C) \qquad\qquad\qquad\qquad\qquad \text{by (9.7)}$$

and $\|h_n * g - g\|_C \to 0$ as required.

In fact $h_n * f$ approximates $f \in L^1[a,b]$ in the L^1-norm, for, choosing $g \in C[a,b]$ close to f, $\|f - g\|_{L^1} < \epsilon$, and n large enough that $\|h_n * g - g\|_C < \epsilon$ holds, then

$$\|h_n * f - f\|_{L^1[a,b]} \leqslant \|h_n * g - g\|_{L^1[a,b]} + \|h_n * (f - g)\|_{L^1[a,b]}$$

$$+ \|f - g\|_{L^1[a,b]}$$

$$< (b-a)\epsilon + 2\epsilon,$$

since $\|h_n * (f - g)\|_{L^1} \leqslant \|h_n\|_{L^1}\|f - g\|_{L^1} < \epsilon$. □

A useful way of generating an approximation of the identity is to start with a single integrable function $h \geqslant 0$, normalized so that $\int h = 1$, and then defining $h_n(t) := nh(nt)$. The conditions of the proposition hold, in particular, $\int_{|t|>\delta} nh(nt)\, dt = \int_{|s|>n\delta} h(s)\, ds \to 0$ as $n \to \infty$.

Typical examples of approximations of the identity are (i) $h_n(t) := \frac{n}{\sqrt{\pi}} e^{-(nt)^2}$, (ii) $h_n(t) = \frac{1}{\pi} \frac{n}{1+(nt)^2}$, and (iii) $h_n(t) := \frac{1}{c_n}(1 - t^2)^n$ supported on $[-1, 1]$, where $c_n = \int_{-1}^{1}(1 - t^2)^n\, dt$.

Corollary 9.31

For any $f \in L^1(A)$,

$$\|f(t+h) - f(t)\|_{L^1} \to 0 \text{ as } h \to 0.$$

Proof Starting from the unit function $h := 1_{[-1/2,1/2]}$, the step functions $h_n(t) := nh(nt)$ clearly form an approximation of the identity, and so $h_n * f \to f$ in $L^1(\mathbb{R})$. But their translations by $T_s f(t) := f(t - s)$, with $s = \pm 1/2n$, namely $h_n^+ := T_s h_n = n1_{[0,1/n]}$ and $h_n^- := T_{-s} h_n = n1_{[-1/n,0]}$, form other approximations of the identity. Since $(T_s h) * f = T_s(h * f)$ and $\|T_s f\|_{L^1} = \|f\|_{L^1}$,

$$\int |f(t-s) - f(t)| \, dt = \|T_s f - f\|_{L^1}$$

$$\leqslant \|T_s f - (T_s h_n) * f\|_{L^1} + \|(T_s h_n) * f - f\|_{L^1}$$

$$= \|f - h_n * f\|_{L^1} + \|h_n^\pm * f - f\|_{L^1}$$

$$\to 0 \text{ as } n \to \infty.$$

$\square$

Exercises 9.32

1. If $\sum_n \|f_n\|_{L^1}$ converges, then $\sum_{n=0}^\infty \int f_n = \int \sum_{n=0}^\infty f_n$.
2. The map $(a_n)_{n\in\mathbb{N}} \mapsto f$ where $f(t) := \sum_{n\in\mathbb{N}} a_n 1_{[n,n+1[}(t)$ isometrically embeds ℓ^p into $L^p(\mathbb{R})$, $1 \leqslant p \leqslant \infty$.
3. Let $f_n := 1_{[n,n+1]}$; if pointwise convergence were induced by a norm, then $f_n/\|f_n\|$ would converge to zero, a contradiction.
4. A simple function on $[0, 1]$ can be approximated by a step function in $L^1[0, 1]$, namely a simple function $\sum_n a_n 1_{I_n}$ where I_n are disjoint intervals. Deduce that the step functions are dense in $L^1(\mathbb{R})$. (Hint: For 1_E, approximate E by a finite union of intervals using Review 9.18(4) and Exercise 2.14(13).)
5. The map $L^1(A) \to \mathbb{C}$, $f \mapsto \int gf$ is linear, and continuous when $g \in L^\infty(A)$. Assuming surjectivity, show $L^1(K)^* \equiv L^\infty(K)$ for $K \subseteq \mathbb{R}$ compact.
6. Show that the functional $\delta_a(f) := f(a)$ on $C[a, b]$ is not integral, that is, it does not correspond to any L^1-function δ in the sense of $\delta_a(f) = \int \delta f$. Hence the dual space of $C[a, b]$ is not $L^1[a, b]$; it consists of functionals called *measures of bounded variation*.
7. *Minkowski's inequality*: Emulate the proof of Proposition 9.12 to show

$$\|f + g\|_{L^p} \leqslant \|f\|_{L^p} + \|g\|_{L^p} \quad (p \geqslant 1).$$

8. Show that $L^p[a, b]$ is a separable space for $1 \leqslant p$.

9. ⋆ Show that convolution on $L^1(\mathbb{R})$ is associative and commutative; but it has no identity, although Gibbs and Dirac audaciously added one and called it δ. Young's inequality is satisfied,

$$\|f * g\|_{L^r} \leqslant \|f\|_{L^p} \|g\|_{L^q}, \quad \frac{1}{p'} + \frac{1}{q'} = \frac{1}{r'}.$$

(Hint: $|f(t)g(x-t)| = |f(t)|^{p/r}|g(x-t)|^{q/r}|f(t)|^{1-p/r}|g(x-t)|^{1-q/r}$.)

10. *Matched Filter*: An electronic *filter* is a circuit acting on a signal $f \in L^2(\mathbb{R})$ and outputting the convolution $g * f$ ($g \in L^1(\mathbb{R})$). Signals often have *white noise* $\eta(t)$, where $\|g * \eta\|_{L^2} = \epsilon \|g\|_{L^2}$. The signal-to-noise ratio is $S/N := \|g * f\|_{L^2}^2 / \|g * \eta\|_{L^2}^2$; show that $S/N \leqslant \|f\|_{L^2}^2/\epsilon^2$, with equality holding when $g(s-t) = \lambda \overline{f}(t)$, for some $s, \lambda \in \mathbb{R}$.

11. The integral operator $Tf(s) := \int_1^\infty t^{-(s+1)} f(t) \, dt$ is continuous as $L^\infty[1, \infty[\to L^\infty[1, \infty[$, satisfying $\|Tf\|_{L^\infty} \leqslant \|f\|_{L^\infty}$.

12. An integral operator $T : L^1[0, 1] \to L^\infty[0, 1]$, with kernel $k \in L^\infty[0, 1]^2$, has $\|T\| \leqslant \|k\|_{L^\infty}$. So if T_n have kernels k_n with $k_n \to k$ in $L^\infty[0, 1]^2$, then $T_n \to T$.

The Fourier Series

We end this chapter with a look at one of the most important operators on $L^1[0, 1]$. Back to the days of Fourier, there arose the question of whether every periodic function f can be built up as a Fourier series $\sum_n a_n \cos nt + b_n \sin nt$. This claim of Fourier was disputed by Lagrange and others; Dirichlet obtained a partial result for the case $f \in C^2$, and Riemann later vastly extended this result. Despite these protests, the use of Fourier series grew, mainly because they actually worked in many examples.

Definition 9.33

The **Fourier coefficients** of an integrable function $f \in L^1[0, 1]$ are the sequence of numbers defined by[1]

$$\mathcal{F}f(n) = \widehat{f}(n) := \int_0^1 e^{-2\pi i n t} f(t) \, dt, \qquad n \in \mathbb{Z}.$$

---

[1] These are a modern version of the classical Fourier coefficients $\frac{1}{\pi} \int_0^{2\pi} \cos(nt) f(t) \, dt$ and $\frac{1}{\pi} \int_0^{2\pi} \sin(nt) f(t) \, dt$.

This section cannot do justice to the immense number of results and applications of Fourier series. It must suffice here to present some of the main results, with the aim of generalizing them later on. Refer to [3] for more details.

Theorem 9.34

$\mathcal{F} : L^1[0, 1] \to c_0(\mathbb{Z})$ **is a 1–1 continuous operator with**

$$\|\widehat{f}\|_{c_0(\mathbb{Z})} \leqslant \|f\|_{L^1[0,1]}$$

Here, $c_0(\mathbb{Z})$ is defined as consisting of those 'sequences' $(a_n)_{n \in \mathbb{Z}}$ such that $a_n \to 0$ as $n \to \pm\infty$.

Proof That $\mathcal{F}$ is linear is easy to show. It is continuous because

$$\|\widehat{f}\|_{\ell^\infty} = \sup_{n \in \mathbb{Z}} \left| \int_0^1 e^{-2\pi i n t} f(t) \, dt \right| \leqslant \int_0^1 |f(t)| \, dt = \|f\|_{L^1[0,1]}.$$

The characteristic function $1_{[a,b]}$, for $[a, b] \subseteq [0, 1]$, has Fourier coefficients

$$\widehat{1}_{[a,b]}(n) = \int_a^b e^{-2\pi i n t} \, dt = \frac{e^{-2\pi i n a} - e^{-2\pi i n b}}{2\pi i n} \to 0 \text{ as } n \to \pm\infty.$$

Hence the vector space of simple functions, as well as its closure $L^1[0, 1]$, are mapped into the complete space c_0 (Exercise 8.6(5)).

$\mathcal{F}$ *is 1–1:* If $\widehat{f}(n) = 0$ for every n, then

$$\int_0^1 e^{-2\pi i n s} f(s) \, ds = 0, \quad \forall n \in \mathbb{Z}.$$

The aim is to show that $f = 0$ a.e. Firstly,

$$\int_0^1 e^{-2\pi i n s} f(t - s) \, ds = \int_0^1 e^{-2\pi i n (t - s)} f(s) \, dy = 0.$$

Secondly, since $(\cos \pi s)^{2n} = (e^{2\pi i s} + e^{-2\pi i s} + 2)^n / 2^{2n}$ is a linear combination of exponentials of various frequencies that are all multiples of $2\pi s$, we have, for $h_n(s) := (\cos \pi s)^{2n} / c_n$,

$$h_n * f(t) = \frac{1}{c_n} \int_0^1 (\cos \pi s)^{2n} f(t - s) \, ds = 0,$$

where $c_n := \int_0^1 (\cos \pi s)^{2n} \, ds = \frac{(2n-1)(2n-3)\cdots}{(2n)(2n-2)\cdots} \geqslant \frac{1}{2n}$.

The functions h_n satisfy the criteria of Proposition 9.30, as they are positive and fall rapidly to 0 for $|s| \geqslant \delta$, as $n \to \infty$. Thus $\|f\|_{L^1} = \|h_n * f - f\|_{L^1} \to 0$, and $f = 0$ a.e. □

Although this Fourier operator is not surjective, and hence not invertible, its coefficients can be used to build up the original function. Note that, in the proof, the convolution of periodic functions is defined by $f * g(t) := \int_0^1 f(t-s)g(s)\,\mathrm{d}s$ where $f(t-s) = f(t-s+1)$ when $t < s$.

Theorem 9.35

The Cesáro sum of $\displaystyle\sum_{n=-\infty}^{\infty} \widehat{f}(n)e^{2\pi i n t}$ converges to f in $C[0,1]$.

Proof Take the finite sum

$$\sum_{k=-n}^{n} \widehat{f}(k)e^{2\pi i k t} = \int_0^1 \sum_{k=-n}^{n} e^{2\pi i k(t-s)} f(s)\,\mathrm{d}s = D_n * f(t),$$

where D_n is the so-called Dirichlet kernel,

$$\begin{aligned}
D_n(t) &= \sum_{k=-n}^{n} e^{2\pi i k t} = \frac{e^{-2\pi i n t} - e^{2\pi i (n+1)t}}{1 - e^{2\pi i t}} \\
&= \frac{(e^{-2\pi i n t} - e^{2\pi i (n+1)t})(1 - e^{-2\pi i t})}{(1 - e^{2\pi i t})(1 - e^{-2\pi i t})} \\
&= \frac{\cos(2\pi n t) - \cos(2\pi (n+1)t)}{1 - \cos(2\pi t)}.
\end{aligned}$$

D_n is not an approximation of the identity, since $|D_n(\frac{1}{2})| = 1 \nrightarrow 0$. But consider the Cesáro sum,

$$\frac{1}{N} \sum_{n=0}^{N} \sum_{k=-n}^{n} \widehat{f}(n)e^{2\pi i k t} = \frac{1}{N} \sum_{n=0}^{N} D_n * f(t) = F_N * f(t)$$

where F_N is the Fejer kernel,

$$F_N = \frac{1}{N} \sum_{n=0}^{N} D_n = \frac{1}{N} \sum_{n=0}^{N} \frac{\cos(2\pi n t) - \cos(2\pi (n+1)t)}{1 - \cos(2\pi t)} = \frac{1}{N} \frac{1 - \cos(2\pi N t)}{1 - \cos(2\pi t)}.$$

It can be verified that F_N is an approximation of the identity: the functions are positive, integrate to 1, and vanish outside a neighborhood of 0,

$$\int_0^1 D_n(t)\,dt = \sum_{k=-n}^n \int_0^1 e^{2\pi ikt}\,dt = \sum_{k\neq 0}[e^{2\pi ikt}/(2\pi ik)]_0^1 + 1 = 1,$$

$$\therefore \int_0^1 F_N(t)\,dt = \frac{1}{N}\sum_{n=0}^N \int_0^1 D_n(t)\,dt = 1.$$

For $\delta < t < \frac{1}{2}$,

$$F_N(t) = \frac{1}{N}\frac{1-\cos(2\pi Nt)}{1-\cos(2\pi t)} \leqslant \frac{1}{N}\frac{1}{1-\cos(2\pi\delta)} \to 0 \text{ as } N \to \infty.$$

Hence by Proposition 9.30, $F_N * f \to f$ both in $L^1[0, 1]$ and uniformly in $C[0, 1]$.

□

The Fourier coefficients have properties that appear remarkable: when f is translated the coefficients rotate in $\mathbb{C}$, at a rate proportional to n, with each $\widehat{f}(n)$ performing n turns as f is translated one whole period; differentiation of f scales the coefficients by a multiple of n; and convolutions are transformed to multiplications.

Proposition 9.36

For periodic functions, with period 1,

$$\widehat{T_a f}(n) = e^{-2\pi ian}\,\widehat{f}(n), \quad \widehat{f'}(n) = 2\pi in\,\widehat{f}(n), \quad \widehat{f * g} = \widehat{f}\,\widehat{g}.$$

Proof A translation $T_a f(t) := f(t - a)$ has the effect

$$\widehat{T_a f}(n) = \int_0^1 e^{-2\pi int} f(t - a)\,dt$$

$$= \int_0^1 e^{-2\pi in(t+a)} f(t)\,dt = e^{-2\pi ina}\,\widehat{f}(n).$$

For the derivative, f', using integration by parts,

$$\widehat{f'}(n) = \int_0^1 e^{-2\pi int} f'(t)\,dt$$

$$= [e^{-2\pi int} f(t)]_0^1 + 2\pi in \int_0^1 e^{-2\pi int} f(t)\,dt = 2\pi in\,\widehat{f}(n),$$

and the convolution of f and g becomes

$$\widehat{f * g}(n) = \int_0^1 e^{-2\pi i n t} \int_0^1 f(t - s) g(s) \, ds \, dt$$

$$= \int_0^1 \int_0^1 e^{-2\pi i n (t+s)} f(t) \, dt \, g(s) \, ds$$

$$= \int_0^1 e^{-2\pi i n t} f(t) \, dt \int_0^1 e^{-2\pi i n s} g(s) \, ds = \widehat{f}(n) \, \widehat{g}(n).$$

$\square$

Exercises 9.37

1. Show

   (a) $\mathcal{F} : 1 \mapsto (\ldots, 0, 0, 1, 0, 0, \ldots)$,
   (b) $\mathcal{F} : t \mapsto \frac{i}{2\pi} (\ldots, -\frac{1}{2}, -1, \frac{\pi}{i}, 1, \frac{1}{2}, \ldots, \frac{1}{n}, \ldots)$,
   (c) $\mathcal{F} : |t - \frac{1}{2}| \mapsto \frac{1}{\pi^2} (\ldots, 0, 1, \frac{1}{4}, 1, 0, \frac{1}{9}, 0, \frac{1}{25}, \ldots)$,
   (d) $\mathcal{F} : t(t - \frac{1}{2})(t - 1) \mapsto \frac{-3i}{4\pi^3} (\ldots, -\frac{1}{8}, -1, 0, 1, \frac{1}{8}, \ldots, \frac{1}{n^3}, \ldots)$.

2. The open mapping theorem implies that a bijective operator is an isomorphism (Corollary 11.2). Use it to show that $\mathcal{F}$ is not onto c_0.

3. The *power spectrum* of a function is a plot of $|\widehat{f}(n)|^2$. It displays the dominant frequencies of f. A better plot is the Nyquist diagram, where $\widehat{f}(n)$ is graphed in three dimensions, with one axis representing n, and the other two representing $\widehat{f} = |\widehat{f}| e^{i\phi}$. Prove that $\mathcal{F} : C^k[0, 1] \to c_k(\mathbb{Z})$, where $C^k[0, 1]$ is the space of k-times continuously differentiable periodic functions, and $c_k(\mathbb{Z}) := \{ (a_n)_{n \in \mathbb{Z}} : n^k a_n \to 0 \}$. Therefore, how fast the power spectrum decays as $n \to \infty$ measures how smooth the function is.

4. The operator $S_a : f(t) \mapsto a^{1/2} f(at)$ $(a > 0)$ stretches or compresses f, while preserving its L^2-norm; prove $\widehat{S_a f}(n) = S_{1/a} \widehat{f}(n)$. This should be familiar: playing a sound clip in half its normal time doubles the frequencies.

5. ▶ The **Fourier transform** of a function $f \in L^1(\mathbb{R})$ is defined to be the function

$$\mathcal{F} f(\xi) = \widehat{f}(\xi) := \int_{-\infty}^{\infty} e^{-2\pi i t \xi} f(t) \, dt.$$

It is an integral operator $\mathcal{F} : L^1(\mathbb{R}) \to L^{\infty}(\mathbb{R})$. Similarly to the Fourier series,

   (a) it is a continuous linear operator $\mathcal{F} : L^1(\mathbb{R}) \to C_0(\mathbb{R})$,
   (b) $\widehat{1_{[-a,a]}}(\xi) = a \sin(\pi a \xi)/(\pi a \xi) =: a \operatorname{sinc}(\pi a \xi) \to 0$ as $\xi \to \pm\infty$,
   (c) $\widehat{T_a f}(\xi) = e^{-2\pi i a \xi} \widehat{f}(\xi)$,
   (d) $\widehat{f'}(\xi) = 2\pi i \xi \widehat{f}(\xi)$,
   (e) $\widehat{f * g} = \widehat{f} \widehat{g}$.

6. $\mathcal{F}\frac{1}{\sqrt{\sigma}}e^{-\pi t^2/\sigma^2} = \sqrt{\sigma}\,e^{-\pi\sigma^2\xi^2}$. Deduce that the convolution of two Gaussian functions is another Gaussian function,

$$e^{-t^2/2\sigma^2} * e^{-t^2/2\tau^2} = \sqrt{2\pi}\,\frac{\sigma\tau}{\sqrt{\sigma^2+\tau^2}}e^{-t^2/2(\sigma^2+\tau^2)}.$$

Notice how there is a trade-off between the 'width' σ of the original Gaussian and that of its Fourier transform, namely $1/\sigma$.

7. *Wiener-Khinchin theorem*: For $f \in L^1(\mathbb{R})$, define $f^*(t) := \overline{f(-t)}$. Show $\widehat{f^*} = \overline{\widehat{f}}$, and that the *auto-correlation* function $f^* * f(t) = \int \overline{f(s)}f(s+t)\,ds$ is transformed to the *power spectrum* $|\widehat{f}(\xi)|^2$. More generally, $f^* * g$ is called the *cross-correlation* function of f and g.

Remarks 9.38

1. The functionals on ℓ^∞ are more difficult to describe. Every sequence $\boldsymbol{y} \in \ell^1$ still acts as a functional on ℓ^∞ via $\boldsymbol{x} \mapsto \boldsymbol{y} \cdot \boldsymbol{x}$, but $(\ell^\infty)^*$ is a complicated non-separable space that includes much more than just ℓ^1 (look up "finitely additive measures" for more).

2. $\ell^\infty = C_b(\mathbb{N})$, so the completeness part of Theorem 9.1 is included in Theorem 6.23.

3. The Fibonacci iteration $a_n := a_{n-1} + a_{n-2}$, starting from $a_0 = 1 = a_1$, is an equation on sequences. It can be expressed in any of the following ways

$$\boldsymbol{x} = R\boldsymbol{x} + R^2\boldsymbol{x} + \boldsymbol{e}_1 + \boldsymbol{e}_0$$

$$(\boldsymbol{e}_0 - \boldsymbol{e}_1 - \boldsymbol{e}_2) * \boldsymbol{x} = \boldsymbol{e}_0 + \boldsymbol{e}_1$$

$$(1, -1, -1, 0, \ldots) * \boldsymbol{x} = (1, 1, 0, \ldots)$$

Convoluting with the inverse of $(1, -1, -1, 0, \ldots)$ gives the terms of the Fibonacci sequence (but note that the inverse is not in ℓ^1). Traditionally, "generating functions" are used to get the same results, the connection being elucidated in Chap. 14.

4. ℓ^1 contains the space of sequences $\ell_s^1 := \{\,(a_n)_{n\in\mathbb{N}} : (n^s a_n)_{n\in\mathbb{N}} \in \ell^1\,\}$, $(s \geqslant 0)$, which in turn contains $\ell_{s+1+\epsilon}^\infty$.

5. The following are some classical criteria for determining that a sequence of measurable functions f_n that converges pointwise a.e. is Cauchy in $L^1(A)$,

   (a) $|f_n|$ are increasing but $\int |f_n|$ are bounded (Monotone Convergence Theorem),
   (b) $|f_n| \leqslant g \in L^1(A)$ (Dominated Convergence Theorem),
   (c) $\int_E f_n$ converges for all measurable sets E (Vitali's theorem).

6. A function on $\mathbb{R}$ has both local and global integrability properties: locally about $t \in \mathbb{R}$, it may belong to some $L^p[t-\delta, t+\delta]$ space, while globally, the sequence

of numbers $a_n := \|f\|_{L^p[n,n+1]}$ may belong to ℓ^q. For example, f is in $L^1(\mathbb{R})$ when it is locally in L^1 and globally in ℓ^1. L^p_{loc} are spaces of functions that are only locally in L^p. For example, the constant function 1 is in all $L^p(\mathbb{R})$ locally, but its sequence of norms is only in ℓ^∞; so $1 \in L^\infty(\mathbb{R})$. Similarly, $1/\sqrt{t}$ is locally in all the L^p spaces for $1 \leqslant p < 2$, but its norm sequence is in ℓ^q, $2 < q$.

7. The Fourier series maps $\mathcal{F} : L^p[0,1] \to \ell^{p'}$ for $1 \leqslant p \leqslant 2$ (see Exercise 10.35(11) for $p = 2$).

Chapter 10
Hilbert Spaces

10.1 Inner Products

There are spaces, such as ℓ^2, whose norms have special properties because they are induced from what are termed *inner products*. Not only do such spaces have a concept of length but also of *orthogonality* between vectors.

Definition 10.1

An **inner product** on a vector space X is a positive-definite sesquilinear form[1], namely a map

$$\langle\,,\,\rangle : X \times X \to \mathbb{F}$$

such that for all $x, y, z \in X, \lambda \in \mathbb{F}$,

$$\langle x, y + z \rangle = \langle x, y \rangle + \langle x, z \rangle, \qquad \langle x, \lambda y \rangle = \lambda \langle x, y \rangle,$$
$$\langle y, x \rangle = \overline{\langle x, y \rangle}, \qquad \langle x, x \rangle \geqslant 0; \quad \langle x, x \rangle = 0 \Leftrightarrow x = 0.$$

Two vectors are said to be *orthogonal* or *perpendicular* when $\langle x, y \rangle = 0$, also written as $x \perp y$. More generally, two subsets are *orthogonal*, $A \perp B$, when any two vectors $a \in A$ and $b \in B$ are orthogonal, $\langle a, b \rangle = 0$.

---

[1] In the mathematical literature, the inner product is often taken to be linear in the *first* variable; this is a matter of convention. The choice adopted here is that of the "physics" community; it makes many formulas, such as the definition $x^*(y) := \langle x, y \rangle$, more natural and conforming with function notation.

© The Author(s), under exclusive license to Springer Nature Switzerland AG 2024   197
J. Muscat, *Functional Analysis*, https://doi.org/10.1007/978-3-031-27537-1_10

Easy Consequences
1. If for all $x \in X$, $\langle x, y \rangle = 0$, then $y = 0$.
2. $\langle x + y, z \rangle = \langle x, z \rangle + \langle y, z \rangle$, but $\langle \lambda x, y \rangle = \bar{\lambda} \langle x, y \rangle$ (conjugate-linear).
3. $\langle x, x \rangle$ is real (and non-negative); its square-root is denoted by $\|x\| := \sqrt{\langle x, x \rangle}$.
4. $\|\lambda x\| = |\lambda| \|x\|$, and $\|x\| = 0 \Leftrightarrow x = 0$.
5. $\|x + y\|^2 = \|x\|^2 + 2 \operatorname{Re} \langle x, y \rangle + \|y\|^2$.
6. (Pythagoras) If $\langle x, y \rangle = 0$ then $\|x + y\|^2 = \|x\|^2 + \|y\|^2$. More generally, if $\langle x_i, x_j \rangle = 0$ for $i \neq j$ then (by induction)

$$\|x_1 + \cdots + x_n\|^2 = \|x_1\|^2 + \cdots + \|x_n\|^2.$$

We will see next that the triangle inequality is also true, making $\| \cdot \|$ a norm, thus inner product spaces are normed spaces.

Examples 10.2

1. The simplest examples are the *Euclidean* spaces $\mathbb{R}^n$ and $\mathbb{C}^n$ with

$$\langle \begin{pmatrix} a_1 \\ \vdots \\ a_n \end{pmatrix}, \begin{pmatrix} b_1 \\ \vdots \\ b_n \end{pmatrix} \rangle := (\overline{a_1} \cdots \overline{a_n}) \begin{pmatrix} b_1 \\ \vdots \\ b_n \end{pmatrix} = \sum_{i=1}^{n} \overline{a_i} b_i.$$

More generally, take any basis $v_1, \ldots, v_n$ of $\mathbb{F}^n$, expand any two vectors x and y as $x = \sum_{i=1}^{n} a_i v_i$, $y = \sum_{i=1}^{n} b_i v_i$, and define $\langle x, y \rangle := \sum_{i=1}^{n} \overline{a_i} b_i$. (The inner product differs depending on the choice of the basis.)
2. The matrices of size $m \times n$ have an inner product given by

$$\langle A, B \rangle := \sum_{i=1}^{m} \sum_{j=1}^{n} \overline{A_{ij}} B_{ij}.$$

The induced norm is the Frobenius norm, not the operator norm (but recall that all norms on a finite-dimensional Banach space are equivalent).
3. ▶ ℓ^2 has the inner product $\langle (a_n), (b_n) \rangle := \sum_{n=0}^{\infty} \overline{a_n} b_n$. The fact that this series converges follows from Cauchy's inequality $| \sum_n \overline{a_n} b_n | \leqslant \|(a_n)\| \|(b_n)\|$.
4. ▶ $L^2(A)$ has the inner product $\langle f, g \rangle := \int_A \bar{f} g$. That this integral has a finite value follows from Hölder's inequality $| \int_A \bar{f} g | \leqslant \|\bar{f} g\|_{L^1} \leqslant \|f\|_{L^2} \|g\|_{L^2}$.
5. The *weighted* ℓ^2 and L^2 spaces generalize these formulae to

$$\langle (a_n), (b_n) \rangle := \sum_n \overline{a_n} b_n w_n, \qquad \langle f, g \rangle := \int \overline{f(x)} g(x) w(x) \, dx,$$

respectively, where w_n and $w(x)$ are called *weights*; what properties do they need to have for the inner product axioms to hold?

Our first proposition generalizes Cauchy's inequality (Proposition 7.5) from ℓ^2 to a general inner product space. It is probably the most used inequality in analysis.

Proposition 10.3 (Cauchy-Schwarz Inequality)

$$|\langle x, y \rangle| \leqslant \|x\|\|y\|$$

Proof The inequality need only be shown for y non-zero. Any other vector x can be decomposed uniquely into two parts, one in the direction of y, and the other perpendicular to it:

$$x = \lambda y + (x - \lambda y), \quad \text{with } \langle y, x - \lambda y \rangle = 0.$$

This yields $\lambda = \langle y, x \rangle / \langle y, y \rangle$. Applying Pythagoras' theorem, we deduce that

$$\|x\|^2 = \|\lambda y\|^2 + \|x - \lambda y\|^2,$$

hence $\|\lambda y\| \leqslant \|x\|$, or $|\lambda| \leqslant \|x\|/\|y\|$, from which follows the assertion.

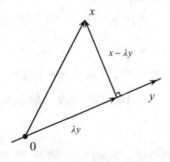

$\square$

Corollary 10.4

$$\|x + y\| \leqslant \|x\| + \|y\|$$

Proof Using the Cauchy-Schwarz inequality, $\mathrm{Re}\,\langle x, y \rangle \leqslant |\langle x, y \rangle| \leqslant \|x\|\|y\|$, so

$$\|x + y\|^2 = \|x\|^2 + 2\,\mathrm{Re}\,\langle x, y \rangle + \|y\|^2 \leqslant \|x\|^2 + 2\|x\|\|y\| + \|y\|^2 = (\|x\| + \|y\|)^2.$$

$\square$

David Hilbert (1862–1943) Hilbert studied invariant theory under Lindemann at Königsberg until 1885. His encyclopedic powers motivated him to explore much of mathematics; in 1899, in Göttingen, he gave rigorous axioms for Euclidean geometry; 1904–1909, he studied Fredholm's integral equations, with his student Schmidt; he defined compact operators, proving they are limits of matrices, with their spectrum of eigenvalues; (Schmidt) defined ℓ^2 with its inner product. On to mathematical physics, quite possibly he inspired Einstein's general relativity. His 1918 'formalist' research programme set out to prove that set axioms are consistent, "*one can solve any problem by pure thought*".

Hence $\| \cdot \|$ is a norm, and all the facts about normed spaces apply to inner product spaces. For example, the norm is continuous.

Proposition 10.5

The inner product is continuous.

Proof Let $x_n \to x$ and $y_n \to y$, then since y_n are bounded (Example 4.3(6)),

$$|\langle x_n, y_n \rangle - \langle x, y \rangle| = |\langle x_n, y_n \rangle - \langle x, y_n \rangle + \langle x, y_n \rangle - \langle x, y \rangle|$$
$$\leqslant |\langle x_n - x, y_n \rangle| + |\langle x, y_n - y \rangle|$$
$$\leqslant \|x_n - x\| \|y_n\| + \|x\| \|y_n - y\| \to 0.$$

$\square$

It follows that taking limits commutes with the inner product:

$$\lim_{n \to \infty} \langle x_n, y_n \rangle = \langle \lim_{n \to \infty} x_n, \lim_{n \to \infty} y_n \rangle.$$

Definition 10.6

A **Hilbert** space is an inner product space which is complete as a metric space.

In the rest of the text, the letter H denotes a Hilbert space.

Examples 10.7

1. $\mathbb{R}^n$, $\mathbb{C}^n$, ℓ^2 and $L^2(\mathbb{R})$ are all Hilbert spaces (Theorems 8.24, 9.8).
2. Every inner product space can be completed to a Hilbert space. In the completion as a normed space (Proposition 7.18), take $\langle \boldsymbol{x}, \boldsymbol{y} \rangle := \lim_{n \to \infty} \langle x_n, y_n \rangle$, for represen-

tative Cauchy sequences $x = [x_n]$, $y = [y_n]$. Note that $\langle x_n, y_n \rangle$ is a Cauchy sequence in $\mathbb{C}$ since

$$|\langle x_n, y_n \rangle - \langle x_m, y_m \rangle| \leqslant |\langle x_n, y_n \rangle - \langle x_m, y_n \rangle| + |\langle x_m, y_n \rangle - \langle x_m, y_m \rangle|$$

$$\leqslant \|x_n - x_m\| \|y_n\| + \|x_m\| \|y_n - y_m\| \to 0$$

as $n, m \to \infty$, with $\|x_m\|$, $\|y_n\|$ bounded.

3. ▶ For an inner product space over $\mathbb{C}$, if $\langle x, Tx \rangle = 0$ for all $x \in X$, then $T = 0$.
Proof: The identities

$$0 = \langle x + y, T(x + y) \rangle = \langle x, Ty \rangle + \langle y, Tx \rangle,$$

$$0 = \langle x + iy, T(x + iy) \rangle = i\langle x, Ty \rangle - i\langle y, Tx \rangle,$$

together imply $\langle x, Ty \rangle = 0$, for any $x, y \in X$, in particular $\|Ty\|^2 = 0$.

4. An alternative proof of the Cauchy-Schwarz inequality is

$$0 \leqslant \|u - \lambda v\|^2 = 1 - 2\,\mathrm{Re}\,\lambda \langle u, v \rangle + |\lambda|^2$$

for $u := x/\|x\|$, $v := y/\|y\|$ unit vectors and all $\lambda \in \mathbb{F}$, in particular for $\lambda = |\langle u, v \rangle|/\langle u, v \rangle$.

5. $\|x\| = \sup_{\|y\|=1} |\langle x, y \rangle|$, with the maximum achieved when $y = x/\|x\|$.

Do all norms on vector spaces come from inner products, and if not, which property characterizes inner product spaces? The answer is given by:

Proposition 10.8 (Parallelogram Law)

A norm is induced from an inner product if, and only if, it satisfies, for all vectors x, y,

$$\|x + y\|^2 + \|x - y\|^2 = 2(\|x\|^2 + \|y\|^2).$$

The statement asserts that the sum of the lengths squared of the diagonals of a parallelogram equals that of the sides.

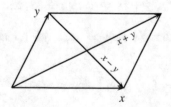

Proof The parallelogram law follows from adding the identities,

$$\|x + y\|^2 = \|x\|^2 + 2\operatorname{Re}\langle x, y \rangle + \|y\|^2,$$
$$\|x - y\|^2 = \|x\|^2 - 2\operatorname{Re}\langle x, y \rangle + \|y\|^2.$$

Subtracting the two gives $4\operatorname{Re}\langle x, y \rangle$. This is already sufficient to identify the inner product when the scalar field is $\mathbb{R}$. Over $\mathbb{C}$, notice that $\operatorname{Im}\langle x, y \rangle = -\operatorname{Re} i\langle x, y \rangle = \operatorname{Re}\langle ix, y \rangle$, so

$$\langle x, y \rangle = \frac{1}{4}\left(\|y + x\|^2 - \|y - x\|^2 + i\|y + ix\|^2 - i\|y - ix\|^2\right). \tag{10.1}$$

This remarkable *polarization identity* expresses the inner product purely in terms of norms. Accordingly, for the converse of the proposition, define

$$\text{for any normed space,}\quad \langle\!\langle x, y \rangle\!\rangle := \tfrac{1}{4}(\|y + x\|^2 - \|y - x\|^2),$$
$$\text{for a complex space,}\quad \langle x, y \rangle := \langle\!\langle x, y \rangle\!\rangle + i\langle\!\langle ix, y \rangle\!\rangle.$$

Two of the inner product axioms follow from $\langle\!\langle y, x \rangle\!\rangle = \langle\!\langle x, y \rangle\!\rangle$ and $\langle x, x \rangle = \langle\!\langle x, x \rangle\!\rangle = \|x\|^2$, as well as $\langle x, 0 \rangle = \langle\!\langle x, 0 \rangle\!\rangle = 0$; $\langle y, x \rangle = \overline{\langle x, y \rangle}$ is readily verified using

$$4\langle\!\langle iy, x \rangle\!\rangle = \|x + iy\|^2 - \|x - iy\|^2 = \|y - ix\|^2 - \|y + ix\|^2 = -4\langle\!\langle ix, y \rangle\!\rangle.$$

To show linearity, let $u = \frac{y+z}{2}$, $v = \frac{y-z}{2}$, then by the parallelogram law,

$$\|x + y\|^2 + \|x + z\|^2 = \|u + v + x\|^2 + \|u - v + x\|^2 = 2\|u + x\|^2 + 2\|v\|^2$$

$$\|y - x\|^2 + \|z - x\|^2 = \|u + v - x\|^2 + \|u - v - x\|^2 = 2\|u - x\|^2 + 2\|v\|^2$$

Subtracting the two equations gives

$$\langle\!\langle x, y \rangle\!\rangle + \langle\!\langle x, z \rangle\!\rangle = 2\langle\!\langle x, \tfrac{1}{2}(y + z) \rangle\!\rangle$$

In particular, putting $z = 0$ gives $\langle\!\langle x, y \rangle\!\rangle = 2\langle\!\langle x, \tfrac{1}{2}y \rangle\!\rangle$ (for any y), reducing the above identity to

$$\langle\!\langle x, y + z \rangle\!\rangle = \langle\!\langle x, y \rangle\!\rangle + \langle\!\langle x, z \rangle\!\rangle. \tag{10.2}$$

By induction, it follows that $\langle\!\langle x, ny \rangle\!\rangle = n\langle\!\langle x, y \rangle\!\rangle$ for $n \in \mathbb{N}$. For the negative integers,

$$\langle\!\langle x, -y \rangle\!\rangle = \|-y + x\|^2 - \|-y - x\|^2 = -\langle\!\langle x, y \rangle\!\rangle$$

while for rational numbers $p = m/n, m, n \in \mathbb{Z}, n \neq 0$,

$$n\langle\!\langle x, \tfrac{m}{n} y \rangle\!\rangle = \langle\!\langle x, my \rangle\!\rangle = m\langle\!\langle x, y \rangle\!\rangle$$

so $\langle\!\langle x, py \rangle\!\rangle = p\langle\!\langle x, y \rangle\!\rangle$. Note that $\langle\!\langle x, y \rangle\!\rangle$ is continuous in x and y since the norm is continuous, so if the rational numbers $p_n \to \alpha \in \mathbb{R}$, then

$$\langle\!\langle x, \alpha y \rangle\!\rangle = \lim_{n\to\infty} \langle\!\langle x, p_n y \rangle\!\rangle = \lim_{n\to\infty} p_n \langle\!\langle x, y \rangle\!\rangle = \alpha \langle\!\langle x, y \rangle\!\rangle.$$

This completes the proof when the scalar field is $\mathbb{R}$. Over the complex numbers, $\langle x, \lambda y \rangle = \lambda \langle x, y \rangle$ for $\lambda \in \mathbb{C}$ is evident from (10.1), (10.2), and

$$\langle x, iy \rangle = -\langle\!\langle ix, y \rangle\!\rangle + i\langle\!\langle x, y \rangle\!\rangle = i\langle x, y \rangle.$$

$\square$

In a sense, it is the presence of orthogonality that distinguishes inner product spaces from normed ones. By the polarization identity, two vectors are perpendicular when $\|x + y\| = \|x - y\|$ and $\|x + iy\| = \|x - iy\|$. Each vector, and more generally each subspace, is complemented by a subspace of those vectors that are perpendicular to it.

Proposition 10.9

The *orthogonal* spaces of subsets $A \subseteq X$,

$$A^\perp := \{ x \in X : \forall a \in A, \ \langle x, a \rangle = 0 \},$$

satisfy

(i) $A \cap A^\perp \subseteq 0$,
(ii) $A \subseteq B \Rightarrow B^\perp \subseteq A^\perp$, **and** $A \subseteq A^{\perp\perp}$,
(iii) $A^\perp$ **is a closed subspace of** X,
(iv) $A^\perp = \overline{[\![A]\!]}^\perp$.

Proof (i) If a vector $a \in A$ is also in $A^\perp$, then it is orthogonal to all vectors in A, including itself, $\langle a, a \rangle = 0$, so $a = 0$.

(ii) If $a \in A \subseteq B$ and $x \in B^\perp$, then $\langle x, a \rangle = 0$, so $x \in A^\perp$. For any $a \in A$ and $x \in A^\perp$, $\langle a, x \rangle = \overline{\langle x, a \rangle} = 0$, so $a \in A^{\perp\perp}$.

(iii) If x and y are in $A^\perp$ and $a \in A$, then

$$\langle \lambda x, a \rangle = \bar{\lambda}\langle x, a \rangle = 0, \qquad \langle x + y, a \rangle = \langle x, a \rangle + \langle y, a \rangle = 0,$$

so $\lambda x, x + y \in A^\perp$. If $x_n \in A^\perp$ and $x_n \to x$, then $0 = \langle x_n, a \rangle \to \langle x, a \rangle$, and $x \in A^\perp$.

(iv) That $\overline{[\![A]\!]}^\perp \subseteq A^\perp$ follows from $A \subseteq \overline{[\![A]\!]}$. Conversely, let $x \in A^\perp$; for any $a, b \in A$,

$$\langle x, a + b \rangle = \langle x, a \rangle + \langle x, b \rangle = 0, \quad \langle x, \lambda a \rangle = \lambda \langle x, a \rangle = 0,$$

so x is orthogonal to the space generated by A, $x \in [\![A]\!]^\perp$. Let $a_n \to y$ with $a_n \in [\![A]\!]$, then $0 = \langle x, a_n \rangle \to \langle x, y \rangle$ and $x \in \overline{[\![A]\!]}^\perp$.   □

Exercises 10.10

1. If $T, S : X \to Y$ are linear maps on inner product spaces such that $\langle y, Tx \rangle = \langle y, Sx \rangle$ for all $x \in X$, $y \in Y$, then $T = S$. Example 10.7(3) is false for real spaces: Find a non-zero 2×2 real matrix T such that $\langle x, Tx \rangle = 0$ for all $x \in \mathbb{R}^2$.

2. The Cauchy-Schwarz inequality becomes an equality if, and only if, $x = \lambda y$ for some scalar λ (or $y = 0$). Similarly, $\|x + y\| = \|x\| + \|y\|$ precisely when $x = \lambda y$, $\lambda \geqslant 0$. More generally, $\| \sum_n x_n \| = \sum_n \|x_n\|$ if, and only if, $x_n = \lambda_n x$ for some $\lambda_n \geqslant 0$.

3. When $T : X \to Y$ is 1–1 and linear, $\langle x, y \rangle_X := \langle Tx, Ty \rangle_Y$ is an inner product on X. What properties does $S : X \to X$ need to have to ensure that $\langle\!\langle x, y \rangle\!\rangle := \langle x, Sy \rangle$ is also an inner product?

4. ⋆ Every inner product on $\mathbb{R}^n$ is of the type $\langle x, Ay \rangle = \sum_{ij} A_{ij} a_i b_j$ where A is a positive symmetric matrix. Deduce that balls have the shape of an ellipse in $\mathbb{R}^2$, and of an ellipsoid in $\mathbb{R}^3$.

5. ▶ The product of two inner product spaces, $X \times Y$, has an inner product defined by

$$\langle \begin{pmatrix} x_1 \\ y_1 \end{pmatrix}, \begin{pmatrix} x_2 \\ y_2 \end{pmatrix} \rangle := \langle x_1, x_2 \rangle_X + \langle y_1, y_2 \rangle_Y.$$

Then the maps $x \mapsto \begin{pmatrix} x \\ 0 \end{pmatrix}$ and $y \mapsto \begin{pmatrix} 0 \\ y \end{pmatrix}$ embed X and Y as orthogonal subspaces of $X \times Y$. Although the induced norm is not the same one we defined for $X \times Y$ as normed spaces (Example 7.4(8)), the two norms are equivalent.

   When X, Y are complete, so is $X \times Y$ with the induced norm (Hint: use $\|x\| \leqslant \| \begin{pmatrix} x \\ y \end{pmatrix} \|$).

6. In any inner product space,

   (a) $\|x - y\|^2 + \|x + y - 2z\|^2 = 2\|x - z\|^2 + 2\|y - z\|^2$.
   (b) $\|x + y + z\|^2 + \|x + y - z\|^2 + \|x - y + z\|^2 + \|x - y - z\|^2$
       $= 4(\|x\|^2 + \|y\|^2 + \|z\|^2)$.
   (c) ⋆ Generalize, by induction, to the sum of n elements,

$$\frac{1}{2^n} \sum_\sigma \| \sum_{i=1}^n \sigma_i x_i \|^2 = \sum_{i=1}^n \|x_i\|^2$$

where σ ranges through all 2^n possible $\pm$ choices for x_i. Deduce that one can always choose the signs such that $\|x_1 \pm x_2 + \cdots \pm x_n\|^2 \geqslant \sum_i \|x_i\|^2$. Deduce further that a random walk of successive vectors $\pm x \pm y \pm x \pm y \pm \cdots$ has an expected root-mean-square distance of $\sqrt{n}\sqrt{\|x\|^2 + \|y\|^2}$.

(d) By first showing that if $\omega = e^{2\pi i/n}$ then $\sum_{k=1}^n \omega^k = 0 = \sum_{k=1}^n \omega^{2k}$ and $\sum_{k=1}^n \omega^k \operatorname{Re}(\omega^{-k}z) = zn/2$, prove

$$\langle x, y \rangle = \frac{1}{n} \sum_{k=1}^n \omega^k \|y + \omega^k x\|^2.$$

7. Verify that the norms for ℓ^2 and $L^2(\mathbb{R})$ satisfy the parallelogram law, and show that the inner product obtained from the polarization identity is the same one defined previously (Examples 10.2(3, 4)).

8. The 1-norm and ∞-norm defined on $\mathbb{R}^2$ are not induced by inner products. Find two vectors that do not satisfy the parallelogram law with these norms.

9. ▶ Similarly, ℓ^1, ℓ^∞, $L^1(\mathbb{R})$ and $L^\infty(\mathbb{R})$ are not inner product spaces. Neither is $B(X, Y)$ in general.

10. A norm $\| \cdot \|$ that satisfies the parallelogram law gives rise to its associated inner product, by the polarization identity. In turn, this inner product induces the norm $\|\|x\|\| := \sqrt{\langle x, x \rangle}$. Show that the two norms are identical.

11. The polynomials t and $2t^2 - 1$ are orthogonal in $L^2[0, 1]$. So are sine and cosine in the space $L^2[-\pi, \pi]$; can you find a function orthogonal to both?

12. $0^\perp = X$, $X^\perp = 0$. In fact, $A^\perp = X \Leftrightarrow A \subseteq \{0\}$. Do you think it is true that $A^\perp = 0 \Leftrightarrow A = X$? What if A is a closed linear subspace of X?

13. Show that (i) $(A + B)^\perp = A^\perp \cap B^\perp$, (ii) $A^{\perp\perp\perp} = A^\perp$. (Hint: Use property (ii) of Proposition 10.9.)

14. Let $d := d(x, [\![y]\!]) = \inf_\lambda \|x + \lambda y\|$, where y is a unit vector; show that (i) $d = \|x + \lambda_0 y\|$ for some λ_0, (ii) $|\langle x, y \rangle|^2 = \|x\|^2 - d^2$, and (iii) $y \perp (x - \lambda_0 y)$.

15. To illustrate the strength of orthogonality, prove that if $M \perp N$ are orthogonal complete subspaces of X, then $M + N$ is also complete (Example 7.12(2)).

16. Suppose a vector space X satisfies all the axioms for an inner product space except that it contains non-zero vectors with $\langle x, x \rangle = 0$. Show that if $\langle x, x \rangle = 0$, then $\forall y$, $\langle x, y \rangle = 0$. (Hint: Expand $\|y - \lambda x\|^2$.)

Deduce that Pythagoras' theorem and Cauchy-Schwarz's inequality remain valid. Show that $Z := \{x : \langle x, x \rangle = 0\}$ is a closed linear subspace, and that there is a well-defined inner product on X/Z, $\langle x + Z, y + Z \rangle := \langle x, y \rangle$.

17. A light 'ray' has a frequency profile $f(\omega)$. Oversimplifying slightly, our eyes convert it to a color vector $(\langle r, f \rangle, \langle g, f \rangle, \langle b, f \rangle)$ where $r(\omega)$, $g(\omega)$, $b(\omega)$ are the absorption profiles of the retinal cone cells. So any two points (rays) in the coset $f + [\![r, g, b]\!]^\perp$ are perceived to have the same color.

10.2   Least Squares Approximation

By Exercise 10.10(14) above, the distance between a point and a line can be minimized by a unique point on the line. This has a generalization with far-reaching consequences:

Theorem 10.11 (Hilbert Projection Theorem)

> **If M is a closed convex subset of a Hilbert space H, then any point in H has a unique point in M which is closest to it,**
>
> $$\forall x \in H,\ \exists! x_* \in M,\ \forall y \in M, \quad \|x - x_*\| \leqslant \|x - y\|.$$
>
> **For any $y \in M$, $\operatorname{Re} \langle x - x_*, y - x_* \rangle \leqslant 0$.**
>
> **The mapping $H \to M, x \mapsto x_*$, is continuous.**

Proof *Existence of x_*:* Let $d := d(x, M) = \inf_{y \in M} \|x - y\|$ be the smallest distance from M to x. Then there is a sequence of vectors $y_n \in M$ such that $\|x - y_n\| \to d$. Using the parallelogram law and the convexity of M, $(y_n)_{n \in \mathbb{N}}$ is a Cauchy sequence,

$$
\begin{aligned}
\|y_n - y_m\|^2 &= 2\|y_n - x\|^2 + 2\|y_m - x\|^2 - \|(y_n + y_m) - 2x\|^2 \\
&= 2\|y_n - x\|^2 + 2\|y_m - x\|^2 - 4\left\|\frac{y_n + y_m}{2} - x\right\|^2 \\
&\leqslant 2\|y_n - x\|^2 + 2\|y_m - x\|^2 - 4d^2 \\
&\to 0, \quad \text{as } n, m \to \infty.
\end{aligned}
$$

But H is complete and M is closed, so y_n converges to some $x_* \in M$. It follows, by continuity of the norm, that $\|x - x_*\| = \lim\limits_{n \to \infty} \|x - y_n\| = d$.

Uniqueness of x_*: Suppose $y \in M$ is another closest point to x, i.e., $\|x - y\| = d$. Then $x_* = y$ since, as in the argument above,

$$
\begin{aligned}
\|x_* - y\|^2 &= 2\|x_* - x\|^2 + 2\|y - x\|^2 - \|(x_* + y) - 2x\|^2 \\
&\leqslant 2\|x_* - x\|^2 + 2\|y - x\|^2 - 4d^2 = 0.
\end{aligned}
$$

Obtuse angle property: Consider the straight line $y(t) := x_* + tv$, where v is a vector pointing inside M. By convexity of M, $y(t) \in M$ for $t \geqslant 0$ small enough; for example, take $v = y - x_*$, with $y \in M, 0 \leqslant t \leqslant 1$. Then

$$\|x - x_*\|^2 \leqslant \|x - y(t)\|^2 = \|x - x_* - tv\|^2$$

$$= \|x - x_*\|^2 - 2t \operatorname{Re} \langle x - x_*, v \rangle + t^2 \|v\|^2$$

$$\therefore \ 2 \operatorname{Re} \langle x - x_*, v \rangle \leqslant t\|v\|^2$$

$$\therefore \ \operatorname{Re} \langle x - x_*, y - x_* \rangle \leqslant 0 \tag{10.3}$$

since t is positive and arbitrarily close to zero.

Continuity of $x \mapsto x_*$: Let x, z be any points in H, with corresponding closest points x_*, z_* in M. Then the map is non-expansive, and thus continuous:

$$\|x - z\|^2 = \|x - x_* + x_* - z_* + z_* - z\|^2$$

$$= \|x - x_* + z_* - z\|^2 + \|x_* - z_*\|^2$$

$$+ 2 \operatorname{Re} \langle x - x_*, x_* - z_* \rangle + 2 \operatorname{Re} \langle z_* - z, x_* - z_* \rangle$$

$$\geqslant \|x_* - z_*\|^2$$

since both inner products are non-negative by (10.3).

$\square$

Let us concentrate on the special case when M is a closed subspace of H.

Theorem 10.12

> **When M is a closed linear subspace of a Hilbert space H, then $y \in M$ is the closest point x_* to $x \in H$ if, and only if,**
>
> $$x - y \in M^{\perp}.$$
>
> **The map $P : x \mapsto x_*$ is an 'orthogonal' projection of norm 1, with im $P = M$ orthogonal to ker $P = M^{\perp}$, so**
>
> $$H = M \oplus M^{\perp}.$$

Proof (i) Let v be any non-zero point of M and let $w := x - (x_* + \lambda v)$ where λ is chosen so that $v \perp w$, that is, $\lambda := \langle v, x - x_* \rangle / \|v\|^2$. By Pythagoras' theorem, we get

$$\|x - x_*\|^2 = \|w + \lambda v\|^2 = \|w\|^2 + \|\lambda v\|^2 \geqslant \|w\|^2$$

making $x_* + \lambda v$ even closer to x than the closest point x_*, unless $\lambda = 0$, i.e., $\langle v, x - x_* \rangle = 0$. Since v is arbitrary, this gives $x - x_* \perp M$.

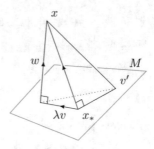

Conversely, if $(x - y) \perp v'$ for any $v' \in M$, then $(x - y) \perp (v' - y)$ and Pythagoras' theorem implies

$$\|x - v'\|^2 = \|x - y\|^2 + \|y - v'\|^2,$$

so that $\|x - y\| \leqslant \|x - v'\|$, making y the closest point in M to x.

(ii) By the above, for any $x \in H$, $P(x)$ is that unique vector in M such that $x - P(x) \in M^\perp$. This characteristic property has the following consequences:

o P is linear since

$$(x + y) - (Px + Py) = (x - Px) + (y - Py) \in M^\perp, \qquad Px + Py \in M,$$

$$\lambda x - \lambda Px = \lambda(x - Px) \in M^\perp, \qquad\qquad\qquad \lambda Px \in M,$$

hence $P(x + y) = Px + Py$ and $P(\lambda x) = \lambda Px$.
o   The closest point in M to $v \in M$ is v itself, i.e., $Pv = v$, so im $P = M$.
o   When $x \in M^\perp$, then $x - 0 \in M^\perp$ and $0 \in M$ so $Px = 0$, i.e., $M^\perp \subseteq \ker P$.

$P^2 = P$ since for any x, $Px \in M$ and so $P^2 x = Px$. P is continuous with $\|P\| = 1$ since $\|x\|^2 = \|x - Px\|^2 + \|Px\|^2$ by Pythagoras' theorem so that $\|Px\| \leqslant \|x\|$; moreover $\|Pv\| = \|v\|$ when $v \in M$.
    $\ker P = M^\perp$ since $Px = 0$ implies $x = x - Px \in M^\perp$.
    Finally, $H = \text{im } P \oplus \ker P = M \oplus M^\perp$, since any vector can be decomposed as $x = Px + (x - Px)$, and $M \cap M^\perp = 0$.   □

Corollary 10.13

For any subset $A \subseteq H$, $A^{\perp\perp} = \overline{[\![A]\!]}$.

Proof Let M be a closed linear subspace of a Hilbert space H. By Proposition 10.9, $M \subseteq M^{\perp\perp}$, so we require the opposite inclusion. Let $x \in M^{\perp\perp}$, then $x = u + v$ where $u \in M$ and $v \in M^{\perp}$, and

$$0 = \langle v, x \rangle = \langle v, u \rangle + \langle v, v \rangle = \|v\|^2,$$

forcing $v = 0$ and $x \in M$; thus $M^{\perp\perp} \subseteq M$. In particular, $A^{\perp\perp} = \overline{[\![A]\!]}^{\perp\perp} = \overline{[\![A]\!]}$.

□

Note that $M^{\perp} = 0 \iff M = M^{\perp\perp} = 0^{\perp} = H$, answering Exercise 10.10(12) in the case of a closed linear subspace of a Hilbert space.

Examples 10.14

1. Let $M := \{ f \in L^2[0, 1] : \int_0^1 f = 0 \}$. To find that function f_0 in M which most closely approximates a given function g, we first note

$$M = \{ f \in L^2[0, 1] : \langle 1, f \rangle = 0 \} = \{1\}^{\perp}, \quad \text{so } M^{\perp} = [\![1]\!].$$

Then f_0 must satisfy $f_0 \in M$ and $g - f_0 \in M^{\perp}$, i.e., $f_0 = g + \lambda$ and $0 = \int_0^1 f_0 = \int_0^1 g + \lambda$, hence $f_0 = g - \int_0^1 g$.
2. The "affine" projection onto a plane with equation $x \cdot n = d$ (n a unit vector) is given by $P(x) := x + (d - x \cdot n)n$.
   Proof: Translate all points $x \mapsto y := x - dn$, so that the plane becomes the subspace M with equation $y \cdot n = 0$, i.e., $M = \{n\}^{\perp}$. The required point satisfies $(y - y_0) \cdot \tilde{y} = 0$ for all $\tilde{y} \in M$, so $y_0 = y + tn$. Dotting with n implies $t = -y \cdot n = d - x \cdot n$, which can be substituted into $x_0 = x + tn$.
3. A projection is orthogonal if, and only if, $\|P\| = 1$ or 0.
   Proof: Using $\langle x - Px, Px \rangle = 0$ and the Cauchy-Schwarz inequality,

$$\|Px\|^2 = \langle x, Px \rangle \leqslant \|x\| \|Px\|,$$

   so $\|Px\| \leqslant \|x\|$; but $Px = x$ for $x \in \operatorname{im} P$, so $\|P\| = 1$ (unless $P = 0$). Conversely, let $u \in \ker P$, $v \in \operatorname{im} P$; then for any λ,

$$\|v\|^2 = \|P(\lambda u + v)\|^2 \leqslant \|\lambda u + v\|^2 = |\lambda|^2 \|u\|^2 + 2\operatorname{Re}\lambda\langle v, u \rangle + \|v\|^2$$

   and after letting $\lambda = |\lambda|e^{i\theta}$ with $|\lambda| \to 0$, we find $\operatorname{Re} e^{i\theta}\langle v, u \rangle \geqslant 0$ for any θ, hence $\langle v, u \rangle = 0$.
4. ▶ $[\![A]\!]$ is dense in H if, and only if, $A^{\perp} = 0$.
   Proof: If $A^{\perp} = 0$, then $\overline{[\![A]\!]} = A^{\perp\perp} = 0^{\perp} = H$. Conversely, if A is dense in H, then $A^{\perp} = \overline{[\![A]\!]}^{\perp} = H^{\perp} = 0$.

Least Squares Approximation

A common problem in mathematical applications is to approximate a generic vector x by one which is more easily handled, such as a linear combination of simpler vectors $y_1, \ldots, y_n$. For Hilbert spaces, there is a guarantee that a unique closest approximation exists, and this lies at the heart of the method of least squares.

Let $M := [\![y_1, \ldots, y_n]\!]$, a closed linear subspace of H; then the closest point in M to x is $x_* = \sum_{j=1}^{n} \alpha_j y_j$ such that $x - x_* \perp M$. Since M is generated by $y_1, \ldots, y_n$, this is equivalent to

$$\langle y_i, x - x_* \rangle = 0, \quad i = 1, \ldots, n,$$

$$\therefore \quad \langle y_i, x \rangle = \langle y_i, x_* \rangle = \sum_{j=1}^{n} \langle y_i, y_j \rangle \alpha_j.$$

These n linear equations in the n unknowns $\alpha_1, \ldots, \alpha_n$, can be recast in matrix form,

$$\begin{pmatrix} \langle y_1, y_1 \rangle & \cdots & \langle y_1, y_n \rangle \\ \vdots & & \vdots \\ \langle y_n, y_1 \rangle & \cdots & \langle y_n, y_n \rangle \end{pmatrix} \begin{pmatrix} \alpha_1 \\ \vdots \\ \alpha_n \end{pmatrix} = \begin{pmatrix} \langle y_1, x \rangle \\ \vdots \\ \langle y_n, x \rangle \end{pmatrix}.$$

Given x, the coefficients α_i can be found by solving these equations. The *Gram* matrix $[\langle y_i, y_j \rangle]$, and possibly its inverse, need only be calculated once, and used to approximate other points.

Example The space of cubic polynomials, $a + bt + ct^2 + dt^3$, is a four-dimensional closed linear subspace of the Hilbert space $L^2[0, 1]$, with basis $1, t, t^2, t^3$. Their Gram matrix and inverse are given by

$$\begin{pmatrix} 1 & \frac{1}{2} & \frac{1}{3} & \frac{1}{4} \\ \frac{1}{2} & \frac{1}{3} & \frac{1}{4} & \frac{1}{5} \\ \frac{1}{3} & \frac{1}{4} & \frac{1}{5} & \frac{1}{6} \\ \frac{1}{4} & \frac{1}{5} & \frac{1}{6} & \frac{1}{7} \end{pmatrix}^{-1} = \begin{pmatrix} 16 & -120 & 240 & -140 \\ -120 & 1200 & -2700 & 1680 \\ 240 & -2700 & 6480 & -4200 \\ -140 & 1680 & -4200 & 2800 \end{pmatrix}.$$

So, to approximate the sine function by a cubic polynomial over the region $[0, 1]$, we first calculate $\langle t^i, \sin t \rangle_{L^2[0,1]}$, which work out to $(0.460, 0.301, 0.223, 0.177)$, and then apply the inverse of the Gram matrix to it, giving

$$p(t) \approx -0.000253 + 1.005t - 0.0191t^2 - 0.144t^3.$$

Notice that the coefficients are close to, but not the same as, the first terms of the MacLaurin expansion of sine. The difference is that, whereas the MacLaurin expansion is accurate at 0 and becomes progressively worse away from it, the L^2-approximation balances out the 'root-mean-square error' throughout the region $[0, 1]$.

Exercises 10.15

1. Find the closest point in the plane $2x + y - 3z = 0$ to a point $x \in \mathbb{R}^3$.
   (Hint: Find $M^\perp$.)

2. Let (i) $M := [\![y]\!]$, or (ii) $M := \{y\}^\perp$, where y is a unit vector. The orthogonal projection P which maps any point x to its closest point in M is (i) $Px = \langle y, x \rangle y$, (ii) $Px = x - \langle y, x \rangle y$.

3. ▶ In the decomposition $x = u + v$ with $u \in M$ and $v \in M^\perp$, u and v are unique. Deduce that if $H = M \oplus N$, where M is a closed linear subspace and $M \perp N$, then $N = M^\perp$.

4. Let $v + M$ be a coset of a closed linear subspace M. Show that there is a unique vector $x \in v + M$ with smallest norm. (Hint: this is equivalent to finding the closest vector in M to $-v$.) Deduce that Riesz's lemma (Proposition 8.22) continues to hold in a Hilbert space even when $c = 1$.

5. If $M \subseteq N$ are both closed linear subspaces, then $M \oplus (M^\perp \cap N) = N$.

6. Let T be a square matrix, and suppose both subspaces M and $M^\perp$ are T-invariant, so that T takes the schematic form $\begin{pmatrix} A & 0 \\ 0 & B \end{pmatrix}$ on $M \oplus M^\perp$. Show that $\|T\| = \max(\|A\|, \|B\|)$. (Hint: If $x = u + v$, then $\|Tx\|^2 = \|Tu\|^2 + \|Tv\|^2$.)

7. ▶ There is a 1–1 correspondence between closed linear subspaces of a Hilbert space and orthogonal projections (onto them). Properties about subspaces are reflected as properties of the projections, e.g., if the orthogonal projections P_M and P_N project onto M and N respectively, then

   (a) $M \subseteq N \Leftrightarrow P_N P_M = P_M \Leftrightarrow P_M P_N = P_M$,
   (b) $M \perp N \Leftrightarrow P_M P_N = 0 \Leftrightarrow P_N P_M = 0$,
   (c) $N = M^\perp \Leftrightarrow I = P_M + P_N$,
   (d) $\operatorname{im} T \subseteq M \Leftrightarrow T = P_M T$, and $M \subseteq \ker T \Leftrightarrow T P_M = 0$,
   (e) M is T-invariant $\Leftrightarrow T P_M = P_M T P_M$,
   (f) M and $M^\perp$ are both T-invariant $\Leftrightarrow T P_M = P_M T$,

8. (a) Let P be a projection onto a closed linear subspace M. Since $\langle x, v \rangle = \langle Px, v \rangle$ for $v \in M$, it follows that $|\langle x, v \rangle| \leqslant \|Px\| \|v\|$. Deduce that in a real Hilbert space, the angle between x and v is at least $\cos^{-1}(\|Px\|/\|x\|)$.

   (b) Let $H = M \oplus N$ with M, N non-zero closed subspaces. Show that there is a minimum distance $d > 0$ between the disjoint closed sets $S_H \cap M$ and $S_H \cap N$, where S_H is the sphere of unit vectors. Thus for any unit vectors $u \in M, v \in N$, $\|u - v\| \geqslant d > 0$. Deduce that $\operatorname{Re} \langle u, v \rangle \leqslant \alpha := 1 - d^2/2$, and hence that

   $$\forall x \in M, \ \forall y \in N, \qquad |\langle x, y \rangle| \leqslant \alpha \|x\| \|y\|.$$

9. The main theorem, which does not refer to inner products, is *not* true in Banach spaces in general.

   (a) In $\mathbb{R}^2$ with the 1-norm, the vector $\binom{1}{1}$ has many closest vectors in the closed ball $\overline{B_1(0)}$.
   (b) In ℓ^∞, there are many sequences in c_0 that have the minimum distance to $(1, 1, \ldots)$.
   (c) Show that, for a convex subset M of a normed space, the set of best approximations to a point x, $\{\, y \in M : \|x - y\| = d(x, M)\,\}$, is convex.
   (d) $\star$ On the other hand, in ℓ^∞, the sequence $\mathbf{0}$ has no closest sequence in the closed convex set $M := \{\, (a_n)_{n \in \mathbb{N}} \in c_0 : \sum_n a_n / 2^n = 1 \,\}$.

10. $\star$ Consider two orthogonal projections P and Q in $\mathbb{R}^N$. Show that the iteration $y_{n+1} := QP y_n$ starting from $y_0 = x$ converges to a point $x_* \in \operatorname{im} P \cap \operatorname{im} Q$.

11. Find

   (a) the best-fitting quadratic and cubic polynomials to the sine function in $[0, 2\pi]$,
   (b) the linear combination of sin and cos which is closest to $1 - t^3$ in $L^2[0, 1]$.

12. (a) The Gram matrix of vectors $y_1, \ldots, y_n$ is $G := A^*A$ where the columns of A are y_j, and the rows of A^* are $\overline{y}_i^\top$. It is invertible when y_j are linearly independent.
   (b) Show that in order to write a vector x as a linear combination of basis vectors $x = \sum_{j=1}^n \alpha_j y_j$, given the numbers $b_j := \langle y_j, x \rangle$, then one needs to solve the matrix equation $G\alpha = b$.
   (c) Given the total mass and moment of inertia of a radially symmetric planar object,

$$M = 2\pi \int_0^R \rho(r) r \, \mathrm{d}r = 2\pi \, \langle r, \rho(r) \rangle_{L^2[0, R]},$$

$$I = 2\pi \int_0^R \rho(r) r^3 \, \mathrm{d}r = 2\pi \, \langle r^3, \rho(r) \rangle_{L^2[0, R]},$$

   find an estimate of $\rho(r)$ as some function $\alpha + \beta r$.

13. $\star$ The symmetric Gram matrix of a set of vectors $x_n \in \mathbb{R}^N$ is useful in other contexts as well. Show how to recover

   (a) the vectors x_n from their Gram matrix, up to an isomorphism (use diagonalization to find A such that $A^2 = G$),
   (b) the Gram matrix of the vectors from the mutual distances between vectors d_{ij}, and their norms r_i,
   (c) the Gram matrix from d_{ij} only, assuming $\sum_n x_n = \mathbf{0}$.

This is essentially what is done in the *Global Positioning System*, when 3–4 distances obtained by time-lags from satellites are converted to a position.

Frigyes Riesz (1880–1956) Riesz was a Hungarian mathematics professor who proved that $L^2(\mathbb{R})$ is complete; in 1907, with E.S. Fischer, he proved that Hilbert's ℓ^2 space is equivalent to $L^2(\mathbb{R})$; he defined compact operators abstractly for more general spaces, including $C[a, b]$ (1918); he introduced the resolvent projection to part of the spectrum and thus $f(T)$ for compact operators.

10.3 Duality $H^* \approx H$

An inner product is a function acting on two variables. But if one input vector is fixed, it becomes a scalar-valued function on vectors, indeed a continuous functional

$$x^* : X \to \mathbb{F}$$

$$y \mapsto \langle x, y \rangle.$$

This is linear by the inner product axioms, while continuity follows from the Cauchy-Schwarz inequality $|x^*y| = |\langle x, y \rangle| \leqslant \|x\| \|y\|$.

Are there any other functionals besides these? Not when the space is complete:

Theorem 10.16 (Riesz Representation Theorem)

Every continuous functional of a Hilbert space H is of the form $x^* := \langle x, \cdot \rangle$,

$$\forall \phi \in H^*, \ \exists! x \in H, \quad \phi = \langle x, \cdot \rangle.$$

The Riesz map

$$J : H \to H^*$$

$$x \mapsto x^*$$

is a bijective conjugate-linear isometry.

Proof (i) Given $\phi \in H^*$, first notice that for any z and y in H,

$$(\phi y)z - (\phi z)y \in \ker \phi.$$

Assuming $\phi \neq 0$, pick a unit vector $z \perp \ker \phi$; this is possible since $\ker \phi \neq H$, so $(\ker \phi)^\perp \neq 0$. Then

$$0 = \langle z, (\phi y)z - (\phi z)y \rangle = (\phi y) - (\phi z)\langle z, y \rangle,$$

$$\therefore \ \phi y = (\phi z)\langle z, y \rangle = \langle x, y \rangle,$$

where $x = (\overline{\phi z})z$. To show that it is unique, suppose $\tilde{x}$ is another such x, then

$$\forall y \in H, \quad \langle x - \tilde{x}, y \rangle = \langle x, y \rangle - \langle \tilde{x}, y \rangle = \phi y - \phi y = 0 \ \Leftrightarrow \ x = \tilde{x}.$$

These considerations prove that J is onto and 1–1.

(ii) Let x and y be two vectors in H. Then for any $z \in H$,

$$(x + y)^*(z) = \langle x + y, z \rangle = \langle x, z \rangle + \langle y, z \rangle = x^*z + y^*z,$$

$$(\lambda x)^*(z) = \langle \lambda x, z \rangle = \bar{\lambda}\langle x, z \rangle = \bar{\lambda}x^*z,$$

showing that $(x + y)^* = x^* + y^*$ and $(\lambda x)^* = \bar{\lambda}x^*$ (conjugate-linear).

J is isometric: Note that

$$\|x^*\|_{H^*} = \sup_{y \neq 0} \frac{|x^*y|}{\|y\|} = \sup_{y \neq 0} \frac{|\langle x, y \rangle|}{\|y\|} = \|x\|,$$

using the Cauchy-Schwarz inequality, in particular with $y = x$. $\square$

Examples 10.17

1. ▶ For $T \in B(X, Y)$ (X, Y Hilbert spaces), $\|T\| = \sup_{\|x\|=1=\|y\|} |\langle y, Tx \rangle|$.

2. The dual space of $\mathbb{R}$ is (isomorphic to) $\mathbb{R}$ itself. Any $\phi : \mathbb{R} \to \mathbb{R}$ that is linear must be of type $\phi(t) = \lambda t$ where $\lambda \in \mathbb{R}$.

3. Functionals are simply row vectors when $H = \mathbb{C}^n$; thus H^* is isometric to $\mathbb{C}^n$ and is generated by the *dual* basis $e_1^\top, \ldots, e_n^\top$.
   Proof: Let $e_1, \ldots, e_n$, be the standard basis for $\mathbb{C}^n$. Then every functional ϕ in $(\mathbb{C}^n)^*$ is of the type $\phi = (b_i)^\top$, where $b_i := \phi e_i$ (Example 8.3(3)). Thus the map $\mathbb{C}^n \to (\mathbb{C}^n)^*$, $y \mapsto y^\top$, where $y^\top x := y \cdot x$, is onto; it is easily seen to be linear, and continuous from Cauchy's inequality $|y \cdot x| \leqslant \|y\|\|x\|$. In fact $\|y^\top\| = \|y\|$ (using $x = (\bar{b}_i)_{i=1}^n$). Note that $y^\top = \sum_{i=1}^n b_n e_i^\top$, and $e_i^\top e_j = \delta_{ij}$.

4. It was noted previously that $\ell^{2*} \equiv \ell^2$ and $L^2(\mathbb{R})^* \equiv L^2(\mathbb{R})$ (Theorem 9.25). These are special cases of the Riesz correspondence.

The Adjoint Map T^*

We now seek to find a generalization of the transpose operation on matrices. In finite dimensions, we have $(A^*v)^* = v^*A$; in terms of inner products, this becomes $\langle A^*v, x \rangle = \langle v, Ax \rangle$. In this form, it can be generalized to any Hilbert space:

Definition 10.18

> The (Hilbert) **adjoint** of an operator $T : X \to Y$ between Hilbert spaces, is the operator $T^* : Y \to X$ uniquely defined by the relation
>
> $$\forall x \in X, \ \forall y \in Y, \qquad \langle T^*y, x \rangle_X = \langle y, Tx \rangle_Y.$$

That T^*y is uniquely defined follows from the Riesz correspondence applied to the functional $x \mapsto \langle y, Tx \rangle$. Linearity and continuity of T^* follow from

$$\langle T^*(y_1 + y_2), x \rangle = \langle y_1 + y_2, Tx \rangle = \langle y_1, Tx \rangle + \langle y_2, Tx \rangle = \langle T^*y_1 + T^*y_2, x \rangle$$

$$\langle T^*(\lambda y), x \rangle = \langle \lambda y, Tx \rangle = \bar{\lambda}\langle y, Tx \rangle = \langle \lambda T^*y, x \rangle$$

$$\|T^*\| = \sup_{\|y\|=1=\|x\|} |\langle T^*y, x \rangle| = \sup_{\|y\|=1=\|x\|} |\langle y, Tx \rangle| = \|T\|$$

The properties of the adjoint map are:

Proposition 10.19

> $$(S + T)^* = S^* + T^*, \qquad (\lambda T)^* = \bar{\lambda}T^*, \qquad (ST)^* = T^*S^*,$$
>
> $$I^* = I, \qquad T^{**} = T, \qquad \|T^*T\| = \|T\|^2$$

Proof These assertions follow from the following identities, valid for all $x \in X$, $y \in Y$:

$$\langle (S + T)^*y, x \rangle = \langle y, (S + T)x \rangle = \langle y, Sx \rangle + \langle y, Tx \rangle = \langle (S^* + T^*)y, x \rangle$$

$$\langle (\lambda T)^*y, x \rangle = \langle y, \lambda Tx \rangle = \lambda\langle T^*y, x \rangle = \langle \bar{\lambda}T^*y, x \rangle$$

$$\langle (ST)^*y, x \rangle = \langle y, STx \rangle = \langle S^*y, Tx \rangle = \langle T^*S^*y, x \rangle$$

$$\langle I^*y, x \rangle = \langle y, Ix \rangle = \langle y, x \rangle = \langle Iy, x \rangle$$

$$\langle y, T^{**}x \rangle = \overline{\langle T^{**}x, y \rangle} = \overline{\langle x, T^*y \rangle} = \langle T^*y, x \rangle = \langle y, Tx \rangle,$$

$$\|T^*T\| = \sup_{x,y \in S_X} |\langle y, T^*Tx \rangle| = \sup_{x,y \in S_X} |\langle Ty, Tx \rangle|$$

$$= \sup_{x,y \in S_X} \|Ty\| \|Tx\| = \|T\|^2,$$

where $S_X := \{ x : \|x\| = 1 \}$, and the equation before the last is valid by the Cauchy-Schwarz inequality, in particular choosing $y = x$.   □

The following proposition reveals an orthogonality between subspaces of adjoint operators. In particular, both M and $M^\perp$ are T-invariant if, and only if, M is T- and T^*-invariant.

Proposition 10.20

For an operator T on Hilbert spaces,

$$\ker T^* = (\operatorname{im} T)^\perp, \quad \overline{\operatorname{im} T^*} = (\ker T)^\perp.$$

If $T \in B(H)$ and M is a closed linear subspace of H,

$$M \text{ is } T\text{-invariant} \iff M^\perp \text{ is } T^*\text{-invariant}.$$

Proof The definition $\langle x, Ty \rangle = \langle T^*x, y \rangle$ implies that

$$x \perp Ty \iff T^*x \perp y,$$

in particular $x \perp \operatorname{im} T \iff T^*x \perp Y \iff x \in \ker T^*$. Consequently, $\ker T^* = (\operatorname{im} T)^\perp$ and thus $\ker T = \ker T^{**} = (\operatorname{im} T^*)^\perp$; furthermore,

$$(\ker T)^\perp = (\operatorname{im} T^*)^{\perp\perp} = \overline{\operatorname{im} T^*}.$$

Suppose M is T-invariant, and let $x \in M^\perp$, $y \in M$, then $\langle T^*x, y \rangle = \langle x, Ty \rangle = 0$, and $T^*x \in M^\perp$. Conversely, if $M^\perp$ is T^*-invariant then $M^{\perp\perp}$ is T^{**}-invariant; but $T^{**} = T$ and $M^{\perp\perp} = M$ for a closed subspace M.   □

Unitary Operators

Definition 10.21

> A **unitary isomorphism** $J : X \to Y$ of inner product spaces is defined as a map which preserves the structure of an inner product space, namely
>
> J is bijective   (preserves the elements),
> J is linear   (preserves vector addition and scalar multiplication),
> $\langle Jx, Jy \rangle_Y = \langle x, y \rangle_X$ (preserves the inner product).

It is obvious that a unitary isomorphism preserves the induced norm (an isometry); the converse is also partly true in Hilbert spaces, because, by the polarization identity, the inner product can be written in terms of norms:

Proposition 10.22

> **An operator $U \in B(X, Y)$ on Hilbert spaces preserves the inner product when U preserves the norm,**
>
> $$\forall x, \tilde{x} \in X, \quad \langle Ux, U\tilde{x} \rangle = \langle x, \tilde{x} \rangle \iff U^*U = I$$
>
> $$\iff \|Ux\| = \|x\| \quad \forall x \in X.$$
>
> **U is unitary when it is also surjective.**

This statement basically says that preserving the inner product (lengths *and* 'angles') is equivalent to preserving lengths.

Proof The first equivalence is trivial

$$\forall x, \tilde{x}, \ \langle x, \tilde{x} \rangle = \langle Ux, U\tilde{x} \rangle = \langle x, U^*U\tilde{x} \rangle \iff U^*U = I.$$

In particular (taking $\tilde{x} = x$), U is isometric. The converse implication from the third statement to the first follows from the polarization identity (10.1),

$$\langle Ux, Uy \rangle = \tfrac{1}{4}(\|Ux + Uy\| + \cdots) = \tfrac{1}{4}(\|x + y\| + \cdots) = \langle x, y \rangle.$$

A superficially different proof of this last fact can be given for complex Hilbert spaces (Example 10.7(3)),

$$\forall x, \ \langle x, x \rangle = \langle Ux, Ux \rangle = \langle x, U^*Ux \rangle \iff U^*U = I.$$

Since isometries are 1–1, we need only require in addition that it is onto for U to be invertible, in which case $U^{-1} = U^*$.   □

Examples 10.23

1. The adjoint of a matrix $A = [A_{ij}]$, with respect to the standard inner product, is the conjugate of its transpose, $\bar{A}^\top$, since

$$\langle x, Ay \rangle = \sum_{i,j} \bar{a}_i A_{ij} b_j = \sum_j \overline{(\sum_i \bar{A}_{ij} a_i)} b_j = \langle \bar{A}^\top x, y \rangle.$$

2. ▶ The adjoint of the left-shift operator (on ℓ^2) is the right-shift, $L^* = R$, since

$$\langle L^* y, x \rangle = \langle y, Lx \rangle = \sum_{n=0}^{\infty} \bar{b}_n a_{n+1} = \sum_{n=1}^{\infty} \bar{b}_{n-1} a_n = \langle Ry, x \rangle$$

and $R^* = L^{**} = L$.

3. The adjoint of an integral operator on $L^2(\mathbb{R})$,

$$Tf(s) := \int k(s,t) f(t) \, dt \quad \text{is} \quad T^* g(t) = \int \overline{k(s,t)} g(s) \, ds.$$

Proof:
$$\langle g, Tf \rangle = \int \overline{g(s)} \int k(s,t) f(t) \, dt \, ds$$

$$= \iint k(s,t) \overline{g(s)} f(t) \, ds \, dt$$

$$= \int \overline{\int \overline{k(s,t)} g(s) \, ds} \; f(t) \, dt = \langle T^* g, f \rangle.$$

4. The unitary[2] isomorphisms of $\mathbb{R}^2$ are the rotations and reflections. More generally, those of $\mathbb{C}^n$ are the matrices whose columns are orthonormal (mutually orthogonal and of unit norm).
   Proof: The column vectors u_i of a unitary matrix U satisfy $u_i = U e_i$, where e_i are the standard basis for $\mathbb{C}^n$. Then, $\langle u_i, u_j \rangle = \langle U e_i, U e_j \rangle = \langle e_i, e_j \rangle = \delta_{ij}$.

5. ▶ By itself, $U^* U = I$ ensures that a linear operator $U : X \to Y$ is isometric (and 1–1), but not that it is onto, that is, it is an isometric embedding of X into Y. For example, the matrix $\begin{pmatrix} 0 & 1 \\ 1 & 0 \\ 0 & 0 \end{pmatrix}$ embeds $\mathbb{R}^2$ into $\mathbb{R}^3$. In general, UU^* is not equal to I but is a projection of Y onto $\operatorname{im} U \subseteq Y$.
   Proof: Clearly, $UU^* UU^* = UU^*$ is a projection from Y to $\operatorname{im} U$. It is onto since $UU^*(Ux) = Ux$.

---

[2] More properly called *orthogonal* isomorphisms when the space is real.

Exercises 10.24

1. The norm of H^* is induced from an inner product, $\langle x^*, y^* \rangle_{H^*} := \langle y, x \rangle_H$. Then the Riesz map is "anti-unitary", that is, $\langle Jx, Jy \rangle = \overline{\langle x, y \rangle}$.

2. A functional $\phi \in H^*$ corresponds to some vector $x \in H$; if M is a closed linear subspace of H, ϕ can be restricted to act on it, $\tilde{\phi} \in M^*$. As M is a Hilbert space in its own right, what vector $v \in M$ corresponds to $\tilde{\phi}$?

3. A second inner product on H which satisfies $|\langle\!\langle x, y \rangle\!\rangle| \leqslant c\|x\|\|y\|$ must be of the type $\langle\!\langle x, y \rangle\!\rangle = \langle Tx, y \rangle = \langle x, Ty \rangle$, where $T \in B(H)$, $\|T\| \leqslant c$.

4. Riesz's representation theorem holds only for complete inner product spaces (it is false for, say, $c_{00} \subset \ell^2$). Where is completeness used in the proof of the theorem?

5. If T is invertible then $(T^{-1})^* = (T^*)^{-1}$.

6. Use $\|T^*T\| = \|T\|^2$ to show $\|T^*\| = \|T\|$.

7. ▶ The adjoint of the multiplier operator in ℓ^2, $x \mapsto ax$, is $y \mapsto \bar{a}y$.

8. Let $a \in \ell^1(\mathbb{Z})$, then Young's inequality (Exercise 9.17(10)) shows that the linear map $x \mapsto a*x$ is continuous on $\ell^2(\mathbb{Z})$. Its adjoint is given by $y \mapsto a^\dagger * y$ where $(a_n)^\dagger := (\bar{a}_{-n})$.

9. The Volterra operator on $L^2[0, 1]$, $Vf(t) := \int_0^t f$, has adjoint $V^*f(t) = \int_t^1 f$.

10. Let $\langle\!\langle x, y \rangle\!\rangle := \langle x, Sy \rangle$ be a new inner product ($S^* = S$), then the adjoint of T with respect to it is $T^\star := S^{-1}T^*S$.

11. If $R \in B(X, Y)$ then $T \mapsto RTR^*$ is an operator $B(X) \to B(Y)$.

12. For any $T \in B(H_1, H_2)$, $\ker(T^*T) = \ker T$ and $\overline{\operatorname{im} T^*T} = \overline{\operatorname{im} T^*}$.

13. A linear map $T : X \to Y$ is said to be *conformal* when it preserves orthogonality,

$$\forall x, \tilde{x} \in X, \qquad \langle x, \tilde{x} \rangle = 0 \Leftrightarrow \langle Tx, T\tilde{x} \rangle = 0.$$

Show that this is the case if, and only if, $T = \lambda U$ for some $\lambda \geqslant 0$, U unitary. Moreover, angles between vectors are preserved (for $\lambda > 0$).

In particular, two inner products on the same vector space are conformal when $\langle\!\langle x, y \rangle\!\rangle = \lambda \langle x, y \rangle$ for some $\lambda > 0$.

14. ⋆ Show that a map between Hilbert spaces which preserves the inner product must be linear. Deduce that isometries on a real Hilbert space must be of the type $f(x) = Ux + v$ where $U^*U = I$ and $v \in H$.

(Hint: Let $g(x) := f(x) - f(0)$, an isometry; show $\langle g(x + y), g(z) \rangle = \langle g(x) + g(y), g(z) \rangle$, so $g(x + y) - g(x) - g(y) \in [\![\operatorname{im} g]\!] \cap (\operatorname{im} g)^\perp$.)

10.4   Inverse Problems

When an operator $T : X \to Y$ is not surjective, the equation $Tx = y$ need not have a solution. The next best thing to ask for is a vector x which *minimizes* $\|Tx - y\|$.

Proposition 10.25

> **For an operator $T : H_1 \to H_2$ between Hilbert spaces and a vector $y \in H_2$, a vector $x \in H_1$ minimizes $\|Tx - y\|$ if, and only if**
>
> $$T^*Tx = T^*y.$$

Proof Suppose $T \in B(X, Y)$, and consider the closed linear subspace $M := \overline{\operatorname{im} T} \subseteq Y$. For each $y \in Y$, there is a unique vector $y_* \in M$ which is closest to it. As proved in Theorem 10.12, a necessary and sufficient condition for $v \in M$ to be y_* is $y - v \in M^\perp = \ker T^*$ (Proposition 10.20), that is, $T^*v = T^*y$. If y_* happens to be in $\operatorname{im} T$, i.e., $y_* = Tx$, then the equation becomes $T^*Tx = T^*y$. □

To continue this discussion, y_* is in $\operatorname{im} T$ only when $y \in \operatorname{im} T \oplus (\operatorname{im} T)^\perp$, a dense subspace of Y. When $\operatorname{im} T$ is closed, e.g., in finite dimensions, this is the case for all $y \in Y$. If $y_* \notin \operatorname{im} T$ then we can only conclude that there is some sequence of vectors $x_n \in X$ such that $Tx_n \to y_*$, and so $T^*Tx_n \to T^*y$. Thus $\|Tx_n - y\|$ converges to $\|y_* - y\|$, but is never equal to it (by uniqueness of y_*).

In the case of finite dimensions, the above situation is typical of an *overdetermined system* of equations, that is, a system $Tx = y$ that represents more equations than there are unknowns. The least squares solution is then found to be

$$x = (T^*T)^{-1}T^*y$$

at least in the generic case when T is 1–1. Then T^*T is also 1–1 since $T^*Tx = 0 \Leftrightarrow \|Tx\|^2 = \langle x, T^*Tx \rangle = 0 \Leftrightarrow x = 0$, so it is invertible at least on $\operatorname{im} T^*$.

The dual problem is that of an *underdetermined system* of equations, $Tx = y$, where there are less equations than unknowns. There is an oversupply of solutions, namely any vector in $x_0 + \ker T$, where x_0 is any single solution of the equation, and $\ker(T^*T) = \ker T \neq 0$. In this case, a unique x that is closest to 0 can be selected from all these solutions, i.e., has the least norm. That is, we seek $x \in (\ker T)^\perp = \operatorname{im} T^*$ (in finite dimensions, every subspace is closed). Thus $x = T^*v$ and $y = Tx = TT^*v$, so the required least norm vector is

$$x = T^*(TT^*)^{-1}y.$$

In the general case, an operator need be neither 1–1 nor onto, so the set of vectors which minimize $\|Tx - y\|$ is a coset, $x + \ker T$. But since $\ker T$ is a closed subspace, it has a unique vector with smallest norm. The mapping from y to this $x \in (\ker T)^{\perp}$ is then well-defined for $y \in \operatorname{im} T + (\operatorname{im} T)^{\perp}$ and is denoted by $T^{\dagger}$, called the Moore-Penrose *pseudo-inverse*. To recap,

$$T^{\dagger} : \operatorname{im} T + (\operatorname{im} T)^{\perp} \subseteq Y \to X,$$

$$y \mapsto x, \text{ where } T^*Tx = T^*y, \ x \in (\ker T)^{\perp}.$$

In the simple case when T is invertible, so $\operatorname{im} T = Y$, it reduces to the usual inverse $T^{\dagger} = (T^*T)^{-1}T^* = T^{-1}$. For example, every $m \times n$ matrix and every vector has a pseudo-inverse, e.g., $x^{\dagger} = x^*/\|x\|^2$, so that $x^{\dagger}x = 1$ (except that $0^{\dagger} = 0$).

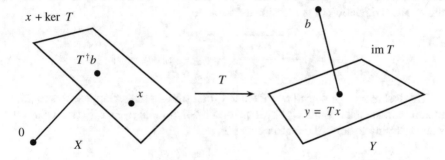

The equations introduced above have found an extremely fertile scope for applications. In many scientific or engineering contexts, an abundant number of measurements of a few variables in general gives an overdetermined system of equations. This also occurs when there is loss of information during measurement, so that the 'space of measurements' ($\operatorname{im} T$) is a proper subspace of the space of variables (H). A small sample of applications is given below:

Regression

To find the best-fitting (least-squares) line $y = mx + c$ to N given points $\binom{x_n}{y_n} \in \mathbb{R}^2$, minimizing the errors in y_n, we require that $mx_n + c$ be collectively as close to y_n as possible. In matrix form, we require

$$\begin{pmatrix} x_1 & 1 \\ \vdots & \vdots \\ x_N & 1 \end{pmatrix} \begin{pmatrix} m \\ c \end{pmatrix} = \begin{pmatrix} y_1 \\ \vdots \\ y_N \end{pmatrix},$$

written as $Am = y$. As this usually has no exact solution, the best alternative is $A^*Am = A^*y$,

$$\begin{pmatrix} x_1 & \cdots & x_N \\ 1 & \cdots & 1 \end{pmatrix} \begin{pmatrix} x_1 & 1 \\ \vdots & \vdots \\ x_N & 1 \end{pmatrix} \begin{pmatrix} m \\ c \end{pmatrix} = \begin{pmatrix} x_1 & \cdots & x_N \\ 1 & \cdots & 1 \end{pmatrix} \begin{pmatrix} y_1 \\ \vdots \\ y_N \end{pmatrix},$$

$$\begin{pmatrix} \sum_n x_n^2 & \sum_n x_n \\ \sum_n x_n & \sum_n 1 \end{pmatrix} \begin{pmatrix} m \\ c \end{pmatrix} = \begin{pmatrix} \sum_n x_n y_n \\ \sum_n y_n. \end{pmatrix}$$

Solving for $m = \begin{pmatrix} m \\ c \end{pmatrix}$ gives the usual regression line as used in statistics. Moreover, the standard deviation of the 'residuals' is

$$\sigma_y = \sqrt{\frac{\sum_{j=1}^n (y_j - \hat{y}_j)^2}{n}} = \frac{1}{\sqrt{n}} \|Am - y\|.$$

This technique is not at all restricted to fitting straight lines. Suppose it is required to approximate data points $\begin{pmatrix} x_n \\ y_n \end{pmatrix}$ by a quadratic polynomial $a + bx + cx^2$. This is the same as trying to solve the matrix equation

$$\begin{pmatrix} 1 & x_1 & x_1^2 \\ 1 & x_2 & x_2^2 \\ & \vdots & \\ 1 & x_N & x_N^2 \end{pmatrix} \begin{pmatrix} a \\ b \\ c \end{pmatrix} = \begin{pmatrix} y_1 \\ y_2 \\ \vdots \\ y_N \end{pmatrix}.$$

Repeating the above procedure gives the solution

$$\begin{pmatrix} a \\ b \\ c \end{pmatrix} = \frac{1}{\Delta} \begin{pmatrix} (S_2^2 - S_1 S_3) \sum_n x_n^2 y_n + (S_1 S_4 - S_2 S_3) \sum_n x_n y_n + (S_3^2 - S_2 S_4) \sum_n y_n \\ (S_0 S_3 - S_1 S_2) \sum_n x_n^2 y_n + (S_2^2 - S_0 S_4) \sum_n x_n y_n + (S_1 S_4 - S_2 S_3) \sum_n y_n \\ (S_1^2 - S_0 S_2) \sum_n x_n^2 y_n + (S_0 S_3 - S_1 S_2) \sum_n x_n y_n + (S_2^2 - S_1 S_3) \sum_n y_n \end{pmatrix}$$

where $S_k = \sum_n x_n^k$, and $\Delta = S_2^3 - 2S_1 S_2 S_3 + S_1^2 S_4 - S_0 S_2 S_4 + S_0 S_3^2$. (Note: In practice, one does not need to program these formulae; multiplying out T^*T as a numerical matrix and solving $T^*Tx = T^*y$ directly is usually a better option.)

In general, one may find the best parameters a_i in the function

$$y = a_1 f_1 + \cdots + a_k f_k$$

to fit data points (x_j, y_j), $1 \leqslant j \leqslant n$, where the functions f_i are given. The corresponding matrix equation is $A\boldsymbol{a} = \boldsymbol{y}$,

$$\begin{pmatrix} f_1(x_1) & \cdots & f_k(x_1) \\ f_1(x_2) & \cdots & f_k(x_2) \\ & \vdots & \end{pmatrix} \begin{pmatrix} a_1 \\ \vdots \\ a_k \end{pmatrix} = \begin{pmatrix} y_1 \\ y_2 \\ \vdots \end{pmatrix}$$

and the best fit parameters a_i found as above.

Tikhonov Regularization

The Moore-Penrose pseudo-inverse is usually either not a continuous operator or has a large condition number; its solutions tend to fluctuate with slight changes in the data (e.g., errors). To address this deficiency, a number of different *regularization* techniques are employed whose aim is to improve the ill-conditioning. One of the more popular techniques is attributed to Tikhonov; it balances out finding the best approximate solution of $Tx = y$ with x having a small norm by seeking the minimum of $\|Tx - y\|^2 + \alpha\|x\|^2$, where $\alpha > 0$ is some pre-determined parameter.

To solve this minimization problem, consider the following more general formulation: Let H be a real Hilbert space and suppose $A \in B(H)$, $b \in H$, and $c \in \mathbb{R}$; to find the minimum of the quadratic function $q : H \to \mathbb{R}$,

$$q(x) := \langle x, Ax \rangle + \langle b, x \rangle + c.$$

Taking small variations of the minimum point x, namely $x + tv$, we deduce

$$\forall t \in \mathbb{R}, \forall v \in H, \qquad q(x) \leqslant q(x + tv) = \langle x + tv, Ax + tAv \rangle + \langle b, x + tv \rangle + c$$

$$\therefore \ 0 \leqslant t\langle v, Ax + A^*x + b \rangle + t^2\langle v, Av \rangle,$$

$$\therefore \ \forall t > 0, \quad -t\langle v, Av \rangle \leqslant \langle v, Ax + A^*x + b \rangle \leqslant t\langle v, Av \rangle.$$

As t and v are arbitrary, it must be the case that x satisfies

$$(A + A^*)x + b = 0.$$

In particular, minimizing $\|Tx - y\|^2 = \langle x, T^*Tx \rangle - 2\langle T^*y, x \rangle + \|y\|^2$ gives the equation inferred previously, $T^*Tx = T^*y$. Similarly, that x which minimizes

$$\|Tx - y\|^2 + \alpha\|x\|^2 = \langle x, (T^*T + \alpha I)x \rangle - 2\langle T^*y, x \rangle + \|y\|^2$$

solves the equation

$$(T^*T + \alpha I)x = T^*y.$$

This is the regularized version of the last proposition. It will be proved later that $T^*T + \alpha I$ is always invertible (regular) for $\alpha > 0$ (Proposition 15.44). This gives an excellent alternative to the Moore-Penrose solution when $y \notin \operatorname{im} T + (\operatorname{im} T)^{\perp}$, although choosing the parameter α may not be straightforward.

Algebraic Reconstruction Technique

ART is an iterative algorithm that generates a solution x of the (real) equation $Ax = b$. The matrix equation can be rewritten as $\langle a_n, x \rangle = b_n$, $n = 1, \dots, N$, where a_n are the rows of A. The iteration is defined in terms of affine projections (Example 10.14(2))

$$ x_n = x_{n-1} + \frac{b_n - \langle a_n, x_{n-1} \rangle}{\|a_n\|^2} a_n, \qquad x_0 \in H. $$

The indices of a_n and b_n are to be understood as modulo N ($a_{N+1} = a_1$, etc.). We show below that starting from any $x_0 \in H$, the iteration converges to the closest point x_* to x_0 that is a solution of $Ax_* = b$. Note that starting from $x_0 = 0$ results in the Moore-Penrose inverse solution.

To see why this works, let $M_n := a_n^{\perp}$ (cycling through $n = 1, \dots, N$), then $M := \bigcap_n M_n$ contains all the solutions of $Av = 0$; let also $v_n := x_n - x_*$. The iteration becomes

$$ v_n = v_{n-1} - \langle \hat{a}_n, v_{n-1} \rangle \hat{a}_n = P_n v_{n-1} \in M_n, $$

where $\hat{a}_n = a_n / \|a_n\|$, and P_n is the projection onto the hyperplane M_n. Notice that $v_0 = x_0 - x_* \in M^{\perp}$, as well as $v_n - v_{n-1} \in M^{\perp}$, so the entire sequence v_n lies in $M^{\perp}$.

Consider the operator $Q := P_N \cdots P_1$ acting on $M^{\perp}$; its norm is bounded by 1 because $\|P_n\| \leqslant 1$ for each n. If $1 = \|Q\| = \sup_{\|w\|=1} \|Qw\|$, then the supremum is achieved by some unit vector $w \in M^{\perp}$ since the unit ball is compact in finite dimensions and $w \mapsto \|Qw\|$ is a continuous function. Denote $w_n := P_n w_{n-1} = w_{n-1} - \langle \hat{a}_n, w_{n-1} \rangle \hat{a}_n$, with $w_0 := w$; then

$$ 1 = \|Qw\| = \|P_N w_{N-1}\| \leqslant \|w_{N-1}\| \leqslant \|w_{N-2}\| \leqslant \cdots \leqslant \|w_1\| \leqslant \|w\| = 1 $$

forces all w_n to have norm 1. But, since $\|w_{n-1}\|^2 = \|w_n\|^2 + |\langle \hat{a}_n, w_{n-1} \rangle|^2$, it follows that $\langle a_n, w_{n-1} \rangle = 0$ and $w_n = w_{n-1}$ for $n = 1, \dots, N$. Hence $w \in M_1 \cap \cdots \cap M_N = M$, yet $w \in M^{\perp}$ is a unit vector.

This contradiction implies $\|Qv\| \leqslant c \|v\|$, $c < 1$, for any $v \in M^{\perp}$. Hence $\|v_{n+N}\| = \|Qv_n\| \leqslant c \|v_n\|$; combined with $\|v_{n+1}\| \leqslant \|v_n\|$, we get $v_n \to 0$. Equivalently, x_n converges to x_*.

The advantages of ART are that it uses less computer memory and is flexible in that it can be used even if there is missing data or newly available data (missing or new rows of A); but, being an iterative procedure, it is generally slower to converge.

Wiener Deconvolution

When a signal $f \in L^2(\mathbb{R})$ passes through a linear modifier (which could be a circuit, some medium such as the atmosphere, say, or a measuring apparatus), it changes in two ways: (1) the signal is distorted slightly to $Kf := k * f$, where $k \in L^1(\mathbb{R})$ is characteristic of the modifier (recall convolution, Example 9.28(3)), (2) random noise in the process adds a little error $\epsilon \in L^2(\mathbb{R})$ to the signal. The net effect is a distorted output signal $y = k * f + \epsilon$. Is it possible to extract the original signal f back again from y? A full reconstruction by solving $Kf = y$ is impossible as lost information cannot be regained; the im K subspace is not the full space $L^2(\mathbb{R})$, and the error displaces the signal off this subspace. But one can use Tikhonov regularization and solve $(K^*K + \alpha)f = K^*y$. The simplest way to do this is to use the properties of the Fourier transform, which converts convolution to multiplication. As in Example 10.23(3), the adjoint of K is given by $K^*g = k^* * g$ where $k^*(t) := \overline{k(-t)}$, since

$$\langle K^*g, f \rangle = \langle g, Kf \rangle = \int \int \overline{g(s)} k(s-t) f(t) \, dt \, ds$$

$$= \int \overline{\int \overline{k(s-t)} g(s) \, ds} \; f(t) \, dt.$$

The Fourier transform of k^* is

$$\widehat{k^*}(\xi) = \int e^{-2\pi i \xi t} \overline{k(-t)} \, dt = \overline{\int e^{2\pi i \xi t} \overline{k(t)} \, dt} = \overline{\widehat{k}(\xi)},$$

so that $(K^*K + \alpha)f = K^*y$ transforms to

$$\widehat{f} = \frac{\overline{\widehat{k}} \, \widehat{y}}{|\widehat{k}|^2 + \alpha}.$$

This is a recipe for finding f from y, called *deconvolution*, that is commonly implemented as a computer program using the Fast Fourier Transform, or directly as an electrical filter circuit.

Image Reconstruction

An image can also be considered as a 'signal', this time in $L^2(\mathbb{R}^2)$, or, when discretized, as a vector of numbers in the form of an array of pixels. Each number represents the brightness of a pixel (neglecting the color content for simplicity). An imaging apparatus transforms the original image x to $y = Ax + \epsilon$, where A is assumed to be a linear operator, as above; examples include a slight spherical aberration or blurring in general. Since such modification incurs a loss of information, the distortion matrix A is not invertible, but the best-fit "regularized" solution of $x = (A^*A + \alpha I)^{-1} A^*y$ restores the image somewhat, as seen in Fig. 10.1.

Fig. 10.1 Image reconstruction. (1) The original image, (2) after it passes an imaging device (exaggerated), (3) the best-fit image

In practice, implementing the reconstruction encounters difficulties that are specific to images. Images are typically in the order of about a million pixels in size; the matrix A would therefore consist of about a trillion coefficients (most of which are zero), and finding the inverse of $A^*A + \alpha I$ is prohibitively time-consuming. Fortunately, blurring is to a good approximation usually independent of the pixel positions; for example, a linear motion blur produces the same streaks everywhere across the picture (but note that this is not true for a rotation blur). In mathematical terms, the transformation A can be taken to be translation invariant, so that it is equivalent to the convolution by some vector $k \in H$. With this simplification, image reconstruction becomes a 2-dimensional version of Wiener deconvolution; the same technique using the Fourier transform can be applied,

$$x = \mathcal{F}^{-1} \frac{\overline{\widehat{k}}\,\widehat{y}}{|\widehat{k}|^2 + \alpha}.$$

Here, $\widehat{y}$ represents the discrete version of the Fourier transform, namely $\widehat{y}_m = \sum_n e^{-2\pi i m n} y_n$. The resulting x may have negative coefficients; these are meaningless and usually replaced by 0.

Tomography

Suppose that instead of a vector x, one is given 'views' of it, $y_n := \langle a_n, x \rangle$, where a_n is a list of known vectors: Is it possible to reconstruct x from these views? If a_n are assembled as rows of a matrix A, one obtains a matrix equation $Ax = y$. In such problems, it may be the case that the number of views is less than the dimension of the vector space, so that the system is under-determined, or that there are a large number of views, making the equation over-determined. In either case, a least-squares solution can be found as above, using the techniques of inverse problem solving (Fig. 10.2).

CT scans: An X-ray passing through a 3-D object of density f diminishes in intensity by an amount $e^{\int f(a+bt)\,dt}$ where $a + bt$ is the straight line followed by the ray. The emitted and received intensity can be measured and, after taking logs, one obtains a 'view' of the object

$$y = \int f(a + bt)\,dt = \langle L_{a,b}, f \rangle,$$

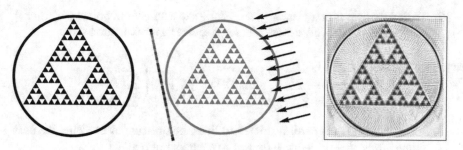

Fig. 10.2 Computed tomography. (1) The original image (360×360 pixels), (2) 80 parallel 'views' of the object, (3) the best-fit reconstruction from 6400 views (80 directions)

where $L_{a,b}$ is the characteristic function of the ray, i.e., a function that is 1 along the ray and 0 outside it (in practice, the ray has a finite width). It should be possible to reconstruct f from a large number of these views. A CT-scan does precisely this: an X-ray source coupled with a detector rotate around the object to produce these views. In one simple configuration, $b = \left(\begin{smallmatrix}\cos\theta\\\sin\theta\end{smallmatrix}\right)$ and $a = s\left(\begin{smallmatrix}-\sin\theta\\\cos\theta\end{smallmatrix}\right)$; the collection of these views, as a function of θ and s, is called the *Radon transform* R of f. The best-fit f that reproduces the data is computed by solving $(R^*R + \alpha)f = R^*y$, either directly in the form of the optimized Filtered Back Projection (FBP) algorithm or by iterative algorithms such as some variants of ART. Other configurations include a fixed source and a rotating detector, producing a fan-shaped collection of rays. In yet other applications, the 'rays' move along curved lines; more generally, the output may depend non-linearly on f and the source (see [21] for an overview of tomography and inverse scattering theory).

The idea obviously has lots of potential: X-ray tomography has revolutionized medical diagnosis, archaeology, and fossil analysis; crystal X-ray diffraction tomography recreates the atomic configuration of molecules in a lattice; impedance tomography takes output currents from input voltages to reconstruct the interior resistance density of an object; seismographs measure the output vibrations after the occurrence of earthquakes to reconstruct the interior density of the Earth; gravity, magnetic, or sound measurements at the Earth's surface can determine rock densities underneath, aiding in the exploration for oil or minerals; ultrasound echoes or scattered light can be used to reconstruct 3-D images of internal organs (or of moths and fish/squid by bats and dolphins). The list is long and increasing!

Exercises 10.26

1. Find best approximate solutions for

(i) $\begin{pmatrix}1 & 4 & 7\\2 & 5 & 8\\3 & 6 & 9\end{pmatrix} x = \begin{pmatrix}4\\-1\\0\end{pmatrix}$,   (ii) $\begin{pmatrix}1 & 4 & 7\\2 & 5 & 8\end{pmatrix} x = \begin{pmatrix}4\\-1\end{pmatrix}$.

2. To find the best-fitting plane $z = ax + by + c$ to a number of points (x_n, y_n, z_n), where z_n is the dependent variable, least squares approximation gives

$$
\begin{pmatrix} \sum_n 1 & \sum_n x_n & \sum_n y_n \\ \sum_n x_n & \sum_n x_n^2 & \sum_n x_n y_n \\ \sum_n y_n & \sum_n x_n y_n & \sum_n y_n^2 \end{pmatrix} \begin{pmatrix} c \\ a \\ b \end{pmatrix} = \begin{pmatrix} \sum_n z_n \\ \sum_n x_n z_n \\ \sum_n y_n z_n \end{pmatrix}.
$$

3. ★ The method is not at all restricted to linear geometric objects. Find the best-fitting *circle* $x^2 + y^2 + ax + by = c$ to a number of points (x_n, y_n).

4. *Weighted Regression*: Suppose, in fitting a least-squares line, that the data points are not equally significant and should be weighted. This can be achieved by a diagonal matrix of weights, S, in the inner product, that is, $\langle x, y \rangle = \bar{x}^\top S y = \sum_i w_i \bar{x}_i y_i$. Show that the new regression equation is

$$
\bar{A}^\top S A m = \bar{A}^\top S y.
$$

5. The pseudo-inverse of the left-shift operator on ℓ^2 is the right-shift operator, and vice versa.

6. For any $T \in B(X, Y)$, $T T^\dagger T = T$, because both x and $T^\dagger T x$ belong to $x +$ ker T. So $T^\dagger T$ and $T T^\dagger$ are projections; which precisely?

7. The transformation $T^\dagger : \text{im } T \oplus \text{im } T^\perp \to \text{ker } T^\perp$ is linear but continuous only when im T is closed (Hint: if $T x_n \to y$ then $T x_n = T T^\dagger T x_n \to T T^\dagger y$).

8. Recall the Volterra 1–1 operator $V f(t) := \int_0^t f$ on $L^2[0, 1]$. If g is differentiable, then $V^\dagger g = g'$, and the Tikhonov regularization solves the equation $f - \alpha f'' = g'$.

9. An oscillating pendulum is captured on video at 25 frames/s. The angle θ (in rad) that the pendulum makes with the vertical, for 1 s worth of frames (1–26), is given in the table below. Theoretically, θ satisfies

$$
\ddot{\theta} + \frac{\kappa r}{m} \dot{\theta}^2 + \frac{g}{r} \sin \theta = 0,
$$

where $g = 9.81 \text{ ms}^{-2}$ and κ/m, and r are unknown numbers. From the data, estimate $\dot{\theta}_n$ by $(\theta_{n+1} - \theta_{n-1})/2\delta t$, and $\ddot{\theta}_n$ by $(\theta_{n+1} - 2\theta_n + \theta_{n-1})/\delta t$, thereby getting equations of the type $a x_n + b y_n = z_n$, where $x_n = \dot{\theta}_n^2$, $y_n = \sin \theta_n$, $z_n = -\ddot{\theta}_n$, and a, b are unknown constants. Use regression to find a, b (hence r and κ/m) that best fit these data.

1	2	3	4	5	6	7	8	9
0.372	0.210	0.043	-0.126	-0.291	-0.447	-0.589	-0.714	-0.816
10	11	12	13	14	15	16	17	18
-0.900	-0.957	-0.988	-0.993	-0.972	-0.923	-0.854	-0.756	-0.640
19	20	21	22	23	24	25	26	
-0.505	-0.353	-0.192	-0.025	0.144	0.308	0.462	0.600	

10. *Phylogeny*: Bioinformaticians can create a score of how far apart two species are genetically. An example is given in the adjoining table, together with the suspected evolutionary tree. Assign constants to each edge in the tree which best match the given scores, i.e., the sum of the edge constants along the path from, say, A to D should be as close to 6.16 as possible.

	A	B	C	D
B	2.22	-		
C	6.12	5.60	-	
D	6.16	5.70	1.70	-
E	5.79	5.06	3.12	3.72

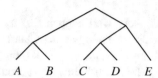

10.5   Orthonormal Bases

Definition 10.27

An **orthonormal basis** of a Hilbert space H is a set of *orthonormal* vectors E whose span is dense,

$$\forall e_i, e_j \in E, \ \langle e_i, e_j \rangle = \delta_{ij}, \qquad \overline{[\![E]\!]} = H.$$

The second condition is equivalent to $E^\perp = 0$ (Example 10.14(4)), i.e.,

$$\forall e \in E, \ \langle e, x \rangle = 0 \Leftrightarrow x = 0.$$

Examples 10.28

1. The sequences $e_n := (0, \ldots, 0, 1, 0, \ldots)$ are an orthonormal basis for ℓ^2.
   Proof: Orthonormality is obvious,

$$\langle e_n, e_m \rangle_{\ell^2} = \langle (0, \ldots, 0, \underset{\uparrow n}{1}, 0, \ldots), (0, \ldots, 0, \underset{\uparrow m}{1}, 0, \ldots) \rangle = \delta_{nm}.$$

If the sequence $x = (a_0, a_1, \ldots)$ is in $[\![e_0, e_1, \ldots]\!]^\perp$, then $a_n = \langle e_n, x \rangle_{\ell^2} = 0$ for any n; hence $x = \mathbf{0}$.

2. In finite dimensions, orthonormal bases span the space, $[\![e_1, \ldots, e_n]\!] = H$.
   In infinite dimensions, an orthonormal basis is *not* a basis in the linear algebra sense (*Hamel* basis), which requires the stronger spanning condition $[\![E]\!] = H$.

3. *Gram-Schmidt orthogonalization*: Any countable number of vectors $\{v_n\}_{n\in\mathbb{N}}$ can be replaced by a set of orthonormal vectors having the same span, using the Gram-Schmidt algorithm:

$$u_0 := v_0, \qquad\qquad\qquad e_0 := u_0/\|u_0\|$$
$$u_n := v_n - \sum_{i=0}^{n-1} \langle e_i, v_n \rangle e_i, \quad e_n := u_n/\|u_n\|.$$

It may very well happen that $u_n = 0$, in which case it and v_n are discarded and v_{n+1} relabeled as v_n. Clearly, the vectors e_n are mutually orthogonal, and $e_n \in [\![e_0, \ldots, e_{n-1}, v_n]\!]$; so, by induction, $[\![e_0, \ldots, e_n]\!] = [\![v_0, \ldots, v_n]\!]$, not taking the discarded v_n into account. Hence $\overline{[\![e_0, e_1, \ldots]\!]} = \overline{[\![v_0, v_1, \ldots]\!]}$.

4. Suppose $x = \sum_m \alpha_m e_m$ for an orthonormal basis $\{e_0, e_1, e_2, \ldots\}$; then taking the inner product with e_n gives the simple formula $\alpha_n = \langle e_n, x \rangle$. The next section discusses whether every x can be so written.

5. The set of basis vectors need not be countable; when uncountable, the Hilbert space is not separable, because the vectors e_n are equally distant from each other $\|e_n - e_m\| = \sqrt{2}$, so that the balls $B_\epsilon(e_n)$ are disjoint for $\epsilon < \sqrt{2}/2$ (Exercise 4.22(4)). Conversely, if $E = \{e_n : n \in \mathbb{N}\}$ is a countable orthonormal basis, then $[\![E]\!]$, and $H = \overline{[\![E]\!]}$, are separable.

6. ⋆ Every Hilbert space has an orthonormal basis.
   Proof: Consider the collection of all orthonormal sets of vectors. It is nonempty for a non-trivial space, so Hausdorff's maximality principle implies that there is a maximal chain of orthonormal sets E_α. But $E := \bigcup_\alpha E_\alpha$ is also an orthonormal set, for pick any two distinct vectors $e_\alpha \in E_\alpha$ and $e_\beta \in E_\beta \subseteq E_\alpha$, say, then $e_\alpha \perp e_\beta$. So E is a maximal set of orthonormal vectors. $E^\perp = 0$ otherwise E can be extended further, so $\overline{[\![E]\!]} = H$.

Fourier Expansion

The utility of orthonormal bases lies in the ease of calculation of the inner product:

Proposition 10.29 (Parseval's Identity)

If $x = \sum_{n\in\mathbb{N}} \alpha_n e_n$ **and** $y = \sum_{n\in\mathbb{N}} \beta_n e_n$, **where** $\{e_n\}_{n\in\mathbb{N}}$ **are orthonormal, then**

$$\langle x, y \rangle = \sum_{n\in\mathbb{N}} \bar{\alpha}_n \beta_n = \sum_{n\in\mathbb{N}} \langle x, e_n \rangle \langle e_n, y \rangle.$$

In particular, $\|x\| = \left(\sum_{n\in\mathbb{N}} |\alpha_n|^2 \right)^{1/2}.$

Proof A simple expansion of the two series in the inner product, making essential use of the linearity and continuity of $\langle\ ,\ \rangle$ as well as orthonormality, gives the result:

$$\langle x, y \rangle = \sum_{n\in\mathbb{N}}\sum_{m\in\mathbb{N}} \overline{\alpha_n}\beta_m \langle e_n, e_m \rangle = \sum_{n\in\mathbb{N}} \overline{\alpha_n}\beta_n.$$

$\square$

Parseval's identity is the generalization of Pythagoras' theorem to infinite dimensions. The question remains: when can a vector be written as a series of orthonormal vectors? The next proposition and theorem give an answer.

Proposition 10.30

> Let $\{e_0, e_1, e_2, \ldots\}$ be a countable orthonormal set of vectors in a Hilbert space H, then
>
> $$\sum_{n\in\mathbb{N}} \alpha_n e_n \text{ converges in } H \iff (\alpha_n)_{n\in\mathbb{N}} \in \ell^2.$$

Proof By Pythagoras' theorem we have

$$\|\alpha_n e_n + \cdots + \alpha_m e_m\|^2 = |\alpha_n|^2 + \cdots + |\alpha_m|^2.$$

This shows that $\sum_{n=1}^{N} \alpha_n e_n$ is a Cauchy sequence in H if and only if $\sum_{n=1}^{N} |\alpha_n|^2$ is Cauchy in $\mathbb{C}$ (Example 7.21(1)). Since H and ℓ^2 are complete, $\sum_n \alpha_n e_n$ converges if, and only if, $(\alpha_n)_{n\in\mathbb{N}}$ is in ℓ^2. $\square$

The convergence of $\sum_n \alpha_n e_n$ need not be absolute in infinite dimensions; for the latter to be true requires that $\sum_n \|\alpha_n e_n\| = \sum_n |\alpha_n|$ converges, that is, $(\alpha_n)_{n\in\mathbb{N}} \in \ell^1 \subset \ell^2$. Nevertheless, a rearrangement σ of an orthonormal basis does not affect the expansion, $\sum_n \alpha_n e_n = \sum_n \alpha_{\sigma(n)} e_{\sigma(n)}$, because $e_{\sigma(n)}$ remain orthonormal and $(\alpha_{\sigma(n)}) \in \ell^2$.

Theorem 10.31 (Bessel's Inequality)

> If $\{e_n\}_{n\in\mathbb{N}}$ is orthonormal in an inner product space, then
>
> $$\sum_{n\in\mathbb{N}} |\langle e_n, x \rangle|^2 \leqslant \|x\|^2.$$
>
> When $\{e_n\}_{n\in\mathbb{N}}$ is an orthonormal basis of a Hilbert space,
>
> $$x = \sum_{n\in\mathbb{N}} \langle e_n, x \rangle e_n.$$

Proof (i) Fix x and let $x_N := \sum_{n=0}^{N} \langle e_n, x \rangle e_n$. Writing $\alpha_n := \langle e_n, x \rangle$, we have

$$0 \leqslant \|x - x_N\|^2 = \|x\|^2 - \langle x_N, x \rangle - \langle x, x_N \rangle + \langle x_N, x_N \rangle$$

$$= \|x\|^2 - 2 \sum_{n=0}^{N} \bar{\alpha}_n \alpha_n + \sum_{n,m=0}^{N} \bar{\alpha}_n \alpha_m \langle e_n, e_m \rangle$$

$$= \|x\|^2 - \sum_{n=0}^{N} |\alpha_n|^2,$$

hence

$$\sum_{n=0}^{N} |\langle e_n, x \rangle|^2 \leqslant \|x\|^2. \tag{10.4}$$

As a bounded increasing series, the left-hand side must converge as $N \to \infty$, and Bessel's inequality holds.

(ii) By the previous proposition, the series $\sum_n \langle e_n, x \rangle e_n$ converges in a Hilbert space, say to $y \in H$. But $x - y \in \{e_0, e_1, e_2, \dots\}^{\perp} = 0$, since for all $N \in \mathbb{N}$,

$$\langle e_N, x - y \rangle = \langle e_N, x \rangle - \sum_{n=0}^{\infty} \langle e_n, x \rangle \langle e_N, e_n \rangle = 0.$$

A countable orthonormal basis is thus a Schauder basis. $\square$

As a matter of fact, even if $\{e_i\}_{i \in I}$ is an uncountable orthonormal set of vectors, the same analysis can be made for any finite subset of them. Inequality (10.4) then shows that there can be at most $N - 1$ vectors e_i with $|\langle e_i, x \rangle|^2 > \|x\|^2/N$, for any positive integer N, and so only a countable number of terms with $\langle e_i, x \rangle \neq 0$. Therefore $\sum_{i \in I} |\langle e_i, x \rangle|^2$ is in fact a countable sum, bounded above by $\|x\|^2$.

Proposition 10.32

Every n-dimensional Hilbert space is unitarily isomorphic to $\mathbb{R}^n$ or $\mathbb{C}^n$.

Every separable infinite-dimensional Hilbert space is unitarily isomorphic to ℓ^2 (real or complex).

Joseph Fourier (1768–1830) A Napoleonic supporter, almost guillotined in the aftermath of the French revolution, Fourier succeeded his teacher Lagrange in 1797. Besides being a government official and an accomplished Egyptologist, his mathematical work culminated in his 1822 book on Fourier series: "sines and cosines as the *atoms* of all functions"; it revolutionized how differential equations were solved. But Lagrange had pointed out that the expansion might not be unique, or even exist. Which functions have a Fourier series? This question led to refined treatments of integration such as Riemann's, and to Cantor's set theory; but also to studies into what convergence of functions is all about, when it is not pointwise.

Proof Suppose H is a separable Hilbert space, with some dense countable subset $A = \{a_0, a_1, a_2, \dots\}$. The Gram-Schmidt process converts this to a list of orthonormal vectors $E = \{e_0, e_1, e_2, \dots\}$, which is then a countable orthonormal basis of H since $\overline{[\![E]\!]} = \overline{[\![A]\!]} \supseteq A = H$. Consider the map

$$J : H \to \ell^2$$
$$x \mapsto (\alpha_n)_{n \in \mathbb{N}}, \qquad \alpha_n := \langle e_n, x \rangle.$$

Bessel's inequality shows that $(\alpha_n)_{n \in \mathbb{N}}$ is indeed in ℓ^2. (If H is a real Hilbert space, α_n are also real.) Linearity of J follows from that of the inner product. Preservation of the inner products and norms, $\langle x, y \rangle_H = \langle Jx, Jy \rangle_{\ell^2}$, is precisely the content of Parseval's identity.

J is surjective: For any $(\alpha_n)_{n \in \mathbb{N}} \in \ell^2$, the series $\sum_n \alpha_n e_n$ converges to some vector x by Proposition 10.30, if E is countably infinite. Then $Jx = (\alpha_n)_{n \in \mathbb{N}}$ since

$$\langle e_n, Jx \rangle = \langle e_n, \sum_{m \in \mathbb{N}} \alpha_m e_m \rangle = \sum_{m \in \mathbb{N}} \alpha_m \langle e_n, e_m \rangle = \alpha_n.$$

The Hilbert space is N-dimensional precisely when E has N vectors; in this case it is a classical basis of H. J remains a surjective isometry, with $\mathbb{R}^N$ or $\mathbb{C}^N$ replacing ℓ^2. $\qquad\square$

Examples of Orthonormal Bases

Orthonormal bases are widely used to approximate functions, and are indispensable for actual calculations. There are various orthonormal bases commonly used for the space of L^2 functions on different domains. Each basis has particular properties that are useful in specific contexts. One should treat these in the same way that one treats bases in finite-dimensional vector spaces—a suitable choice of basis may make a problem amenable. For example, for a problem that has spherical symmetry, it would probably make sense to use an orthonormal basis adapted to spherical symmetry.

Consider the simplest domain, the real line. There are three different classes of non-empty closed intervals (up to a homeomorphism): $[a, b]$, $[a, \infty[$, and $\mathbb{R}$. Various orthonormal bases have been devised for each, with the most popular being listed here.

$L^2[a, b]$—Fourier Series

The classical Fourier series were the original impetus for much of this theory of orthonormal bases. C. Sturm and J. Liouville extended the concept substantially, and F. Riesz showed in 1907 that there exists a function f with Fourier coefficients equal to a given sequence a_n iff $(a_n)_{n \in \mathbb{Z}} \in \ell^2$.

Proposition 10.33

> **The functions $e^{2\pi i n t}$ ($n \in \mathbb{Z}$), form an orthonormal basis for $L^2[0, 1]$.**

Proof Orthonormality of the functions is trivial to establish,

$$\langle e^{2\pi i n t}, e^{2\pi i m t} \rangle = \int_0^1 e^{2\pi i t(m-n)} \, \mathrm{d}t = \delta_{nm}.$$

Suppose $f \in \{e^{2\pi i n t}\}_{n \in \mathbb{Z}}^\perp$, i.e., $\int_0^1 e^{-2\pi i n t} f(t) \, \mathrm{d}t = 0$ for all $n \in \mathbb{Z}$. Recall that the Fourier coefficients give a 1–1 operator $\mathcal{F} : L^1[0, 1] \to c_0(\mathbb{Z})$ (Theorem 9.34) (note: $L^2[0, 1] \subset L^1[0, 1]$), so $\mathcal{F}f = \mathbf{0}$ implies $f = 0$ and hence $\{e^{2\pi i n t} : n \in \mathbb{Z}\}^\perp = 0$. $\qquad\square$

Of course, there is nothing special about the interval $[0, 1]$. Any other interval $[a, b]$ has a modified Fourier basis, namely $\frac{1}{\sqrt{L}}e^{2\pi i n t/L}$, where $L = b - a$. For example, $\frac{1}{\sqrt{2\pi}}e^{i n t}$ ($n \in \mathbb{Z}$), is an orthonormal basis for $L^2[-\pi, \pi]$.

Examples 10.34

1. ▶ The Fourier expansion becomes, for $f \in L^2[0, 1]$,

$$f(t) = \sum_{n=-\infty}^{\infty} \alpha_n e^{2\pi i n t}$$

where $\alpha_n = \langle e^{2\pi i n t}, f \rangle = \int_0^1 e^{-2\pi i n t} f(t)\, dt$ are the Fourier coefficients of f. Note carefully that the convergence of the sum is to be understood as $\| \sum_n \alpha_n e^{2\pi i n t} - f \|_{L^2[0,1]} \to 0$, not necessarily pointwise for each $t \in [0, 1]$. (However, a lengthy proof [30] shows that there is pointwise convergence a.e.; see also Example 11.31(5)).

2. The classical Parseval identity is

$$\int_{-\pi}^{\pi} |f(t)|^2\, dt = \sum_{n=-\infty}^{\infty} |a_n|^2 + |b_n|^2,$$

where $a_n - i b_n = \frac{1}{\sqrt{2\pi}} \int_{-\pi}^{\pi} e^{-i n t} f(t)\, dt$ are the $L^2[-\pi, \pi]$-Fourier coefficients.

3. Fourier series have a wide range of applications, especially in signal processing. For example, the operator $\mathcal{F}^* 1_{[-n,n]} \mathcal{F}$ is called a *low(frequency)-pass filter*: Given a signal f, $1_{[-n,n]}$ discards the higher-frequency terms from the Fourier coefficients $\mathcal{F}f$; $\mathcal{F}^*$ then builds a function from the remaining coefficients, resulting in a smoothed out low frequency band signal (for example, without a high frequency hiss).

$L^2[-1, 1]$—Legendre Polynomials

We've seen that the set of polynomials is dense in the space $L^2[a, b]$ (Proposition 9.29) but the simplest basis, namely $1, t, t^2, \ldots$, is not orthogonal, as can be easily verified by calculating, say, $\langle 1, t^2 \rangle = (b^3 - a^3)/3$. This can be rectified by applying the Gram-Schmidt algorithm. On the interval $[-1, 1]$, the resulting polynomials are called the (normalized) *Legendre* polynomials (Fig. 10.3). The first few are

$$\tfrac{1}{\sqrt{2}}, \quad \sqrt{\tfrac{3}{2}}\, t, \quad \tfrac{3}{2}\sqrt{\tfrac{5}{2}} \left(t^2 - \tfrac{1}{3} \right), \ldots$$

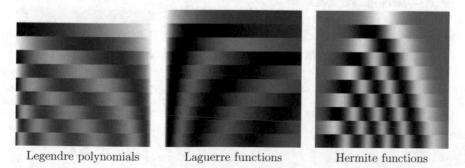

Legendre polynomials   Laguerre functions   Hermite functions

Fig. 10.3 Orthonormal bases (The first ten functions of each basis are plotted as rows in each image; brightness is proportional to the value of the function, mid-grey being 0.)

with the general formula being

$$p_n(t) = \left(\frac{\sqrt{n + \frac{1}{2}}}{2^n n!}\right) D^n (t^2 - 1)^n,$$

where $D = \frac{d}{dt}$. These polynomials satisfy the differential equation

$$Lp_n = -n(n+1)p_n, \quad \text{where } L = D(1 - t^2)D = (1 - t^2)D^2 - 2tD.$$

$L^2[0, \infty[$—Laguerre Functions

This Hilbert space contains, *not* the polynomials t^n, but their modified versions $t^n e^{-t/2}$. A Gram-Schmidt orthonormalization of them gives the *Laguerre* functions, the first few terms of which are

$$e^{-t/2}, \quad (1 - t)e^{-t/2}, \quad (1 - 2t + \tfrac{1}{2}t^2)e^{-t/2}, \ldots$$

and the general formula is

$$l_n(t) = \frac{1}{n!}e^{t/2}D^n(t^n e^{-t}).$$

The Laguerre functions satisfy (prove!)

$$Sl_n = -(n + \tfrac{1}{2})l_n, \quad \text{where } S := DtD - t/4.$$

The Laguerre polynomials (the polynomial part of l_n) can also be thought of as an orthonormal basis for $L^2_w(\mathbb{R}^+)$ with the weight e^{-t}.

$L^2(\mathbb{R})$—Hermite Functions

Here, orthonormalization is performed on the functions $t^n e^{-t^2/2}$ (equivalently, take t^n in $L^2_w(\mathbb{R})$ with the weight e^{-t^2}) to get the *Hermite* functions,

$$\frac{1}{\pi^{1/4}}e^{-t^2/2}, \quad \frac{2}{\sqrt{2\pi}^{1/4}}te^{-t^2/2}, \quad \frac{1}{\sqrt{2\pi}^{1/4}}(2t^2 - 1)e^{-t^2/2}, \ldots.$$

$$h_n(t) = \frac{(-1)^n}{\sqrt{2^n n!}\,\pi^{1/4}}e^{t^2/2}D^n e^{-t^2}.$$

To prove orthogonality, first show that $D(e^{t^2} D^n e^{-t^2}) = -2n e^{t^2} D^{n-1} e^{-t^2}$, and deduce that $\langle h_n, h_m \rangle = 2n \langle h_{n-1}, h_{m-1} \rangle$. The Hermite functions satisfy

$$Rh_n = -(2n+1)h_n, \quad \text{where } R := D^2 - t^2.$$

Moreover, they are eigenvectors of the Fourier transform: $\frac{1}{\sqrt{2\pi}} \int_{-\infty}^{\infty} e^{-it\xi} h_n(t) \, dt = (-i)^n h_n(\xi)$.

Some other useful orthogonal bases on $L^2(A)$ spaces on other domains are, in brief:

Circle. Since the circle $\mathbb{S}^1$ is essentially the interval $[0, 2\pi]$ as far as L^2-functions are concerned, the periodic Fourier functions $e^{in\theta}$ form an orthogonal basis for it.

The *Chebyshev* polynomials, $T_n(\cos\theta) := \cos n\theta$, are the projection of the $\cos n\theta$ part of this Fourier basis, from the unit semi-circle to the x-axis $[-1, 1]$. They are thus orthogonal on $L^2_w[-1, 1]$ with the weight $1/\sqrt{1-t^2}$ (since $d\theta = -dt/\sqrt{1-t^2}$).

There are many other orthonormal bases adapted to $L^2_w[a, b]$. Rodrigues' formula describes orthogonal functions on $L^2_w[a, b]$,

$$f_n(t) := w(t)^{-1} D^n(w(t)p(t)^n)$$

for a quadratic polynomial p with roots at the endpoints a, b, and weight function w: the Legendre, Laguerre, Hermite, and Chebyshev functions are all of this type.

Plane $\mathbb{R}^2$. An orthonormal basis for the plane can be obtained by multiplying Hermite functions $h_n(x)h_m(y)$. In general, if $e_n(x)$ and $\tilde{e}_n(y)$ are orthonormal bases for $L^2(A)$ and $L^2(\tilde{A})$, then $e_n(x)\tilde{e}_m(y)$ form an orthonormal basis of $L^2(A \times \tilde{A})$.

Disk $B_{\mathbb{C}}$—Bessel Functions. The functions on the unit disk taking the value zero at the boundary have an orthogonal basis $J_n(\lambda_{m,n}r)e^{in\theta}$, where $\lambda_{m,n}$ are the zeros of the Bessel function $J_n(r) := \sum_{m=0}^{\infty} \frac{(-1)^m}{m!(n+m)!}(r/2)^{2m+n}$ (Fig. 10.4).

Sphere $\mathbb{S}^2$—Spherical Harmonics.

$$Y_m^l(\theta, \phi) := \sqrt{\frac{(2l+1)(l-m)!}{4\pi(l+m)!}} P_m^l(\cos\theta)e^{im\phi},$$

where $P_m^l(t) = (-1)^m(1-t^2)^{m/2} D^m P_l(t)$ are the "associated Legendre functions". They depend on two indices, $l \in \mathbb{N}$ and $m = -l, \ldots, +l$. These are the spherical projection of the atomic orbitals.

Exercises 10.35

1. Orthonormal vectors must be linearly independent.
2. Comparing coefficients: If $\sum_n \alpha_n e_n = \sum_n \beta_n e_n$, then $\alpha_n = \beta_n$.

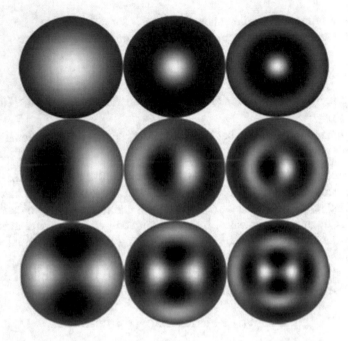

Fig. 10.4 Bessel's functions, $J_n(\lambda_{m,n} r) \cos(n\theta)$, $n, m = 0, 1, 2$

3. If $\{e_n\}_{n \in \mathbb{N}}$ and $\{\tilde{e}_m\}_{m \in \mathbb{N}}$ are orthonormal bases for Hilbert spaces X and Y respectively, then $\{(e_n, 0)\}_{n \in \mathbb{N}} \cup \{(0, \tilde{e}_m)\}_{m \in \mathbb{N}}$ form an orthonormal basis for $X \times Y$ (Exercise 10.10(5)).

4. Let $E := \{e_0, e_1, e_2, \dots\}$ be a set of orthonormal vectors, with $\overline{[\![E]\!]} = M \subset H$. For any $x \in H$, the sum $\sum_{n \in \mathbb{N}} \langle e_n, x \rangle e_n$ gives the closest point x_* in M to x.

5. ▶ An operator $U \in B(H_1, H_2)$ is a unitary isomorphism if, and only if, it maps orthonormal bases to orthonormal bases.

6. Expand the function t on $[0, 1]$ as a Fourier series.

   (a) Assuming pointwise convergence, derive Gregory's formula

   $$1 - \frac{1}{3} + \frac{1}{5} - \frac{1}{7} + \cdots = \frac{\pi}{4}.$$

   (b) Use Parseval's identity to obtain Euler's formula

   $$1 + \frac{1}{2^2} + \frac{1}{3^2} + \cdots = \frac{\pi^2}{6}.$$

   Deduce that $\left(\sum_{n \in \mathbb{Z}} \frac{(-1)^n}{2n+1} \right)^2 = \frac{\pi^2}{4} = \sum_{n \in \mathbb{Z}} \frac{1}{(2n+1)^2}$.

(c) Similarly find the Fourier coefficients for the functions t^2, $4t^3 - t$, and $2t^4 - t^2$ on $L^2[-\frac{1}{2}, \frac{1}{2}]$, to obtain

$$\sum_{n=1}^{\infty} \frac{1}{n^4} = \frac{\pi^4}{90}, \qquad \sum_{n=1}^{\infty} \frac{1}{n^6} = \frac{\pi^6}{945}, \qquad \sum_{n=1}^{\infty} \frac{1}{n^8} = \frac{\pi^8}{9450}.$$

7. When $f \in L^2[-\frac{1}{2}, \frac{1}{2}]$ is an even function, meaning $f(-t) = f(t)$, then $\alpha_{-n} = \alpha_n$ and

$$\sum_{n=-\infty}^{\infty} \alpha_n e^{2\pi int} = \alpha_0 + \sum_{n=1}^{\infty} 2\alpha_n \cos(2\pi nt).$$

What if f is odd, or neither odd nor even?

8. Show that $\cos n\pi t$, $n = 0, 1, \ldots$, is an orthogonal basis for the real space $L^2[0, 1]$.

9. $\star$ It is quite possible for $x = \sum_{n\in\mathbb{N}} \langle e_n, x \rangle e_n$ to hold true for all x in a Hilbert space, without e_n being orthonormal. Find three such vectors e_1, e_2, e_3, in $\mathbb{R}^2$. But if Parseval's identity $\|x\|^2 = \sum_n |\langle e_n, x \rangle|^2$ holds for all $x \in H$, and $\|e_n\| = 1$ for all n, then the vectors e_n form an orthonormal basis.

10. Show that $Uf(t) := \frac{1}{\sqrt{b-a}} f(\frac{t-a}{b-a})$ is a unitary operator $L^2[0, 1] \to L^2[a, b]$. Hence find an orthonormal basis for $L^2[a, b]$.

11. ▶ The Fourier operator $\mathcal{F} : L^2[0, 1] \to \ell^2$ is a unitary isomorphism between Hilbert spaces. Its adjoint maps $x = (a_n)_{n\in\mathbb{N}} \in \ell^2$ to $\mathcal{F}^*(x) = \sum_{n=-\infty}^{\infty} a_n e^{2\pi int}$.

12. Prove that the Legendre polynomials are orthonormal in $L^2[-1, 1]$, as follows: Define $u_n(t) := (t^2 - 1)^n$, and $q_n := D^n u_n$; show by induction that

(a) $D^k u_n(\pm 1) = 0$, for $k < n$,
(b) $\langle D^n u_n, D^m u_m \rangle = -\langle D^{n-1} u_n, D^{m+1} u_m \rangle$,
(c) $\langle q_n, q_m \rangle = 0$ unless $n = m$.

13. $\star$ The Legendre polynomials $P_n := p_n / \sqrt{n + \frac{1}{2}}$ have the property,

$$\frac{1}{\|u - y\|} = \sum_{n=0}^{\infty} r^n P_n(\cos\theta)$$

where u is a unit vector, $r := \|y\| < 1$, and θ is the angle between u and y. (Hint: Show $f_r(t) := 1/\sqrt{1 + r^2 - 2rt}$ satisfies $Lf_r = r\frac{\partial^2}{\partial r^2}(rf_r)$, then write $f_r(t) = \sum_n \alpha_n(r) p_n(t)$.)

14. ▶ A *frame* is a sequence of vectors $e_n \in H$ (not necessarily linearly independent) for which the mapping $J : x \mapsto (\langle e_n, x \rangle)_{n\in\mathbb{N}}$ is an embedding

$H \to M \subseteq \ell^2$. By Proposition 8.15, this is equivalent to there being positive constants $a, b > 0$, $a\|x\| \leqslant \|Jx\|_{\ell^2} \leqslant b\|x\|$, i.e.,

$$\exists c > 0, \ \forall x \in H, \qquad \frac{1}{c}\|x\|^2 \leqslant \sum_n |\langle e_n, x\rangle|^2 \leqslant c\|x\|^2.$$

Let $\delta_k(a_n) := a_k$ and $L := (J^*)^{-1}$; then $x \mapsto \delta_k L x$ is a continuous functional, hence there is a unique vector $\tilde{e}_k$ such that $\delta_k L x = \langle \tilde{e}_k, x\rangle$.

(a) The two sets of vectors e_n and $\tilde{e}_n$ are bi-orthogonal, that is, $\langle \tilde{e}_m, e_n\rangle = \delta_{mn}$.
(b) $J^*L = I = L^*J$, so

$$x = \sum_n \langle \tilde{e}_n, x\rangle e_n = \sum_n \langle e_n, x\rangle \tilde{e}_n.$$

Applications of Orthonormal Bases

Frequency-Time Orthonormal Bases

An improvement on the classical orthonormal bases for functions $t \mapsto f(t)$ in $L^2(\mathbb{R})$ are bases that give information in both 'frequency' and 'time'. In contrast, the Fourier coefficients, for example, only give information about the frequency content of the function. A large nth Fourier coefficient means that there is a substantial amount of the term $e^{2\pi i n t}$, *somewhere* in the function $f(t)$ without indicating at all where. The aim of frequency-time bases is to have coefficients $a_{m,n}$ that depend on two parameters n and m, one of which is a frequency index, the other a "time" index. The $a_{m,n}$ coefficients, much like musical notes placed on a score, indicate how much of the frequency corresponding to n, is "played" at the time corresponding to m; they are able to track the change of frequency content of f with time. Of course, the reference to t as time is not of relevance here; t can represent any other varying real quantity.

Windowed Fourier Bases (Short Time Fourier Transform): A basic way to achieve this is to define the basis functions by

$$h_{m,n}(t) := e^{2\pi i n t} h(t - m),$$

where h is a carefully chosen (real) *window* function, with $\|h\|_{L^2} = 1$, such that $h_{m,n}$ are orthonormal. The simplest choice of window function is $h = 1_{[-\frac{1}{2}, \frac{1}{2}]}$; other popular possibilities, such as the Hann window $\cos^2(\pi t)$ ($-\frac{1}{2} \leqslant t \leqslant \frac{1}{2}$) and the Gaussian $c_\sigma e^{-t^2/2\sigma^2}$ do not give orthonormal bases but are useful nonetheless.

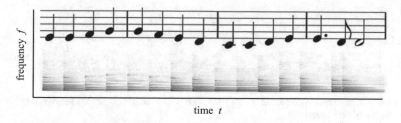

Fig. 10.5 Spectrogram of a piano piece, showing clearly the duration, frequency, and harmonics of each note

One can then obtain a picture of f spread out in time and frequency, called a *spectrogram* (Fig. 10.5), by plotting the coefficients $|\langle h_{m,n}, f \rangle|^2$ (often interpolating in m and n to get a smooth picture).

Note that the coefficients $\langle h_{m,n}, f \rangle$ are really just $(a_{m,n}) = \mathcal{F}(h(t - m)f(t))$. So summing the coefficients in n, keeping the position m fixed, gives the windowed function:

$$\sum_{n \in \mathbb{Z}} a_{m,n} e^{2\pi i n t} = h(t - m)f(t)$$

and similarly, when $\sum_{m \in \mathbb{Z}} h(t - m) = 1$,

$$\widehat{f}(n) = \int e^{-2\pi i n t} \sum_{m \in \mathbb{Z}} h(t - m)f(t)\,\mathrm{d}t = \sum_{m \in \mathbb{Z}} a_{m,n}$$

The greatest disadvantage of these bases is that the window 'width' is predetermined; it ought to be large enough to contain the low frequency oscillations, but then the time localization of the high frequencies is lost. The aim of the windowed Fourier basis is only achieved over a limited range of frequencies. To circumvent this, one can make the window width decrease with the frequency parameter n— this is the idea of wavelets.

Wavelet Bases: The basis in this case consists of the following functions in $L^2[0, 1]$

$$\psi_{m,n}(t) := T_m S_{2^n} \psi(t) = 2^{n/2} \psi(2^n t - m), \quad (m, n \in \mathbb{Z})$$

where ψ is a carefully chosen 'mother' function in $L^2(\mathbb{R})$. It serves both as a window (ideally with compact support) and an oscillation. The basis functions $\psi_{m,n}$ are thus scaled and translated versions of ψ. They have the advantage that the resolution in 'time' is better for higher frequencies than the windowed Fourier bases, and so require less coefficients to represent a function to the same level of detail. One example is the classical *Haar basis*, generated by $\psi(t) := 1_{[0,1]} - 1_{[1,2]}$ (prove orthogonality of $\psi_{m,n}$). Other wavelets, generated by continuous functions, are more popular, e.g., Mexican-hat $((1 - t^2)e^{-t^2/2})$, Gabor/Morlet $(e^{2\pi i f t} e^{-t^2/2},$

Fig. 10.6 Three wavelets: Haar, Mexican hat (with a translated and scaled version), and Morlet (real and imaginary parts)

usually $f = 1$; Fig. 10.6). The analogue of the spectrogram is the *scalogram*, which is a plot of the coefficients $Wf(a, b) := \langle \psi_{a,b}, f \rangle$ where $\psi_{a,b}(t) = \frac{1}{\sqrt{a}} \psi(\frac{t-b}{a})$.

In a *multi-resolution* wavelet scheme, a subspace V_k of the Hilbert space $L^2(\mathbb{R})$ is split recursively into low and high resolution parts as $V_{n+1} = V_n \oplus W_n$, where $W_n = V_n^\perp \cap V_{n+1}$

$$V_k = V_{k-1} \oplus W_{k-1} = \cdots = V_0 \oplus W_0 \oplus W_1 \oplus \cdots \oplus W_{k-1}$$

If we suppose V_n and W_n to be spanned by orthonormal bases $\{ \phi_{m,n} : m = 0, \ldots, N - 1 \}$ and $\{ \psi_{m,n} : m = 1, \ldots, N - 1 \}$, that are generated by scaling and translation from a "father" and "mother" wavelets ϕ and ψ respectively, then, by recursion, one need only ensure $V_1 = V_0 \oplus W_0 = [\![\phi]\!] \oplus [\![\psi]\!]$ for this scheme to work. Therefore the requirements are that $\phi, \psi \in V_1$ be orthonormal. For N even, the following "refinement equations" are sufficient,

$$\phi(t) = a_0 \phi(2t) + a_1 \phi(2t - 1) + \cdots + a_{N-1} \phi(2t - N + 1),$$

$$\psi(t) = a_{N-1} \phi(2t) - a_{N-2} \phi(2t - 1) + \cdots - a_0 \phi(2t - N + 1)$$

$$a_0^2 + \cdots + a_{N-1}^2 = 2.$$

Recall here that $\phi(2t - m) = 2^{-1/2} \phi_{m,1}(x)$ has norm $1/\sqrt{2}$, so $\|\phi\|^2 = \sum_m a_m^2 / 2$. For example, the Haar basis has $\phi = 1_{[0,2]}$ and ψ which satisfy

$$\phi(t) = \phi(2t) + \phi(2t - 1), \quad \psi(t) = \phi(2t) - \phi(2t - 1).$$

The Daubechies wavelet basis of order N is a multi-resolution scheme with an optimal choice of coefficients a_i, in which the wavelet ψ is taken to be of compact support and 'smooth' (more precisely, with N zero moments; see [27]).

Solving Linear Equations

Orthonormal expansions can be used to solve linear equations $Tx = y$, where x and y are elements of some (separable) Hilbert space, and T an operator on it. Given an orthonormal basis $\{e_n\}$, the vectors x and y can be written in terms of it as $x = \sum_n a_n e_n$ and $y = \sum_n b_n e_n$. Of these, the scalar coefficients $a_n := \langle e_n, x \rangle$

are unknown and to be determined, but $b_n := \langle e_n, y \rangle$ can be calculated explicitly. Substituting into $Tx = y$ we get

$$\sum_{n \in \mathbb{N}} a_n T e_n = T\left(\sum_{n \in \mathbb{N}} a_n e_n\right) = \sum_{n \in \mathbb{N}} b_n e_n.$$

$$\therefore \ b_m = \langle e_m, y \rangle = \sum_{n \in \mathbb{N}} a_n \langle e_m, T e_n \rangle = \sum_{n \in \mathbb{N}} t_{m,n} a_n,$$

where $t_{m,n} = \langle e_m, T e_n \rangle$. This is a matrix equation in ℓ^2, representing $Tx = y$, i.e., written in terms of the coefficients of T, x and y in the orthonormal basis e_n. Effectively, the problem has been transferred from one in H to one in ℓ^2, via the isomorphism $J : H \to \ell^2$.

Converting to matrices is especially useful if the orthonormal basis elements e_n are *eigenvectors* of T, that is, $T e_n = \lambda_n e_n$. This makes the matrix of T diagonal.

$$\begin{pmatrix} \lambda_1 & 0 & \cdots \\ 0 & \lambda_2 & \\ \vdots & & \ddots \end{pmatrix} \begin{pmatrix} a_1 \\ a_2 \\ \vdots \end{pmatrix} = \begin{pmatrix} b_1 \\ b_2 \\ \vdots \end{pmatrix}$$

The equation is easily solved, $a_n = b_n/\lambda_n$, unless $\lambda_n = 0$. If $\lambda_n = 0$ (i.e., $Tx = 0$ has non-trivial solutions) there are no solutions of $0 a_n = b_n$ unless $b_n = 0$, in which case the a_n are arbitrary. Thus there will be a solution x if, and only if, b_n vanishes whenever λ_n does, or equivalently, $y \perp \ker T$. Separating the vectors e_m that satisfy $T e_m = 0$ from the rest, the complete solution is

$$x = \sum_{m : \lambda_m = 0} \alpha_m e_m + \sum_{n : \lambda_n \neq 0} \frac{b_n}{\lambda_n} e_n,$$

where α_m are arbitrary constants. The first series is a solution of the "homogeneous equation" $Tx = 0$, while the second series is a "particular solution" of $Tx = y$.

For the case of the Hilbert space $L^2(A)$, with e_n and $y = f$, all functions, the particular solution can be rewritten as

$$\sum_n \frac{b_n}{\lambda_n} e_n = \sum_n \frac{\langle e_n, f \rangle}{\lambda_n} e_n = \int_A \left(\sum_n \frac{\overline{e_n(s)} e_n(t)}{\lambda_n}\right) f(s) \, ds = \int_A G(t, s) f(s) \, ds,$$

where the kernel $G(t, s) := \sum_n \overline{e_n(s)} e_n(t)/\lambda_n$ is called the *Green's function* of the operator T.

Gaussian Quadrature

A central problem in numerical analysis is to find an approximation for the integral of a real function, in the form

$$\int_a^b f \approx a_1 f(t_1) + \cdots + a_n f(t_n) =: \phi(f),$$

where a_i, t_i are fixed numbers; note that ϕ is a functional acting on f. The familiar trapezoid rule and Simpson's rule are of this type, where the t_i are equally spaced along $[a, b]$. The question arises as to whether we can do better by choosing t_i in some other optimal way.

Let $e_i(t)$ be real orthonormal polynomials of degree i in the space $L^2[a, b]$, obtained from $1, t, t^2, \ldots$, by the Gram-Schmidt algorithm. By orthogonality, their integrals vanish since $\int_a^b e_i = \langle 1, e_i \rangle = 0$, except for $\int_a^b e_0 = \|1\|_{L^2[a,b]}$. Certainly, for $\phi(e_i)$ to agree with the integral $\int_a^b e_i$ for $i = 0, \ldots, n-1$, we must require

$$\begin{pmatrix} e_0(t_1) & \cdots & e_0(t_n) \\ e_1(t_1) & \cdots & e_1(t_n) \\ \vdots & & \vdots \\ e_{n-1}(t_1) & \cdots & e_{n-1}(t_n) \end{pmatrix} \begin{pmatrix} a_1 \\ \vdots \\ a_n \end{pmatrix} = \begin{pmatrix} \|1\| \\ 0 \\ \vdots \\ 0 \end{pmatrix},$$

which can be solved for a_i when t_i are known. The main point of Gaussian quadrature is that if t_i are chosen to be the n roots of the polynomial $e_n(t)$ (assuming they lie in $[a, b]$), we also get $\int_a^b e_i = 0 = \phi(e_i)$ for $i \leqslant 2n - 1$.

For consider the polynomial division of any $e := e_m$ ($1 \leqslant m \leqslant 2n - 1$) by e_n, $e = q e_n + r$ where q and r are real polynomials of degree at most $n - 1$. Then, as e_0 is proportional to 1, and $q \in [\![1, t, \ldots, t^{n-1}]\!] = [\![e_0, \ldots, e_{n-1}]\!]$,

$$0 = \langle 1, e \rangle = \int_a^b q e_n + r = \langle q, e_n \rangle + \langle 1, r \rangle = \langle 1, r \rangle.$$

Hence $r = \sum_{k=1}^{n-1} b_k e_k$ for some scalars b_k. So by the choice of the coefficients a_i, and $e_n(t_i) = 0$,

$$e(t_i) = q(t_i) e_n(t_i) + r(t_i) = r(t_i),$$

so $\phi(e) = \sum_{i=1}^{n} a_i e(t_i) = \sum_{i=1}^{n} a_i r(t_i) = \sum_{k=1}^{n-1} b_k \sum_{i=1}^{n} a_i e_k(t_i) = 0 = \int_a^b e.$

Thus the integral of any $f = \sum_i \alpha_i e_i \in L^2[a, b]$ agrees with $\phi(f)$ up to order $i = 2n - 1$,

$$\int_a^b f = \sum_i \alpha_i \int_a^b e_i \approx \sum_{i=1}^{2n-1} \alpha_i \phi(e_i) \approx \phi(f).$$

The residual error can be made as small as needed by taking a larger n.

For example, using the Legendre polynomials, (prove!)

$$\int_{-1}^{1} f(t)\, dt \approx 0.35 f(-0.86) + 0.65 f(-0.34) + 0.65 f(0.34) + 0.35 f(0.86).$$

All of this applies equally well for weighted $L_w^2(A)$ spaces; for example, using Laguerre polynomials,

$$\int_{0}^{\infty} f(t) e^{-t}\, dt \approx 0.60 f(0.32) + 0.36 f(1.75) + 0.039 f(4.5) + 0.00054 f(9.4).$$

In practice, the algorithm of choice of most mathematics software is currently the Gauss-Kronrod algorithm, which performs Gaussian quadrature but refines it adaptively by taking more evaluation points if necessary.

Signal Processing

Sounds, images, and signals in general can be thought of as vectors in $L^2(\mathbb{R})$, $L^2(\mathbb{R}^2)$, and $L^2(A)$ respectively. They can thus be decomposed into orthonormal sums with all the advantages that entails. Three applications are:

(a) Storing only the "largest" coefficients $\alpha_n := \langle e_n, x \rangle$ of an orthonormal expansion leads to a useful compressed form of the vector x. Compression ratios of about 100 are quite typical. A close copy of x can easily be regenerated from these coefficients using $x = \sum_n \alpha_n e_n$. Although not identical to the original (because the small terms are omitted), it may be good enough for the purpose, especially since the smallest coefficients are usually unappreciated fine detail or noise.

(b) A vector can be altered intentionally by manipulating its coefficients. For example, it can be improved by filtering out noise coefficients, or particular features in a function may be picked out, e.g., image contrast may be enhanced if certain coefficients are weighted more than others.

(c) A vector may be matched with a database of other vectors, by taking the inner product with each of them, using Parseval's identity $\langle x, y \rangle = \sum_n \bar{\alpha}_n \beta_n$. That vector with the largest *correlation* $\langle x, y \rangle$ gives the best match and can be selected for further investigation.

Consequently, the storage, transmission, rapid retrieval, and comparison of images and sounds have seen a tremendous change since 1990, in part feeding the growth not only of the internet and mobile phones, but also of new scientific tools. For example, speech-, handwriting-, and face-recognition software find phonemes, characters, and faces that best match the given input; an E.C.G./E.K.G. or E.E.G. signal may be compared to a database for the early detection of cardiac arrest or epileptic fits; countless mobile phones perform fingerprint matches daily; satellite and medical imagery are usually enhanced to assist analysis, etc.

To see one application in some detail, let us look at one popular image format—JPEG (1992 standard). Color images consist of an array of pixels, each digitized into three numbers $(R, G, B) \in [0, 1]^3$ representing the red, green, and blue content. In the JPEG algorithm, the three RGB color bytes for each pixel are usually first converted to brightness, excess red, and excess blue,

$$Y := rR + gG + bB,$$

$$C_r := \frac{1}{2} + \frac{1}{2(g + b)}(R - Y),$$

$$C_b := \frac{1}{2} + \frac{1}{2(r + g)}(B - Y),$$

where $r \approx 0.25$, $g \approx 0.65$, $b \approx 0.1$ are agreed-upon constants such that $r + g + b = 1$. This is done to avoid effects due to color-shifts and because the brightness picture carries most of the visible information; in fact the excess red/blue pixels are reduced in number by a factor of 4 because the eye is not sensitive to fine detail in pure color.

The image is then split into 8×8 blocks, and each block is expanded with respect to the cosine basis $\cos(\pi n(x + \frac{1}{2})/8) \cos(\pi m(y + \frac{1}{2})/8)$ (the cosine transform is preferred for positive functions in general because the first few coefficients are larger; however it is not so good for sharp lines). The resulting 64 coefficients for each block are discretized (by multiplying by a user-defined weight, and taking the integer part). Most are now zero, and the rest are squeezed further using the standard Huffman compression algorithm. This way, a 4 Mpixel image, that normally requires 12 million bytes in raw formats, can easily be reduced a 100-fold in file-size without any visible loss of quality. JPEG 2000 uses wavelets instead but works in essentially the same way; MPEG is JPEG 1992 adapted to video.

Similarly a 5 min CD-quality stereo sound clip, sampled at 44,000 times 16 bits a second, would normally need at least 52 Mbytes. It can be compressed to about 10% of that by MP3, an algorithm that works in an analogous way as JPEG, but adapted to sound signals.

Remarks 10.36

1. Re $\langle x, y \rangle$ is a real-valued inner product (over the reals), but Im $\langle x, y \rangle$ fails the last two axioms.
2. A real inner product on the real vector space X can be uniquely extended to its *complexification* $X + iX$, by

$$\langle x_1 + ix_2, y_1 + iy_2 \rangle := (\langle x_1, y_1 \rangle + \langle x_2, y_2 \rangle) + i(\langle x_1, y_2 \rangle - \langle x_2, y_1 \rangle).$$

   Thus an inner product on $\mathbb{R}^n$ can extend in several ways to $\mathbb{R}^{2n}$, but in only one way to $\mathbb{C}^n$.
3. There is an interesting analogy between linear subspaces and logic: Think of subspaces as "statements", with $A \Rightarrow B$ meaning $A \subseteq B$, and FALSE, TRUE, A AND B, A OR B, NOT A, corresponding to 0, X, $A \cap B$, $A + B$, and $A^\perp$, respectively. What are the logical rules that correspond to Proposition 10.9? Are all classical logic rules true in this sense?

4. A normed space with a conjugate-linear "isomorphism" $J : X \to X^*$, has a sesquilinear product $\langle x, y \rangle := (x^*y + \overline{y^*x})/2$ (where $x^* := Jx$). The additional property $x^*x = \|x\|^2$ turns it into an inner product space, compatible with the norm of X.

5. The *conjugate gradient* method is an iteration to solve $T^*Tx = y$, used especially when T is a very large matrix. Note that $\langle\langle x, y \rangle\rangle := \langle x, T^*Ty \rangle$ is an inner product when T is 1–1. If e_i were an orthonormal basis with respect to *this* inner product, and $x = \sum_j \alpha_j e_j$, then

$$\alpha_j = \langle\langle e_j, x \rangle\rangle = \langle e_j, T^*Tx \rangle = \langle e_j, y \rangle,$$

and x can be found. The iteration is essentially the Gram-Schmidt process applied to the residual vectors $r_n = y - T^*Tx_n$, while calculating the approximate solutions x_n on the go, ($\|x\|^2 := \langle\langle x, x \rangle\rangle$)

$$
\begin{aligned}
e_0 &:= y/\|y\|, & u_{n+1} &:= r_n - \langle\langle e_n, r_n \rangle\rangle e_n, \\
& & e_{n+1} &:= u_{n+1}/\|u_{n+1}\|, \\
x_0 &= \langle e_0, y \rangle e_0, & x_{n+1} &:= x_n + \langle e_{n+1}, y \rangle e_{n+1}, \\
r_0 &:= y - T^*Tx_0, & r_{n+1} &:= y - T^*Tx_{n+1}.
\end{aligned}
$$

6. *QR decomposition*: Any operator $T : X \to Y$ between Hilbert spaces maps an orthonormal basis $e_i \in X$ to a sequence of vectors $Te_i \in Y$. If these are orthonormalized to e_i' using the Gram-Schmidt process, then $Te_i = \sum_{j=1}^i \alpha_{ij} e_j'$. This means that, with respect to the bases e_i and e_j', T has the upper-triangular matrix R. If Q represents the change of bases in Y from e_j' to the original one, then the matrix of T is QR.

7. A continuous function $f : [0, 2\pi] \to \mathbb{C}$, $f(0) = f(2\pi)$, traces out a looped path or 'orbit' in the complex plane. If the Fourier coefficients are written in polar form, it is clear that each term $\alpha_n e^{in\theta} = r_n e^{i(n\theta + \phi_n)}$ describes a circle; and the sum of two terms describes the motion along a circle whose center also moves in a circle. The whole Fourier sum then represents a motion along regressively smaller circles. Ptolemy and other Greek astronomers were the first to describe a periodic motion in terms of these cycles within cycles.

8. A non-separable Hilbert space is still isomorphic to an $\ell^2(A)$ space, one with an uncountable number of orthonormal basis vectors. For example every Hilbert space with an orthonormal basis $\{e_t\}_{t \in [0,1]}$ is isomorphic to the space $\ell^2[0, 1]$ consisting of functions α_t for which $\|\alpha\|^2 := \sum_t |\alpha_t|^2 < \infty$ (Note: α can take only a countable number of non-zero values.)

9. The first important application of the least-squares method was by Gauss. In 1801, G. Piazzi found the long-sought 'missing' planet between the orbits of Jupiter and Mars, but could not observe it again after it went behind the Sun. Gauss managed to recover its orbital parameters from Piazzi's observations, and Ceres was relocated almost a year after its discovery. Essentially the same techniques were used in 1846 to predict the location of a new planet, Neptune, from the irregularities in the observed positions of Uranus.

Chapter 11
Banach Spaces

In this chapter, we explore deeper into the properties of operators and functionals on general Banach spaces. We will find that several definitions and propositions that hold for Hilbert spaces generalize to Banach spaces. As Hilbert spaces are, in many ways, very special and non-typical examples of Banach spaces, these results need to be modified in several technical ways: There are no orthonormal bases, or Riesz correspondence, or orthogonal projections available in Banach spaces.

11.1 The Open Mapping Theorem

The following theorem holds the key to several unanswered questions that were raised earlier.

Theorem 11.1 (The Open Mapping Theorem)

> **A continuous linear map between Banach spaces is an open mapping if, and only if, it is surjective.**

Recall that an open mapping is one that maps open sets to open sets.

Proof Let $T : X \to Y$ be a surjective operator between the Banach spaces X and Y. Let U be an open subset of X, and let $x \in U$, so that $x \in B_\epsilon(x) \subseteq U$. If it can be shown that the image of the unit ball $T B_X$ contains a ball $B_\delta(0)$, then

$$Tx \in B_{\delta\epsilon}(Tx) = Tx + \epsilon B_\delta(0) \subseteq Tx + \epsilon T B_X = T B_\epsilon(x) \subseteq TU$$

implies that TU is an open set in Y, proving the theorem.

© The Author(s), under exclusive license to Springer Nature Switzerland AG 2024
J. Muscat, *Functional Analysis*, https://doi.org/10.1007/978-3-031-27537-1_11

Now $X = \bigcup_{n=1}^{\infty} B_n(0)$, so $TX = \bigcup_{n=1}^{\infty} T B_n(0)$. But $TX = Y$ is complete, so by Baire's category theorem, not all the sets $T B_n(0)$ are nowhere dense: there must be an N such that $\overline{T B_N(0)}$ contains a ball. By re-scaling we find that $\overline{T B_X}$ contains a ball $B_r(v)$. It follows that for every $y \in B_r(0)$ we have

$$v + y = \lim_{n \to \infty} T x_n, \quad \text{for some } x_n \in B_X,$$

$$v - y = \lim_{n \to \infty} T x'_n, \quad \text{for some } x'_n \in B_X,$$

$$\therefore \quad y = \lim_{n \to \infty} T \left(\frac{x_n - x'_n}{2} \right) \in \overline{T B_X}$$

since $\|x_n - x'_n\| < 2$. Consequently we have that $B_r(0) \subseteq \overline{T B_X}$.

$\overline{T B_X} \subseteq T B_3(0)$ Let $y \in \overline{T B_X}$, so that there must be an $x_0 \in B_X$ such that $\|y - T x_0\| < r/2$; that is, $\|x_0\| < 1$ and $y - T x_0 \in B_{r/2}(0) \subseteq \overline{T B_{1/2}(0)}$. But this implies that there is an $x_1 \in B_{1/2}(0)$ such that $\|y - T x_0 - T x_1\| < r/4$. Continuing in this fashion, we get a sequence x_n such that

$$\|x_n\| < \frac{1}{2^n}, \quad \|y - T(x_1 + \cdots + x_n)\| < \frac{r}{2^n}.$$

We can conclude that $x := \sum_{n \in \mathbb{N}} x_n$ converges absolutely, with $\|x\| \leqslant \sum_{n=0}^{\infty} \frac{1}{2^n} = 2$, and that $y = Tx \in T \overline{B_2(0)} \subset T B_{2+\epsilon}(0)$.

Re-scaling the vectors in $B_r(0) \subseteq T B_3(0)$ gives $B_{r/3}(0) \subseteq T B_X$ and closes the argument.

The converse was shown in Example 8.5(3).   □

Corollary 11.2

A bijective operator between Banach spaces is an isomorphism.

With this fact, we are ready for the analogue of the first isomorphism theorem of vector spaces, which is a generalization of the corollary.

Proposition 11.3

For any operator $T : X \to Y$ between Banach spaces,

$$X/\ker T \cong \operatorname{im} T \quad \Leftrightarrow \quad \operatorname{im} T \text{ is closed in } Y$$

$$\Leftrightarrow \quad \exists c > 0, \ \forall x \in X, \ \|x + \ker T\| \leqslant c \|Tx\|.$$

Proof The mapping $J : x + \ker T \mapsto Tx$ is well-defined and 1–1 because

$$x + \ker T = y + \ker T \iff x - y \in \ker T \iff Tx = Ty,$$

and it is obviously onto $\operatorname{im} T$. It is trivially linear, and continuous since

$$\forall u \in \ker T, \quad \|Tx\| = \|T(x + u)\| \leqslant \|T\|\|x + u\|,$$

$$\therefore \|Tx\| \leqslant \|T\| \inf_{u \in \ker T} \|x + u\| = \|T\|\|x + \ker T\|.$$

So J is an isomorphism precisely when J^{-1} is continuous, i.e., when the stated inequality holds (Proposition 8.15).

If the range of J, namely $\operatorname{im} T$, is closed in Y, then it is a Banach space (Proposition 4.7), so that Corollary 11.2 implies that J is an isomorphism. For the converse, $X/\ker T$ is complete (Proposition 8.20), as must be any isomorphic copy such as $\operatorname{im} T$ (Exercise 4.18(5)).   □

Examples 11.4

1. ▶ It is important that Y be complete for the open mapping theorem to be valid. The identity map $\ell^1 \to \ell^\infty$ is continuous and 1–1, but ℓ^1 is not isomorphic to its image, because the latter is not complete (in the ∞-norm). For example, $x_n := (1, \frac{1}{2}, \dots, \frac{1}{n}, 0, \dots)$ converge in the ∞-norm, but not to an ℓ^1-sequence.

2. ▶ Let $T : X \to Y$ be a linear map between Banach spaces; its graph $M := \{ (x, Tx) : x \in X \}$ is a linear subspace of $X \times Y$, and the map $J : M \to X$, defined by $J(x, Tx) := x$ is bijective, linear, and continuous.

    Closed Graph Theorem: If M is also closed in $X \times Y$, then it is a Banach subspace, and the open mapping theorem implies that J is an isomorphism, so that

$$\|Tx\|_Y \leqslant \|(x, Tx)\|_{X \times Y} \leqslant c\|x\|_X$$

    and T must be continuous.

3. If X has two complete norms, and $\|x\| \leqslant c\|x\|$ for some fixed $c > 0$, then the two norms are equivalent: the identity map $X_{\|\|} \to X_{\|\|}$ is continuous by hypothesis, and obviously linear and bijective; so its inverse is also continuous. Put differently, if two complete norms on X are inequivalent, then one can find vectors x_n which are unit with respect to one norm, but growing indefinitely with respect to the other. Clearly, this can only happen in infinite dimensions.

4. ⋆ A *Banach theorem*: Any separable Banach space is isomorphic to a quotient of ℓ^1.

    Proof: Let e_n be dense in B_X and let $T : \ell^1 \to X$ be defined by $T(a_n) := \sum_{n \in \mathbb{N}} a_n e_n$; it satisfies $\|T\| = 1$. Moreover, $B_X \subseteq \overline{T B_{\ell^1}}$ since $e_n = T e_n$. The last part of the proof of the open mapping theorem then shows that $B_{1/3}(0) \subseteq T B_{\ell^1}$ and hence T is surjective, with $T B_{\ell^1} = B_X$. The proposition above then shows $\ell^1 / \ker T \cong X$.

Complementary Subspaces

We are now in a position to answer an earlier question about projections: Unlike the case of Hilbert spaces, it is *not* always possible to project continuously to a closed subspace of a Banach space. The following proposition determines exactly when such a projection exists:

Proposition 11.5

> **There is a continuous projection P onto a closed linear subspace M of a Banach space X if, and only if,**
>
> $$X = M \oplus N$$
>
> **for some closed linear subspace N. In this case, $M = \operatorname{im} P$, $N = \ker P$, and $M \oplus N \cong M \times N$.**

We say that M, N are *complementary* closed subspaces.

Proof The forward implication has already been proved (Example 8.18(3)).
  Conversely, suppose $X = M \oplus N$, so that any $x = u + v$ for some $u \in M$, $v \in N$. Uniqueness of u, v follows from

$$u_1 + v_1 = x = u_2 + v_2 \Rightarrow u_1 - u_2 = v_2 - v_1 \in M \cap N = 0,$$

$$\Rightarrow u_1 = u_2 \text{ AND } v_1 = v_2.$$

This allows us to define the function $P : X \to X$ by $P(x) := u$. It is linear since

$$P(\lambda x_1 + x_2) = P(\lambda u_1 + \lambda v_1 + u_2 + v_2) = \lambda u_1 + u_2 = \lambda P(x_1) + P(x_2).$$

When x belongs to M or N, we get the special cases

$$\forall u \in M, \ Pu = P(u + 0) = u; \quad \forall v \in N, \ Pv = P(0 + v) = 0,$$

so $\operatorname{im} P = M$ and $N \subseteq \ker P$; moreover, any $x \in \ker P$ satisfies $0 = Px = u$ implying $x = v \in N$.
P is a continuous projection: For any $x = u + v \in M \oplus N$, $P^2 x = Pu = u = Px$, so $P^2 = P$. Finally, the map $J : M \times N \to X$, $J(u, v) := u + v$, between Banach spaces, is 1–1, onto and continuous

$$\|u + v\|_X \leqslant \|u\|_X + \|v\|_X = \|(u, v)\|_{M \times N}$$

and so is an isomorphism by the open mapping theorem. Therefore

$$\|Px\| = \|u\| \leqslant \|u\| + \|v\| = \|(u, v)\|_{M \times N} \leqslant c\|u + v\|_X = c\|x\|.$$

$\square$

Every subspace M can be extended by another subspace N such that $X = M \oplus N$ (by extending a basis for M to span X) but complementarity requires M, N to be closed.

Examples 11.6

1. Finite-dimensional subspaces are always complemented.
   Proof: The projection to $M = [\![e_1, \ldots, e_n]\!]$ is simply

$$x \mapsto \delta_1(x)e_1 + \cdots + \delta_n(x)e_n,$$

   where δ_j are the dual basis for M^* $(\delta_j(e_i) := \delta_{in})$. Although δ_j are defined on M, they can be extended to X^* as seen later (Theorem 11.19).
2. Finite-codimensional closed subspaces are complemented.
   Proof: Let $e_1 + M, \ldots, e_n + M$ be a basis for X/M, and let $N := [\![e_1, \ldots, e_n]\!]$ (complete). Then, for any $x \in X$,

$$x + M = \sum_{i=1}^n \alpha_i(e_i + M) = \sum_{i=1}^n \alpha_i e_i + M =: v + M$$

   which shows $x - v \in M$, $v \in N$, so $x \in M + N$. If $x \in M \cap N$, then $x = \sum_i \alpha_i e_i$ and the above identity gives $M = x + M = \sum_i \alpha_i(e_i + M)$, so $\alpha_i = 0$ (by linear independence of $e_i + M$) and $x = 0$.
3. Let $T \in B(X, Y)$ be an operator on Banach spaces.

   (a) If $Y = \operatorname{im} T \oplus M$ for some closed linear subspace M of Y, then $\operatorname{im} T$ is closed in Y.
   (b) If $X = \ker T \oplus M$ and $\operatorname{im} T$, M are closed, then $M \cong \operatorname{im} T$.

   Proof: (a) The mapping $X/\ker T \to \operatorname{im} T$ defined in the proof of Proposition 11.3 can be extended to $(X/\ker T) \times M \to Y$ by $(x + \ker T, v) \mapsto Tx + v$; it is continuous and bijective, hence an isomorphism. The conclusion follows since it sends the closed set $(X/\ker T) \times \{0\}$ to $\operatorname{im} T$.
   (b) $M \cong X/\ker T \cong \operatorname{im} T$.
4. If X is a separable Banach space, then there is a surjective operator $T : \ell^1 \to X$ (Example 11.4(4)). By Example 3(b) above, either X is embedded in ℓ^1 or $\ker T$ is not complemented.

Exercises 11.7

1. For a continuous projection $P : X \to X$, im P is closed and $X/\operatorname{im} P \cong \ker P$, while $\|x + \ker P\| \leqslant \|Px\|$.
2. *Second isomorphism theorem*: If M, N, and $M + N$ are closed subspaces of a Banach space, then $(M+N)/N \cong M/M \cap N$, using the map $M \to (M+N)/N$, $x \mapsto x + N$.
3. *Third isomorphism theorem*: Let $M \subseteq N$ be closed subspaces of X, then $\frac{X/M}{N/M} \cong \frac{X}{N}$ using the map $X/M \to X/N$, $x + M \mapsto x + N$. If M is finite-codimensional then codim $N \leqslant$ codim M.
4. Let $T : X \to Y$ and $S : X \to Z$ be operators on Banach spaces.

   (a) If M is a closed linear subspace of $\ker T$, then $x + M \mapsto Tx$ is well defined, linear, and continuous.
   (b) If S is onto and $Sx = 0 \Rightarrow Tx = 0$, then $Sx \mapsto Tx$ is a well-defined operator in $B(Z, Y)$.

5. Let M, N be closed subspaces of a Banach space, with $M \cap N = 0$. Then $M + N$ is closed $\Leftrightarrow P : M + N \to M$, $x + y \mapsto x$, is continuous.
6. If $\phi : X \to \mathbb{F}$ is linear with $\ker \phi$ closed, then ϕ is continuous.
7. If M is a complemented closed subspace of X, then $X \cong \frac{X}{M} \times M$.
8. If $X = M \oplus N$ with M, N closed, then there is a minimum separation $\|u - v\| \geqslant c$ between any unit vectors $u \in M$, $v \in N$.
9. ⋆ Suppose the Banach space X has a Schauder basis e_n (of unit norm). For $x = \sum_n \alpha_n e_n$, it can be shown that $\|\|x\|\| := \sup_n \| \sum_{i=1}^{n} \alpha_i e_i \|$ exists and is a complete norm. Show $\|x\| \leqslant \|\|x\|\|$ and deduce that the map $\phi_n : x \mapsto \alpha_n$ is in X^*. These functionals form a Schauder basis for X^*, called the *bi-orthogonal* or *dual* basis, and satisfy $\phi_n(e_m) = \delta_{nm}$.

11.2   Compact Operators

A linear map is continuous when it maps bounded sets to bounded sets. There is a special subclass of linear maps that go further:

Definition 11.8

A linear mapping between Banach spaces is called **compact** when it maps bounded sets to totally bounded sets.

Easy Consequences
1. Compact linear maps are continuous.
2. If T, S are compact operators, then so are $T+S$ and λT (since B bounded implies $\lambda T B$ and subsets of $T B + S B$ are totally bounded (Proposition 7.14)).
3. The identity map $I : X \to X$ is not compact when the Banach space is infinite dimensional (it cannot convert the unit ball to a totally bounded set (Proposition 8.25)).
4. It is enough to show that T maps the unit ball to a totally bounded set for T to be compact (since $B \subseteq B_r(0) \Rightarrow T B \subseteq r T B_X$).

Proposition 11.9

> **If T is compact and S continuous linear, then ST and TS are compact (when defined).**
>
> **If T_n are compact and $T_n \to T$ then T is compact.**
>
> **For a compact operator T, im T is separable, and is complete only when finite-dimensional.**

Proof (i) Starting from a bounded set, T maps it to a totally bounded set and S, being Lipschitz, maps this to another totally bounded set (Proposition 6.7); or starting with a bounded set, S maps it to another bounded set (Exercise 4.18(3)), which is then mapped by T to a totally bounded set.

(ii) Let B be a bounded set, with its vectors having norm at most c. Then for any $x \in B$, $Tx = T_n x + (T - T_n)x$, and

$$\|(T - T_n)x\| \leqslant \|T - T_n\|\|x\| \leqslant c\|T - T_n\| \to 0.$$

Hence for n large enough, independent of $x \in B$, $\|(T - T_n)x\| < \epsilon/2$; in other words $(T - T_n)B \subseteq B_{\epsilon/2}(0)$. Moreover $T_n B$ is totally bounded and so,

$$T B \subseteq T_n B + (T - T_n)B \subseteq \bigcup_{i=1}^{N} B_{\epsilon/2}(x_i) + B_{\epsilon/2}(0) = \bigcup_{i=1}^{N} B_{\epsilon}(x_i).$$

Thus $T B$ is totally bounded and T is compact.

(iii) Totally bounded sets are separable (Example 6.6(3)), so the image of T,

$$\text{im } T = TX = T[\bigcup_{n=1}^{\infty} B_n(0)] = \bigcup_{n=1}^{\infty} T[B_n(0)],$$

being the countable union of separable sets, is separable (Exercise 4.22(3)).

Suppose im T to be complete, then it would be a Banach space in its own right. The open mapping theorem can be used to conclude that B_X is mapped to an open and totally bounded set TB_X. As 0 is an interior point of it, there is a totally bounded ball $B_r(0) \subseteq TB_X$. This can only be the case if im T is finite dimensional (Proposition 8.25).   □

Examples 11.10

1. A *finite rank* operator, i.e., whose image has finite dimension, is compact. The reason is that, in a finite-dimensional space, bounded sets are necessarily totally bounded (Proposition 8.25, Exercise 6.9(5)). For example, matrices and functionals are compact operators of finite rank.
2. ▶ A common way of showing that an operator is compact is to show that it is the limit of operators of finite rank.
   For example, let $T : \ell^2 \to \ell^2$ be defined by $T(a_n) := (a_n/n)_{n=1}^\infty$. First cleave the operator to T_N defined by $T_N(a_n) := (a_1/1, a_2/2, \dots, a_N/N, 0, 0, \dots)$. This maps ℓ^2 linearly to an N-dimensional space. Showing it is continuous would imply it is compact of finite rank:

$$\|T_N(a_n)\|_{\ell^2}^2 = \sum_{n=1}^N |a_n/n|^2 \leqslant \sum_{n=1}^N |a_n|^2 \leqslant \|(a_n)\|_{\ell^2}^2.$$

   Furthermore, $T_N \to T$:

$$\|(T - T_N)(a_n)\|_{\ell^2}^2 = \sum_{n=N+1}^\infty |a_n/n|^2 \leqslant \tfrac{1}{N^2} \sum_{n=N+1}^\infty |a_n|^2 \leqslant \tfrac{1}{N^2}\|(a_n)\|_{\ell^2}^2.$$

   Hence $\|T - T_N\| \leqslant 1/N \to 0$ as $N \to \infty$ as required.
   Note that in this example, im T contains c_{00} which is dense in ℓ^2.
3. $T_N f(x) := \sum_{n=-N}^N \widehat{f}(n)e^{2\pi i n x}$ is an example of an operator of finite rank on $L^1[0, 1]$.
4. ▶ An operator T on Banach spaces is compact iff for every sequence $(x_n)_{n\in\mathbb{N}}$ that is bounded, $(Tx_n)_{n\in\mathbb{N}}$ has a convergent subsequence.
   Proof: The sequence $(Tx_n)_{n\in\mathbb{N}}$ is totally bounded, hence has a Cauchy subsequence, which converges by virtue of the completeness of the codomain.
   Conversely, if B is bounded, then any sequence $Tx_n \in TB$ has a convergent, hence Cauchy, subsequence; thus TB is totally bounded (Proposition 6.8).

An important source of examples of compact operators is the following:

Proposition 11.11

If the kernel k is a continuous function $[c, d] \times [a, b] \to \mathbb{C}$, then the integral operator $T : C[a, b] \to C[c, d]$,

$$Tf(s) := \int_a^b k(s, t) f(t) \, \mathrm{d}t$$

is compact.

Proof Let F be the unit ball of functions in $C[a, b]$. For any $s \in [c, d]$, and $f \in F$,

$$|Tf(s)| \leqslant (b-a)\|k\|_{L^\infty}\|f\|_{L^\infty} \leqslant (b-a)\|k\|_{L^\infty},$$

so $(TF)[c, d]$ is bounded in $\mathbb{C}$, hence totally bounded.

As k is continuous on the compact set $[a, b] \times [c, d]$, it is uniformly continuous (Proposition 6.17). So for any $\epsilon > 0$ there is a $\delta > 0$ such that for $|s_1 - s_2| < \delta$,

$$|Tf(s_1) - Tf(s_2)| \leqslant \int_a^b |k(s_1, t) - k(s_2, t)||f(t)| \, \mathrm{d}t \leqslant \epsilon(b-a).$$

This implies that Tf is continuous and, as δ is independent of f, TF is equicontinuous. By the Arzelà-Ascoli theorem (Theorem 6.26), TF is totally bounded in $C[c, d]$, and the integral operator T is compact. $\square$

Note that without the compactness of $[a, b]$ and $[c, d]$, the proposition need not hold; e.g., the Fourier transform has a continuous kernel but is not compact.

Fredholm Operators

Recall that in the linear equation $Tx = y$, the image space and kernel determine to what extent solutions exist and are (non-)unique. In infinite dimensions, the next best thing to an invertible operator is one which misses out from being injective and surjective by finite dimensional spaces.

Definition 11.12

A **Fredholm operator** is one whose kernel is finite-dimensional and whose image space has finite codimension. The **index** of a Fredholm operator is the difference

$$\text{index}(T) := \dim \ker T - \text{codim im } T.$$

Examples 11.13

1. (a) Any invertible operator is Fredholm with index 0.
   (b) A continuous projection with a finite-dimensional null space is Fredholm with index 0. This follows immediately from $X = \ker P \oplus \text{im } P$.
   (c) The left-shift operator is Fredholm with index $+1$ (since $\ker L = \{e_0\}$, $\text{im } L = X$).
2. Let $T : \ell^1 \to \ell^1$ be the diagonal operator defined by $T(a_n)_{n \in \mathbb{N}} := (b_n a_n)_{n \in \mathbb{N}}$ where $0 < c \leqslant |b_n| \leqslant d$ except for a finite number of indices only; then T is Fredholm of index 0.
   Proof: Let J be the finite set of indices for which $b_n = 0$. Then $\ell^1 = M \oplus N$ where M and N consist of those sequences with non-zero coefficients in J and J^c respectively. Thus $\ker T = M$, and $\text{im } T = N$ since for any $(a_n) \in N$, let $u_n := a_n/b_n$ when $n \in J^c$ and 0 otherwise; then $(u_n) \in \ell^1$ and $T(u_n) = (a_n)$.
3. A Fredholm operator $T : X \to Y$ between Banach spaces gives rise to decompositions

$$X = \ker T \oplus M, \quad Y = \text{im } T \oplus N,$$

for some closed linear subspaces M, N by Examples 11.6(1,2,3). The restricted operator $R : M \to \text{im } T$, $x \mapsto Tx$ is then bijective and continuous, and thus an isomorphism by the open mapping theorem.

Proposition 11.14 (Index Theorem)

The composition of Fredholm operators is again Fredholm, and

$$\text{index}(ST) = \text{index}(S) + \text{index}(T).$$

Proof Let $T \in B(X, Y)$, $S \in B(Y, Z)$, both Fredholm, with $k := \dim(\ker T)$, $l := \text{codim}(\text{im } S)$. Y decomposes as

$$Y = N \oplus \text{im } T = \ker S \oplus M = A \oplus B \oplus C \oplus D$$

where

$$A := \ker S \cap N \text{ of dimension } a,$$
$$B := \operatorname{im} T \cap \ker S \text{ of dimension } b,$$
$$C := M \cap N \text{ of dimension } c,$$
$$D := M \cap \operatorname{im} T.$$

Then $\dim \ker ST = k + b$, $\operatorname{codim} \operatorname{im} ST = c + l$, both finite, and the index of ST is $k + b - c - l = (a + b - l) + (k - a - c) = \operatorname{index}(S) + \operatorname{index}(T)$. $\qquad\square$

What is the connection with compact operators, one might ask?

Proposition 11.15

> **Let $T, K \in B(X, Y)$ be operators on Banach spaces, with T invertible and K compact, then $T + K$ is Fredholm.**

It is shown in Proposition 14.18 that $\operatorname{index}(T + K) = \operatorname{index} T (I + T^{-1}K) = 0$.

Proof $\ker(T + K)$ *is finite-dimensional*: Let $S := T + K$. On $\ker S$, $K = -T$, so $-T^{-1}K = I$, but the identity map is compact only in finite dimensions. By Example 11.6(1), $X = \ker S \oplus M$, with M a closed subspace. The restriction map $R : M \to Y, x \mapsto Sx$ is injective since $\ker S \cap M = 0$.

R *satisfies* $c\|x\| \leqslant \|Rx\|$: Suppose, to the contrary, that there are unit vectors $x_n \in M$ such that $Rx_n \to 0$. Then there is a convergent subsequence, $Kx_{n_i} \to y$, by the compactness of K, and therefore

$$Tx_{n_i} = Rx_{n_i} - Kx_{n_i} \to -y$$
$$Rx_{n_i} \to -RT^{-1}y = 0$$

so $y = 0$, contradicting that x_n are unit. It follows from Example 8.16(3) that $\operatorname{im} S = \operatorname{im} R$ is closed.

S *has a finite codimensional image*: Consider the map $Y \to Y/\operatorname{im} S$,

$$y \mapsto KT^{-1}y + \operatorname{im} S = (S - T)T^{-1}y + \operatorname{im} S = -y + \operatorname{im} S.$$

It is compact (for any bounded sequence y_n, Ky_n has a convergent subsequence) and surjective. By Proposition 11.9, $Y/\operatorname{im} S$ is finite dimensional.

$\qquad\square$

Proposition 11.16

An operator $T : X \to Y$ on Banach spaces is Fredholm, if and only if T is invertible "up to compact operators", that is, there exist $K_1 \in B(X)$, $K_2 \in B(Y)$ compact and $S \in B(Y, X)$, such that

$$ST = I + K_1, \quad TS = I + K_2.$$

The operator S is also Fredholm with $\mathrm{index}(S) = -\mathrm{index}(T)$.

In fact, K_1, K_2 can be taken to be of finite rank.

Proof Suppose T is Fredholm, so $X = \ker T \oplus M$, $Y = \operatorname{im} T \oplus N$, for some closed subspaces M, N, with accompanying finite-rank projections P onto $\ker T$ with kernel M, and Q onto N along $\operatorname{im} T$. The restriction $R : M \to \operatorname{im} T$ is an isomorphism. Define $S : Y \to X$ by $Sy := R^{-1}(I - Q)y \in M \subseteq X$.

Starting with $y \in Y$, $y = Tu + v \in \operatorname{im} T \oplus N$ (u, v unique), so

$$TSy = TR^{-1}Tu = Tu = (I - Q)y.$$

Similarly, starting with $x \in X$, $x = w + z \in M \oplus \ker T$ (w, z unique), then

$$STx = R^{-1}Tx = w = (I - P)x,$$

so $ST = I - P$, $TS = I - Q$. Note that $\ker S = N$ and $\operatorname{im} S = M$, so S is Fredholm and

$$\mathrm{index}(S) = \dim N - \operatorname{codim} M = \operatorname{codim}(\operatorname{im} T) - \dim \ker T = -\mathrm{index}(T)$$

Conversely, suppose $ST = I + K_1$, $TS = I + K_2$, then

$$\ker T \subseteq \ker ST = \ker(I + K_1).$$

Since $I + K_1$ is Fredholm, its kernel, and thus $\ker T$, are finite-dimensional. Similarly,

$$\operatorname{im} T \supseteq \operatorname{im} TS = \operatorname{im}(I + K_2).$$

Since $I + K_2$ is Fredholm, $Y/\operatorname{im}(I + K_2)$, and by implication $Y/\operatorname{im} T$, are finite-dimensional.

□

Exercises 11.17

1. The multiplication operator $(a_n)_{n \in \mathbb{N}} \mapsto (b_n a_n)_{n \in \mathbb{N}}$ (on ℓ^1, ℓ^2, or ℓ^∞) is compact $\Leftrightarrow b_n \to 0$.

2. The operator $V(a_n) := (0, a_0, a_1/2, a_2/3, \ldots)$ (on ℓ^1, say) is compact. But the shift operators are not.

3. The operator $Tx := \sum_{n=1}^{N}(\phi_n x)y_n$, for any $\phi_n \in X^*$, $y_n \in Y$, is of finite rank. In the limit $N \to \infty$ it gives a compact operator if $\sum_{n=1}^{\infty}\|\phi_n\|\|y_n\| < \infty$.
   In fact, any operator of finite rank must be of this type $Tx = \sum_{n=1}^{N}(\phi_n x)e_n$ with $\phi_n \in X^*$ and e_n a basis for im T.

4. If S, T are linear of finite rank, then so are λT and $S + T$; if S is any linear map, then ST and TS are of finite rank, when defined.

5. If $T : X \to Y$ is compact, then so is its restriction to a subspace $M \subseteq X$, $T|_M : M \to Y$.

6. A compact operator between infinite dimensional Banach spaces is not surjective. Indeed, its image cannot contain an infinite dimensional complete subspace, for then, the operator can be restricted to a compact operator onto it.

7. The index of an $m \times n$ matrix is $n - m$.

8. The right-shift operator R (on ℓ^∞ say) is Fredholm with index -1.

11.3 The Dual Space X^*

Functionals provide very useful tools in converting vectors to numbers, and vector sequences to more amenable numerical sequences. Thus if we are uncertain whether $x_n \to x$ then we might try to see if $\phi x_n \to \phi x$ for some continuous functional—if it does not converge, neither does x_n. Moreover, X^* is a sort of mirror-image, or dual, of X: Just as a vector in $\mathbb{R}^n$ can be thought of as a one-column matrix, every vector in X can be represented as a linear operator $x : \mathbb{F} \to X, \lambda \mapsto \lambda x$; dually, functionals are linear operators $\phi : X \to \mathbb{F}, x \mapsto \phi x$. It turns out that the space X^* is at least as "rich" as the normed space X, in the sense that X can be recovered from X^* as a subspace of X^{**}.

Examples 11.18

1. The functionals of a Hilbert space are in 1–1 correspondence with the vectors by the Riesz representation theorem.

2. Recall that $\ell^{1*} \equiv \ell^\infty$, $\ell^{2*} \equiv \ell^2$, and $c_0^* \equiv \ell^1$ (Propositions 9.3, 9.6, and 9.9).

3. We will see later that every functional on $B(\mathbb{C}^n)$ is of the type $\phi T = \text{tr}(ST)$ where tr S is the trace of the matrix S (Theorems 15.32 and 10.16).

4. $(X \times Y)^* \cong X^* \times Y^*$, via the isomorphism $(\phi, \psi) \mapsto \omega$ where $\omega(x, y) := \phi x + \psi y$.

5. For $\phi_i, \psi \in X^*$, $\psi[\bigcap_{i=1}^{n} \ker \phi_i] = 0 \Leftrightarrow \psi \in [\![\phi_1, \ldots, \phi_n]\!]$.
   Proof: Consider the map $(\phi_1 x, \ldots, \phi_n x) \mapsto \psi x$, which is well defined since $(\phi_i x)_{i=1}^{n} = \mathbf{0} \Rightarrow \psi x = 0$; extend it, by linearity, to a functional $\xi : \mathbb{F}^n \to \mathbb{F}$. Then ξ is some row vector $(\alpha_i)_{i=1}^{n}$, that is, $\sum_{i=1}^{n} \alpha_i \phi_i x = \psi x$.

One of the main questions that arise in functional analysis is to find a function f that satisfies

$$\int_A f g_n = a_n$$

where g_n are given functions and a_n are given coefficients. Two important examples, both in practice and historically, are the following:

The moment problem: to find a probability distribution when all the moments $\int_{\mathbb{R}} p(t) t^n \, dt = a_n$ are given.

The Fourier coefficient problem: to find a periodic function with given Fourier coefficients $\frac{1}{2\pi} \int_0^{2\pi} e^{-int} f(t) \, dt = a_n$.

The problem can be made abstract and more general and potent by thinking of f as a *functional* rather than a function:

To find a functional ϕ which satisfies $\phi(x_i) = a_i$ for given linearly independent elements x_i and scalars a_i.

Written this way, ϕ would be determined on the linear subspace $Y = [\![x_i]\!]$ by $\phi(\sum_{i=1}^n \alpha_i x_i) = \sum_{i=1}^n \alpha_i a_i$. So the question becomes that of extending the functional further to cover all of X, starting from a "fragment" of it on Y. The secondary issue of whether such a functional corresponds to a function or not, has been positively answered for several classical spaces.

The next result is a powerful theorem which asserts the existence of such an extension, but like many abstract existence-type theorems, the path to construct such an extension is not straightforward.

Theorem 11.19 (The Hahn-Banach Theorem)

Let Y be a subspace of a normed space X. Then every functional $\phi \in Y^*$ can be extended to some $\tilde{\phi} \in X^*$, with $\|\tilde{\phi}\|_{X^*} = \|\phi\|_{Y^*}$.

Proof Let us try to extend ϕ from a functional on Y to a functional $\tilde{\phi}$ on $Y + [\![v]\!]$, for a vector $v \notin Y$, by selecting a number $\tilde{\phi}v := c$. Once c is chosen, we are forced to set $\tilde{\phi}(y + \lambda v) := \phi y + \lambda c$, for any $\lambda v \in [\![v]\!]$, to make $\tilde{\phi}$ a linear extension of ϕ; and to retain continuity with $\|\tilde{\phi}\| = \|\phi\|$, we need, for any $y \in Y$ and $\lambda \in \mathbb{F}$ ($\lambda \neq 0$),

$$|\phi y + \lambda c| = |\tilde{\phi}(y + \lambda v)| \leqslant \|\phi\| \|y + \lambda v\|$$

$$\Leftrightarrow \qquad |\phi(y/\lambda) + c| \leqslant \|\phi\| \|y/\lambda + v\|$$

$$\Leftrightarrow \qquad |\phi y + c| \leqslant \|\phi\| \|y + v\|, \qquad\qquad (11.1)$$

(since the vectors y/λ account for all of Y). To proceed, we consider first the case of real scalars and then generalize to the complex field.

Real normed space: Let us suppose that ϕ is real-valued. Thus we are required to find a $c \in \mathbb{R}$ that satisfies inequality (11.1)

$$-\phi y - \|\phi\|\|y + v\| \leqslant c \leqslant -\phi y + \|\phi\|\|y + v\|, \quad \forall y \in Y.$$

Is this possible? Yes, because for any $y_1, y_2 \in Y$,

$$\phi y_1 - \phi y_2 \leqslant |\phi(y_1 - y_2)| \leqslant \|\phi\|\|y_1 - y_2\|$$
$$\leqslant \|\phi\|(\|y_1 + v\| + \|y_2 + v\|)$$
$$\Rightarrow \quad -\phi y_2 - \|\phi\|\|y_2 + v\| \leqslant -\phi y_1 + \|\phi\|\|y_1 + v\|.$$

Since y_1, y_2 are arbitrary vectors in Y, there must be a constant c separating the two sides of the inequality, as sought. Choosing any such c gives an extended functional with $\|\tilde{\phi}\| \leqslant \|\phi\|$ (inequality (11.1)); but $\tilde{\phi}$ extends ϕ, so $\|\tilde{\phi}\| = \|\phi\|$.

Complex normed space: Now consider the case when the functional is complex-valued. It decomposes into its real and imaginary parts $\phi = \phi_1 + i\phi_2$, but the two are not independent of each other because

$$\phi_1(iy) + i\phi_2(iy) = \phi(iy) = i\phi y = i\phi_1(y) - \phi_2(y)$$

so that $\phi_2(y) = -\phi_1(iy)$. Being real-valued, they cannot possibly belong to Y^*, but they *do* qualify as functionals on Y when restricted to the real scalars,

$$\phi_1(y_1 + y_2) = \mathrm{Re}(\phi(y_1) + \phi(y_2)) = \phi_1(y_1) + \phi_1(y_2),$$
$$\forall \lambda \in \mathbb{R}, \quad \phi_1(\lambda y) = \mathrm{Re}\,\phi(\lambda y) = \lambda\phi_1(y),$$
$$|\phi_1(y)| = |\mathrm{Re}\,\phi y| \leqslant |\phi y| \leqslant \|\phi\|\|y\|$$

(for ϕ_2, substitute Re with Im). So they have real-valued extensions $\tilde{\phi}_i$ to $Y + [\![v]\!]$ that are linear over the real scalars; actually, extending ϕ_1 to $\tilde{\phi}_1$ automatically gives the extension for ϕ_2. That is, define $\tilde{\phi}(x) := \tilde{\phi}_1(x) - i\tilde{\phi}_1(ix)$. This is obviously linear over the real scalars since $\tilde{\phi}_1$ is. It is also linear over the complex scalars because

$$\tilde{\phi}(ix) = \tilde{\phi}_1(ix) - i\tilde{\phi}_1(-x) = i(-i\tilde{\phi}_1(ix) + \tilde{\phi}_1(x)) = i\tilde{\phi}(x).$$

Moreover it is continuous since, using the polar form $\tilde{\phi}x = |\tilde{\phi}x|e^{i\theta_x}$,

$$|\tilde{\phi}x| = e^{-i\theta_x}\tilde{\phi}x = \tilde{\phi}(e^{-i\theta_x}x) = \tilde{\phi}_1(e^{-i\theta_x}x) \leqslant \|\tilde{\phi}_1\|\|x\| = \|\phi_1\|\|x\| \leqslant \|\phi\|\|x\|,$$

so that $\|\tilde{\phi}\| \leqslant \|\phi\|$; in fact, equality holds because the domain of $\tilde{\phi}$ includes that of ϕ.

Extending to X: If X can be generated from Y and a countable number of vectors v_n, then ϕ can be extended in steps, first to some ϕ_1 acting on $Y + [\![v_1]\!]$, then to ϕ_2 acting on $Y + [\![v_1]\!] + [\![v_2]\!]$, etc. The final extension is then $\tilde{\phi}x := \phi_n x$ for $x \in X$, whenever $x \in Y + [\![v_1, \ldots, v_n]\!]$. If these vectors are only dense in X (e.g., when X is separable), $\tilde{\phi}$ can be extended further with the same norm via $\tilde{\phi}(x) := \lim_{n \to \infty} \tilde{\phi}(x_n)$ when $x_n \to x$, as a special case of extending a linear continuous function to the completion spaces (Example 8.9(5)).

But even if X needs an uncountable number of generating vectors, then "Hausdorff's maximality principle" can be applied to conclude that the extension goes through to X. Let $\mathcal{M}$ be the collection of functionals ϕ_M acting on linear subspaces M containing Y and extending ϕ with the same norm

$$\mathcal{M} := \{ \phi_M \in M^* : \forall y \in Y, \phi_M y = \phi y, \text{ AND } \|\phi_M\| = \|\phi\| \}.$$

By Hausdorff's maximality principle, $\mathcal{M}$ contains a maximal chain of subspaces $\{M_\alpha\}$, where ϕ_α extends ϕ_β whenever $M_\beta \subseteq M_\alpha$. But $E := \bigcup_\alpha M_\alpha$ also allows an extension of ϕ, namely $\psi(x) := \phi_\alpha x$ for $x \in M_\alpha$. It is well-defined because $x \in M_\alpha \cap M_\beta$ implies $M_\alpha \subseteq M_\beta$ say, so $\phi_\alpha x = \phi_\beta x$. It is linear and continuous with the same norm as ϕ,

$$|\psi x| = |\phi_\alpha x| \leqslant \|\phi_\alpha\| \|x\| = \|\phi\| \|x\|.$$

Hence ψ is a maximal extension in $\mathcal{M}$; in fact, $E = X$, for were it to exclude any vector v, the first part of the proof assures us of an extension that includes v, contradicting the maximality of ψ. $\square$

The next proposition is used repeatedly throughout the rest of the book.

Proposition 11.20

For any $x \neq 0$, there is a unit $\phi \in X^*$ with $\phi x = \|x\|$.

More generally, if M is a closed linear subspace and $x \notin M$, then there is a functional $\phi \in X^*$ with $\|\phi\| = 1$, such that

$$\phi M = 0, \quad \phi x \neq 0.$$

Proof If $x \neq 0$, there are non-zero functionals on $[\![x]\!]$, such as $\psi(\lambda x) := \lambda c$ ($c \neq 0$); in particular, to satisfy the requirement $\|\phi\| = 1$, choose $\phi(\lambda x) := \lambda \|x\|$. By the Hahn-Banach theorem, it has an extension to all of X, with the same norm.

More generally, given $x \notin M$, form the linear subspace

$$Y := [\![x]\!] + M = \{ \lambda x + v : \lambda \in \mathbb{C}, v \in M \}.$$

Y^* contains the functional defined by $\psi(\lambda x + v) := \lambda \|x + M\|$. It clearly satisfies $\phi M = 0$ ($\lambda = 0$) and is linear and continuous since, for $v_i, v \in M, \lambda_i, \lambda \in \mathbb{F}$,

$$\psi(\lambda_1 x + v_1 + \lambda_2 x + v_2) = (\lambda_1 + \lambda_2)\|x + M\| = \psi(\lambda_1 x + v_1) + \psi(\lambda_2 x + v_2),$$

$$\psi(\mu(\lambda x + v)) = \lambda\mu\|x + M\| = \mu\psi(\lambda x + v),$$

$$|\psi(\lambda x + v)| = |\lambda|\|x + M\| = \|\lambda x + M\| \leqslant \|\lambda x + v\|$$

and in fact $\|\psi\| = 1$,

$$\frac{|\psi(x + v_n)|}{\|x + v_n\|} = \frac{\|x + M\|}{\|x + v_n\|} \to 1$$

for $v_n \in M$ chosen so that convergence of $\|x + v_n\| \to \|x + M\|$ occurs (Proposition 8.20). So ψ can be extended to a functional ϕ on all of X with the same norm.

□

The Hahn-Banach theorem and its corollaries show that there is a ready supply of functionals on normed spaces; admittedly, this does not sound exciting, but consider that there are vector spaces (not normed), such as $L^p(\mathbb{R})$ with $p < 1$, that have only trivial continuous functionals. For our purposes, its greater importance lies in its ability to show a certain duality between X and its space of functionals X^*. For example, the dual of the statement $\|\phi\| = \sup_{\|x\|=1} |\phi x|$ is:

Proposition 11.21

$$\|x\| = \sup_{\|\phi\|=1} |\phi x|, \qquad \|T\| = \sup_{\|\phi\|=1=\|x\|} |\phi T x|$$

Proof $|\phi x| \leqslant \|x\|$ for all unit $\phi \in X^*$. But the functional just constructed satisfies $\phi x = \|x\|$ and $\|\phi\| = 1$, so $\sup_{\|\phi\|=1} |\phi x| = \|x\|$.

This in turn allows us to deduce

$$\|T\| = \sup_{\|x\|=1} \|Tx\| = \sup_{\|x\|=1} \sup_{\|\phi\|=1} |\phi T x|.$$

□

Proposition 11.22 (Separating Hyperplane Theorem)

> **If $x \in X$ does not lie in the closed ball $\overline{B_r(0)}$, then there is a hyperplane $\phi^{-1}\alpha$ which separates the two, that is,**
>
> $$\exists \phi \in X^*, \ \exists \alpha > 0, \ \forall y \in \overline{B_r(0)}, \quad |\phi y| < \alpha < |\phi x|.$$

Proof Let $\phi : [\![x]\!] \to \mathbb{F}$, $\phi(\lambda x) := \lambda \|x\|$; its norm is 1 and $\phi x = \|x\| > r$. It can be extended to a functional on X with the same norm. Hence for any y in the closed ball, $|\phi y| \leqslant \|\phi\| \|y\| \leqslant r$. With $\alpha := \lambda_0 \|x\|$ and $r/\|x\| < \lambda_0 < 1$, the hyperplane is then $\phi^{-1}\alpha = \lambda_0 x + \ker \phi$. $\qquad\square$

Note: The proof remains valid when $\overline{B_r(0)}$ is replaced by a closed balanced convex set C since $C + B_\epsilon(0)$ determines a semi-norm in which it is the open unit ball (Exercise 7.8(10)).

Examples 11.23

1. The Hahn-Banach theorem and its corollaries are evident for Hilbert spaces:

   (a) Any functional ϕ on a closed subspace M corresponds to a vector $x \in M$, and hence has the obvious extension $\tilde{\phi} := \langle x, \cdot \rangle$ on H.
   (b) $x = 0 \Leftrightarrow \forall y \in H, \ \langle y, x \rangle = 0$.
   (c) $\|x\| = \sup_{\|y\|=1} |\langle x, y \rangle| = \sup_{\|y^*\|=1} |y^* x|, \|T\| = \sup_{\|x\|=1=\|y\|} |\langle y, Tx \rangle|$
   (Exercise 10.17(1)).
   (d) One hyperplane separating x from $\overline{B_r(0)}$ is $\alpha x + x^\perp$, $\frac{r}{\|x\|} < \alpha < 1$.

2. Operators do not extend automatically as functionals do:

   (a) If M is a complemented closed subspace of X, then every operator $T : M \to Y$ can be extended continuously to $X \to Y$.
   (b) If the identity map I on the closed subspace M can be extended to $X \to M$, then the extension is a projection and M is complemented in X.

   Proof: Let $X = M \oplus N$ with M, N closed subspaces, and define $\tilde{T}(u + v) := Tu$ for $u \in M$, $v \in N$. Then $\|\tilde{T}(u + v)\| = \|Tu\| \leqslant c\|T\| \|u + v\|$ (Proposition 11.5).
   If $\tilde{I} : X \to M$ is an extension of $I : M \to M$, then $\tilde{I}^2 x = I\tilde{I}x = \tilde{I}x$, so it is a projection in $B(X)$. X then splits up as $\ker \tilde{I} \oplus \mathrm{im}\, \tilde{I}$, where $\mathrm{im}\, \tilde{I} = M$.

3. If X is not separable then neither is X^*. But recall that the separable space ℓ^1 has the non-separable dual ℓ^∞.
   Proof: Assume X^* separable, with $\phi_1, \phi_2, \ldots$ dense in it. By definition of their norm, there must be unit vectors x_n such that for a fixed $\epsilon > 0$,

   $$|\phi_n x_n| > \|\phi_n\| - \epsilon.$$

The claim is that $M := \overline{[\![x_n]\!]}$ is equal to X, making X separable. For if not, then there is a unit functional $\psi \in X^*$ such that $\psi M = 0$; and there is a ϕ_n close to it, $\|\psi - \phi_n\| < \epsilon$, so

$$|\phi_n x_n| = |(\psi - \phi_n)x_n| \leqslant \|\psi - \phi_n\| < \epsilon.$$

Combining the two inequalities yields $\|\phi_n\| < 2\epsilon$, and this contradicts that ϕ_n is within ϵ of the unit functional ψ.

4. A Banach space, whose dual is separable, is embedded in ℓ^∞.
   Proof: Let ϕ_n be dense in B_{X^*} and let $T : X \to \ell^\infty$ be defined by $Tx := (\phi_n x)_{n \in \mathbb{N}}$. It is linear, and an isometry:

   $$\|(\phi_n x)\|_\infty = \sup_n |\phi_n x| = \sup_{\|\phi\| \leqslant 1} |\phi x| = \|x\|.$$

   Note: The Banach-Mazur theorem states that *every* real separable Banach space is embedded in $C[0, 1]$, and thus in ℓ^∞.

5. *Banach Limits*. The functional Lim on c (Exercise 9.4(1)) can be extended (non-uniquely) to a functional on ℓ^∞. Even better, let $Y := \overline{\text{im}(L - I)}$, where L is the left-shift operator. Note that Y contains c_{00} (prove!) and hence c_0, but not $\mathbf{1}$. Extend the functional $\text{Lim}\,\mathbf{1} := 1$ to Y by zero, i.e., $\text{Lim}(a\mathbf{1} + y) = a$ for $y \in Y$, and then to all of ℓ^∞ by the Hahn-Banach theorem. Such a Banach limit also satisfies $\text{Lim}(a_{n+1}) = \text{Lim}(a_n)$, as well as $\mathbf{0} \leqslant x \Rightarrow 0 \leqslant \text{Lim}\,x$. For example, taking $x := (0, 1, 0, 1, \ldots)$, then $\text{Lim}\,x = \frac{1}{2}(\text{Lim}\,x + \text{Lim}\,Lx) = \frac{1}{2}\text{Lim}\,\mathbf{1} = \frac{1}{2}$.

Annihilators

Let us explore the duality between X and X^* more closely. The connection between the two is the following construction, which allows us to shuttle between subspaces of X and those of X^*. It is the generalization of the orthogonal subspaces in Hilbert spaces which, under the Riesz correspondence J, can be rewritten in terms of functionals,

$$A^\perp = \{ x \in H : \forall a \in A, \ \langle x, a \rangle = 0 \} \xrightarrow{\ J\ } \{ \phi \in H^* : \forall a \in A, \ \phi a = 0 \}.$$

Definition 11.24

The **annihilator** of a set of vectors $A \subseteq X$ is the set of functionals

$$A^\perp := \{ \phi \in X^* : \forall x \in A, \ \phi x = 0 \}.$$

Similarly, given a set of functionals $\Phi \subseteq X^*$ then the *pre-annihilator* is

$${}^\perp\Phi := \{ x \in X : \forall \phi \in \Phi, \ \phi x = 0 \}.$$

Easy Consequences
1. $0^\perp = X^*$, $X^\perp = 0$.
2. $A \subseteq B \Rightarrow B^\perp \subseteq A^\perp$.
3. $A \subseteq {}^\perp\Phi \Leftrightarrow \Phi A = 0 \Leftrightarrow \Phi \subseteq A^\perp$.

The properties of $A^\perp$ generalize those for Hilbert spaces, such as Proposition 10.9 and Example 10.14(4).

Proposition 11.25

$A^\perp$ is a closed linear subspace of X^* with the following properties:

(i) $(A \cup B)^\perp = A^\perp \cap B^\perp$ **and** $A^\perp + B^\perp \subseteq (A \cap B)^\perp$,
(ii) ${}^\perp(A^\perp) = \overline{\llbracket A \rrbracket}$,
(iii) $\llbracket A \rrbracket$ **is dense in** $X \Leftrightarrow A^\perp = 0$.

Proof That $A^\perp$ is a linear subspace is evident from

$$\forall \phi, \psi \in A^\perp, a \in A, \lambda \in \mathbb{F}, \quad (\lambda\phi + \psi)a = \lambda\phi a + \psi a = 0.$$

Let $\phi_n \to \phi$ with $\phi_n \in A^\perp$; for any $a \in A$, $0 = \phi_n a \to \phi a$, so $\phi \in A^\perp$ and $A^\perp$ is closed in X^*.

(i) Clearly, $(A \cup B)^\perp$ is a subset of $A^\perp$ and $B^\perp$, while $\phi A = 0 = \phi B$ imply $\phi(A \cup B) = 0$. If $\phi \in A^\perp$, $\psi \in B^\perp$, and $x \in A \cap B$, then $(\phi + \psi)x = \phi x + \psi x = 0$.

(ii) ${}^\perp(A^\perp)$ is a closed linear subspace of X (Exercise 8 below), and it contains A, since for $a \in A$ and any $\phi \in A^\perp$, $\phi a = 0$, so $a \in {}^\perp(A^\perp)$. Thus $\overline{\llbracket A \rrbracket} \subseteq {}^\perp(A^\perp)$ (Proposition 7.11).

Conversely, let $x \notin \overline{\llbracket A \rrbracket}$. Then by Proposition 11.20, there is a functional ϕ satisfying both $\phi\overline{\llbracket A \rrbracket} = 0$, hence $\phi \in A^\perp$, and $\phi x \neq 0$, hence $x \notin {}^\perp(A^\perp)$.

(iii) Consequently, $\llbracket A \rrbracket$ is dense precisely when ${}^\perp(A^\perp) = \overline{\llbracket A \rrbracket} = X$, and this is equivalent to $A^\perp = 0$ ($\forall x \in X, \phi x = 0 \Leftrightarrow \phi = 0$). $\square$

The Double Dual X^{**}

A functional ϕ is an assignment of numbers ϕx as the vectors x vary in X. Suppose we fix x and vary ϕ instead, $\phi \mapsto \phi x$, what kind of object do we get? It is a mapping from X^* to $\mathbb{F}$, which is a possible candidate for a "double" functional in X^{**}.

Proposition 11.26

> **For any $x \in X$, the map $x^{**}\phi := \phi x$ is a functional on X^*, and $x \mapsto x^{**}$ is a linear isometry, embedding X in X^{**}.**

The map x^{**} is not $(x^*)^*$, as is the case in Hilbert spaces. There is no correspondence between X and X^* in general Banach spaces.

Proof The mapping $x^{**} : X^* \to \mathbb{F}$, $\phi \mapsto \phi x$, is clearly linear in ϕ, and continuous with $|x^{**}\phi| = |\phi x| \leqslant \|x\|\|\phi\|$, i.e., $x^{**} \in X^{**}$ with $\|x^{**}\| \leqslant \|x\|$.

Hence we can form the map $J : X \to X^{**}$, defined by $J(x) := x^{**}$. It is linear, since for any $\phi \in X^*$, $x, y \in X$, $\lambda \in \mathbb{F}$,

$$(x + y)^{**}(\phi) = \phi(x + y) = \phi x + \phi y = x^{**}(\phi) + y^{**}(\phi),$$

$$(\lambda x)^{**}(\phi) = \phi(\lambda x) = \lambda \phi x = \lambda x^{**}(\phi).$$

J is isometric by Proposition 11.21, $\|x^{**}\| = \sup_{\|\phi\|=1} |x^{**}\phi| = \sup_{\|\phi\|=1} |\phi x| = \|x\|$.   □

Examples 11.27

1. Given any normed space X, the double dual X^{**} is a Banach space. Hence the closure $\overline{JX}$, being a closed linear subspace of X^{**}, is itself a Banach space. It is isomorphic to the *completion* of X, denoted by $\widetilde{X}$.
2. Several Banach spaces, called *reflexive* spaces, have the property that the mapping $x \mapsto x^{**}$ is an isomorphism. Examples include ℓ^p $(p > 1)$ and all Hilbert spaces (Proposition 10.16).
3. But in general, X need not be isomorphic to X^{**}, even if X is complete. For example, some elements of $(\ell^1)^{**}$ are not of the type x^{**} for any $x \in \ell^1$.
4. In this embedding, $A \subsetneq A^{\perp\perp}$ (since for $x \in A$ and $\phi \in A^\perp$, $x^{**}\phi = \phi x = 0$, so $x^{**} \in A^{\perp\perp}$). Note that $A^{\perp\perp}$ is always a closed linear subspace even if A isn't. Question: if M is a closed linear subspace is it necessarily true that $M \cong M^{\perp\perp}$?
5. Since a functional is determined by its values on the unit sphere, we can think of the double-functional x^{**} as a continuous function on the unit sphere in X^*; its norm is none other than its maximum value there, $\|x^{**}\| = \sup_{\|\phi\|=1} |\phi x|$. Hence the vectors of any normed space can be thought of as continuous functions on a (possibly infinite-dimensional) sphere.

Exercises 11.28

1. X^* *distinguishes points*: If $x \neq y$ then there is a $\phi \in X^*$ such that $\phi x \neq \phi y$.
2. If $x \notin [\![y]\!]$, find a functional on X with $\phi x = 1$ and $\phi y = 0$.
3. For normed spaces, $X^* = 0 \Leftrightarrow X = 0$.
4. Given the functional $\phi x := x, x \in \mathbb{R}$, find all equal-norm extensions to $\mathbb{R}^2$ with the 1-norm.
5. Given $x \in X$, the set $\{\phi \in X^* : \phi x = \|x\|\}$ is non-empty and convex.
6. In a normed space of dimension bigger than n, for any vectors $x_1, \ldots, x_n$, there exists a unit vector y such that $\|y - x_i\| \geqslant \|x_i\|, i = 1, \ldots, n$.
   (Hint: Consider unit $\phi_i, \phi_i x_i = \|x_i\|$.)
7. Show that if $\{x\}^\perp = X^*$ then $x = 0$, and if $\{x\}^\perp = 0$ then $X \cong \mathbb{F}$ or $X = 0$.
8. Show $^\perp \Phi$ is a closed linear subspace of X.
9. $(^\perp \Phi)^\perp$ need not equal $\overline{[\![\Phi]\!]}$. For example, take $\Phi := \{\delta_n : n \in \mathbb{N}\}$ in ℓ^{1*}.
10. Let M be a closed subspace of a normed space X. The following maps are isomorphisms

$$M^\perp \to (X/M)^* \qquad\qquad X^*/M^\perp \to M^*$$
$$\phi \mapsto \psi \qquad\qquad \phi + M^\perp \mapsto \phi|_M.$$
$$\psi(x + M) := \phi x,$$

   Hence, $\dim M^\perp = \operatorname{codim} M$ and $\operatorname{codim} M^\perp = \dim M$, when finite.
11. ⋆ Let Y be a closed subspace of the Banach space X. Let $j_X : x \mapsto x^{**} \in X^{**}$, $j_Y : y \mapsto y^{**} \in Y^{**}$ and let $J : Y^{**} \to X^{**}$ be defined by $J\Psi(\phi) := \Psi\phi|_Y$.

    (a) Use the Hahn-Banach theorem to show that J is an isometry, such that $j_X(y) = J \circ j_Y(y)$ for $y \in Y$.
    (b) If $x \notin Y$, then $j_X(x) \notin \operatorname{im} J$ (use Proposition 11.20).
    (c) If X is reflexive then so is Y, since for any $\Psi \in Y^{**}$, $J\Psi = j_X(y)$.
    (d) Deduce that ℓ^1 and c_0 are not embedded in ℓ^2.

11.4   The Adjoint $T^\top$

Recall the adjoint of an operator on Hilbert spaces $T^* : Y \to X$ defined by the identity $\langle T^*y, x \rangle = \langle y, Tx \rangle$. Is there an analogous definition that can be applied to Banach spaces? First, one needs to recast the defining relation, replacing inner products by functionals, $(T^*y)^*x = y^*Tx$. Although not exactly the same thing, the definition $(T^\top \phi)x := \phi Tx$ captures the essentials of this identity in terms of functionals. The relation between them is $T^* : y \mapsto y^* \mapsto T^\top y^* = u^* \mapsto u$. More formally, using the Riesz correspondences $J_Y : Y \to Y^*$ and $J_X : X \to X^*$, $T^* := J_X^{-1} T^\top J_Y$.

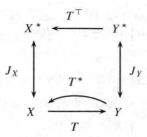

T^* is sometimes called the Hilbert adjoint to distinguish it from the adjoint $T^\top$.

Definition 11.29

The **adjoint**[1] of an operator $T : X \to Y$ is $T^\top : Y^* \to X^*$ defined by

$$(T^\top \phi)x := \phi(Tx) \text{ for any } \phi \in Y^* \text{ and } x \in X.$$

That $T^\top \phi : X \to \mathbb{F}$ is linear and continuous can be seen from

$$(T^\top \phi)(x + y) = \phi T(x + y) = \phi Tx + \phi Ty = (T^\top \phi)(x) + (T^\top \phi)(y)$$
$$(T^\top \phi)(\lambda x) = \phi T(\lambda x) = \lambda \phi Tx = \lambda (T^\top \phi)(x)$$
$$|(T^\top \phi)x| = |\phi(Tx)| \leqslant \|\phi\| \|Tx\| \leqslant \|\phi\| \|T\| \|x\|. \tag{11.2}$$

Proposition 11.30

$T^\top$ **is linear and continuous when** T **is, and the map** $T \mapsto T^\top$ **is a linear isometry from** $B(X, Y)$ **into** $B(Y^*, X^*)$,

$$(S + T)^\top = S^\top + T^\top, \quad (\lambda T)^\top = \lambda T^\top, \quad \|T^\top\| = \|T\|.$$

When defined, $(ST)^\top = T^\top S^\top$.

Proof Linearity of $T^\top$: For all $x \in X$, $\phi, \psi \in Y^*$, $\lambda \in \mathbb{F}$,

$$T^\top(\phi + \psi)(x) = (\phi + \psi)(Tx) = \phi Tx + \psi Tx = (T^\top \phi)x + (T^\top \psi)x,$$
$$T^\top(\lambda \phi)x = \lambda \phi \, Tx = (\lambda T^\top \phi)x.$$

---

[1] There is no standard name or notation for the adjoint operator. It has also been called the *dual* or *transpose* and denoted by various symbols such as T', T^*, $T^\times$, $T^\sharp$ etc.

That $T^\top$ is continuous follows from $\|T^\top\phi\| \leqslant \|T\|\|\phi\|$ by (11.2).

The other assertions are implied by the following statements, true for all $x \in X$ and all $\phi \in Y^*$:

$$(S+T)^\top\phi x = \phi(Sx+Tx) = \phi Sx + \phi Tx = (S^\top\phi + T^\top\phi)x,$$

$$(\lambda T)^\top\phi x = \phi(\lambda Tx) = \lambda\phi Tx = (\lambda T^\top\phi)x.$$

Using Proposition 11.21,

$$\|T\| = \sup_{\|x\|=1}\ \sup_{\|\phi\|=1}\ |\phi Tx| = \sup_{\|\phi\|=1}\ \sup_{\|x\|=1}\ |(T^\top\phi)x| = \sup_{\|\phi\|=1}\ \|T^\top\phi\| = \|T^\top\|.$$

Finally, when $T \in B(X, Y)$, $S \in B(Y, Z)$, and any $\psi \in Z^*$,

$$(ST)^\top\psi = \psi ST = (S^\top\psi)T = T^\top S^\top\psi.$$

$\square$

Examples 11.31

1. $0^\top = 0$, $I^\top = I$.
2. The adjoint of a (complex) matrix is its transpose, with the columns becoming the rows, $\phi Tx = y \cdot Tx = (T^\top y) \cdot x$, e.g.,

$$\begin{pmatrix} y_1 \\ y_2 \end{pmatrix} \cdot \begin{pmatrix} a & b & c \\ d & e & f \end{pmatrix}\begin{pmatrix} x_1 \\ x_2 \\ x_3 \end{pmatrix} = \begin{pmatrix} a & d \\ b & e \\ c & f \end{pmatrix}\begin{pmatrix} y_1 \\ y_2 \end{pmatrix} \cdot \begin{pmatrix} x_1 \\ x_2 \\ x_3 \end{pmatrix}, \quad \therefore \begin{pmatrix} a & b & c \\ d & e & f \end{pmatrix}^\top = \begin{pmatrix} a & d \\ b & e \\ c & f \end{pmatrix},$$

and generally, $\sum_i \left(y_i \sum_j T_{ij}x_j\right) = \sum_j \left(\sum_i T_{ij}y_i\right)x_j$, so $T^\top_{ji} = T_{ij}$.

3. ▶ To find the adjoint of an operator T on the sequence spaces ℓ^1, ℓ^2, or c_0, the effect of T on a vector x needs to reevaluated as an effect on a functional ϕ. But, identifying $(\ell^1)^*$ with ℓ^∞, etc., the adjoint $T^\top$ can be thought of as a mapping on sequences y:

$$\phi Tx = y \cdot Tx = (T^\top y) \cdot x.$$

For example, to show that the adjoint of the operator $T(a_n) := (a_1, 0, 0, \ldots)$ in $B(\ell^1)$ is $T^\top(b_n) = (0, b_0, 0, \ldots)$ in $B(\ell^\infty)$, consider

$$y \cdot Tx = (b_0, b_1, b_2, \ldots) \cdot (a_1, 0, 0, \ldots) = b_0 a_1$$

$$= (0, b_0, 0, \ldots) \cdot (a_0, a_1, a_2, \ldots).$$

4. ► The adjoint of the left-shift operator is the right-shift operator, on ℓ^p or c_0:

$$\phi Lx = y \cdot Lx = \sum_{n=0}^{\infty} b_n a_{n+1} = (0, b_0, b_1, \ldots) \cdot (a_0, a_1, a_2, \ldots) = (Ry) \cdot x.$$

5. The adjoint of the Fourier coefficients operator $\mathcal{F} : L^1[0, 1] \to c_0(\mathbb{Z})$ is $\mathcal{F}^\top :$ $\ell^1(\mathbb{Z}) \to L^\infty[0, 1]$ defined by $\mathcal{F}^\top(a_n) = \sum_n a_n e^{-2\pi i n t}$. (Compare with $\mathcal{F}^*$, Exercise 10.35(11))
   Proof: For $y = (a_n)_{n \in \mathbb{Z}} \in \ell^1(\mathbb{Z})$,

$$y \cdot \mathcal{F}f = \sum_{n \in \mathbb{Z}} a_n \int_0^1 e^{-2\pi i n t} f(t) \, dt$$

$$= \int_0^1 \left(\sum_{n \in \mathbb{Z}} a_n e^{-2\pi i n t} \right) f(t) \, dt = (\mathcal{F}^\top y) \cdot f$$

with the placement of the sum in the integral justified by $\sum_n a_n e^{-2\pi i n t} \in L^\infty[0, 1]$.

★ Note that $\ell^1 \subset c_0$, so the composition $\mathcal{F}^\top \mathcal{F}$ is not defined on all of $L^1[0, 1]$, i.e., rebuilding an L^1-function from its Fourier coefficients is not guaranteed to converge uniformly back to the function, vividly demonstrated by the Gibbs phenomenon. However, with this machinery in place, it is now easy to prove part of Dirichlet's assertion for periodic functions (see Exercise 9.37(3)):

$$\mathcal{F}^\top \mathcal{F} : C^2[0, 1] \to c_2(\mathbb{Z}) \subset \ell^1 \to L^\infty[0, 1].$$

6. ★ If the codomain of $T : X \to Y$ is reduced to the linear subspace $\operatorname{im} T$, the image of $T^\top$ remains the same.
   Proof: Let $M := \operatorname{im} T$ and let $\tilde{T} : X \to M$, $\tilde{T}x := Tx$, be the new operator; then $\tilde{T}^\top : M^* \to X^*$. Any functional $\phi \in M^*$ can be extended to $\tilde{\phi} \in Y^*$, and for all $x \in X$,

$$(T^\top \tilde{\phi})x = \tilde{\phi}Tx = \phi Tx = \phi \tilde{T}x = (\tilde{T}^\top \phi)x.$$

Hence $\operatorname{im} \tilde{T}^\top \subseteq \operatorname{im} T^\top$. Conversely, any $\phi \in Y^*$ can be restricted to $\operatorname{im} T$, and the same reasoning shows the opposite inclusion.

7. For a Hilbert space H, every operator $T \in B(H)$ is paired up with its adjoint $T^* \in B(H)$. This fact makes $B(H)$ much more special than spaces of operators on Banach spaces, as we shall see later in Chap. 15 on C^*-algebras.

The Hilbert space identity $\ker T^* = (\operatorname{im} T)^\perp$ generalizes to Banach spaces, but the closure of $\operatorname{im} T^\top$ is *not* always $(\ker T)^\perp$.

Proposition 11.32 (Closed Range Theorem)

If X, Y are **Banach spaces** and $T \in B(X, Y)$, **then**

$$\ker T^\top = (\operatorname{im} T)^\perp, \quad \ker T = {}^\perp \operatorname{im} T^\top,$$

$$\overline{\operatorname{im} T} = {}^\perp \ker T^\top, \quad \operatorname{im} T^\top \subseteq (\ker T)^\perp.$$

Moreover, $\operatorname{im} T^\top = (\ker T)^\perp \Leftrightarrow \operatorname{im} T$ **is closed** $\Leftrightarrow \operatorname{im} T^\top$ **is closed.**

Proof The central statement is, for $T \in B(X, Y)$,

$$\phi T x = (T^\top \phi) x.$$

If these quantities vanish for all $x \in X$, then the two sides of the equation state $\phi \in (\operatorname{im} T)^\perp$ and $\phi \in \ker T^\top$, which must therefore be logically equivalent. If they vanish for all $\phi \in Y^*$, then they state $x \in \ker T$ and $x \in {}^\perp \operatorname{im} T^\top$ respectively.

We have already seen that $\Phi \subseteq A^\perp \Leftrightarrow A \subseteq {}^\perp \Phi$; so the statements in the second line of the proposition follow from the identities in the top line, using first $\Phi = \ker T^\top$, $A = \operatorname{im} T$, and secondly $A = \ker T$, $\Phi = \operatorname{im} T^\top$. Moreover, by Proposition 11.25,

$$\overline{\operatorname{im} T} = {}^\perp (\operatorname{im} T^\perp) = {}^\perp (\ker T^\top).$$

$\operatorname{im} T$ *closed* $\Rightarrow \operatorname{im} T^\top$ *closed:* To show that equality holds in $\operatorname{im} T^\top \subseteq (\ker T)^\perp$, let $\phi \in (\ker T)^\perp$, i.e., $Tx = 0 \Rightarrow \phi x = 0$. T can be considered as a surjective operator $T : X \to \operatorname{im} T$, so the mapping $\tilde{\phi} : Tx \mapsto \phi x$ is a well-defined functional on $\operatorname{im} T$ (Exercise 11.7(4)). It can be extended to a functional $\psi \in Y^*$ by the Hahn-Banach theorem.

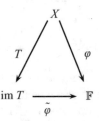

Then, for all $x \in X$,

$$\phi x = \tilde{\phi} T x = \psi T x = (T^\top \psi) x,$$

so $\phi = T^\top \psi$ and $\operatorname{im} T^\top$ is equal to the closed subspace $(\ker T)^\perp$.

im $T^\top$ *closed* $\Rightarrow$ im T *closed:* Define $\widetilde{T} : X \to \overline{\text{im } T} =: M$, $\widetilde{T}x := Tx$; by Example 11.31(6) above and the fact that the annihilator of im $\widetilde{T}$ in M is 0, it follows that $\widetilde{T}^\top$ is 1–1 and has the closed image im $T^\top$. Hence, for all $\phi \in M^*$, $\|\widetilde{T}^\top \phi\| \geqslant c\|\phi\|$ (Proposition 11.3). Now $C := \overline{\widetilde{T} B_X}$ is a closed balanced convex subset of Y, so by the separating hyperplane theorem, any $y \notin C$ can be separated from it by means of a functional $\psi \in Y^*$,

$$\forall x \in \overline{B_1(0)}, \quad |\psi \widetilde{T} x| \leqslant r < |\psi y|.$$

Note that $\|\widetilde{T}^\top \psi\| \leqslant r$. Then

$$r < |\psi y| \leqslant \|\psi\| \|y\| \leqslant \tfrac{1}{c} \|\widetilde{T}^\top \psi\| \|y\| \leqslant \tfrac{r}{c} \|y\|$$

and $\|y\| > c$. This implies that $\overline{\widetilde{T} B_X}$ contains the ball $B_c(0)$. But we have already seen in the proof of the open mapping theorem that when this is the case, then $\widetilde{T} B_X$ contains some open ball $B_\epsilon(0)$ of M. This can only be true if $\widetilde{T}$ is onto, that is, im $\widetilde{T} = $ im T is equal to the closed space M.

$\square$

Proposition 11.33 (Schauder's Theorem)

If T is compact then so is its adjoint $T^\top$.

Proof Let $T : X \to Y$ be a compact operator, so the image of the unit ball $T B_X$ is totally bounded in Y, that is, for arbitrarily small $\epsilon > 0$, it can be covered by a finite number of balls $B_\epsilon(Tx_i)$ where $x_1, \dots, x_n \in B_X$. We want to show that $T^\top$ maps the unit ball of functionals $B_{Y^*} \subset Y^*$ to a totally bounded set of functionals in X^*.

The linear map $S : Y^* \to \mathbb{F}^n$ defined by $S\psi := (\psi T x_1, \dots, \psi T x_n)$ is continuous (because T is, and n is finite), so compact of finite rank. Hence $S B_{Y^*}$ is totally bounded in $\mathbb{F}^n$ and can be covered by balls $B_\epsilon(S\psi_j)$ for a finite number of $\psi_j \in B_{Y^*}$.

We now show that balls of radius 4ϵ centered at $T^\top \psi_j$ cover $T^\top B_{Y^*}$. For any $\psi \in B_{Y^*}$ and any $x \in B_X$, there are Tx_i and $S\psi_j$ close to Tx and $S\psi$ respectively, resulting in

$$|\psi T x - \psi_j T x| \leqslant |\psi T x - \psi T x_i| + |\psi T x_i - \psi_j T x_i| + |\psi_j T x_i - \psi_j T x|$$

$$\leqslant \|\psi\| \|T x - T x_i\| + \|S\psi - S\psi_j\|_{\mathbb{F}^n} + \|\psi_j\| \|T x_i - T x\|$$

$$< \|\psi\| \epsilon + \epsilon + \|\psi_j\| \epsilon$$

$$< 3\epsilon$$

So $\|T^\top \psi - T^\top \psi_j\| \leqslant 3\epsilon$, and $T^\top B_{Y^*} \subseteq \bigcup_j B_{4\epsilon}(T^\top \psi_j)$.

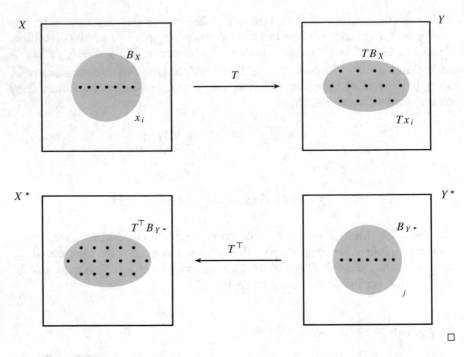

□

Exercises 11.34

1. The adjoint of a multiplier operator $M_y(x) := yx$, where $M_y \in B(\ell^1)$, is $M_y \in B(\ell^\infty)$.

2. The adjoint of a finite-rank operator $Tx := \sum_{n=1}^{N}(\phi_n x)e_n$ is another finite-rank operator $T^\top \psi = \sum_{n=1}^{N}(\psi e_n)\phi_n$.

3. Taking the adjoint is continuous: If $T_n \to T$ then $T_n^\top \to T^\top$.

4. T maps a linear subspace M onto TM; show $T^\top$ maps $(TM)^\perp$ into $M^\perp$. So, if M is T-*invariant*, i.e., $TM \subseteq M$, then $M^\perp$ is $T^\top$-invariant.

5. ⋆ In the embedding of X in X^{**}, show that $T^{\top\top} : X^{**} \to Y^{**}$ is an extension of $T : X \to Y$ in the sense that $T^{\top\top}x^{**} = (Tx)^{**}$.

6. $T^\top$ is 1–1 $\Leftrightarrow$ im T is dense in Y; and im $T^\top$ is dense in $X^* \Rightarrow T$ is 1–1.

7. Let $T \in B(X, Y)$, with X, Y Banach spaces,

$$\begin{aligned} T \text{ is an embedding} &\quad\Leftrightarrow\quad T^\top \text{ is surjective,} \\ T^\top \text{ is an embedding} &\quad\Leftrightarrow\quad T \text{ is surjective,} \\ T \text{ is an isomorphism} &\quad\Leftrightarrow\quad T^\top \text{ is an isomorphism, with } (T^\top)^{-1} = (T^{-1})^\top. \end{aligned}$$

8. A necessary condition for the equation $Tx = y$ to have a solution in x is that y have the property $T^\top \phi = 0 \Rightarrow \phi y = 0$. When is it also sufficient?

9. If P is a projection, then so is $P^\top$, with kernel $(\text{im } P)^\perp$ and image $(\ker P)^\perp$. Deduce that for closed complemented subspaces M, N,

$$X = M \oplus N \Rightarrow X^* = N^\perp \oplus M^\perp.$$

10. If T is a Fredholm operator, then so is $T^\top$ and

$$\text{index}(T^\top) = -\text{index}(T).$$

Moreover, $\text{index}(T) = \dim \ker T - \dim \ker T^\top$. (Hint: Exercise 11.28(10).)

11.5 Strong and Weak Convergence

We have already encountered two types of convergence for operators $T_n \in B(X, Y)$, to which can be added yet another, weaker, type:

(i) **Convergence in norm**

$$T_n \to T \quad \Leftrightarrow \quad \|T_n - T\| \to 0,$$

(ii) **Strong, or pointwise, convergence**

$$\forall x \in X, \quad T_n x \to T x \quad \Leftrightarrow \quad \forall x \in X, \quad \|T_n x - T x\|_Y \to 0,$$

(iii) **Weak convergence**

$$T_n \rightharpoonup T \quad \Leftrightarrow \quad \forall x \in X, \quad \forall \phi \in Y^*, \quad \phi T_n x \to \phi T x.$$

Examples 11.35

1. ▶ Convergence in norm is "stronger" than pointwise convergence, since for each $x \in X$,

$$\|T_n x - T x\| = \|(T_n - T)x\| \leqslant \|T_n - T\| \|x\| \to 0.$$

But the converse is false: it is possible to have strong convergence without convergence in norm. For example, let $\delta_k : \ell^1 \to \mathbb{C}$ be defined by $\delta_k(a_n) := a_k$; then $\delta_k x \to 0$ as $k \to \infty$ for each $x \in \ell^1$, but $\|\delta_k\| = 1$.
Similarly, when defined on $c \subset \ell^\infty$, δ_k converge pointwise to Lim yet $\delta_k \not\to$ Lim, since $\delta_k(a_n) = a_k \to \lim_{n\to\infty} a_n$, but $\delta_k = e_k^\top$ can converge only if e_k converge in ℓ^1.

2. Another example is the projection operator defined by n left shifts followed by n right shifts, $T_n := R^n L^n : \ell^1 \to \ell^1$. It converges pointwise to the 0 operator, since for each $x = (a_i)_{i\in\mathbb{N}} \in \ell^1$, $\|R^n L^n x\| = \sum_{i=n}^\infty |a_i| \to 0$. However there are sequences, such as $x := e_n$, for which $T_n x = x$, so that $\|T_n\| = 1 \not\to 0$.

3. If T_n converge pointwise, $\forall x, T_n x \to T x$, it does not follow that $T_n^\top$ converge pointwise, $\forall \phi, T_n^\top \phi \to T^\top \phi$. For example, in ℓ^1, $L^n x \to \mathbf{0}$ for the left-shift operator L; but $R^n x \not\to \mathbf{0}$ in ℓ^∞. Another example is $T_n(a_i) := (a_n, 0, 0, \ldots)$.

It often happens that a map is defined as the pointwise limit of a sequence of operators, $T(x) := \lim_{n \to \infty} T_n x$, assuming this is defined for all $x \in X$. It is then natural to ask what properties does T enjoy: That it is linear is easy to prove, but is it also necessarily continuous? The answer is yes when X is a Banach space, as implied by the following stronger assertion, one of the pillars of Banach space theory:

Theorem 11.36 (Uniform Boundedness Theorem)

For a Banach space X, a normed space Y, and $T_i \in B(X, Y)$,

$$(\forall x \in X,\ \exists C_x > 0,\ \forall i,\ \|T_i x\| \leqslant C_x) \Rightarrow \forall i,\ \|T_i\| \leqslant C.$$

The index set of i need not be countable.

Proof Suppose that T_i are not uniformly bounded; then there is a sequence of operators from this family, $(T_n)_{n \in \mathbb{N}}$, such that $\|T_n\| \geqslant a^n$ ($a > 1$). For each such operator, there are unit vectors $\pm x_n$ such that $\|T_n x_n\| > \|T_n\| - 1$. Whatever signs are chosen for x_n, the series

$$x_0 + \frac{x_1}{r} + \cdots + \frac{x_n}{r^n}$$

converges absolutely to some vector y, when $r > 1$; in fact, the remainder term is at most

$$\|z_n\| = \left\| \sum_{k>n} \frac{x_k}{r^k} \right\| \leqslant \sum_{k>n} \frac{1}{r^k} = \frac{1}{(r-1)r^n}.$$

Now, for any vectors u, v, either $\|u + v\|$ or $\|u - v\|$ is larger than $\|v\|$ (Exercise 7.8(1)), so the sign of x_n can be chosen so that

$$\| \underbrace{(T_n x_0 + \cdots + r^{-(n-1)} T_n x_{n-1})}_{u} + \underbrace{r^{-n} T_n x_n}_{v} \| \geqslant \| \underbrace{r^{-n} T_n x_n}_{v} \|.$$

$$\therefore\ \|T_n y\| = \|T_n x_0 + \cdots + r^{-n} T_n x_n + T_n z_n\|$$

$$\geqslant r^{-n} \|T_n x_n\| - \|T_n\| \|z_n\|$$

$$\geqslant \frac{\|T_n\| - 1}{r^n} - \frac{\|T_n\|}{(r-1)r^n}$$

$$\geqslant \frac{r-2}{r-1} \frac{a^n}{r^n} - \frac{1}{r^n} \to \infty \text{ as } n \to \infty$$

for $r = 3$ and $a = 4$, say. Thus there exists a $y \in X$ such that $\|T_n y\| \to \infty$, which is the negation of the hypothesis. $\square$

Corollary 11.37

If $T_n \in B(X, Y)$ with X a Banach space, and $T_n x \to T(x)$ for all x, then T is linear and continuous,

$$\|T\| \leqslant \liminf_n \|T_n\|.$$

Proof T is necessarily linear, by continuity of addition and scalar multiplication (see the proof of Theorem 8.7). Any convergent sequence is bounded, so $\|T_n x\|$ is bounded for each x, from which follows that $\forall n, \|T_n\| \leqslant C$, by the uniform boundedness theorem.

If we now choose a subsequence of T_n, for which $\|T_n\| \to \alpha := \liminf_n \|T_n\|$, and take the limit $n \to \infty$ of $\|T_n x\| \leqslant \|T_n\| \|x\|$, we get $\|T x\| \leqslant \alpha \|x\|$ and $\|T\| \leqslant \alpha$. $\qquad\square$

Examples 11.38

1. If the coefficients a_n are such that $\sum_{n=0}^{\infty} a_n b_n$ converges for any sequence $x = (b_n)_{n \in \mathbb{N}} \in \ell^p$, then $(a_n)_{n \in \mathbb{N}} \in \ell^{p'}$.
   Proof: The numbers $\phi_N x := \sum_{n=0}^{N} a_n b_n$ converge to $\phi(x) := (a_n) \cdot (b_n)$ as $N \to \infty$; hence by the corollary, ϕ is a functional on ℓ^p.
2. A common error is to define or prove $T x = \sum_{n \in \mathbb{N}} T_n x$ for all x and then deduce $T = \sum_{n \in \mathbb{N}} T_n$. It is true that two functions are the same, $f = g$, when $f(x) = g(x)$ for all $x \in X$, but the point is that the meaning of the limit in the sum $\sum_n$ differs in the two expressions, the first occurring in Y and the second in $B(X, Y)$.
3. ⋆ Recall the Fourier sum $S_n f := \sum_{k=-n}^{n} \hat{f}(k) e^{2\pi i k t} = D_n * f$, as an operator $C[0, 1] \to C[0, 1]$, where D_n is the Dirichlet kernel

$$D_n(t) := \sum_{k=-n}^{n} e^{2\pi i k t} = \frac{\sin(2n + 1)\pi t}{\sin \pi t}.$$

Let $\phi_n f := S_n f(0)$, which has norm $\|\phi_n\| = \|D_n\|_{L^1[0,1]}$. Assuming that $\|D_n\|_{L^1} \to \infty$, then the uniform boundedness theorem shows that there is a continuous function f for which $S_n f(0) \to \infty$ as $n \to \infty$.

Weak Convergence

Let us now consider weak convergence of operators

$$T_n \rightharpoonup T \quad \Leftrightarrow \quad \phi T_n x \to \phi T x \quad \forall x \in X, \ \forall \phi \in Y^*.$$

For vectors (considered as operators $\mathbb{F} \to X, \lambda \mapsto \lambda x$), weak convergence takes the form

$$x_n \rightharpoonup x \quad \Leftrightarrow \quad \phi x_n \to \phi x, \quad \forall \phi \in X^*.$$

For functionals ($X \to \mathbb{F}$), this convergence is called *weak-∗ convergence*, sometimes denoted $\phi_n \overset{*}{\rightharpoonup} \phi$ for emphasis; it coincides with their pointwise convergence,

$$\phi_n \rightharpoonup \phi \quad \Leftrightarrow \quad \phi_n x \to \phi x, \quad \forall x \in X.$$

One must guard against a possible source of confusion: the weak convergence of functionals, *when thought of as vectors in X^**, is different:

$$\phi_n \rightharpoonup \phi \quad \Leftrightarrow \quad \Psi \phi_n \to \Psi \phi, \quad \forall \Psi \in X^{**},$$

hence the need for a new name.

Examples 11.39

1. Strong convergence implies weak convergence because, by continuity of ϕ,

$$T_n x \to T x \;\Rightarrow\; \phi T_n x \to \phi T x.$$

2. ▶ But the converse is false in general: For example, in c_0, $R^n \rightharpoonup 0$, since for any $x = (a_i)_{i \in \mathbb{N}} \in c_0$, and $y = (b_i)_{i \in \mathbb{N}} \in \ell^1 \equiv c_0^*$,

$$|y \cdot R^n x| = \left| \sum_{i=0}^{\infty} b_{i+n} a_i \right| \leqslant \sum_{i=n}^{\infty} |b_i| \|x\| \to 0 \text{ as } n \to \infty,$$

yet $R^n x \not\to 0$, since $\|R^n x\| = \|x\| \not\to 0$.

3. To prove weak convergence, $x_n \rightharpoonup x$, given that $(x_n)_{n \in \mathbb{N}}$ is bounded in X, it is enough to check $\psi x_n \to \psi x$ for ψ in a *dense* subset of X^*.
   Proof: Any $\phi \in X^*$ can be approximated by functionals $\psi_n \to \phi$, by their density in X^*. For $y_n := x_n - x$ (bounded), it is not hard to show that $\psi_n y_n \to 0$, so

$$\phi y_n = \psi_n y_n + (\phi - \psi_n) y_n \to 0 \quad \text{as } n \to \infty.$$

4. Weak convergence of vectors and operators in an inner product space become

$$x_n \rightharpoonup x \Leftrightarrow \langle y, x_n \rangle \to \langle y, x \rangle \text{ as } n \to \infty, \quad \forall y \in X,$$
$$T_n \rightharpoonup T \Leftrightarrow \langle y, T_n x \rangle \to \langle y, T x \rangle \text{ as } n \to \infty, \quad \forall x, y \in X.$$

5. In an inner product space,

$$x_n \rightharpoonup x \text{ AND } \|x_n\| \to \|x\| \Leftrightarrow x_n \to x.$$

Proof: When $x_n \rightharpoonup x$, we get $\langle x, x_n \rangle \to \langle x, x \rangle$ since x^* is a functional, so

$$\|x - x_n\|^2 = \|x\|^2 - 2\operatorname{Re}\langle x, x_n \rangle + \|x_n\|^2 \to 0.$$

Proposition 11.40

In finite dimensions, all three convergence types are equivalent.

Proof Let $A_n \rightharpoonup A$ where A_n, A are $M \times N$ matrices. This means that for any $\phi \in (\mathbb{F}^M)^*$ and $x \in \mathbb{F}^N$, $\phi(A_n - A)x \to 0$ as $n \to \infty$. In particular if we let $\phi = \tilde{e}_i^\top, x = e_j$ be basis vectors for $\mathbb{F}^{M*}$ and $\mathbb{F}^N$ respectively, then each component of A_n converges to the corresponding component in A:

$$A_{n,ij} = \tilde{e}_i^\top A_n e_j \to \tilde{e}_i^\top A e_j = A_{ij}, \text{ as } n \to \infty.$$

This then implies that $\|A_n - A\| \leqslant c\sum_{j=1}^N \sum_{i=1}^M |A_{n,ij} - A_{ij}| \to 0$ (Proposition 8.10). □

The analogous result of the uniform boundedness theorem for weak convergence is also true, but more care is needed: Although every convergent sequence is bounded (Example 4.3(6)), that fact was proved using a metric, whereas weak convergence $T_n \rightharpoonup T$ is not equivalent, in general, to such a strong type of convergence as $d(T_n, T) \to 0$ for any distance function.

Proposition 11.41

If $T_n \rightharpoonup T$ where $T_n \in B(X, Y)$, X a Banach space, then

(i) $\{T_n : n \in \mathbb{N}\}$ **is bounded, and**
(ii) $T \in B(X, Y)$ **with $\|T\| \leqslant \liminf \|T_n\|$.**

Proof (i) Let $T_n \rightharpoonup T$; the set $\{T_1 x, T_2 x, \dots\}$ is *weakly bounded* in the sense that for all $n \in \mathbb{N}$, $\phi \in X^*$, $|\phi T_n x| \leqslant C_{\phi,x}$, since $(\phi T_n x)_{n \in \mathbb{N}}$ is a convergent sequence in $\mathbb{C}$. But an application of the uniform boundedness theorem twice shows first that $\|T_n x\| \leqslant C_x$, and then that T_n is bounded. Of course, a simplified version of this argument applies equally well to weakly convergent sequences of vectors $x_n \rightharpoonup x$ and to weak-*-convergent sequences of functionals $\phi_n \rightharpoonup \phi$.

(ii) Take the limit of $\phi T_n(x + y) = \phi T_n x + \phi T_n y$ and $\phi T_n(\lambda x) = \lambda \phi T_n x$ to show linearity of T. Similarly, the bounded set $\{ \|T_n\| : n \in \mathbb{N} \}$ possesses a smallest limit point α, so taking a subsequence of $\|T_n\|$ which converges to it, we obtain

$$\forall x \in X, \phi \in Y^*, \qquad |\phi T_n x| \leqslant \|\phi\| \|T_n\| \|x\|$$
$$\downarrow \qquad\qquad \downarrow$$
$$|\phi T x| \qquad \alpha \|\phi\| \|x\|$$

and $\|T\| \leqslant \alpha$ follows. Thus $B(X, Y)$ is closed under weak convergence. $\square$

As a partial converse there is:

Theorem 11.42

> **When X is a separable Banach space, every bounded sequence in X^* has a weak-$*$-convergent subsequence.**
>
> **If $x_1, x_2, \ldots \in X$ are dense in the unit ball, then X^* has a norm**
>
> $$\|\phi\|_w := \sum_{n=1}^{\infty} \frac{1}{2^n} |\phi x_n| \leqslant \|\phi\|$$
>
> **such that for ϕ_n bounded,**
>
> $$\phi_n \rightharpoonup \phi \Leftrightarrow \|\phi_n - \phi\|_w \to 0.$$
>
> **Thus the unit closed ball of X^* is a compact metric space with this norm.**

This theorem can be generalized to non-separable spaces (see [10]), when it is known as the Banach-Alaoglu theorem: *The unit closed ball of X^* is a compact topological space.*

Proof (i) Let $\{x_m\}_{m \in \mathbb{N}}$ be a countable dense subset of X, and suppose $\|\phi_n\| \leqslant c$. Then the sequence of complex numbers $\phi_n x_1$ is bounded, $|\phi_n x_1| \leqslant c \|x_1\|$, and so must have a convergent subsequence (Exercise 6.9(6)), which we shall denote by $\phi_{1,n} x_1 \to \psi(x_1)$. *This* subsequence is also bounded on x_2, $|\phi_{1,n} x_2| \leqslant c \|x_2\|$, and so we can extract, by the same means, a convergent sub-subsequence, $\phi_{2,n} x_2$. Notice that, not only does $\phi_{2,n} x_2 \to \psi(x_2)$ but also $\phi_{2,n} x_1 \to \psi(x_1)$. Continuing this way, we get subsequences $\phi_{m,n}$ and numbers $\psi(x_m)$ such that $\phi_{m,n} x_i \to \psi(x_i)$, for $i \leqslant m$, and $|\psi(x_m)| \leqslant c \|x_m\|$.

$$\phi_n \begin{vmatrix} \phi_1 & \phi_2 & \phi_3 & \phi_4 & \phi_5 & \dots \end{vmatrix}$$

$$\phi_{1,n} \begin{vmatrix} \phi_1 & & \phi_3 & \phi_4 & \phi_5 & \dots \end{vmatrix} \phi_{1,n} x_1 \to \psi(x_1)$$

$$\phi_{2,n} \begin{vmatrix} \phi_1 & & \phi_3 & & \phi_5 & \dots \end{vmatrix} \phi_{2,n} x_2 \to \psi(x_2)$$

$$\phi_{3,n} \begin{vmatrix} & & & & \phi_5 & \dots \end{vmatrix} \phi_{3,n} x_3 \to \psi(x_3)$$

$$\dots$$

$$\phi_{k,k} \begin{vmatrix} \phi_1 & & \phi_3 & & & \dots \end{vmatrix} \phi_{k,k} x_m \to \psi(x_m)$$

Let $\psi_k := \phi_{k,k}$, a subsequence of the original sequence ϕ_n. In fact, ψ_k is a subsequence of every $\phi_{m,n}$ from some point onward ($k \geqslant m$), so $\psi_k x_m \to \psi(x_m)$, as $k \to \infty$. This implies that the function ψ is Lipschitz on the dense set $\{x_m\}$,

$$|\psi(x_i) - \psi(x_j)| = \lim_{k \to \infty} |\psi_k x_i - \psi_k x_j)| = \lim_{k \to \infty} |\phi_{k,k}(x_i - x_j)| \leqslant c\|x_i - x_j\|$$

and so can be extended uniformly to a continuous function on X (Theorem 4.14), and still satisfying $|\psi(x)| \leqslant c\|x\|$. It is linear, as seen by taking the limit $k \to \infty$ of $\psi_k(x + y) = \psi_k x + \psi_k y$ and $\psi_k(\lambda x) = \lambda \psi_k x$.

Now, for any $\epsilon > 0$, there is an x_m close to x, $\|x_m - x\| < \epsilon$, so that

$$\exists K \in \mathbb{N}, \ k \geqslant K \implies |\psi_k x_m - \psi x_m| < \epsilon$$
$$\implies \quad |\psi_k x - \psi x| \leqslant |\psi_k x - \psi_k x_m| + |\psi_k x_m - \psi x_m|$$
$$+ |\psi x_m - \psi x|$$
$$\leqslant (2c + 1)\epsilon,$$

in other words $\psi_k x \to \psi x$ for all x, or $\psi_k \rightharpoonup \psi$, as $k \to \infty$.

(ii) That $\|\phi\|_w$ is well-defined and bounded by $\|\phi\|$ follows from $|\phi x_n| \leqslant \|\phi\|\|x_n\| \leqslant \|\phi\|$; that it is a norm follows from $|\phi x_n + \psi x_n| \leqslant |\phi x_n| + |\psi x_n|$ and $|\lambda \phi x_n| = |\lambda||\phi x_n|$, as well as

$$0 = \|\phi\|_w = \sum_{n=1}^{\infty} \frac{1}{2^n} |\phi x_n| \Leftrightarrow \forall n, \ |\phi x_n| = 0 \Leftrightarrow \phi = 0$$

since $\{x_n\}_{n \in \mathbb{N}}$ is dense in B_X.

(iii) *When* $\|\phi_n\| \leqslant c, \phi_n \rightharpoonup \phi \Leftrightarrow \|\phi_n - \phi\|_w \to 0$: It is enough to consider functionals ϕ_n such that $\phi_n \rightharpoonup 0$. Let $\epsilon > 0$ and M large enough that $1/2^M < \epsilon$. For all m, $\phi_n x_m \to 0$ as $n \to \infty$; this convergence may not be uniform in m, but it will be for the first M points $x_1, \dots, x_M$, i.e.,

$$\exists N, \ n \geqslant N \implies |\phi_n x_m| < \epsilon, \ \forall m = 1, \dots, M.$$

So $\|\phi_n\|_w \to 0$, because for $n \geqslant N$,

$$\|\phi_n\|_w = \sum_{m=1}^{\infty} \frac{1}{2^m} |\phi_n x_m| < \sum_{m=1}^{M} \frac{1}{2^m} \epsilon + \sum_{m=M+1}^{\infty} \frac{1}{2^m} \|\phi_n\| \|x_m\| < (1+c)\epsilon.$$

Conversely, let ϕ_n be bounded functionals such that $\|\phi_n\|_w \to 0$. This implies that for any fixed m,

$$\frac{1}{2^m} |\phi_n x_m| \leqslant \sum_{m=1}^{\infty} \frac{1}{2^m} |\phi_n x_m| \to 0, \text{ as } n \to \infty, \tag{11.3}$$

so $\phi_n x_m \to 0$. For any $x \in X$, choose x_m close to within ϵ of $y := x/\|x\|$. This is possible because $\{x_m\}$ are dense in the unit ball. Then, for n large enough,

$$|\phi_n y| \leqslant |\phi_n x_m| + |\phi_n (x_m - y)| \leqslant \epsilon + c\epsilon$$

$$\Rightarrow \quad |\phi_n x| \leqslant (1+c)\|x\|\epsilon$$

Hence $\phi_n x \to 0$ for any x and so $\phi_n \rightharpoonup 0$.

(iv) $\overline{B}_{X*}$ *is compact with respect to* $\| \cdot \|_w$: Every sequence ϕ_n in $\overline{B_1(0)}$ has a weak-$*$-convergent subsequence by (i), i.e., $\|\phi_n - \phi\|_w \to 0$. For any $x \in X$,

$$|\phi x| = \lim_{n \to \infty} |\phi_n x| \leqslant \|\phi_n\| \|x\| \leqslant \|x\|,$$

so $\|\phi\| \leqslant 1$, and $\overline{B_1(0)}$ has the Bolzano-Weierstraß property of compactness (Theorem 6.21). $\qquad\qquad\square$

Note carefully that the unit ball of X^* is not necessarily compact in the standard norm of X^* because the "unit" is measured in one norm and the "compact" in another; only in finite-dimensions are balls totally bounded (Proposition 8.25).

The next proposition characterizes compact operators in terms of weak convergence, at least for reflexive spaces.

Proposition 11.43

> **A linear $T : X \to Y$ is continuous iff, $x_n \rightharpoonup x \Rightarrow Tx_n \rightharpoonup Tx$.**
>
> **For any compact operator $T : X \to Y, x_n \rightharpoonup x \Rightarrow Tx_n \to Tx$, with the converse being true when X is a (separable) reflexive space.**

Proof (i) Given $x_n \rightharpoonup x$, and any functional $\phi \in Y^*$, then

$$\phi T x_n = (T^\top \phi) x_n \to (T^\top \phi) x = \phi T x.$$

Conversely, if $\|Tx\|/\|x\|$ is unbounded, then one can find vectors $x_n \to 0$ while $\|Tx_n\| \to \infty$ (prove!). But $x_n \rightharpoonup 0$ implies $Tx_n \rightharpoonup 0$ and so Tx_n are bounded, a contradiction.

(ii) For any Banach space, if $y_n \rightharpoonup y$ and $\{\, y_n : n \in \mathbb{N} \,\}$ is totally bounded, then $y_n \to y$. This is because there is a convergent subsequence $y_{n_i} \to z$; but $y_{n_i} \rightharpoonup y$, and therefore $z = y$ (Exercise 11.49(3)). This holds for any convergent subsequence, and thus $y_n \to y$.

Let $T : X \to Y$ be a compact operator and suppose $x_n \rightharpoonup x$ in X; then $\{x_n\}_{n \in \mathbb{N}}$ is bounded by Proposition 11.41, and $\{Tx_n\}_{n \in \mathbb{N}}$ is totally bounded by virtue of T being compact. Since $Tx_n \rightharpoonup Tx$, it follows by the above that $Tx_n \to Tx$.

Conversely, let Tx_n be any sequence in TB for a bounded subset B of a reflexive space. There is a weakly convergent subsequence $x_n \rightharpoonup x$ by the previous theorem, since $X \equiv X^{**}$. Recall that $x^{**}\phi = \phi x$, so the weak-$*$ convergence of x_n^{**} is equivalent to the weak convergence of x_n. Therefore $Tx_n \to Tx$ for this subsequence and hence TB is totally bounded.

$\square$

Examples 11.44

1. A subset $A \subseteq X$ is said to be *weakly bounded* when $\forall \phi \in X^*$, ϕA is bounded. It turns out that A is weakly bounded $\Leftrightarrow$ A is bounded.
   Proof: Given that $|\phi a| \leqslant R_\phi$ for all $a \in A$ and $\phi \in X^*$, then $\phi a = a^{**}\phi$, so the uniform boundedness theorem can be used to yield $\|a\| = \|a^{**}\| \leqslant C$.

   The idea of using functionals to transfer sets in X to sets in $\mathbb{F}$ is so convenient and useful that it is applied, not just to convergence, but to various other properties. In a general sense, we say that a set $A \subseteq X$ is *weakly* $\mathcal{P}$ when for all $\phi \in X^*$, ϕA has the property $\mathcal{P}$.

2. A vector x is a *weak limit point* of a subset A when for any $\phi \in X^*$, every open ball in $\mathbb{F}$ which contains ϕx also contains another point ϕa for $a \in A$, $a \neq x$. A is said to be *weakly closed* when it contains all its weak limit points. Every weakly closed set is closed, since $x_n \to x \Rightarrow x_n \rightharpoonup x$, but not conversely: $E := \{e_n : n \in \mathbb{N}\}$ is closed in ℓ^2 but $e_n \rightharpoonup 0 \notin E$.

3. If T is linear and ϕT is continuous for each $\phi \in Y^*$ (i.e., $x_n \to x \Rightarrow Tx_n \rightharpoonup Tx$), then in fact T is continuous.
   Proof: For every bounded set B, ϕTB is bounded by continuity. So TB is weakly bounded, which is the same as bounded.

4. A Hilbert space is weakly complete: if $(\phi x_n)_{n \in \mathbb{N}}$ is Cauchy in $\mathbb{F}$ for each $\phi \in H^*$, then $x_n \rightharpoonup x$ for some x.

Proof: Let $f(y) := \lim_{n \to \infty} \langle x_n, y \rangle$; f is linear and continuous by the uniform boundedness theorem, so must be of the form $f = \langle x, \cdot \rangle$ and $\langle x_n, y \rangle \to \langle x, y \rangle$ for each y. Can you extend this to reflexive spaces?

5. Closed and bounded sets of a reflexive space are weakly sequentially compact, meaning any bounded sequence has a weakly convergent subsequence.
   Proof: x_n^{**} has a weak-$*$ convergent subsequence, $x_{n_k}^{**} \rightharpoonup x^{**}$; but, as noted in the proof of the proposition above, this simply means $x_{n_k} \rightharpoonup x$.

6. $\star$ The "Hilbert Projection Theorem" 10.11 can be generalized to when M is weakly closed. (Note that closed convex subsets are weakly closed.)
   Proof: The sequence $(y_n)_{n \in \mathbb{N}}$ of the theorem is bounded, hence has a weakly convergent subsequence $y_{n_i} \rightharpoonup y_* \in M$. Moreover $\|y_n - x\| \to d$. Taking the limit of $|\langle y_{n_i} - x, y_* - x \rangle| \leqslant \|y_{n_i} - x\| \|y_* - x\|$ gives $\|y_* - x\| \leqslant d$.

Weak Convergence in ℓ^p

We now turn our attention to the difference between weak convergence and convergence in norm in ℓ^p ($p \geqslant 1$).

Proposition 11.45

> **For ℓ^p, $1 < p < \infty$, a sequence $x_n = (a_{n,i})_{i \in \mathbb{N}}$ converges weakly to some $x = (a_i)_{i \in \mathbb{N}}$ if, and only if, it is bounded and each component converges,**
>
> $$x_n \rightharpoonup x \quad \Leftrightarrow \quad (\exists c > 0, \forall n, \|x_n\| \leqslant c) \text{ AND } (\forall i \in \mathbb{N}, \lim_{n \to \infty} a_{n,i} = a_i).$$

Proof A weakly convergent sequence, $x_n \rightharpoonup x$, is bounded as noted in Proposition 11.41. Consider the functional $e_i^\top \in \ell^{p*}$; then

$$a_{n,i} = e_i^\top x_n \to e_i^\top x = a_i.$$

Conversely, by subtracting $(a_i)_{i \in \mathbb{N}}$, it is enough to consider a sequence x_n in ℓ^p whose components converge to 0. Let $y = (b_i)_{i \in \mathbb{N}} \in \ell^{p'}$ act as a functional on ℓ^p. The proof hinges on the fact that both the tail part of y and the leading part of x_n are small. For any $\epsilon > 0$, there are integers k and N_i for each $i \leqslant k$, beyond which

$$\left(\sum_{i > k} |b_i|^{p'} \right)^{1/p'} < \epsilon \quad \text{AND} \quad \forall i \leqslant k, \exists N_i, \ n \geqslant N_i \Rightarrow |a_{n,i}| < \epsilon.$$

Then for $b := |b_1| + \cdots + |b_k|$ and $n \geqslant \max(N_1, \ldots, N_k)$, by Hölder's inequality,

$$|y^\top x_n| \leqslant \sum_{i \leqslant k} |b_i||a_{n,i}| + \sum_{i > k} |b_i||a_{n,i}|$$

$$\leqslant b\epsilon + \Big(\sum_{i > k} |b_i|^{p'}\Big)^{1/p'} \|x_n\|_p$$

$$\leqslant b\epsilon + \epsilon c$$

Hence $y^\top x_n \to 0$; since $y^\top$ is arbitrary in ℓ^{p*}, it follows that $x_n \rightharpoonup 0$. $\qquad \square$

Note that the proof is still valid for the space c_0, with minor modifications. But the proposition is false for ℓ^1: the bounded sequence e_n converges component-wise to $\mathbf{0}$ but not weakly, since $\mathbf{1}^\top e_n = 1 \not\to 0$.

Consider now a weakly convergent sequence, $x_n \rightharpoonup \mathbf{0}$, which does *not* converge to $\mathbf{0}$ in norm, in effect, $\|x_n\|_{\ell^p} \geqslant c > 0$ (for a subsequence). As proved above, each component converges to 0, yet the sequence as a whole is not diminishing in size. The example sequences e_n and $(0, \ldots, 0, \underset{\uparrow n}{1}, \ldots, \underset{\uparrow n+k}{1}, 0, \ldots)$ turn out to be quite typical. The following *gliding hump* argument shows that there must exist a subsequence whose terms are approximately non-overlapping.

Proposition 11.46

> **A sequence $x_n \in \ell^p$ ($1 \leqslant p < \infty$) which converges weakly to 0, with norm bounded below,**
>
> $$x_n \rightharpoonup \mathbf{0}, \qquad \|x_n\| \geqslant c > 0,$$
>
> **has a gliding hump subsequence y_n, such that**
>
> $$\sum_{n=0}^{\infty} \alpha_n y_n \in \ell^p \quad \Leftrightarrow \quad (\alpha_n)_{n \in \mathbb{N}} \in \ell^p.$$

With minor modifications, the proof is also valid for the space c_0.

Proof Let ϵ_n be an arbitrary sequence of (small) positive numbers.

(i) Starting with $y_1 = x_1 = (a_{1,i})_{i=0}^{\infty}$, there is some i_1 such that

$$\|(0, \ldots, 0, a_{1,i_1}, a_{1,i_1+1}, \ldots)\|_p < \epsilon_1.$$

Since each of the first i_1 components are converging to 0, there must be some sequence, x_{n_1}, which we'll relabel as $y_2 = (a_{2,i})_{i\in\mathbb{N}}$ such that

$$\|(a_{2,0}, \ldots, a_{2,i_1-1}, 0, \ldots)\|_p < \epsilon_2.$$

So the bulk of the norm occurs after i_1; as before there is some i_2 such that

$$\|(0, \ldots, 0, a_{2,i_2}, \ldots)\|_p < \epsilon_2.$$

Repeating the argument we can find a sequence $y_n = (a_{n,i})_{i\in\mathbb{N}}$ such that

$$\|(a_{n,0}, \ldots, a_{n,i_{n-1}-1}, 0, \ldots, \qquad \ldots, 0, 0, \ldots, \ldots)\|_p < \epsilon_n$$
$$\|(0, \ldots, \qquad \ldots, 0, a_{n,i_{n-1}}, \ldots, a_{n,i_n-1}, 0, \ldots, \ldots)\|_p > c - 2\epsilon_n,$$
$$\|(0, \ldots, \qquad \ldots, 0, 0, \ldots, \qquad \ldots, 0, a_{n,i_n}, \ldots)\|_p < \epsilon_n$$

By construction, the middle 'humps' of y_n, call them b_n, occur on consecutive and non-overlapping intervals $\{i_{n-1}, \ldots, i_n - 1\}$.

(ii) The weakly convergent sequence x_n is bounded by, say, d. Let y_n be a gliding hump subsequence with $\epsilon_n = \epsilon^n$. Then

$$\left\| \sum_n \alpha_n y_n - \sum_n \alpha_n b_n \right\| \leqslant \sum_n |\alpha_n| \|y_n - b_n\| \leqslant 2\sum_n |\alpha_n|\epsilon^n \leqslant \frac{2\epsilon}{1-\epsilon} \sup_n |\alpha_n|.$$

Hence $\sum_n \alpha_n y_n$ converges iff $\sum_n \alpha_n b_n$ does. Now, because the sequences b_n have non-overlapping supports,

$$\left\| \sum_{n=M}^N \alpha_n b_n \right\|_p^p = \sum_{n=M}^N \left(|\alpha_n|^p \sum_{i=i_{n-1}}^{i_n-1} |a_{n,i}|^p \right)$$

$$\therefore (c - 2\epsilon^M) \left(\sum_{n=M}^N |\alpha_n|^p \right)^{1/p} \leqslant \left\| \sum_{n=M}^N \alpha_n b_n \right\|_p \leqslant d \left(\sum_{n=M}^N |\alpha_n|^p \right)^{1/p}$$

The result then follows: $\sum_{n\in\mathbb{N}} |\alpha_n|^p$ is Cauchy in $\mathbb{R}$ iff $\sum_n \alpha_n b_n$ is Cauchy in ℓ^p, and so converges iff $\sum_n \alpha_n y_n$ does so in ℓ^p. $\square$

Proposition 11.47 (Schur's Property)

In ℓ^1, weak convergence of sequences is equivalent to convergence in norm,

$$x_n \rightharpoonup x \Leftrightarrow x_n \to x.$$

Proof To obtain a contradiction, we may suppose, without loss of generality, $x_n \rightharpoonup 0$ but $x_n \nrightarrow 0$. There is a subsequence such that $\|x_n\| \geqslant c > 0$, and a further gliding hump subsequence $y_n = (a_{n,i})_{i \in \mathbb{N}}$ with $\epsilon_n = \epsilon < c/4$. Consider the sequence z built up from the concatenation of the humps of y_n, as $z := (|a_{n,i}|/a_{n,i})_{i \in \mathbb{N}} \in \ell^\infty$ where, for each i, n is such that $i_{n-1} \leqslant i < i_n$ (note: use 1 instead of $|a_{n,i}|/a_{n,i}$ if $a_{n,i} = 0$). Then

$$|z \cdot y_n| \geqslant \sum_{i=i_{n-1}}^{i_n - 1} z_i a_{n,i} - 2\epsilon = \sum_{i=i_{n-1}}^{i_n - 1} |a_{n,i}| - 2\epsilon \geqslant (c - 4\epsilon) > 0,$$

contradicting $y_n \rightharpoonup 0$. The converse follows by continuity of functionals (Example 11.39(1)). □

A second application of the gliding hump argument is to compact operators on ℓ^p spaces. Proposition 11.43 immediately implies that any operator $T : \ell^p \to \ell^1$ is compact, since if $x_n \rightharpoonup x$ then $Tx_n \rightharpoonup Tx$ which is equivalent to $Tx_n \to Tx$ in ℓ^1. The following proposition is a strengthening of this result:

Proposition 11.48 (Pitt's Theorem)

Any operator $T : \ell^p \to \ell^q$, with $q < p < \infty$, is compact.

Proof Let $x_n \rightharpoonup 0$ in ℓ^p; then $Tx_n \rightharpoonup 0$ in ℓ^q. Suppose $Tx_n \nrightarrow 0$, so must have a subsequence bounded below, $\|Tx_n\| \geqslant c > 0$. Since T is continuous, it follows that $x_n \nrightarrow 0$, and therefore for a further subsequence $\|x_n\| \geqslant d > 0$. Hence x_n and Tx_n have joint 'gliding hump' subsequences, which we'll call y_n and Ty_n, such that

$$(\alpha_n)_{n \in \mathbb{N}} \in \ell^p \Leftrightarrow \sum_n \alpha_n y_n \in \ell^p \Rightarrow \sum_n \alpha_n T y_n \in \ell^q \Leftrightarrow (\alpha_n)_{n \in \mathbb{N}} \in \ell^q$$

But, of course, $\ell^p \nsubseteq \ell^q$, implying that $x_n \rightharpoonup 0 \Rightarrow Tx_n \to 0$ and therefore that T is compact, since ℓ^p is reflexive.

□

Exercises 11.49

1. Show $e_n \rightharpoonup 0$ in c_0 or ℓ^p $(p > 1)$, yet $e_n \not\to 0$. Hence the norm is not continuous with respect to weak convergence; or the inner product for Hilbert spaces.
   In ℓ^1, $e_n \not\rightharpoonup 0$, but when thought of as functionals on c_0, $e_n \overset{*}{\rightharpoonup} 0$.

2. In $L^1[0, 1]$, the functions $f_n(t) := e^{2\pi int}$ converge weakly $f_n \rightharpoonup 0$, but not in norm and not pointwise $f_n(t) \not\to 0$ at any t (see Theorem 9.34). Examples of functions that converge weakly but not in norm are typically rapidly oscillating.

3. ▶ The weak limit of T_n, if it exists, is unique. A subsequence of T_n also converges weakly to the same limit.

4. The map $T : c_{00} \to c_{00}$ defined by $(a_n)_{n \in \mathbb{N}} \mapsto (na_n)_{n \in \mathbb{N}}$ is linear but not continuous; yet $\|Tx\|_\infty \leqslant c_x$. Does this contradict the uniform boundedness theorem?

5. In a Hilbert space with an orthonormal basis e_n,

   (a) $e_n \rightharpoonup 0$,
   (b) $\sum_n \alpha_n e_n \rightharpoonup x \Leftrightarrow \sum_n \alpha_n e_n \to x$.

   (Hint: The series is bounded, by Proposition 11.41, i.e., $\|\alpha_1 e_1 + \cdots + \alpha_n e_n\|^2 \leqslant c$ and so $(\alpha_n)_{n \in \mathbb{N}} \in \ell^2$; or use Example 11.39(5).)

6. $\phi_n \to \phi \Rightarrow \phi_n \rightharpoonup \phi$; but show that the converse is not true for $e_n^\top \in c_0^*$.

7. Addition and scalar multiplication are continuous with respect to weak convergence, that is, if $T_n \rightharpoonup T$ and $S_n \rightharpoonup S$ then $T_n + S_n \rightharpoonup T + S$, and $\lambda T_n \rightharpoonup \lambda T$. Of course, they are also continuous with respect to norm-wise and strong convergence.

8. Continuous functions in general do not preserve weak convergence. For example, multiplication does not: $e^{2\pi int} e^{-2\pi int} = 1$ though $e^{\pm 2\pi int} \rightharpoonup 0$ in $L^1[0, 1]$.
   The most that can be said regarding the multiplication of operators is:

   (a) if $T_n \rightharpoonup T$ then $T_n S \rightharpoonup TS$ and $ST_n \rightharpoonup ST$,
   (b) if $T_n \rightharpoonup T$ and $S_n x \to Sx$ for all x, then $T_n S_n \rightharpoonup TS$,
   (c) if $\forall \phi \in X^*$, $\phi S_n \to \phi S$ and $T_n \rightharpoonup T$ then $S_n T_n \rightharpoonup ST$.

9. (a) For Banach spaces, if $T_n^\top \rightharpoonup T^\top$ then $T_n \rightharpoonup T$ (but not conversely).
   (b) For Hilbert spaces, if $T_n \rightharpoonup T$ then $T_n^* \rightharpoonup T^*$ (weakly continuous).

10. If $x_n \rightharpoonup x$ in X, then $\phi \mapsto (\phi x_n)_{n \in \mathbb{N}}$ maps X^* into $c \subset \ell^\infty$. For example, when X is ℓ^1, this map converts bounded sequences to convergent ones.

11. Every closed linear subspace is weakly closed (by Proposition 11.20). Thus, if $x_n \rightharpoonup x$, then there is a sequence $y_n \in [\![x_1, x_2, \dots]\!]$ which converges in norm, $y_n \to x$.

12. A set in X^* is weak-∗-closed when it contains all weak-∗-limit points; for example, $A^\perp$.

13. The strong limit of unitary isomorphisms U_n between two Hilbert spaces is an isometry U. But U need not be unitary; e.g., let U_n be defined on ℓ^2 by

$$U_n(a_1, a_2, \ldots) := (a_{n+1}, a_1, a_2, \ldots, a_n, a_{n+2}, a_{n+3}, \ldots).$$

Then U_n converges strongly to the right-shift operator R.

14. Use Pitt's theorem to show that $\ell^p \ncong \ell^q$ for $p \neq q$.

15. Justify the following step used in the proof of Pitt's theorem: If both $x_n \in \ell^p$ and $y_n \in \ell^q$ converge weakly to $\mathbf{0}$ and are bounded from below by $c > 0$, then they have joint gliding hump subsequences, agreeing in their indices and gliding hump positions.

16. The *Hadamard* matrices are defined recursively by $T_1 := \begin{pmatrix} 1 & 1 \\ 1 & -1 \end{pmatrix}$, $T_{n+1} :=$ $\begin{pmatrix} T_n & T_n \\ T_n & -T_n \end{pmatrix}$. $S_n := T_n/2^{n/2}$ are $2^n \times 2^n$ unitary matrices; they can be extended to unitary operators on ℓ^2 by $U_n x := S_n x$ when $x \in M_n := [\![e_0, \ldots, e_{2^n-1}]\!]$, and $U_n x := x$ when $x \in M_n^\perp$, and then $U_n \rightharpoonup 0$.

17. If a sequence of unitary isomorphisms U_n converges weakly to U, then $\|U\| \leqslant 1$. If U is known to be unitary, then the convergence is pointwise.
   (Hint: Expand $\|U_n x - U x\|^2$.)

Remarks 11.50

1. Not every closed subspace of a Banach space need be "complemented", e.g., the space $\ell^\infty \neq c_0 \oplus M$ for any closed linear subspace M (see [39]). Indeed there exist infinite-dimensional Banach spaces whose only complemented subspaces are the finite-dimensional or codimensional closed ones [34].

2. It is a theorem that Hilbert spaces are the only Banach spaces in which every closed subspace is complemented [38].

3. Weak convergence does not obey all the convergence properties of metric spaces. For example, not every weak limit point of a set M need have a sequence in M that converges weakly to it.

4. There are yet other types of convergence. For example, $B(X, Y)$ is itself a Banach space, and so there is *weak convergence* with respect to $B(X, Y)^*$, meaning $\Phi T_n \to \Phi T$ for all $\Phi \in B(X, Y)^*$.

Chapter 12
Differentiation and Integration

12.1 Differentiation

Although continuous linear transformations are stressed throughout the book—with good reason, for they are the morphisms of normed spaces—they represent, of course, a very special part of all the functions from one normed space to another. To put things in perspective, recall that the linear maps on $\mathbb{R}$ are $x \mapsto \lambda x$, a very restricted set of functions in comparison with the non-linear real continuous functions. However, the linear maps are still relevant for one class of continuous functions: maps that are 'locally linear', meaning that they can be approximated by linear operators up to second-order errors:

Definition 12.1

A function $f : X \to Y$ between normed spaces (over the same field) is said to be (Fréchet) **differentiable** at $x \in X$ when there is a continuous linear map $f'(x) \in B(X, Y)$ such that for h in a neighborhood of 0,

$$f(x + h) = f(x) + f'(x)h + o(h)$$

where $\|o(h)\|/\|h\| \to 0$ as $h \to 0$.

Note that f need not be defined on all of X but only on a neighborhood of x. The set of functions $f : U \subseteq X \to Y$, where U is an open subset of a normed space X and f is differentiable at all points $x \in U$, is here denoted $D(U, Y)$.

J. Muscat, *Functional Analysis*, https://doi.org/10.1007/978-3-031-27537-1_12

Proposition 12.2

The set $C^1(\bar{U}, Y) := \{ f \in D(U, Y) : f, f' \in C_b(\bar{U}) \}$ **is a vector space with norm**

$$\|f\|_{C^1} := \|f\|_C + \|f'\|_C.$$

Differentiation $D : f \mapsto f'$ **is an operator** $C^1(\bar{U}, Y) \to C(\bar{U}, Y)$, **which takes composition of functions to operator products,**

$$(f + g)' = f' + g', \quad (\lambda f)' = \lambda f',$$
$$(f \circ g)'(x) = f'(g(x))g'(x).$$

The last identity is called the *chain rule* of differentiation.

Proof The following identities and inequalities demonstrate the closure of C^1 as a vector space, linearity of D, and the chain rule:

$$(f + g)(x + h) = f(x + h) + g(x + h)$$
$$= f(x) + f'(x)h + o_f(h) + g(x) + g'(x)h + o_g(h)$$
$$= f(x) + g(x) + (f' + g')(x)h + (o_f(h) + o_g(h))$$

$$\lambda f(x + h) = \lambda f(x) + \lambda f'(x)h + \lambda o(h)$$

$$f \circ g(x + h) = f\big(g(x + h)\big)$$
$$= f\big(g(x) + g'(x)h + o_g(h)\big)$$
$$= f\big(g(x)\big) + f'\big(g(x)\big)\big(g'(x)h + o_g(h)\big) + o_f(h)$$
$$= f\big(g(x)\big) + f'\big(g(x)\big)g'(x)h + (f'(g(x))o_g(h) + o_f(h))$$

$$\|o_f(h) + o_g(h)\| \leqslant \|o_f(h)\| + \|o_g(h)\|,$$
$$\|\lambda o(h)\| = |\lambda| \|o(h)\|,$$
$$\|T o_g(h) + o_f(h)\| \leqslant \|T\| \|o_g(h)\| + \|o_f(h)\|, \quad \text{for any } T \in B(X, Y).$$

The norm axioms of $\| \cdot \|_{C^1}$ are easy to verify:

$$\|f + g\|_{C^1} \leqslant \|f\|_C + \|g\|_C + \|f'\|_C + \|g'\|_C = \|f\|_{C^1} + \|g\|_{C^1},$$
$$\|\lambda f\|_{C^1} = \|\lambda f\|_C + \|\lambda f'\|_C = |\lambda| \|f\|_C + |\lambda| \|f'\|_C = |\lambda| \|f\|_{C^1},$$
$$0 = \|f\|_{C^1} = \|f\|_C + \|f'\|_C \Rightarrow \|f\|_C = 0 \Leftrightarrow f = 0.$$

The continuity of D results from the inequality:

$$\|Df\|_C = \|f'\|_C \leqslant \|f\|_C + \|f'\|_C = \|f\|_{C^1}.$$

$\square$

Examples 12.3

1. The constant functions $f(x) := y_0$ are differentiable with $f' = 0$.
2. In $\mathbb{R}$ or $\mathbb{C}$, the functions $f(x) := x^n$ are differentiable with

$$f(x + h) = (x + h)^n = x^n + nx^{n-1}h + o(h),$$

so $f'(x) = nx^{n-1}$. Polynomials are thus differentiable.
3. Continuous linear maps are differentiable, $T(x+h) = Tx+Th$, so $T'(x) = T$. A special case of the composition law is $(T \circ f)' = T \circ f'$ when T is a fixed operator.
4. The derivative of $\boldsymbol{F} : \mathbb{R} \to \mathbb{R}^2$, $\boldsymbol{F}(t) := \binom{f(t)}{g(t)} = f(t)\binom{1}{0} + g(t)\binom{0}{1}$ is $\boldsymbol{F}'(t) = \binom{f'(t)}{g'(t)}$. A differentiable path $r : \mathbb{R} \to X$ is called a *curve*. The direction of its derivative r' is called its *tangent*. The *arclength* of a curve is $\int_r ds := \int |r'(t)| \, dt$.
5. Define $f : \mathbb{R}^2 \to \mathbb{R}$ by $f(x, y) := x^2 - y$. Then $f'(x, y) : \mathbb{R}^2 \to \mathbb{R}$ is its *gradient* $f'(x, y) = (2x, -1)$ since

$$f(x + h, y + k) = (x + h)^2 - (y + k) = (x^2 - y) + (2x - 1)\binom{h}{k} + h^2.$$

The map $(h, k) \mapsto (x^2 - y) + 2xh - k$ gives the *tangent plane* to the surface $z = f(x, y)$ at the point (x, y, z).
6. A real inner product $\langle \cdot, \cdot \rangle : X^2 \to \mathbb{R}$ is differentiable,

$$\langle x + h, y + k \rangle = \langle x, y \rangle + (\langle x, k \rangle + \langle h, y \rangle) + \langle h, k \rangle.$$

The middle term is linear in (h, k), and the last term is $o(h, k)$ by the Cauchy-Schwarz inequality,

$$\frac{|\langle h, k \rangle|}{\|h\| + \|k\|} \leqslant \frac{\|h\|\|k\|}{\|h\| + \|k\|} \leqslant \|h\| \to 0 \text{ as } (h, k) \to (0, 0).$$

7. We often write $D_v f(x) := f'(x)v$. Note that

$$D_{v+w}f = D_v f + D_w f, \quad D_{\lambda v} f = \lambda D_v f.$$

Because of this last property, v is usually taken to be a unit vector. It is the *directional derivative* of f along v.

When $X = \mathbb{R}$ there are only two unit vectors, $v = \pm 1$, and the notation used is $\frac{d}{dx} := D_1$ for the derivative in the positive direction. Similarly, for $\mathbb{C}$, $\frac{d}{dz} := D_1$. In $\mathbb{R}^n$, the standard basis consists of n unit vectors e_k, and we define $\partial_k := D_{e_k}$.

8. For $X = \mathbb{R}$, the derivative can be taken to be a function $f' : \mathbb{R} \to Y$, since $B(\mathbb{R}, Y) \equiv Y$.

9. ▶ Differentiable functions are continuous in x, in fact are Lipschitz in a neighborhood of any point

$$\| f(y) - f(x) \| = \| f'(x)(y - x) + o(y - x) \| \leqslant c \| y - x \|.$$

In particular, $f(y) \to f(x)$ as $y \to x$. But there are Lipschitz functions, such as $x \mapsto |x|$ on $\mathbb{R}$, that are not differentiable.

10. ⋆ $C^1(\mathbb{R})$ is a non-closed linear subspace of $C_b(\mathbb{R})$.

Proof: The functions $\sin nt$ have unit norms in $C_b(\mathbb{R})$, but their derivatives $n \cos nt$ have arbitrarily large ∞-norm. Let us define

$$f(t) := \sum_{n=0}^{\infty} \frac{1}{2^n} \sin 4^n t$$

with the partial sums f_N converging absolutely in $C_b(\mathbb{R})$. But this is an example of a nowhere-differentiable function (check it is not differentiable at 0 at least), so although $f_N \in C^1(\mathbb{R})$ and $f_N \to f$ uniformly, $f \notin C^1(\mathbb{R})$.

Proposition 12.4

> **The kernel of D on $D(X, Y)$ consists of the constant functions,**
>
> $$Df = 0 \Rightarrow f \text{ is constant.}$$

Proof We first identify the kernel when the differentiable functions are real valued, $g : \mathbb{R} \to \mathbb{R}$. Suppose $g'(t) = 0$ for all $t \in [a, b]$, and let

$$G(t) := g(t) - \frac{(t - a)g(b) + (b - t)g(a)}{b - a}$$

also differentiable, with $G(a) = 0 = G(b)$, and

$$G(t + h) - G(t) = G'(t)h + o(h) = -\frac{g(b) - g(a)}{b - a} h + o(h), \qquad t \in]a, b[.$$

$$(12.1)$$

G is continuous on the compact set $[a, b]$, so it must have maximum and minimum points. We can assume one of them to be inside $]a, b[$, for if they are at a and b, then trivially G is 0 throughout $[a, b]$.

Now, on any minimum of G within $]a, b[$, as h changes sign from negative to positive, $G(t_0 + h) - G(t_0)$ remains positive; on a maximum it remains negative. From (12.1), this can only hold if $g(a) = g(b)$. As a and b are arbitrary, this shows that g is constant.

For $f' = 0$ on X, we can use functionals to reduce it to a real-valued function: let $g(t) := \phi \circ f(tx)$ for any non-zero $x \in X$ and $\phi \in Y^*$. It is differentiable,

$$g(t + h) = \phi \circ f(tx + hx) = \phi\big(f(tx)\big) + \phi\big(f'(tx)hx\big) + o(hx) = g(t) + o(hx),$$

with derivative $g'(t) = 0$. By the first part, $g(t) = g(0) = \phi \circ f(0)$ constant. But with ϕ and x arbitrary, this shows that $f = f(0)$, a constant function.   □

Exercises 12.5

1. For differentiable functions $\lambda : \mathbb{R} \to \mathbb{F}$, $f, g : \mathbb{R} \to X$, $F : X^2 \to X$, $T : \mathbb{R} \to B(X, Y)$,

   (a) $\frac{\mathrm{d}}{\mathrm{d}t}(\lambda(t)f(t)) = \lambda'(t)f(t) + \lambda(t)f'(t)$,
   (b) $\langle f, g \rangle' = \langle f', g \rangle + \langle f, g' \rangle$,
   (c) $\frac{\mathrm{d}}{\mathrm{d}t}F(f(t), g(t)) = \partial_1 F(f(t), g(t))f'(t) + \partial_2 F(f(t), g(t))g'(t)$,
   (d) $\frac{\mathrm{d}}{\mathrm{d}t}T(t)f(t) = T'(t)f(t) + T(t)f'(t)$.

2. For a curve on the sphere of a real Hilbert space $r : [0, 1] \to S_H$, the tangent t at any point satisfies $\langle t, r \rangle = 0$.
3. ▶ For a differentiable function $y : \mathbb{R}^n \to \mathbb{R}^m$, y' is the Jacobian matrix $[\partial_i y_j]$.
4. The derivative itself, $f'(x)$, need not be continuous in x. For example, show that $f(x) := x^2 \sin(1/x)$ (and $f(0) := 0$) is differentiable at all points, yet its derivative is not continuous at 0.
5. If $f : X \to \mathbb{R}$ is differentiable and has a maximum/minimum at x in some open set $U \subseteq X$, then $f'(x) = 0$.
6. *L'Hôpital's rule*: If $f : \mathbb{R} \to X$, $g : \mathbb{R} \to \mathbb{R}$ are differentiable functions satisfying $f(a) = 0$, $g(a) = 0$, but $g'(a) \neq 0$, then

$$\lim_{x \to a} \frac{f(x)}{g(x)} = \frac{f'(a)}{g'(a)}.$$

12.2   Integration for Vector-Valued Functions

The construction of $L^1(\mathbb{R})$ can be extended to include functions $f : \mathbb{R} \to X$, where X is a Banach space, as done in Sect. 9.2. Briefly,

- a vector-valued *characteristic* function $x1_E$ maps t to $x \in X$ when $t \in E \subseteq \mathbb{R}$ and to 0 otherwise;

- a simple function is a linear combination of vector characteristic functions on sets of finite measure, in which case,

$$\int \sum_{n=1}^{N} 1_{E_n} x_n := \sum_{n=1}^{N} \mu(E_n) x_n.$$

The set of simple functions is a normed space with $\|s\| := \int \|s(t)\|_X \, dt$.
- a function $f : \mathbb{R} \to X$ is **integrable** when it is the ae-limit of a Cauchy sequence of simple functions $s_n \to f$ a.e., $\|s_n - s_m\| \to 0$ a.e., $n, m \to \infty$; its **integral** is

$$\int f := \lim_{n \to \infty} \int s_n.$$

- on a measurable set $A \subseteq \mathbb{R}$, $\int_A f := \int f 1_A$, e.g., $\int_a^b f = \int_{[a,b]} f$ for $a \leqslant b$.

Quoting the results of Sect. 9.2,

Proposition 12.6

For $f, g : \mathbb{R} \to X$ integrable,

  (i) $\int f + g = \int f + \int g$, $\quad \int \lambda f = \lambda \int f \quad (\lambda \in \mathbb{F})$,
  (ii) $\| \int f \| \leqslant \int \|f(t)\| \, dt$,
  (iii) $\int \lambda(t) x \, dt = (\int \lambda) x$ **for** $\lambda \in L^1(\mathbb{R}), x \in X$,
  (iv) $\int T f = T \int f$ **for** $T \in B(X, Y)$.

Examples 12.7

1. $\displaystyle \int \begin{pmatrix} f(t) \\ g(t) \end{pmatrix} \, dt = \int f(t) \begin{pmatrix} 1 \\ 0 \end{pmatrix} + g(t) \begin{pmatrix} 0 \\ 1 \end{pmatrix} \, dt = \begin{pmatrix} \int f \\ \int g \end{pmatrix}$, when $f, g : \mathbb{R} \to \mathbb{R}$ are integrable. Similarly, $\displaystyle \int_0^1 \begin{pmatrix} 1 & t \\ t^2 & t^3 \end{pmatrix} \, dt = \begin{pmatrix} 1 & \frac{1}{2} \\ \frac{1}{3} & \frac{1}{4} \end{pmatrix}$.

2. Any continuous function $f : [a, b] \to X$ is integrable, since

$$\int_a^b \|f(t)\| \, dt \leqslant (b - a) \|f\|_C.$$

3. If $f_n(t) \to f(t)$ in X, uniformly in $t \in [a, b]$, then $\int_a^b f_n \to \int_a^b f$ in X, since

$$\left\| \int_a^b (f_n - f) \right\| \leqslant \int_a^b \|f_n(t) - f(t)\| \, dt \leqslant (b - a) \|f_n - f\|_{L^\infty[a,b]}.$$

The connection between differentiation and integration is one of the cornerstones of classical mathematics. It remains valid for vector-valued functions:

Theorem 12.8 (Fundamental Theorem of Calculus)

If $f : [a, b] \to X$ is integrable, and continuous at $t \in \,]a, b[$, then its integral is differentiable at t, and

$$\frac{\mathrm{d}}{\mathrm{d}t} \int_a^t f = f(t).$$

If $f' : [a, b] \to X$ is continuous, then

$$\int_a^b f' = f(b) - f(a).$$

Proof (i) The first part is a consequence of

$$\int_a^{t+h} f = \int_a^t f + f(t)h + \left(\int_t^{t+h} f - f(t)h \right)$$

and

$$\left\| \frac{1}{h} \int_t^{t+h} f(\tau)\, \mathrm{d}\tau - f(t) \right\| = \left\| \int_t^{t+h} \frac{f(\tau) - f(t)}{h}\, \mathrm{d}\tau \right\|$$

$$\leqslant \left| \int_t^{t+h} \frac{\|f(\tau) - f(t)\|}{|h|}\, \mathrm{d}\tau \right|$$

$$< \frac{\epsilon}{|h|} \left| \int_t^{t+h} \mathrm{d}\tau \right| = \epsilon$$

for arbitrary $\epsilon > 0$ and $|h|$ sufficiently small, since f is continuous at t.

(ii) For the second part, let $F(t) := \int_a^t f'$. By (i) we obtain $F' = f'$ on $]a, b[$, so their difference $F(t) - f(t)$ must be a constant c. As $F(a) = 0$, $c = -f(a)$. ∎

Proposition 12.9 (Mean Value Theorem)

For a continuous function $f : [a, b] \to X$, the mean value

$$\frac{1}{b - a} \int_a^b f(t)\, \mathrm{d}t$$

belongs to the closed convex hull of $f[a, b]$.

Proof The function is uniformly continuous (Proposition 6.17), so splitting $[a, b]$ into small enough intervals $[t_n, t_{n+1}]$ of size $h = (b - a)/N$ each ($t_n := a + nh$), ensures that $\| f(t) - f(\tilde{t}) \| < \epsilon$ whenever $t, \tilde{t}$ are in the same sub-interval. This means that f can be approximated uniformly by a simple function which takes the value $f(t_n)$ on the interval $[t_n, t_{n+1}[$, and its integral $\int_a^b f$ can be approximated to within $\epsilon(b - a)$ by the sum

$$\big(f(a) + f(t_1) + \cdots + f(t_{N-1})\big)h.$$

Thus $\frac{1}{b-a} \int_a^b f$ is within ϵ of $(f(a) + f(a+h) + \cdots + f(b-h))/N$ which belongs to the convex hull of $f[a, b]$. Since ϵ is arbitrarily small, the result follows. $\square$

Corollary 12.10

For a continuously differentiable function $f : [a, b] \to X$,

$$\frac{f(b) - f(a)}{b - a}$$

belongs to the closed convex hull of $f'[a, b]$.

Proof

$$\frac{f(b) - f(a)}{b - a} = \frac{1}{b - a} \int_a^b f'.$$

$\square$

Recall that f' is a function $U \to B(X, Y)$; it may itself be differentiable, with derivative denoted by $f''(x) \in B(X, B(X, Y))$. This Banach space is actually isomorphic to the space of bilinear maps $B(X^2, Y)$ via the identification $T_{x_1 x_2} = T(x_1, x_2)$. Because of this, $f''(x)$ is akin to an operator that converts a pair of vectors of X into a vector in Y; in particular, $f''(x)(h, h)$ makes sense, and is often shortened into the form $f''(x)h^2$ (though no vectors are squared).

More generally, $f^{(n)}$ is the nth derivative of f: it takes n vectors in X and outputs a vector in Y. The set of n-times differentiable functions $f : \mathbb{R} \to X$, with $f^{(n)}$ continuous, is denoted by $C^n(\mathbb{R}, X)$.

Theorem 12.11 (Taylor's Theorem)

For $f \in C^n(\mathbb{R}, X)$ ($n = 1, 2, \ldots$),

$$f(t + h) = f(t) + f'(t)h + \cdots + \frac{f^{(n)}(t)h^n}{n!} + o(h^n).$$

Proof As expected the proof proceeds by induction on n. To illustrate the idea behind the inductive step, we only consider how the statement for $n = 2$ follows from that for $n = 1$. Let $f \in C^2(\mathbb{R}, X)$, and let

$$F(s) := f(t + s) - f(t) - f'(t)s - f''(t)s^2/2!$$

We wish to show $F(h) = o(h^2)$. F is continuously differentiable in s because it consists of sums and products of continuously differentiable functions, in fact

$$F'(s) = f'(t + s) - f'(t) - f''(t)s = o(s),$$

since f' is differentiable. Using the above corollary, it follows that $\frac{F(h)-F(0)}{h}$ belongs to the closed convex hull of $F'[0, h]$, whose values are at most of order $o(h)$. Since $F(0) = 0$, we have $F(h) = o(h^2)$ as required.

The reader is invited to adapt this proof to show that if the statement is correct for n then it is also true for $n + 1$. The case $n = 1$ is, of course, part of the definition of the derivative. $\square$

Exercises 12.12

1. *Integration by parts*: $\int_a^b f(t)F'(t)\,dt = [fF]_a^b - \int_a^b f'(t)F(t)\,dt$, where $f : \mathbb{R} \to \mathbb{F}$ and $F : \mathbb{R} \to X$ have continuous derivatives.
2. *Change of variables*: $\int_a^b f(x)\,dx = \int_{y(a)}^{y(b)} F(y)\frac{dx}{dy}\,dy$, where $y : \mathbb{R} \to \mathbb{R}$ has an invertible continuous derivative, and $F(y(x)) = f(x)$.
3. If $f : [a, b] \to M$ is continuous, where M is a closed linear subspace of X, then $\int_a^b f \in M$.
4. The symbol $o(h)$ satisfies $\|o(h)\| \leqslant c\|h\|$ for h small enough, but not necessarily $\|o(h)\| \leqslant c\|h\|^2$. However show that the latter inequality is true if $f'(y)$ is Lipschitz in y in some ball about x, by evaluating

$$\left\| \int_0^1 \frac{d}{dt} f(x + th) - f'(x)h\,dt \right\|.$$

5. A bounded set B in $C^1[a, b]$ is uniformly bounded and equicontinuous.
   It follows by the Arzelà-Ascoli theorem that a sequence of bounded functions in $C^1[a, b]$ has a convergent subsequence.
   (Hint: Use the mean value theorem and the fact that $\|f'\|_C \leqslant k$.)

Application: The Newton-Raphson Algorithm

It would be no exaggeration to claim that a large proportion of real-world problems reduce to solving some (non-linear) equation $f(x) = y$ where x and y belong to some Banach spaces. From designing whole electronic circuits, finding the right image in some feature space, solving partial differential equations, to finding the right parameters in data models, such equations are ubiquitous in "continuous" models.

When f is differentiable, we can hope to approximate f by some affine map, and thereby solve the resulting equation by inverting the operator. In detail, we might start with a first estimate x and find a better approximation from

$$y = f(x + h) \approx f(x) + f'(x)h,$$

namely $h = f'(x)^{-1}(y - f(x))$. This suggests the following iteration:

Proposition 12.13 (The Newton-Raphson Method)

Let $f(\tilde{x}) = y$ and suppose that f is differentiable in a neighborhood of $\tilde{x}$, with $f'(x)$ Lipschitz in x and $\| f'(x)^{-1} \| \leqslant c$. Then if x_0 is sufficiently close to $\tilde{x}$, the iteration

$$x_{n+1} := x_n + f'(x_n)^{-1}(y - f(x_n))$$

converges to $\tilde{x}$.

Proof The differentiability of f at $\tilde{x}$ states that for $h = x_n - \tilde{x}$, $|h| < \epsilon$,

$$f(x_n) = f(\tilde{x} + h) = f(\tilde{x}) + f'(\tilde{x})h + o(h),$$
$$\therefore \qquad f(x_n) = y + f'(x_n)h + (f'(\tilde{x}) - f'(x_n))h + o(h),$$
$$\therefore \; f'(x_n)^{-1}(f(x_n) - y) = x_n - \tilde{x} + f'(x_n)^{-1}((f'(\tilde{x}) - f'(x_n))h + o(h))$$
$$\therefore \qquad x_{n+1} - \tilde{x} = -f'(x_n)^{-1}((f'(\tilde{x}) - f'(x_n))h + o(h))$$
$$\therefore \qquad \|x_{n+1} - \tilde{x}\| \leqslant \tfrac{3ck}{2}\|h\|^2 = \tilde{c}\|x_n - \tilde{x}\|^2,$$

where k is the Lipschitz constant of f' and $\|o(h)\| \leqslant \tfrac{1}{2}k\|h\|^2$ (Exercise 12.12(4)). If $\epsilon < 1/\tilde{c}$ then it implies firstly that if x_n belongs to $B_\epsilon(\tilde{x})$, then so does x_{n+1}, and secondly by induction it follows that $\|x_n - \tilde{x}\| \leqslant (\tilde{c}\|x_0 - \tilde{x}\|)^{2^n}/\tilde{c} \to 0$ as $n \to \infty$. $\qquad \square$

This algorithm is very effective since it converges quadratically, as long as x_0 is already sufficiently close to $\tilde{x}$. In practice, other algorithms are utilized to perform a broad search for a solution, and Newton's method is then used to rapidly home in on it. Another caveat is that it may be computationally expensive: at each step, one has to calculate not only the derivative $f'(x)$ but effectively also its inverse. The methods that are most often used employ modified iterations like $x_{n+1} := x_n + H_n(y - f(x_n))$, where H_n are operators that approximate $f'(x_n)^{-1}$ but are easier to calculate.

Examples 12.14

1. To solve for $e^{iz} = 1$ close to $z = 6$, apply Newton's iteration:

$$z_{n+1} := z_n + i(1 - e^{-iz_n})$$
$$z_0 = 6$$
$$z_1 = 6.27942 + 0.03983i$$
$$z_2 = 6.28334 - 0.00080i$$
$$z_3 = 6.28319 - 0.0i$$

Examples of other equations whose solutions are routinely found using this method are (a) roots of polynomials, e.g., $x^3 = 2$, (b) transcendental equations such as $x - \sin x = 1$ or $x \tan x = 1$.

2. Find the points on the orbit of Mars whose distance from Earth is 1 a.u. on 1 Jan. In appropriate coordinates, the problem reduces to solving the equations,

$$f(x, y) := x^2 + 1.009y^2 - 0.284x = 2.302,$$

$$g(x, y) := x^2 + y^2 + 1.11x - 1.622y = 0.0343.$$

Setting up the Newton-Raphson iteration gives

$$\begin{pmatrix} x_{n+1} \\ y_{n+1} \end{pmatrix} = \begin{pmatrix} x_n \\ y_n \end{pmatrix} + \frac{1}{\Delta} \begin{pmatrix} 2y_n - 1.622 & -2.018y_n \\ -2x_n - 1.11 & 2x_n - 0.284 \end{pmatrix} \begin{pmatrix} 2.302 - f(x_n, y_n) \\ 0.0343 - g(x_n, y_n) \end{pmatrix},$$
$$\Delta = 0.461 - 3.244x_n - 2.808y_n - 0.036x_ny_n.$$

Depending on the starting point, it may converge either to $(0.153, 1.517)$ or to $(-1.365, 0.225)$. But most of the time it does not converge—hence the need to perform a rough search for solutions before zeroing in using the algorithm.

3. The method can be used to find the minimum of a scalar differentiable function, which is equivalent to finding zeros of its derivative. For example, if a function were exactly quadratic

$$f(x) = c + b \cdot x + \tfrac{1}{2}x^\top A x$$

(here A is a symmetric matrix) then the minimum occurs when $Ax + b = 0$, and Newton's method, starting from $x_0 = 0$, finds the minimum point in one step: $x_1 = -A^{-1}b$. The more undulating a function is, the more demanding it becomes to find the true minimum. Two challenging functions that have served as benchmarks are the following

(a) $(1 - x)^2 + 100(y - x^2)^2$ (Rosenbrock's valley),
(b) $(x^2 + y - 11)^2 + (x + y^2 - 7)^2$ (Himmelblau's function).

4. ⋆ To align two real-valued functions f and g as best as possible, one may find a that minimizes $\int (f(t-a) - g(t))^2 \, dt$. Expanding this out, then differentiating in a, gives up to order $o(a^2)$,

$$\int (f(t) - g(t))^2 - 2(f(t) - g(t))f'(t)a + (f'(t)a)^2$$

$$+ (f(t) - g(t))f''(t)a^2 \, dt,$$

$$\therefore \int (f(t) - g(t))f'(t) - \left(f'(t)^2 + (f(t) - g(t))f''(t) \right)a \, dt + o(a) = 0$$

The Newton-Raphson estimate of a is

$$a = \frac{\langle f - g, f' \rangle}{\|f'\|^2 + \langle f - g, f'' \rangle}.$$

Letting $f_{n+1}(t) := f_n(t-a)$, $f_0(t) := f(t)$, and iterating aligns the two functions. (You can try this out with $f(t) = \cos t$ and $g(t) = \cos(t+1)$ over the interval $[0, 2\pi]$.) This method, modified to $\mathbb{R}^2$, has been implemented to align images, for example to compensate for video camera jitter from one frame to the next.

12.3 Complex Differentiation and Integration

Let X be a complex Banach space, then a differentiable function $f : \mathbb{C} \to X$ is also called **analytic**, i.e., for all z, h,

$$f(z+h) = f(z) + f'(z)h + o(h).$$

The set of functions $f : \mathbb{C} \to \mathbb{C}$ which are analytic at all points z in an open set $U \supseteq A$, is denoted by $C^\omega(A)$.

A function $f : \mathbb{C} \to X$ is *integrable* along a differentiable path $w : [t_0, t_1] \to \mathbb{C}$, when the composition $f \circ w : [t_0, t_1] \to \mathbb{C} \to X$ is integrable. Its **integral** is then

$$\int_w f(z) \, dz := \int_{t_0}^{t_1} f(w(t))w'(t) \, dt.$$

Notice that dz/i is along the normal to a path. Proposition 12.6 remains true, for example property (ii) becomes

$$\left\| \int_w f(z) \, dz \right\| \leqslant \int \|f(w(t))\| \, ds, \quad \text{where} \ ds := |w'(t)| \, dt.$$

Examples 12.15

1. Along any curve w which starts at $w(0) = a + bi$ and ends at $w(1) = c + di$, $\int_{a+bi}^{c+di} 1 \, dz = \int_0^1 w'(t) \, dt = [w(t)]_0^1 = [z]_{a+bi}^{c+di}$. More generally,

$$\int_{a+bi}^{c+di} f'(z) \, dz = \int_0^1 f'(w(t)) w'(t) \, dt = [f(z)]_{a+bi}^{c+di}$$

for f analytic (with f' continuous). Thus one can integrate complex function derivatives in the same manner as real-valued functions.

2. The map $z \mapsto \frac{1}{z}$ is analytic except at $z = 0$. On a circular path $w(t) := re^{it}$, $0 \leqslant t \leqslant 2\pi$,

$$\int_{\circ} \frac{1}{z} \, dz = \int_0^{2\pi} \frac{1}{r} e^{-it} i r e^{it} \, dt = 2\pi i$$

(independent of the radius). Thus the integral $\int_{-1}^{1} \frac{1}{z} \, dz$ does not have a unique answer, but depends on whether one traverses a path that passes above or below the origin, and how often it loops around it. But otherwise $\int_{\circ} \frac{1}{z^n} \, dz = 0$.

3. *Cauchy-Riemann equations*: An analytic function $f : \mathbb{C} \to \mathbb{C}$, $x + iy \mapsto u(x, y) + iv(x, y)$ satisfies the equations

$$\frac{\partial u}{\partial x} = \frac{\partial v}{\partial y}, \qquad \frac{\partial u}{\partial y} = -\frac{\partial v}{\partial x},$$

since $f'(z) = \frac{\partial u}{\partial x} + i\frac{\partial v}{\partial x} = \frac{\partial v}{\partial y} - i\frac{\partial u}{\partial y}$, which can be obtained by comparing

$$f(z + h) = u(x, y) + \frac{\partial u}{\partial x} h + iv(x, y) + i\frac{\partial v}{\partial x} h + o(h)$$

$$= f(z) + f'(z)h + o(h),$$

$$f(z + ih) = u(x, y) + \frac{\partial u}{\partial y} h + iv(x, y) + i\frac{\partial v}{\partial y} h + o(h)$$

$$= f(z) + f'(z)ih + o(h).$$

4. The conjugate map $z \mapsto \bar{z}$ is not analytic, since $\overline{z + h} = \bar{z} + \bar{h}$. Therefore, $\text{Re}(z) = (z + \bar{z})/2$, $\text{Im}(z) = (z - \bar{z})/2i$, and $|z| = z\bar{z}$, are not analytic. Indeed the Cauchy-Riemann equations can be written symbolically as $\frac{\partial f}{\partial \bar{z}} = 0$, and interpreted as f being independent of $\bar{z}$.

Analytic functions $f : \mathbb{C} \to X_{\mathbb{C}}$ are profoundly different from similar-looking functions $f : \mathbb{R}^2 \to X_{\mathbb{R}}$ that are simply differentiable over the reals. This is borne out by a string of results discovered by Augustin Cauchy in the 19th century. We will only present here the essential theorems (See [20] for a more thorough presentation).

Theorem 12.16 (Cauchy's Theorem)

> Let $\Omega \subset \mathbb{C}$ be a bounded open set having a finite number of differentiable curves as boundary. Let f be a function from $\mathbb{C}$ into a Banach space, which is analytic on and in Ω, then along these boundary curves,
>
> $$\oint f(z)\,dz = 0.$$

Warning: the curves must be traversed in a consistent manner, say with the region Ω to the left of each curve. A fully rigorous proof requires results that are too technical to be presented in a simplified form (see [10]). These details will be disregarded in favor of a more intuitive approach, both for this theorem and its corollaries.

Proof At any analytic point, $f(z+h) = f(z) + f'(z)h + o(h)$, where $o(h)/h \to 0$ as $h \to 0$. So for any $\epsilon > 0$ and $|h| < \delta$ small enough, we have $\|o(h)\| < \epsilon\delta$. For any closed curve $\square$ inside a disk $B_\delta(z_0) \subseteq \Omega$ we get, using Example 12.15(1) above,

$$\int_\square f(w)\,dw = \int_\square f(z_0 + z)\,dz$$

$$= \int_\square f(z_0) + f'(z_0)z + o(z)\,dz$$

$$= f(z_0)\int_\square 1\,dz + f'(z_0)\int_\square z\,dz + \int_\square o(z)\,dz$$

$$= \int_\square o(z)\,dz$$

$$\therefore \ \left\|\int_\square f(w)\,dw\right\| \leqslant \int_\square \|o(h)\|\,ds < \epsilon\delta \times \text{Perimeter}(\square) \tag{12.2}$$

Each point $z_0 \in \Omega$ might need a different δ, but since $\overline{\Omega}$ is compact, there is a minimum δ that works at all points (as in Proposition 6.17).

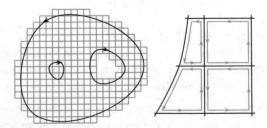

The region Ω can be covered by an array of squares of side δ, as shown in the diagram. The integral on the boundary $\partial\Omega$ can be split up into a sum of integrals along the squares that are within Ω, except that when a square intersects the boundary $\partial\Omega$, the integral is partly along the square and partly along the boundary. Each tiny loop has perimeter at most $4\delta + l$, where l is the length of that part of the boundary curve which lies inside the square.

If Ω is enclosed in a square of side L, there are at most $(L/\delta)^2$ squares in all, so the sum of the integrals is at most

$$\left\| \int_{\partial\Omega} f(w)\,dw \right\| \leqslant \sum_i \left\| \int_{\square_i} f(w)\,dw \right\|$$

$$\leqslant \sum_i \epsilon\delta(4\delta + l_i) \text{ by } (12.2)$$

$$\leqslant \left(4L^2 + \text{Perimeter}(\Omega)\delta\right)\epsilon$$

With ϵ arbitrarily small, the integral must vanish. $\qquad\square$

Corollary 12.17

> If $f : \mathbb{C} \to X$ is analytic in the interior Ω of a simple closed curve w, then the integral $\int_a^b f(z)\,dz$ is well-defined when $a, b \in \Omega$, independent of the path taken (within Ω).

Proof Any two paths inside Ω, from a to b, together form one or more simple closed paths, inside which f is analytic. Hence the integral of f on this closed loop is 0. $\qquad\square$

One of the surprising results of Cauchy's theorem is that the value of the integral $\oint f(z)\,dz$ is independent of the bounding curve itself, but only on interior "distant" regions!

Corollary 12.18 (Cauchy's Residue Theorem)

> The integral of f over a closed simple curve depends only on those regions inside where f is not analytic,
>
> $$\frac{1}{2\pi i} \oint f(z)\,dz = \sum_i \textbf{Residue}_i(f).$$

Proof Enclose the non-analytic parts by a finite number of curves w_i—the outer boundary curve γ already does this, but it may be possible to further isolate the non-analytic parts—to form one analytic region, around which the integral is zero,

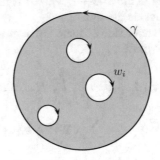

$$\frac{1}{2\pi i} \oint_\gamma f + \frac{1}{2\pi i} \sum_i \int_{w_i} f = 0$$

traversing each curve w_i in a clockwise direction. The value of the integral around each non-analytic region in a counter-clockwise direction may be called a 'residue' of f.   □

Because of this, the integral around a closed simple curve is often denoted by $\oint f(z)\,dz$, without reference to the (counter-clockwise) path taken, as long as it is clear from the context which non-analytic regions are included.

   The simplest cases in which a function fails to be analytic are of isolated points, called *isolated singularities*. An example of an isolated singularity a is a *pole* of order n when the function is of the type $f(z)/(z - a)^n$ with f analytic in a neighborhood of a and $f(a) \neq 0$. A *simple pole* is a pole of order 1. All other isolated singularities are called *essential singularities*. We shall see later that the residue of a function at a pole of order $n + 1$ is $f^{(n)}(a)/n!$, but what can be proved here is the case for a simple pole:

Proposition 12.19 (Cauchy's Integral Formula)

If $f : \mathbb{C} \to X$ is analytic inside a simple closed path that contains a, then

$$f(a) = \frac{1}{2\pi i} \oint \frac{f(z)}{z - a}\,dz.$$

Agustin Louis Cauchy (1789–1857) Cauchy studied under Lagrange and Laplace as a military engineer, but decided to continue with mathematics. A staunch royalist, he replaced Monge at the Académie des Sciences after the fall of Napoleon. Although he published important papers in the fields of elasticity and waves, he became famous for his taught courses on analysis and calculus in the 1820s, in which he proved the diagonalization of real quadratic forms and pushed forward the new standards of rigor, e.g. limits, continuity, convergence.

Proof The integrand $f(z)/(z-a)$ is analytic except at $z = a$, so by Cauchy's theorem the path of integration can be taken to be a small circle of radius r about a. As f is analytic at a, we know $f(a+w) = f(a) + f'(a)w + o(w)$, so

$$\frac{f(z)}{z-a} = \frac{f(a+w)}{w} = \frac{f(a)}{w} + f'(a) + \frac{o(w)}{w}.$$

Integrating around a closed simple path eliminates the constant function $f'(a)$, and

$$\left| \frac{1}{2\pi i} \oint \frac{o(w)}{w} \, dz \right| \leqslant \frac{1}{2\pi} \int_0^{2\pi} \frac{|o(w)|}{|w|} r \, dt < r\epsilon$$

if r is small enough that $|o(w)|/|w| < \epsilon$. Thus in the limit as we take smaller circles, only the term $\frac{1}{2\pi i} \oint \frac{f(a)}{w} \, dw = f(a)$ remains.   □

Examples 12.20

1. Interpreting the residue theorem in actual examples often yields integration results that would be harder to obtain otherwise. For example, the function e^{iz}/z has a simple pole at 0 with residue 1. So using a contour as shown in the diagram, we obtain

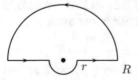

$$2\pi i = \int_r^R \frac{e^{ix}}{x} \, dx + \int_{-R}^{-r} \frac{e^{ix}}{x} \, dx + \int_0^\pi e^{-R(\sin\theta - i\cos\theta)} i \, d\theta + \int_\pi^{2\pi} e^{ire^{i\theta}} i \, d\theta$$

As $R \to \infty$ and $r \to 0$, the imaginary part is $2 \int_0^\infty \frac{\sin x}{x} \, dx + \pi = 2\pi$, which gives $\lim\limits_{\substack{r \to 0 \\ R \to \infty}} \int_r^R \frac{\sin x}{x} \, dx = \pi/2$.

2. *Maximum modulus principle*: If $f : \mathbb{C} \to \mathbb{C}$ is analytic and has a local maximum (or minimum) at a, then f is constant in a neighborhood of a. It follows that on a compact subset K, $|f|$ attains its maximum and minimum at the boundary ∂K.
   Proof: Using a circular path of any radius r, centered at a,

$$|f(a)| = \left| \frac{1}{2\pi i} \oint \frac{f(z)}{z-a} \, dz \right| \leqslant \frac{1}{2\pi} \int_0^{2\pi} |f(a+re^{i\theta})| \, d\theta \leqslant |f(a)|$$

$$\therefore \int_0^{2\pi} |f(a)| - |f(z)| \, d\theta = 0$$

so $|f(z)| = |f(a)|$ within the disk, which in turn implies $f(z)$ is constant (Exercise 12.21(5)). Let $f^{-1}M$ be the subset of the interior of K where $|f|$ attains the maximum $M := \max_{z \in K^\circ} |f(z)|$. It is open by the above, and closed in K° (Exercises 3.13(11)), hence must contain whole components of K°, unless empty. By continuity, f takes the same value M on the boundary.

3. We say that a function f has a *zero* of order n at a when $f(z) = (z-a)^n g(z)$, with $g(a) \neq 0$, g analytic in a neighborhood of a.
   If $f : \mathbb{C} \to \mathbb{C}$ has a zero (or pole) of order n at a, then f'/f has a simple pole at a with residue n (resp. $-n$)

$$\frac{f'(z)}{f(z)} = \frac{n(z-a)^{n-1}g(z) + (z-a)^n g'(z)}{(z-a)^n g(z)} = \frac{n}{z-a} + \frac{g'(z)}{g(z)},$$

(g'/g is analytic at a). Thus $\frac{1}{2\pi i} \oint \frac{f'}{f} = n$; more generally it equals the difference between the number of zeros and poles (counted with their order) inside the curve of integration.

4. *Rouché's theorem*: If $p_n \to f$ inside a closed simple curve γ, with f non-zero on γ, then f and p_n have the same number of zeros inside γ, from some n onwards.
   Proof: As $|f|$ has a non-zero minimum on γ, there is an n such that $|\frac{p_n}{f} - 1| < 1$ on γ. Let $F := p_n/f$ then $\oint_\gamma \frac{F'}{F} = \oint_{F\circ\gamma} \frac{1}{z} \, dz = 0$, since $F \circ \gamma$ is a closed curve that excludes 0. By the previous example, this implies that F has the same number of zeros as poles, that is, the zeros of p_n and of f are the same in number.

Exercises 12.21

1. The function $x+iy \mapsto (x^2-y^2)+2xyi$ is analytic, but $x+iy \mapsto (x^2-y^2)+xyi$ is not.
2. Show that, along any closed curve $\square$ in $\mathbb{C}$, $\int_\square 1 \, dz = 0$ and $\int_\square z \, dz = 0$, but on a unit circle centered at the origin, $\int_\circ \operatorname{Re}(z) \, dz = \pi i$.
3. If $f_n(z) \to f(z)$ in X for all z on a simple closed curve w, on which f_n and f are continuous, then $\int_w f_n(z) \, dz \to \int_w f(z) \, dz$.

4. Assuming u and v are sufficiently differentiable, deduce from the Cauchy-Riemann equations that the real and imaginary parts of an analytic function $f = u + iv$ are 'harmonic',

$$\frac{\partial^2 u}{\partial x^2} + \frac{\partial^2 u}{\partial y^2} = 0, \quad \frac{\partial^2 v}{\partial x^2} + \frac{\partial^2 v}{\partial y^2} = 0.$$

5. Let $f : \mathbb{C} \to \mathbb{C}$ be analytic. Suppose $|f|$ is constant in some open set, then f is constant. (Hint: Differentiate $|f|^2 = u^2 + v^2$.)

6. Find the poles and residues of (a) $e^{iz}/(z^2 + 1)$, (b) $\frac{1}{z^3-1}\left(\frac{e^z}{e^{-z}}\right)$, (c) $(\sin z)/z^2$ (First show $(\sin z)/z$ is analytic at 0).

7. $\dfrac{1}{2\pi i} \oint \dfrac{z^2 + 2}{z(z^2 - 1)}\, dz = -\dfrac{1}{2}$ along a simple closed counter-clockwise path that includes 0, 1, but not -1.

8. Show

(a) $\displaystyle\int_0^{2\pi} \frac{1}{2 + \cos\theta}\, d\theta = \frac{2\pi}{\sqrt{3}}$ using $f(z) := \dfrac{1}{z^2 + 4z + 1}$, $z(\theta) = e^{i\theta}$;

(b) $\displaystyle\int_{-\infty}^{\infty} \frac{\cos x}{x^2 + 1}\, dx = \frac{\pi}{e}$ using $f(z) := \dfrac{e^{iz}}{z^2 + 1}$;

(c) $\displaystyle\int_0^{\infty} \frac{1 - \cos x}{x^2}\, dx = \frac{\pi}{2}$ using $f(z) := \dfrac{1 - e^{iz}}{z^2}$.

9. By applying Example 12.20(3) to $f = e^g$, prove that the order of any of its poles must be zero. As this is impossible, the isolated singularities of f must be essential singularities.

10. Use Rouché's theorem to show that $\cosh z - 2\cos z$ has 2 zeros in the unit disk, assuming it equals its MacLaurin series.

Remarks 12.22

1. The first use of the Newton-Raphson method was by the "Babylonians" who used the iteration $x_{n+1} = \frac{1}{2}(x_n + n/x_n)$ to find square roots, $x^2 = n$. Newton's method was initially restricted to finding roots of polynomials, and it was Simpson (1740) who described the iteration we use today.

2. Cauchy's theorem for analytic functions is a special case of Green's or Stoke's theorem $\oint F \cdot d\mathbf{r} = \iint \nabla \times F \cdot d\mathbf{A}$. In this case, using the Cauchy-Riemann equations,

$$\oint f(z)\, dz = \oint (u + iv)(dx + i\, dy) = \oint u\, dx - v\, dy + i \oint v\, dx + u\, dy$$

$$= -\iint \left(\frac{\partial v}{\partial x} + \frac{\partial u}{\partial y}\right) dA + i \iint \left(\frac{\partial u}{\partial x} + \frac{\partial v}{\partial y}\right) dA$$

$$= 0.$$

Part III
Banach Algebras

Chapter 13
Banach Algebras

13.1 Introduction

We now turn our attention to the space of operators $B(X)$. We have seen that it is a Banach space when X is one, but additionally, one can compose, or *multiply*, operators in $B(X)$. This extra structure turns the vector space $B(X)$ into what is called an *algebra*. We shall mostly study these spaces as abstract algebras $\mathcal{X}$ without specific reference to them being spaces of operators, in order to include other examples of algebras and to make some of the proofs clearer. Nonetheless, $B(X)$ remains our primary interest, and accordingly, the elements of an algebra will be denoted in general by upper-case letters $T, S, \ldots$ to remind us of operators and to distinguish them from mere vectors x.

Definition 13.1

> A unital **Banach algebra** $\mathcal{X}$ is a Banach space over $\mathbb{C}$ that has an associative multiplication of vectors with unity 1, such that for all $R, S, T \in \mathcal{X}, \lambda \in \mathbb{C}$,
>
> $$(R + S)T = RT + ST, \quad R(S + T) = RS + RT,$$
>
> $$(\lambda S)T = \lambda(ST) = S(\lambda T),$$
>
> $$\|ST\| \leqslant \|S\|\|T\|, \quad \|1\| = 1.$$

Throughout this book, a Banach algebra will mean a unital Banach algebra over $\mathbb{C}$. Of course, Banach algebras over $\mathbb{R}$ are also of interest, and all the results in this chapter apply to them in modified form; but complex scalars are necessary for an adequate spectral theory of $\mathcal{X}$.

© The Author(s), under exclusive license to Springer Nature Switzerland AG 2024
J. Muscat, *Functional Analysis*, https://doi.org/10.1007/978-3-031-27537-1_13

Easy Consequences

1. 1 is unique, because $1' = 1'1 = 1$ for any other unity $1'$.
2. T is said to be *invertible* (or *regular*) when there is an element S, called its *inverse*, such that $ST = 1 = TS$. The inverse of T is unique when it exists, and is denoted T^{-1}. If $AT = 1 = TB$ then $A = A(TB) = (AT)B = B$ so $A = T^{-1}$.
3. $(S + T)^2 = S^2 + ST + TS + T^2$, and more generally,

$$(S + T)^n = S^n + (ST^{n-1} + TST^{n-2} + \cdots + T^{n-1}S) + \cdots + T^n.$$

4. $\|T^n\| \leqslant \|T\|^n$.

Proposition 13.2

Multiplication, $(T, S) \mapsto TS$, is a differentiable map.

Proof In the identity

$$(T + H)(S + K) = TS + (TK + HS) + HK, \tag{13.1}$$

the map $(H, K) \mapsto TK + HS$, $\mathcal{X}^2 \to \mathcal{X}$, is linear and continuous, and HK is of lower order, since

$$\|TK + HS\| \leqslant \|T\|\|K\| + \|S\|\|H\| \leqslant \max(\|T\|, \|S\|)(\|H\| + \|K\|)$$

$$\|HK\| \leqslant \|H\|\|K\| \leqslant (\|H\| + \|K\|)^2 = \|(H, K)\|^2.$$

$\square$

Needless to say, every differentiable map is continuous.

Examples 13.3

1. $\mathbb{C}^n$ with the ∞-norm and the following pointwise multiplication and unity:

$$\begin{pmatrix} a_1 \\ \vdots \\ a_n \end{pmatrix} \begin{pmatrix} b_1 \\ \vdots \\ b_n \end{pmatrix} := \begin{pmatrix} a_1 b_1 \\ \vdots \\ a_n b_n \end{pmatrix}, \quad 1 = \begin{pmatrix} 1 \\ \vdots \\ 1 \end{pmatrix}.$$

2. ▶ ℓ^∞ with pointwise multiplication xy, and unity $1 = (1, 1, \ldots)$ (Exercise 9.4(3)).
3. $C(K)$, the space of continuous functions on a compact set K, with pointwise multiplication $fg(x) := f(x)g(x)$, and unity being the constant function 1. For example, $C[0, 1]$ is a space of paths in the complex plane.

4. ▶ ℓ^1 with the convolution product; unity is $e_0 = (1, 0, \ldots)$ (Exercise 9.7(2)).
5. The space $L^1(\mathbb{R})$ with convolution as a product; although it does not have a unity, we can artificially add a δ, called Dirac's "function", such that $\delta * f := f =: f * \delta$. (To make this rigorous, one needs to consider $L^1(\mathbb{R}) \times \mathbb{C}$ with elements (f, a) representing $f + a\delta$.)

   The above examples happen to be *commutative*, i.e., $ST = TS$ holds. But this is not assumed in general. For example, $T^2 - S^2 \neq (T - S)(T + S)$ in general.
6. ▶ $B(X)$ for any Banach space X; the product is operator composition and unity is the identity operator I (Proposition 8.8).
7. ▶ If $\mathcal{X}$ and $\mathcal{Y}$ are Banach algebras, then so is $\mathcal{X} \times \mathcal{Y}$ with

$$\begin{pmatrix} S_1 \\ T_1 \end{pmatrix} \begin{pmatrix} S_2 \\ T_2 \end{pmatrix} := \begin{pmatrix} S_1 S_2 \\ T_1 T_2 \end{pmatrix}, \quad 1 = \begin{pmatrix} 1_{\mathcal{X}} \\ 1_{\mathcal{Y}} \end{pmatrix}, \quad \left\| \begin{pmatrix} S \\ T \end{pmatrix} \right\| := \max(\|S\|_{\mathcal{X}}, \|T\|_{\mathcal{Y}}).$$

8. Every normed algebra can be completed to a Banach algebra.
   Proof: Using the notation of Proposition 7.18, if $T = [T_n]$ and $S = [S_n]$, let $ST := [S_n T_n]$ and $1 := [1]$. Note that $S_n T_n$ is a Cauchy sequence by

$$\|S_n T_n - S_m T_m\| \leqslant \|S_n T_n - S_n T_m\| + \|S_n T_m - S_m T_m\|$$
$$\leqslant \|S_n\| \|T_n - T_m\| + \|S_n - S_m\| \|T_m\|$$
$$\leqslant c(\|S_n - S_m\| + \|T_n - T_m\|).$$

Hence

$$R(ST) = [R_n(S_n T_n)] = [(R_n S_n) T_n] = (RS)T,$$
$$\lambda(ST) = [\lambda(S_n T_n)] = (\lambda S)T = S(\lambda T),$$
$$\|ST\| = \lim_{n \to \infty} \|S_n T_n\| \leqslant \lim_{n \to \infty} \|S_n\| \|T_n\| = \|S\| \|T\|.$$

9. The polynomials $\mathbb{C}[z]$ on $\bar{B}_{\mathbb{C}}$ with the ∞-norm form an incomplete algebra. As we shall see shortly, its completion is the space of analytic functions $C^\omega(\bar{B}_{\mathbb{C}})$. More general is the *tensor algebra*, consisting of polynomials and series in a number of non-commuting variables.
10. ▶ If $ST = 0$ and S is invertible, then $T = 0$. But there may exist non-zero non-invertible elements S, T, called *divisors of zero*, for which $ST = 0$. Note that TS need not also be 0, so S and T are more precisely called *left* and *right* divisors of zero, respectively.
11. ▶ The product of invertible elements is invertible, with $(ST)^{-1} = T^{-1} S^{-1}$. Also, $(T^{-1})^{-1} = T$. If T^n is invertible, for some $n \geqslant 1$, then so is T.
    But it is possible for two non-invertible elements to have an invertible product, i.e., ST invertible $\not\Rightarrow T$ invertible (unless TR is also invertible for some R). In particular, $ST = 1$ by itself is not enough to ensure T and S are invertible.

For example, in $B(\ell^1)$, the product of the (non-invertible) shift-operators is $LR = I$.

12. (a) $(1 + TS)^{-1} = 1 - T(1 + ST)^{-1}S$,

(b) *Woodbury's formula* $(A + TBS)^{-1} = A^{-1} - A^{-1}T(B^{-1} + SA^{-1}T)^{-1}SA^{-1}$.

Proof: (a) Starting from the identities $(1 + TS)T = T(1 + ST)$ and $S(1 + TS) = (1 + ST)S$, we deduce

$$1 + (1 + TS)T(1 + ST)^{-1}S = 1 + TS = 1 + T(1 + ST)^{-1}S(1 + TS)$$

from which the result follows.

(b) Use (a) with $(A + TBS)^{-1} = (1 + A^{-1}TBS)^{-1}A^{-1}$.

13. Suppose an element satisfies some non-zero polynomial, $p(T) = 0$. The unique such polynomial of minimum degree and leading coefficient 1 is called its *minimal polynomial* p_m. It divides all other polynomials p such that $p(T) = 0$. *Proof*: There cannot be two minimal polynomials, p_m and p, otherwise $p_m - p$ has a lesser degree than both and $p_m(T) - p(T) = 0$. If $p(T) = 0$, then $p = qp_m + r$ by the division algorithm of polynomials. As r has a strictly smaller degree than p_m, yet $r(T) = p(T) - q(T)p_m(T) = 0$, it must be the zero polynomial.

14. The derivative of the map $T \mapsto ST$ is S. Similarly the derivative of $T \mapsto T^n$ is

$$H \mapsto HT^{n-1} + THT^{n-2} + \cdots + T^{n-1}H.$$

Because of commutativity, this simplifies to $(z^n)' = nz^{n-1}$ in $\mathbb{C}$. Thus, any polynomial in T is differentiable in T.

Subalgebras and Ideals

Definition 13.4

A **subalgebra** of an algebra $\mathcal{X}$ is a subset which is itself an algebra with the same (induced) addition, scalar multiplication, product, and unity. It is a *Banach subalgebra* when, additionally, the induced norm is complete.

An **ideal** is a linear subspace $\mathcal{I}$ such that $ST, TS \in \mathcal{I}$ for any $T \in \mathcal{X}, S \in \mathcal{I}$.

To show that a non-empty subset $\mathcal{A}$ is a subalgebra of $\mathcal{X}$, one need only show closure of the various operations, i.e., for any $S, T \in \mathcal{A}, S + T \in \mathcal{A}, \lambda T \in \mathcal{A}, ST \in \mathcal{A}$, $1 \in \mathcal{A}$. The required properties of the induced operations are obviously inherited from those of $\mathcal{X}$.

Examples 13.5

1. $\mathbb{C}$ is embedded in every (complex) Banach algebra as $\mathbb{C}1 = \{z1 : z \in \mathbb{C}\}$. In fact, it is customary to write z when we mean $z1$.
2. An element T generates the subalgebra of *polynomials*

$$\mathbb{C}[T] := \{a_0 + a_1 T + \cdots + a_n T^n : a_1, \ldots, a_n \in \mathbb{C}, n \in \mathbb{N}\}.$$

More generally, a finite number of commuting elements $T_1, \ldots, T_n$ generate the commutative algebra $\mathbb{C}[T_1, \ldots, T_n]$, which may contain, for example, the element $1 - 2T_2 + T_1^2 T_2$.
3. The algebra ℓ^∞ contains the closed ideal c_0.
   Proof: That c_0 is a closed linear subspace of ℓ^∞ is proved in Proposition 9.2. Let $(a_n)_{n\in\mathbb{N}} \in \ell^\infty$, $(b_n)_{n\in\mathbb{N}} \in c_0$, then $(a_n b_n)_{n\in\mathbb{N}} \in c_0$ since

$$|\lim_{n\to\infty} a_n b_n| \leqslant \sup_n |a_n| \lim_{n\to\infty} |b_n| = 0.$$

We will see later that every commutative Banach algebra, except $\mathbb{C}$, has non-trivial ideals (Example 14.5(4)).
4. The *center* $\mathcal{X}' := \{T : ST = TS, \forall S \in \mathcal{X}\}$ is a commutative closed subalgebra of $\mathcal{X}$.
   Proof: If $T_n \in \mathcal{X}'$, then

$$S(T_1 + \lambda T_2) = ST_1 + \lambda ST_2 = T_1 S + \lambda T_2 S = (T_1 + \lambda T_2)S,$$

$$S(T_1 T_2) = T_1 ST_2 = (T_1 T_2)S, \quad SI = S = IS,$$

$$T_n \to T \Rightarrow ST = \lim_{n\to\infty} ST_n = \lim_{n\to\infty} T_n S = TS.$$

The algebra is commutative by definition of $\mathcal{X}'$.
5. ▶ Proper ideals do not contain 1, or any other invertible element T, otherwise it would have to contain every element $S = ST^{-1}T$. (However, as remarked in Example 13.3(11), the set of non-invertible elements need not be an ideal, or even a subspace.)
6. A closed ideal gives rise to a quotient algebra $\mathcal{X}/\mathcal{I}$ with multiplication and unity defined by

$$(S + \mathcal{I})(T + \mathcal{I}) := ST + \mathcal{I}, \quad 1 + \mathcal{I}.$$

7. A *maximal* ideal is a proper ideal $\mathcal{I}$ for which the only other ideal containing it is $\mathcal{X}$ itself,

$$\mathcal{I} \subseteq \mathcal{J} \subseteq \mathcal{X} \Rightarrow \mathcal{J} = \mathcal{I} \text{ OR } \mathcal{J} = \mathcal{X}.$$

Maximal ideals are necessarily closed, assuming that the closure of a proper ideal is also a proper ideal (Example 13.22(3)).

8. ⋆ Every proper ideal is contained in a maximal ideal.

Proof: Let $\mathcal{C}$ be the collection of all proper ideals that contain the proper ideal $\mathcal{I}$. By Hausdorff's maximality principle, $\mathcal{C}$ contains a maximal chain of nested ideals $\mathcal{I}_\alpha$. Then $\mathcal{M} := \bigcup_\alpha \mathcal{I}_\alpha$ is an ideal, since if $T \in \mathcal{I}_\alpha$ and $S \in \mathcal{I}_\beta \subseteq \mathcal{I}_\alpha$, say, then $S + T \in \mathcal{I}_\alpha \subseteq \mathcal{M}$, and for any $S \in \mathcal{X}$, both ST and TS are in $\mathcal{I}_\alpha \subseteq \mathcal{M}$. It is obvious that $\mathcal{M}$ is proper and contains $\mathcal{I}$ since $1 \notin \mathcal{I}_\alpha \supseteq \mathcal{I}$ for every α, and that $\mathcal{M}$ is maximal since the chain $\mathcal{I}_\alpha$ is maximal.

Morphisms

Definition 13.6

A **morphism** $\Phi : \mathcal{X} \to \mathcal{Y}$ of Banach algebras is a continuous linear map (preserving limits, addition, and scaling) which preserves multiplication and the unity,

$$\Phi(ST) = \Phi(S)\Phi(T), \qquad \Phi(1_{\mathcal{X}}) = 1_{\mathcal{Y}}.$$

A **character** is a Banach algebra morphism $\phi : \mathcal{X} \to \mathbb{C}$. The set of characters, denoted by Δ, is called the *character space*, or *spectrum*, of $\mathcal{X}$.

Examples 13.7

1. Invertible elements of $\mathcal{X}$ are mapped by algebra morphisms to invertible elements of $\mathcal{Y}$,

$$\Phi(T)^{-1} = \Phi(T^{-1}),$$

   since $\Phi(T)\Phi(T^{-1}) = \Phi(TT^{-1}) = \Phi(1) = 1$ and similarly, $\Phi(T^{-1})\Phi(T) = 1$.

2. ▶ The kernel of a Banach algebra morphism, $\ker \Phi := \{T : \Phi(T) = 0\}$, is a closed ideal. It is maximal when $\Phi \in \Delta$.
   Proof: If $\Phi(T) = 0$, then $\Phi(ST) = \Phi(S)\Phi(T) = 0$; similarly, $\Phi(TS) = 0$.
   Maximality: Let $\Phi : \mathcal{X} \to \mathbb{C}$ be a morphism, and let the ideal $\mathcal{I}$ contain $\ker \Phi$ as well as some $T \notin \ker \Phi$. Then $\Phi(T) = \lambda \neq 0$, and $\Phi(\lambda - T) = 0$; so $\lambda = (\lambda - T) + T \in \mathcal{I}$, and $\mathcal{I}$ must equal $\mathcal{X}$ (Example 13.5(5) above).
   (Every maximal ideal of a commutative Banach algebra is of the type $\ker \phi$ with $\phi \in \Delta$, but the proof requires Exercise 13.10(21) and Example 14.5(4); see the proof of Theorem 14.39.)

3. An *isomorphism* of Banach algebras is defined to be an invertible morphism $\Phi :$ $\mathcal{X} \to \mathcal{Y}$ such that Φ^{-1} is also a morphism. In fact, an invertible morphism is automatically an isomorphism.
4. An *automorphism* of a Banach algebra $\mathcal{X}$ is an isomorphism from $\mathcal{X}$ to itself. For example, the *inner automorphisms* $T \mapsto S^{-1}TS$, for any fixed invertible S.
5. Since $\mathbb{C}$ is commutative, *commutators* $[S, T] := ST - TS$ are mapped to 0 by characters (if they exist).

Representation in $B(X)$

Some mathematical theories contain a set of theorems stating that any abstract model of the theory can be *represented* concretely. For example, every group can be represented by a permutation group, and every smooth manifold is embedded as a smooth "surface" of a Euclidean space. In this regard, every finite-dimensional Banach algebra can be embedded, or "faithfully represented", as a matrix algebra, and more generally, we have the following representation theorem:

Theorem 13.8

Every Banach algebra can be embedded as a closed subalgebra of $B(X)$, for some Banach space X.

Proof The Banach space X can be taken to be the Banach algebra $\mathcal{X}$ itself without the product (although there may well be 'smaller' Banach spaces that fit the job). That is, the claim is that $\mathcal{X}$ is embedded in $B(\mathcal{X})$. To avoid confusion, we temporarily denote elements of $\mathcal{X}$ by lower-case letters, and the operators on them by upper-case letters.

Let $L_a(x) := ax$ be left-multiplication by a. Then $L_a \in B(\mathcal{X})$ since multiplication is distributive and continuous:

$$L_a(x + y) = a(x + y) = ax + ay = L_a(x) + L_a(y),$$

$$L_a(\lambda x) = a(\lambda x) = \lambda(ax) = \lambda L_a(x),$$

$$\|L_a(x)\| = \|ax\| \leqslant \|a\|\|x\|,$$

so that $\|L_a\| \leqslant \|a\|$. Furthermore,

$$L_{a+b}(x) = (a + b)x = ax + bx = L_a(x) + L_b(x), \qquad L_1(x) = 1x = x = I(x),$$
$$L_{\lambda a}(x) = (\lambda a)x = \lambda L_a(x), \qquad\qquad\qquad L_a(1) = a1 = a,$$
$$L_{ab}(x) = (ab)x = a(bx) = L_a L_b(x),$$

so $\|a\| = \|L_a 1\| \leqslant \|L_a\|\|1\| = \|L_a\|$ and $\|L_a\| = \|a\|$. These show that the mapping $L : \mathcal{X} \to B(\mathcal{X})$ defined by $L : a \mapsto L_a$ is an isometric morphism of Banach algebras. In fact, the space of such operators, $\operatorname{im} L$, is a closed subalgebra of $B(\mathcal{X})$ since isometries preserve completeness (Exercise 4.18(5)). Note that all the Banach algebra axioms have been used.   □

As one may anticipate, $B(X)$ and $B(Y)$ are not isomorphic as Banach algebras, when X and Y are not isomorphic as Banach spaces. The proof, however, is not as obvious as one might expect.

Theorem 13.9

> **Let X and Y be Banach spaces. A Banach algebra isomorphism J : $B(X) \to B(Y)$ induces a Banach space isomorphism $L : X \to Y$, such that**
>
> $$J(T) = LTL^{-1}.$$
>
> **Thus, every automorphism of $B(X)$ is inner.**

Proof The idea is to establish a 1–1 correspondence between vectors $x \in X$ and certain projection-like operators $P_x \in B(X)$, and similarly $y \leftrightarrow R_y$ for Y; using the given mapping $J : T \mapsto \widetilde{T}$, the sought isomorphism would then be

$$L : x \mapsto P_x \overset{J}{\mapsto} R_y \mapsto y.$$

The correspondence $x \leftrightarrow P_x$: For the remainder of the proof, fix a vector $u \in X$, $u \neq 0$, and a functional $\phi \in X^*$ such that $\phi u = 1$. Multiplying any $x \in X$ by ϕ gives an operator $P_x := x\phi : z \mapsto (\phi z)x$; conversely, multiplying P_x with u gives back the vector $P_x u = x\phi u = x$. The crucial characteristics of these operators are, for any $T \in B(X)$ (including scalar multiplication),

$$T P_x = T x\phi = (Tx)\phi = P_{Tx}, \qquad P_{x_1+x_2} = (x_1 + x_2)\phi = P_{x_1} + P_{x_2}.$$

In particular, $P_x P_u = x\phi u\phi = x\phi = P_x$. Note that $\|P_x\| = \|x\phi\| \leqslant \|x\|\|\phi\|$ and $\|x\| = \|P_x u\| \leqslant \|P_x\|\|u\|$. Thus, $P : X \to B(X), x \mapsto P_x$ is an embedding.

The isomorphism J maps $P_x \in B(X)$ to a similar operator $R_y \in B(Y)$: The relation $P_u^2 = P_u$ is preserved by J, so $\widetilde{P}_u := J(P_u)$ is a non-zero projection in $B(Y)$. Pick $v \in \operatorname{im} \widetilde{P}_u$ and $\psi \in Y^*$ such that $\psi v = 1$ and $\psi \ker \widetilde{P}_u = 0$ (Proposition 11.20), and define $R_y := y\psi$ for any $y \in Y$. R_y satisfies analogous properties as P_x, such as $R_y v = y$ and $\widetilde{T} R_y = R_{\widetilde{T}y}$. First we show that $J(P_u) = R_v$: for suppose $w \in \operatorname{im} \widetilde{P}_u$, and let $T \in B(X)$ correspond to $R_w \in B(Y)$ under J; then J transforms the identity

$$P_u T P_u = u(\phi Tu)\phi = \lambda P_u, \qquad \text{where } \lambda = \phi Tu,$$

to $\widetilde{P}_u R_w \widetilde{P}_u = \lambda \widetilde{P}_u$, so im $\widetilde{P}_u = [\![v]\!]$ since

$$w = \widetilde{P}_u w = \widetilde{P}_u R_w v = \widetilde{P}_u R_w \widetilde{P}_u v = \lambda \widetilde{P}_u v = \lambda v.$$

Thus the projections $\widetilde{P}_u$ and R_v have the same image and the same kernel, and we can conclude that they are equal to each other.

Hence, the identity $P_x = P_x P_v$ becomes, in $B(Y)$,

$$J(P_x) = \widetilde{P}_x = \widetilde{P}_x R_v = R_{\widetilde{P}_x v} = R_y, \quad \text{where } y = \widetilde{P}_x v.$$

The map $L : x \mapsto y = J(P_x)v$ *is an isomorphism*: That L is linear, continuous, and 1–1 follow from:

$$L(x_1 + x_2) = J(P_{x_1+x_2})v = J(P_{x_1} + P_{x_2})v = L(x_1) + L(x_2),$$

$$L(\lambda x) = J(P_{\lambda x})v = J(\lambda P_x)v = \lambda L(x),$$

$$\|Lx\| = \|J(P_x)v\| \leqslant \|J\| \|\phi\| \|x\| \|v\|,$$

$$Lx = 0 \Leftrightarrow y = J(P_x)v = 0 \Leftrightarrow J(P_x) = R_y = 0 \Leftrightarrow P_x = 0 \Leftrightarrow x = 0.$$

Given any $y \in Y$, J^{-1} maps the identity $R_y = R_y R_v$ to $S = S P_u = P_{Su}$. So for $x := Su$,

$$Lx = J(P_{Su})v = J(S)v = R_y v = y,$$

and L is onto. By the open mapping theorem (Theorem 11.1), L is an isomorphism. $\widetilde{T} = L T L^{-1}$: J maps the identity $T P_x = P_{Tx}$ to $\widetilde{T} R_{L(x)} = R_{L(Tx)}$. Multiplying by v to get the vector form, this reads $\widetilde{T} Lx = LTx$ for all $x \in X$.

When $X = Y$, then $L \in B(X)$, and J is an inner automorphism. □

Exercises 13.10

1. Banach algebras of square matrices abound: the sets of matrices of type $\left(\begin{smallmatrix} a & 0 \\ b & c \end{smallmatrix}\right)$, $\left(\begin{smallmatrix} a & b \\ 0 & a \end{smallmatrix}\right)$, $\left(\begin{smallmatrix} a & b \\ b & a \end{smallmatrix}\right)$, or $\left(\begin{smallmatrix} a & b \\ -b & a \end{smallmatrix}\right)$ are each closed under addition and multiplication, and are Banach subalgebras of $B(\mathbb{C}^2)$. The last three examples can be written as $a + bJ$ where $J^2 = 0, 1, -1$ respectively.

2. $\mathcal{C} := \mathbb{C}^2$ with $\left(\begin{smallmatrix} a \\ b \end{smallmatrix}\right)\left(\begin{smallmatrix} c \\ d \end{smallmatrix}\right) := \left(\begin{smallmatrix} ac \\ ad+bc+bd \end{smallmatrix}\right)$ is a Banach algebra, with unity $\left(\begin{smallmatrix} 1 \\ 0 \end{smallmatrix}\right)$. (Hint: it is a matrix algebra in disguise.)

3. Find examples of 2×2 matrix divisors of zero, $ST = 0 \neq TS$.

4. In an n-dimensional algebra, every element has a minimal polynomial of degree at most n; e.g., every square matrix A has a minimal polynomial. Show also how the Gram-Schmidt process (with respect to the inner product of Example 10.2(2)) can be applied to the sequence $I, A, A^2, \ldots$ to construct this minimal polynomial.

5. An *idempotent* satisfies $P^2 = P$. They are the projections in $B(X)$; what are they in $\mathbb{C}^n$ and ℓ^∞? The idempotents of $C[0, 1]$ are trivial. Show further that $P\mathcal{X}P$ is an algebra with unity P, called a "reduced algebra".

6. A *nilpotent* satisfies $Q^n = 0$ for some n, e.g., $\begin{pmatrix} 0 & 1 \\ 0 & 0 \end{pmatrix}$ and $\begin{pmatrix} 1 & i \\ i & -1 \end{pmatrix}$. In $\mathbb{C}^N$, ℓ^∞, and $C[0, 1]$, there are no nilpotents except zero. Find all the 2×2 matrix nilpotents of index 2, i.e., $Q^2 = 0$, $Q \neq 0$.

7. An element is *cyclic* when $T^n = 1$ for some n, e.g., $\begin{pmatrix} 1 & 0 \\ 0 & i \end{pmatrix}$. In $\mathbb{C}^N$ and ℓ^∞ they are sequences whose terms are of the type $e^{2\pi i m/n}$ for a fixed n.

8. The product of differentiable functions is again differentiable, with

$$(fg)'(T)H = [f'(T)H]g(T) + f(T)[g'(T)H].$$

This can be written in short as the familiar *product rule* $(fg)' = f'g + fg'$, provided it is remembered that the vector H is acted upon by each derivative.

9. If $F : \mathbb{R} \to \mathcal{X}$ is integrable and $T \in \mathcal{X}$, then $\int F(t)T \, dt = (\int F)T$ (First show it is true for simple functions).

10. ⋆ *Group Algebra*: Let G be a finite group of order n, and $\{ e_g : g \in G \}$ be an orthonormal basis for $\mathbb{C}^n$; define $e_g * e_h := e_{gh}$, and extend the product to all other vectors by distributivity. The result is a Banach algebra, denoted $\mathbb{C}^G$ or $\ell^1(G)$, with unity e_1 and the 1-norm. Every basis element is cyclic.
For example, the cyclic group $\{ 1, g : g^2 = 1 \}$, gives rise to an algebra generated by $e_1 := \begin{pmatrix} 1 \\ 0 \end{pmatrix}$ and $e_g := \begin{pmatrix} 0 \\ 1 \end{pmatrix}$, and the product

$$\begin{pmatrix} a \\ b \end{pmatrix} * \begin{pmatrix} c \\ d \end{pmatrix} := (ae_1 + be_g) * (ce_1 + de_g) = \begin{pmatrix} ac + bd \\ bc + ad \end{pmatrix}.$$

11. The closure of a subalgebra is an algebra (use continuity of the product).

12. If $\mathcal{I}$ and $\mathcal{J}$ are ideals, then so are $\mathcal{I} + \mathcal{J}$ and $\overline{\mathcal{I}}$.

13. The center of $B(X)$ is $\mathbb{C}$. (Hint: Consider projections $x\phi$, for any $x \in X$, $\phi \in X^*$.)

14. ▶ The *centralizer* or *commutant* of a subset $\mathcal{A} \subseteq \mathcal{X}$,

$$\mathcal{A}' := \{ T : AT = TA, \ \forall A \in \mathcal{A} \}$$

is a closed subalgebra of $\mathcal{X}$. (In fact, when $\mathcal{X} = B(H)$, $\mathcal{A}'$ is weakly closed by Exercise 11.49(8a).) Prove:

(a) $\mathcal{A} \subseteq \mathcal{B} \Rightarrow \mathcal{B}' \subseteq \mathcal{A}'$,
(b) $\mathcal{A} \subseteq \mathcal{A}''$ and $\mathcal{A}''' = \mathcal{A}'$,
(c) If $T \in \mathcal{A}'$ is invertible in $\mathcal{X}$ then $T^{-1} \in \mathcal{A}'$,
(d) If elements of $\mathcal{A}$ commute, then $\mathcal{A} \subseteq \mathcal{A}'$ and $\mathcal{A}''$ is a commutative Banach algebra.

15. A *left*-ideal is a linear subspace $\mathcal{I} \subseteq \mathcal{X}$ such that $T\mathcal{I} \subseteq \mathcal{I}$ for any $T \in \mathcal{X}$. Similarly, for a *right*-ideal, $\mathcal{I}T \subseteq \mathcal{I}$. For example, $\mathcal{X}S$ is a left-ideal, and $S\mathcal{X}$ is a right-ideal, but $\mathcal{X}S\mathcal{X}$ need not be an ideal. Instead, the ideal *generated* by S is $[\![\mathcal{X}S\mathcal{X}]\!]$.

16. The set of compact operators in $B(X)$ form a closed ideal $\mathcal{K}(X)$; the set of finite rank operators form an ideal (but note that the closure of this ideal need not be $\mathcal{K}(X)$.)

17. Show that any closed ideal of c_0 consists of those sequences that vanish on some specific set of indices. What are the maximal ideals?

18. Let A be a closed subset of $[0, 1]$, then

$$\mathcal{I}_A := \{ f \in C[0, 1] : \forall x \in A, f(x) = 0 \}$$

is a closed ideal of $C[0, 1]$. Conversely, given a closed ideal $\mathcal{I}$ of $C[0, 1]$, let

$$A := \{ x \in [0, 1] : \forall f \in \mathcal{I}, f(x) = 0 \},$$

then $\mathcal{I} = \mathcal{I}_A$. What are the maximal ideals?

19. Let $\mathcal{I}_A$ be a closed ideal of $C[0, 1]$, where A is a closed subset of $[0, 1]$. Then the mapping $f + \mathcal{I}_A \mapsto f|_A$ is an isomorphism $C[0, 1]/\mathcal{I}_A \equiv C(A)$.

20. An algebra morphism $\Phi : \mathcal{X} \to \mathcal{Y}$ 'pulls' ideals $\mathcal{I}$ in $\mathcal{Y}$ to ideals $\Phi^{-1}\mathcal{I}$ in $\mathcal{X}$.

21. If $\mathcal{I}$ is a closed ideal, then $\Phi(T) := T + \mathcal{I}$ gives a Banach algebra morphism $\Phi : \mathcal{X} \to \mathcal{X}/\mathcal{I}$ with kernel $\ker \Phi = \mathcal{I}$.

22. The mapping $\sum_n a_n z^n \mapsto (a_n)_{n \in \mathbb{N}}$ from the set of power series converging absolutely on the closed unit disk $\bar{B}_{\mathbb{C}}$ of $\mathbb{C}$, considered as a subspace of $C(\bar{B}_{\mathbb{C}})$, to ℓ^1 is a 1–1 Banach algebra morphism.

23. Let σ be a permutation of $1, \ldots, n$; then the mapping defined by $(z_1, \ldots, z_n) \mapsto (z_{\sigma(1)}, \ldots, z_{\sigma(n)})$ is an automorphism of $\mathbb{C}^n$.

24. For the group algebra $\mathbb{C}^G$, let σ be an automorphism of the group G; then $e_g \mapsto e_{\sigma(g)}$ induces an automorphism on $\mathbb{C}^G$.

25. The algebra $\mathbb{C}^n$ is embedded in $B(\mathbb{C}^n)$ as diagonal matrices. $\mathcal{C}$ is represented by the matrices $\begin{pmatrix} a & 0 \\ b & a+b \end{pmatrix}$. The group algebra $\mathbb{C}^G$ is generated by the Cayley matrices of G.

26. Show that every Banach algebra of dimension 2 (over $\mathbb{C}$) can be represented by the matrices generated from I and $\begin{pmatrix} 0 & \alpha \\ 1 & \beta \end{pmatrix}$, where α is a fixed number and β is 0 or 1. What are α and β for the group algebra generated by $\{ 1, g : g^2 = 1 \}$?

27. Let $\mathcal{X}$ be a Banach algebra contained in $B(X)$. Its unity $P = P^2$ is a projection, so $X = M \oplus N$ where $M = \text{im } P$. For every $T \in \mathcal{X}$, $PT = T = TP$ implies M is T-invariant and $TN = 0$, hence $\mathcal{X}$ acts on M.

13.2 Power Series

Definition 13.11

A **power series** is a series $\sum_n a_n T^n$ where $a_n \in \mathbb{C}$ and $T \in \mathcal{X}$.

Recall that the root test can help determine whether such a series converges or not: If $\|a_n T^n\|^{1/n} = |a_n|^{1/n}\|T^n\|^{1/n}$ converges to a number less than 1, then the power series converges. It is important to know that $\|T^n\|^{1/n}$ converges:

Proposition 13.12

For any T in a Banach algebra, the sequence $\|T^n\|^{1/n}$ converges to a number denoted by $\rho(T)$, where

$$\forall n \in \mathbb{N}, \quad \rho(T) \leqslant \|T^n\|^{1/n} \leqslant \|T\|.$$

Proof It is clear that $0 \leqslant \|T^n\|^{1/n} \leqslant \|T\|$. Let $\rho(T)$ be the infimum value of $\|T^n\|^{1/n}$, meaning that $\|T^n\|^{1/n}$ is bounded below by $\rho(T)$ and

$$\forall \epsilon > 0, \ \exists N, \quad \rho(T) \leqslant \|T^N\|^{1/N} < \rho(T) + \epsilon.$$

Although the sequence $\|T^n\|^{1/n}$ is not necessarily decreasing towards $\rho(T)$, notice that $\|T^{qm}\|^{1/qm} \leqslant \|T^m\|^{1/m}$. For any n, let $n = q_n N + r_n$ with $0 \leqslant r_n < N$ (by the remainder theorem), then $0 \leqslant r_n/n < N/n \to 0$ and $q_n/n = \frac{1}{N}(1 - \frac{r_n}{n}) \to \frac{1}{N}$ as $n \to \infty$, so that

$$\rho(T) \leqslant \|T^n\|^{1/n} = \|T^{q_n N}T^{r_n}\|^{1/n} \leqslant \|T^N\|^{q_n/n}\|T\|^{r_n/n} \to \|T^N\|^{1/N} < \rho(T)+\epsilon.$$

Since ϵ is arbitrarily small, this shows that $\|T^n\|^{1/n} \to \rho(T)$ from above. $\square$

Examples 13.13

1. ▶ (a) $\rho(1) = 1$, (b) $\rho(\lambda T) = |\lambda|\rho(T)$, (c) $\rho(ST) = \rho(TS)$, (d) $\rho(T^n) = \rho(T)^n$, since

$$\|1^n\|^{1/n} = 1, \quad \|\lambda^n T^n\|^{1/n} = |\lambda|\|T^n\|^{1/n},$$

$$\rho(ST) \leqslant \|(ST)^n\|^{1/n} \leqslant \|S\|^{1/n}\|(TS)^{n-1}\|^{1/n}\|T\|^{1/n} \to \rho(TS),$$

$$\|(T^n)^m\|^{1/m} = \|T^{nm}\|^{\frac{n}{nm}} \to \rho(T)^n \text{ as } m \to \infty.$$

But $\rho(T)$ may be 0 without $T = 0$; and $\rho(S + T) \not\leqslant \rho(S) + \rho(T)$ in general, e.g., $\left(\begin{smallmatrix} 0 & 1 \\ 0 & 0 \end{smallmatrix}\right)$, $\left(\begin{smallmatrix} 0 & 0 \\ 1 & 0 \end{smallmatrix}\right)$. So ρ is not usually a *norm* on $\mathcal{X}$.

2. $\rho(T) = \|T\| \Leftrightarrow \forall n \in \mathbb{N}, \ \|T^n\| = \|T\|^n$, since $\|T\| = \rho(T) \leqslant \|T^n\|^{1/n} \leqslant \|T\|$.

3. ▶ If $\rho(T) < 1$, then $T^n \to 0$ (even though $\|T\|$ may be bigger than 1). If $\rho(T) > 1$, then $T^n \to \infty$.
   Proof: For ϵ small enough and n large enough,

$$\|T^n\|^{1/n} \leqslant \rho(T) + \epsilon < 1 \Rightarrow \|T^n\| \leqslant (\rho(T) + \epsilon)^n \to 0, \text{ as } n \to \infty,$$

$$\|T^n\|^{1/n} \geqslant \rho(T) > 1 + \epsilon \Rightarrow \|T^n\| \geqslant (1+\epsilon)^n \to \infty.$$

For example, if $A := \frac{1}{3}\begin{pmatrix} 0 & 1 \\ 6 & 0 \end{pmatrix}$ then $\|A\begin{pmatrix} 1 \\ 0 \end{pmatrix}\| = 2 > 1$ yet $A^2 = \frac{2}{3}I$, so $A^n \to 0$. On the other hand, (i) $B := \begin{pmatrix} 1 & 1 \\ 0 & -1 \end{pmatrix}$ satisfies $B^n = I$ or B, so $\rho(B) = 1$ and B^n does not converge; (ii) $C := \begin{pmatrix} 1 & 1 \\ 0 & 1 \end{pmatrix}$ satisfies $C^n = \begin{pmatrix} 1 & n \\ 0 & 1 \end{pmatrix} \to \infty$ as $n \to \infty$, yet $\rho(C) = 1$; (iii) $D := \begin{pmatrix} 1 & 0 \\ 0 & 1/2 \end{pmatrix}$ satisfies $\rho(D) = 1$ and $D^n = \begin{pmatrix} 1 & 0 \\ 0 & 1/2^n \end{pmatrix} \to \begin{pmatrix} 1 & 0 \\ 0 & 0 \end{pmatrix}$ as $n \to \infty$.

4. If $\|\!|\cdot|\!\|$ is an equivalent norm, then $\rho(T) = \lim_{n\to\infty} \|\!|T^n|\!\|^{1/n}$; for example, for matrices, one can use the Frobenius norm, which is easier to calculate than the standard norm, although the convergence rate may differ.

Theorem 13.14 (Cauchy-Hadamard Theorem)

The power series $\sum_{n=0}^{\infty} a_n T^n$, where $a_n \in \mathbb{C}, T \in \mathcal{X}$,

- **converges absolutely when** $\rho(T) < r$, **and**
- **diverges when** $\rho(T) > r$,

where $r := 1/\limsup_n |a_n|^{1/n}$ **is called the** *radius of convergence* **of the series.**

Proof This is a simple application of the root test. The nth root of the general term satisfies

$$\limsup_n \|a_n T^n\|^{1/n} = \limsup_n |a_n|^{1/n} \rho(T) = \rho(T)/r.$$

Thus, if $\rho(T) < r$, then the series converges absolutely, while if $\rho(T) > r$, then it diverges. Assuming $\mathcal{X}$ is complete, the power series converges or diverges accordingly. $\square$

Examples 13.15

1. *Ratio test*: If $|a_n|/|a_{n+1}| \to r$ then so does $|a_n|^{-1/n}$ (Sect. 7.5), hence r would be the radius of convergence of $\sum_n a_n T^n$.
2. Some aspects of power series may seem mysterious from the point of view of real numbers: The series $1 - t^2 + t^4 - t^6 + \cdots$ has a radius of convergence of 1 yet converges to $(1 + t^2)^{-1}$ which takes a finite value at all $t \in \mathbb{R}$ (but not at $t = i$). Moreover the same function can also be written as $(5 - (4 - t^2))^{-1} = \frac{1}{5} \sum_{n=0}^{\infty} \left(\frac{4-t^2}{5}\right)^n$, but in this form it converges in the larger range $-3 < t < 3$.
3. The theorem also applies to power series $\sum_n A_n z^n$, where A_n is a sequence of elements in $\mathcal{X}$. The radius of convergence is then $1/\limsup_n \|A_n\|^{1/n}$.
4. When $|a_n| \leqslant c$ for all n, then $a_2 T^2 + a_3 T^3 + \cdots = o(T)$ for small T, since it is bounded above by $c\|T\|^2/(1 - \|T\|)$.

When can a function be written as a power series? We wish to establish that being analytic in a neighborhood of 0 is a necessary and sufficient condition. The necessity part is the content of the following proposition, but sufficiency will be shown later (Theorem 13.26).

Proposition 13.16

A power series $f(z) := \sum_{n=0}^{\infty} a_n z^n$ **is analytic strictly within its radius of convergence, and**

$$f'(z) = \sum_{n=1}^{\infty} n a_n z^{n-1}.$$

Proof First of all, the power series $\sum_n a_n n z^{n-1}$ converges, with the same radius of convergence R as $\sum_n a_n z^n$,

$$\limsup_n |n a_n|^{1/n} = \lim_{n \to \infty} n^{1/n} \limsup_n |a_n|^{1/n} = 1/R \qquad \text{(Exercise 3.6(1d))}.$$

For each individual term of the given power series,

$$(z + h)^n = z^n + n z^{n-1} h + o_n(h).$$

It needs to be shown that $|\sum_n a_n o_n(h)|/|h| \to 0$ as $h \to 0$. One trick is to find an alternative way of expanding $(z + h)^n$ as follows:

$$
\begin{aligned}
(z + h)^n &= (z + h)^{n-1} h + (z + h)^{n-1} z \\
&= (z + h)^{n-1} h + (z + h)^{n-2} z h + (z + h)^{n-2} z^2 \\
&= (z + h)^{n-1} h + \cdots + (z + h)^{n-k} z^{k-1} h + \cdots \\
&\quad + z^{n-1} h + z^n \\
\Rightarrow \qquad |(z + h)^n - z^n| &\leqslant (|z + h|^{n-1} + \cdots + |z|^{n-1}) |h| \\
&\leqslant n r^{n-1} |h|, \qquad\qquad\qquad\qquad (13.2)
\end{aligned}
$$

where r is larger than $|z| + |h|$ but smaller than R. Now,

$$
\begin{aligned}
o_n(h) &= (z + h)^n - z^n - n z^{n-1} h \\
&= (z + h)^{n-1} h + \cdots + (z + h)^{n-k} z^{k-1} h + \cdots + z^{n-1} h \\
&\quad - z^{n-1} h - \cdots - z^{n-k} z^{k-1} h - \cdots - z^{n-1} h
\end{aligned}
$$

$$= \sum_{k=1}^{n} \left((z+h)^{n-k} - z^{n-k} \right) z^{k-1} h$$

so $|o_n(h)| \leqslant (n-1)r^{n-2}|h|^2 + \cdots + r^{n-2}|h|^2$ $\qquad$ by (13.2)

$$= \frac{n(n-1)}{2} r^{n-2} |h|^2$$

But the series $c := \sum_{n=2}^{\infty} |a_n| \frac{n(n-1)}{2} r^{n-2}$ converges for $r < R$, so

$$\left| \sum_{n=0}^{\infty} a_n o_n(h) \right| \leqslant \sum_{n=2}^{\infty} |a_n| |o_n(h)| \leqslant c|h|^2$$

which proves that the remainder term $\sum_n a_n o_n(h)$ is $o(h)$. $\qquad\qquad$ □

There are two important consequences: Since differentiating a power series gives another power series with the same radius of convergence, then we can differentiate repeatedly. Secondly, we know that polynomials are distinct as functions on $\mathbb{C}$ when they have different coefficients; this property remains valid for power series: If a function can be written as a power series, then its coefficients are unique to it.

Proposition 13.17

Assuming a strictly positive radius of convergence,

(i) **a power series $f(z) := \sum_{n=0}^{\infty} a_n z^n$ is infinitely many times differentiable, and**

$$a_n = \frac{f^{(n)}(0)}{n!}$$

(ii) **distinct power series are not equal as functions.**

Proof (i) By induction on n, $f^{(n)}$ has the power series

$$f^{(n)}(z) = n!a_n + (n+1)!a_{n+1}z + \frac{(n+2)!}{2}a_{n+2}z^2 + \cdots$$

Substituting $z = 0$ gives the stated formula.

(ii) Suppose $\sum_n b_n T^n = \sum_n c_n T^n$ for all T such that $\rho(T) < r$, the smaller of their radii of convergence. By taking the difference of the two series, it is enough to show that if $f(z) := \sum_n a_n z^n = 0$ for all $z \in B_r(0)$, then $a_n = 0$ for all n. But this is immediate from (i) since $f^{(n)}(0) = 0$ in this case. $\qquad\qquad$ □

The Exponential and Logarithm Maps

There are a couple of power series of supreme importance. As motivation, consider the possibility of converting addition in a Banach algebra to multiplication,

$$f(x + y) = f(x)f(y), \quad f(0) = 1.$$

Apart from the constant function $f = 1$, are there any others? If f exists, it would have to satisfy a number of properties:

(a) $f(nx) = f(x)^n$, $f(-x) = f(x)^{-1}$,
(b) When the algebra is $\mathbb{R}$, $f(m/n) = a^{m/n}$ where $a := f(1) > 0$ (Hint: $f(n/n) = f(1/n)^n$),
(c) f is uniformly continuous on $\mathbb{Q} \cap [0, 1]$, so it can be extended to a continuous function on $\mathbb{R}$, usually denoted by $f(x) = a^x$,
(d) $f'(x) = f'(0)f(x)$ if f is differentiable at 0, since $f(h) = 1 + f'(0)h + o(h)$ so

$$f(x + h) = f(x)f(h) = f(x) + f(x)f'(0)h + o(h);$$

consequently f is infinitely many times differentiable with $f^{(n)}(x) = f'(0)^n f(x)$. Taking the simplest case $f'(0) = 1$ (so $f^{(n)}(0) = 1$) leads to the following definition:

The **exponential** function is defined by

$$e^T := 1 + T + \frac{T^2}{2!} + \frac{T^3}{3!} + \cdots = \sum_{n=0}^{\infty} \frac{1}{n!} T^n.$$

Its radius of convergence is $\liminf_n |a_n|^{-1/n} = \lim_{n \to \infty} \frac{1/n!}{1/(n+1)!} = \infty$ by the ratio test, so e^T exists for any T and satisfies $\|e^T\| \leqslant e^{\|T\|}$.

Similarly, starting with $f(xy) = f(x) + f(y)$, we are led to the **logarithm** function, defined by

$$\log(1 + T) := T - \frac{T^2}{2} + \frac{T^3}{3} + \cdots = \sum_{n=1}^{\infty} \frac{(-1)^{n+1}}{n} T^n,$$

with radius of convergence $\liminf_n |a_n|^{-1/n} = \lim_{n \to \infty} \frac{1/n}{1/(n+1)} = 1$.

Proposition 13.18

When S, T commute, $e^{S+T} = e^S e^T$. For $\rho(T) < 1$, $e^{\log(1+T)} = 1 + T$.

Proof (i) The product $e^S e^T$ can be obtained in table form as,

$$e^S = 1 + S + \tfrac{1}{2!}S^2 + \cdots$$

e^T			
$=$			
1	1	S	$\tfrac{1}{2}S^2$
$+$			
T	T	ST	$\tfrac{1}{2}S^2T$
$+$			
$\tfrac{1}{2!}T^2$	$\tfrac{1}{2}T^2$	$\tfrac{1}{2}ST^2$	$\tfrac{1}{4}S^2T^2$
$+$			
$\vdots$			

The general term in this array is $\frac{1}{n!}\frac{1}{m!}S^n T^m = \frac{1}{N!}\binom{N}{n}S^n T^{N-n}$ where $N := n + m$ is the Nth diagonal from the top left corner. This is precisely the nth term of the expansion of $\frac{1}{N!}(S + T)^N$ when S and T commute, so the array sum is e^{S+T}.

(ii) The second part can be (tediously) proved by making a power series expansion as above (Exercise 13.19(8)). We defer the proof until we have better tools available (Example 13.30(3)). □

Exercises 13.19

1. Calculate $\rho(T)$ for the following matrices

   (a) $\begin{pmatrix} 0 & 1 \\ 0 & 0 \end{pmatrix}$, (b) $\begin{pmatrix} 1 & a \\ 0 & 0 \end{pmatrix}$, (c) $\begin{pmatrix} a & 0 \\ 0 & b \end{pmatrix}$, (d) $\begin{pmatrix} a & 1 \\ 0 & a \end{pmatrix}$.

   Only one of these examples satisfies $\rho(T) = \|T\|$.
2. Every idempotent P, except 0, satisfies $\rho(P) = 1$; every nilpotent Q has $\rho(Q) = 0$, and every cyclic element T has $\rho(T) = 1$.
3. For any invertible S, $\rho(S^{-1}TS) = \rho(T)$, yet $\|S^{-1}TS\|$ may be much larger than $\|T\|$. For example, let $P := \begin{pmatrix} 1 & c \\ 0 & 0 \end{pmatrix}$ and $S := \begin{pmatrix} 1 & 0 \\ 0 & a \end{pmatrix}$, then $S^{-1}PS = \begin{pmatrix} 1 & ac \\ 0 & 0 \end{pmatrix}$ has norm $\sqrt{1 + |ac|^2}$.
4. If $ST = TS$, then $\rho(ST) \leqslant \rho(S)\rho(T)$. Deduce $\rho(T^{-1})^{-1} \leqslant \rho(T)$, and find examples of non-commuting matrices such that $\rho(ST) > \rho(S)\rho(T)$.
5. The equation $T - ATB = C$ has a solution $T = \sum_{n=0}^{\infty} A^n C B^n$ if $\rho(A)\rho(B) < 1$.

6. The radii of convergence of

$$\sum_{n=0}^{\infty} n^n T^n, \quad \sum_{n=0}^{\infty} n T^n, \quad \sum_{n=1}^{\infty} T^n/n, \quad \sum_{n=0}^{\infty} T^n/n!$$

are $0, 1, 1, \infty$, respectively. A quick way of estimating the radius of convergence r is to judge how fast the coefficients grow: if $c_0 r_0^n \leqslant |a_n| \leqslant c_1 r_1^n$ then $\frac{1}{r_1} \leqslant r \leqslant \frac{1}{r_0}$.

7. How are the radii of convergence of $\sum_n (a_n + b_n) T^n$ and $\sum_n a_n b_n T^n$ related to those of $\sum_n a_n T^n$ and $\sum_n b_n T^n$?

8. Let $f(T) := \sum_{n=0}^{\infty} a_n T^n$ and $g(T) := \sum_{n=0}^{\infty} b_n T^n$. Find the first few terms of the power series expansions of $f + g$, fg and $f \circ g$; in particular, find $-f(T)$, $f(T)^{-1}$, $f^{-1}(T)$.

9. Let $f(T) := \sum_n a_n T^n$ be a power series, and $F(T) := \sum_n |a_n| T^n$; they have the same radius of convergence r. If $\|T\| < r$, then $\|f(T)\| \leqslant F(\|T\|)$; e.g., $\|e^T\| \leqslant e^{\|T\|}$.

10. The convergence of a power series is uniform in T on $\overline{B_s}(0)$, for $s < r$.

11. When T satisfies a polynomial $p(T) = 0$, then every (convergent) power series on T reduces to a polynomial in T.

12. (a) $e^0 = 1$, (b) the inverse of e^T is e^{-T}, (c) $e^{nT} = (e^T)^n$.

13. By analogy with the complex case, define the *hyperbolic* and *trigonometric* functions of T as power series, and show (a) $\exp \left(\begin{smallmatrix} 0 & -1 \\ 1 & 0 \end{smallmatrix} \right) t = \left(\begin{smallmatrix} \cos t & -\sin t \\ \sin t & \cos t \end{smallmatrix} \right)$, (b) $\cos \left(\begin{smallmatrix} 1 & 1 \\ 0 & 1 \end{smallmatrix} \right) t = \left(\begin{smallmatrix} \cos t & -t \sin t \\ 0 & \cos t \end{smallmatrix} \right)$, (c) $e^T = \cosh T + \sinh T$, (d) $e^{iT} = \cos T + i \sin T$.

14. Prove that there is a non-zero complex number τ such that $e^\tau = 1$. Thus the exponential function has a *period*, $e^{T + n\tau} = e^T$. The 'smallest' such number is $6.283\ldots i =: 2\pi i$.

15. $\star$ $(1 + T/n)^n \to e^T$ as $n \to \infty$.
    (Hint: Each component in the series is $\frac{1}{n^k} \binom{n}{k} T^k \to \frac{1}{k!} T^k$, then use Exercise 9.7(1).)

16. $\star$ The product of n terms, $(1 + S/n)(1 + T/n)(1 + S/n) \cdots (1 + T/n) \to e^{S+T}$ as $n \to \infty$. (At least show convergence for each power term.)

17. $\star$ Trotter formula: $e^{S/n} e^{T/n} e^{S/n} \cdots e^{T/n} \to e^{S+T}$. For example,

$$e^{S+T} \approx e^{S/2} e^{T/2} e^{S/2} e^{T/2}.$$

Find the exact coefficients used in the Trotter-Suzuki approximation

$$e^{0.293S} e^{0.707T} e^{0.707S} e^{0.293T},$$

that make it the best possible to second order. These formulas are very useful to approximate e^{S+T} whenever S and T do not commute.

13.3 The Group of Invertible Elements

Among the invertible elements of a Banach algebra, one finds all the exponentials e^T (including all non-zero complex numbers) and all their products, as well as the unit ball around 1, as the next key theorem proves:

Theorem 13.20

> **If $\rho(T) < 1$ then $1 - T$ is invertible:** $(1 - T)^{-1} = 1 + T + T^2 + \cdots$

Proof The radius of convergence of the series $\sum_n T^n$ is 1, by Hadamard's formula. For $\rho(T) < 1$, let $S_N := 1 + T + \cdots + T^N \to \sum_{n=0}^{\infty} T^n$. Then, remembering that $\rho(T) < 1 \Rightarrow T^N \to 0$ as $N \to \infty$ (Example 13.13(3)),

$$
\begin{aligned}
S_N &= 1 + T + \cdots + T^N \\
T S_N &= \quad\ T + \cdots + T^N + T^{N+1} \\
\Rightarrow (1 - T)S_N &= 1 \qquad\qquad\qquad - T^{N+1} \to 1.
\end{aligned}
$$

Similarly, $S_N(1 - T) \to 1$ as $N \to \infty$. This shows that $\sum_{n=0}^{\infty} T^n$ is the inverse of $1 - T$. $\qquad\square$

Theorem 13.21

> **The invertible elements of a Banach algebra $\mathcal{X}$ form a group $\mathcal{G}(\mathcal{X})$ with the operation of multiplication. $\mathcal{G}(\mathcal{X})$ is an open set in $\mathcal{X}$, and the map $T \mapsto T^{-1}$ is differentiable on it.**

Proof Multiplication in a Banach algebra is associative and has a unity $1 \in \mathcal{G}(\mathcal{X})$. To prove $\mathcal{G}(\mathcal{X})$ is a group, it needs to be shown that if $S, T \in \mathcal{G}(\mathcal{X})$, then ST and T^{-1} are invertible, a fact that is evident from

$$(ST)^{-1} = T^{-1}S^{-1}, \quad (T^{-1})^{-1} = T.$$

Let T be any invertible element of $\mathcal{X}$, and consider any neighboring element

$$T + H = T(1 + T^{-1}H)$$

with $\|H\| < \|T^{-1}\|^{-1}$. Then $\rho(T^{-1}H) \leqslant \|T^{-1}\|\|H\| < 1$, so that $1 + T^{-1}H$, and by implication $T + H$, are invertible. As the neighboring points of T are invertible, T is an interior point of $\mathcal{G}(\mathcal{X})$ and the group is open in $\mathcal{X}$.
In fact, writing $T + H = T(I + T^{-1}H)$,

$$(T + H)^{-1} = (1 + T^{-1}H)^{-1}T^{-1} = T^{-1} - T^{-1}HT^{-1} + T^{-1}HT^{-1}HT^{-1} + \cdots$$

This shows that $T \mapsto T^{-1}$ is differentiable with derivative $H \mapsto -T^{-1}HT^{-1}$, by verifying

$$\|T^{-1}HT^{-1}\| \leqslant \|T^{-1}\|^2 \|H\|$$

$$\|T^{-1}HT^{-1}HT^{-1} + \cdots\| \leqslant \sum_{n=0}^{\infty} \|H\|^{n+2}\|T^{-1}\|^{n+3} = \frac{\|H\|^2\|T^{-1}\|^3}{1 - \|T^{-1}\|\|H\|} = o(H).$$

□

A group, for which the acts of multiplication and taking the inverse are differentiable, is called a 'Lie group', a topic that has a vast literature devoted to it.

A particular case of the above, for $H = z1$, is the following series:

$$(T + z)^{-1} = T^{-1} - zT^{-2} + z^2T^{-3} + \cdots, \tag{13.3}$$

Note that the map $z \mapsto (T - z) \mapsto (T - z)^{-1}$ is analytic wherever the inverse exists; its derivative is $(T - z)^{-2}$.

Examples 13.22

1. The group of $n \times n$ invertible complex matrices is often denoted $GL(n, \mathbb{C})$. It has a group-morphism, the *determinant* $\det : GL(n, \mathbb{C}) \to \mathbb{C}^\times = \mathcal{G}(\mathbb{C})$,

$$\det AB = \det A \det B$$

   whose kernel is the normal subgroup $SL(n, \mathbb{C})$ of 'special matrices' with determinant 1.

2. In $\mathbb{C}$, when z is large, z^{-1} is small. But for general Banach algebras there is no such relation between $\|T^{-1}\|$ and $\|T\|$, e.g., the inverse of $(10, 0.01) \in \mathbb{C}^2$ is $(0.1, 100)$.

3. The set of non-invertible elements is closed in $\mathcal{X}$. So the closure of a proper ideal is a proper ideal.
   Proof: By Example 13.5(5), $\mathcal{I} \subset \mathcal{G}(\mathcal{X})^c$, so $\overline{\mathcal{I}} \subseteq \mathcal{G}(\mathcal{X})^c$ and $1 \notin \overline{\mathcal{I}}$.

4. If T is invertible, then $B_\epsilon(TS) \subseteq TB_{\epsilon\|T^{-1}\|}(S)$. Consequently, multiplication by T is an open mapping.
   Proof: Let $\|A - TS\| < \epsilon$; then $\|T^{-1}A - S\| \leqslant \|T^{-1}\|\|A - TS\| < \|T^{-1}\|\epsilon$, as required. If U is an open set in $\mathcal{X}$ and $S \in U$, then $S \in B_\epsilon(S) \subseteq U$, so

$$TS \in B_{\epsilon/\|T^{-1}\|}(TS) \subseteq TB_\epsilon(S) \subseteq TU$$

   and TU is open in $\mathcal{X}$.

5. The set of non-invertible elements is path-connected (to the origin, say), and may disconnect the group of invertible elements, e.g., $GL(2, \mathbb{R})$ disconnects into the

two open sets of matrices whose determinants are strictly positive and strictly negative, respectively.

The following proposition confirms that as an invertible operator R approaches the boundary of $\mathcal{G}(\mathcal{X})$, $\|R^{-1}\|$ grows to infinity, as expected.

Proposition 13.23

Let T be on the boundary of the group of invertible elements.

(i) **For any invertible element R, $\|R^{-1}\| \geqslant 1/\|R - T\|$,**

(ii) **T is a *topological divisor of zero*, meaning there are unit elements S_n such that**

$$T S_n \to 0 \text{ AND } S_n T \to 0, \text{ as } n \to \infty.$$

Proof (i) Since T is at the boundary of the open set of invertible elements, it cannot be invertible, whereas R and all elements in its surrounding ball of radius $\|R^{-1}\|^{-1}$ are invertible, by the proof of the previous theorem. Thus $\|R - T\| \geqslant \|R^{-1}\|^{-1}$ as claimed.

(ii) Let invertible elements R_n converge to a boundary element T, and let $S_n :=$ $R_n^{-1}/\|R_n^{-1}\|$; then

$$T S_n = \frac{T R_n^{-1}}{\|R_n^{-1}\|} = (T - R_n)\frac{R_n^{-1}}{\|R_n^{-1}\|} + \frac{I}{\|R_n^{-1}\|} \to 0$$

since $R_n \to T$ and $\|R_n^{-1}\|^{-1} \leqslant \|R_n - T\| \to 0$. Similarly $S_n T \to 0$ as well. $\qquad\square$

As remarked earlier, the group $\mathcal{G}(\mathcal{X})$ need not be a connected set, but splits into connected components, with, say, $\mathcal{G}_1$ being the component containing 1. Recall that a component is maximal connected, so if $\mathcal{G}_1$ contains part of a connected subset of $\mathcal{G}(\mathcal{X})$, it must contain all of it (Theorem 5.12).

Proposition 13.24

The component of invertible elements containing 1 is an open normal subgroup, generated by e^T for all T.

Proof $\mathcal{G}_1$ *is open in* $\mathcal{G}(\mathcal{X})$: Any $T \in \mathcal{G}_1$ is an interior point of $\mathcal{G}(\mathcal{X})$, so $T \in B_\epsilon(T) \subseteq \mathcal{G}(\mathcal{X})$. But the ball $B_\epsilon(T)$ is (path-)connected and intersects $\mathcal{G}_1$, so $B_\epsilon(T) \subseteq \mathcal{G}_1$.

$\mathcal{G}_1$ *is a subgroup of* $\mathcal{G}(\mathcal{X})$: Multiplication by T is a continuous operation, so $T\mathcal{G}_1$ is connected (Proposition 5.6). When $T \in \mathcal{G}_1$, then $T = T1 \in T\mathcal{G}_1 \subseteq \mathcal{G}(\mathcal{X})$, so $\mathcal{G}_1$ contains part, and therefore all, of $T\mathcal{G}_1$. Hence $T, S \in \mathcal{G}_1 \Rightarrow TS \in T\mathcal{G}_1 \subseteq \mathcal{G}_1$. Similarly, inversion is a continuous mapping, so $\mathcal{G}_1^{-1}$ is connected; it contains 1, so must be a subset of $\mathcal{G}_1$, i.e., $T \in \mathcal{G}_1 \Rightarrow T^{-1} \in \mathcal{G}_1$.

$\mathcal{G}_1$ *is a normal subgroup*: By the same reasoning, for any invertible T, $T^{-1}\mathcal{G}_1 T$ is a connected subset of $\mathcal{G}(\mathcal{X})$ and contains 1, so it is a subset of $\mathcal{G}_1$ (in fact it must equal it).

$\mathcal{G}_1$ *is generated by the exponentials*: Let $\mathcal{E}$ be the group generated by the exponentials e^T for all $T \in \mathcal{X}$; its elements are finite products $e^T \cdots e^S$. $\mathcal{E}$ is clearly closed under multiplication and inversion, $(e^T \cdots e^S)^{-1} = e^{-S} \cdots e^{-T}$, so $\mathcal{E} \subseteq \mathcal{G}$. It contains $1 = e^0$, and is connected since there is a continuous path from 1 to every element $e^T \cdots e^S$, namely $t \mapsto e^{tT} \cdots e^{tS}$ for $t \in [0, 1]$. We can conclude that $\mathcal{E}$ lies inside $\mathcal{G}_1$.

The elements near to 1 are all exponentials, since for $\tilde{H}$ small, $\log(1 + \tilde{H})$ exists as a power series and hence[1] $1 + \tilde{H} = e^{\log(1+\tilde{H})}$. So a small enough neighborhood around $E := e^T \cdots e^S \in \mathcal{E}$ consists of elements

$$E + H = E(1 + E^{-1}H) = e^T \cdots e^S e^{\log(1+E^{-1}H)} \in \mathcal{E}$$

at least for $\|H\| < e^{-\|S\|} \cdots e^{-\|T\|}$. This means that E is an interior point of $\mathcal{E}$, which is thus open. Its complement in $\mathcal{G}_1$ is also open, since $\mathcal{G}_1 \setminus \mathcal{E} = \bigcup_{T \in \mathcal{G}_1 \setminus \mathcal{E}} T\mathcal{E}$ (prove!) and each $T\mathcal{E}$ is open (Example 13.22(4)). $\mathcal{E}$, being open and closed in $\mathcal{G}_1$, must equal $\mathcal{G}_1$ (Proposition 5.4). □

Exercises 13.25

1. The invertible elements of $\mathbb{C}^n$ are $(z_1, \ldots, z_n)$ such that none of the components are zero.

2. In ℓ^∞, a sequence $(a_n)_{n \in \mathbb{N}}$ is invertible if, and only if, it is bounded away from $\mathbf{0}$, i.e., $0 < c \leqslant |a_n|$. Paths $t \mapsto w(t)$ in $C[0, 1]$ are invertible when they do not pass through 0.

3. In $B(X)$, the invertible elements are the automorphisms of X.

4. In $B(X)$, $\|T^{-1}\| = 1/\inf_{\|x\|=1} \|Tx\|$.

5. In $\mathcal{X} \times \mathcal{Y}$, (S, T) is invertible if, and only if, both S and T are invertible.

6. The integral operator on $C[a, b]$, $Tf(s) := \int_a^b k(s, t) f(t) \, dt$ has norm satisfying $\|T\| \leqslant \|k\|_{L^\infty} |b - a|$. Deduce that when $\|k\|_{L^\infty} < 1/|b - a|$, the equation $Tf + g = f$ has the unique solution $f = \sum_{n=0}^\infty T^n g$.

7. If T is invertible and $Tx = y$, $(T + H)(x + x_\epsilon) = y$, then $\dfrac{\|x_\epsilon\|}{\|x\|} \leqslant \dfrac{\|T^{-1}\|\|H\|}{1-\|T\|\|H\|}$.

8. The map $t \mapsto e^{tT}$ is a differentiable group-morphism $\mathbb{R} \to \mathcal{G}(\mathcal{X})$; its derivative at t is Te^{tT}.

9. ⋆ Conversely, every differentiable group-morphism $A : \mathbb{R} \to \mathcal{G}(\mathcal{X})$, meaning $A_{t+s} = A_t A_s$, is of this type:

---

[1] This was stated, not proved, in Proposition 13.18, but the argument is not circular.

(a) $\exists h > 0$, $\int_0^h A_t\, dt$ is invertible, by the mean value theorem (Proposition 12.9), and $\int_t^{t+h} A = (\int_0^h A)A_t$;

(b) Let $T := (A_h - 1)(\int_0^h A)^{-1}$, so that $A_{t+h} = A_t + hTA_t + o(h)$;

(c) $\frac{d}{dt}(A_t e^{-tT}) = (\frac{d}{dt}A_t)e^{-tT} - A_t T e^{-tT} = 0$, so $A_t = A_0 e^{tT} = e^{tT}$.

10. Verify Proposition 13.23 for $\begin{pmatrix} 1 & 1 \\ 0 & \frac{1}{n} \end{pmatrix} \to \begin{pmatrix} 1 & 1 \\ 0 & 0 \end{pmatrix}$.

11. A topological divisor of zero, also called a *generalized* divisor of zero, does not have right or left inverses.

12. The right-shift operator R on ℓ^∞ is a right divisor of zero but not a topological divisor of zero.

13. In finite dimensions, there is no distinction between divisors of zero and topological ones. (Hint: $S_n \in \overline{B}_{\mathcal{X}}$, which is compact.)

14. An isomorphism between Banach algebras preserves topological divisors of zero.

15. If R is invertible, then $\|R^{-1}\| \geqslant 1/d(R, \partial \mathcal{G}(\mathcal{X}))$.
   (Hint: By the definition of $d(\mu, \partial \mathcal{G}(\mathcal{X}))$ (Exercise 2.20(9)), there is a sequence $T_n \in \partial \mathcal{G}(\mathcal{X})$ such that $\|T_n - R\| \to d(R, \mathcal{G}(\mathcal{X}))$.)

16. Every invertible $n \times n$ matrix has a logarithm (over $\mathbb{C}$; see Example 14.27(1)), so $\mathcal{G} = \mathcal{G}_1$ for $B(\mathbb{C}^n)$. But over the reals, any diagonal matrix with some negative components are not exponentials; they have no real logarithms.

13.4 Analytic Functions

There are two ways of connecting the coefficients of a power series to its function

$$f(z) = a_0 + a_1 z + a_2 z^2 + \cdots,$$

(i) by differentiation

$$f^{(n)}(z) = n!a_n + (n+1)!a_{n+1}z + \cdots \quad \Rightarrow \quad f^{(n)}(0) = n!\, a_n.$$

(ii) by integration

$$\frac{f(z)}{z^n} = \frac{a_0}{z^n} + \cdots + \frac{a_{n-1}}{z} + a_n + \cdots \quad \Rightarrow \quad \oint \frac{f(z)}{z^n}\, dz = 2\pi i\, a_{n-1}.$$

These formulas raise the possibility of *creating* a power series from a given function, by defining the coefficients in these ways. The latter one is more useful because it does not assume f to be differentiable infinitely often.

Theorem 13.26 (Taylor Series)

If $f : \mathbb{C} \to \mathbb{C}$ is analytic in a disk $B_R(0)$, then it is a power series inside the disk. For $\rho(T) < R$,

$$f(T) := \frac{1}{2\pi i} \oint f(z)(z - T)^{-1}\, dz = \sum_{n=0}^{\infty} a_n T^n,$$

where for all $n \in \mathbb{N}$

$$a_n = \frac{1}{2\pi i} \oint f(z) z^{-1-n}\, dz = \frac{f^{(n)}(0)}{n!},$$

and

$$\forall r < R,\ \exists c_r,\ \forall n \in \mathbb{N}, \quad |a_n| \leqslant \frac{c_r}{r^n}.$$

To justify the use of the notation $f(T)$, note that when $T = a1$, the two uses of the symbol f agree, i.e., $f(a1) = f(a)1$, by Cauchy's integral formula.

Proof The path of integration is along a circle with center 0 and radius r just less than R but larger than $\rho(T)$. For z on this circle, $\rho(T/z) = \rho(T)/r < 1$, so

$$(z - T)^{-1} = z^{-1}(1 - T/z)^{-1} = \sum_{n=0}^{\infty} z^{-1-n} T^n, \text{ and}$$

$$\frac{1}{2\pi i} \oint f(z)(z - T)^{-1}\, dz = \sum_{n=0}^{\infty} \frac{1}{2\pi i} \oint f(z) z^{-1-n}\, dz\, T^n = \sum_{n=0}^{\infty} a_n T^n.$$

However we need to justify the swap of the summation with the integral. Recall that $z \mapsto (z - T)^{-1}$ is continuous in z by (13.3), and the circle is a compact set, so $\|f(z)(z - T)^{-1}\| \leqslant C$ for z on the circle (Corollary 6.16). It follows that

$$\left\| \sum_{n=N}^{\infty} f(z) T^n / z^{n+1} \right\| = \|T^N f(z)(z - T)^{-1}/z^{N+1}\| \leqslant C\|T^N\|/r^{N+1} \to 0$$

uniformly in z. So $\sum_{n=0}^{N} \oint f(z) T^n / z^{n+1}\, dz \to \oint \sum_n f(z) T^n / z^{n+1}\, dz$.
    Note that

$$|a_n| \leqslant \frac{1}{2\pi} \oint c / r^{n+1}\, dt = c / r^n,$$

where c is the maximum value of f on the compact disk $\overline{B_r(0)} \subset \mathbb{C}$. The radius of convergence of this power series is at least R since for any $0 < r < R$,

$$\liminf_n |a_n|^{-1/n} \geqslant \lim_{n\to\infty} \frac{r}{c^{1/n}} = r.$$

$\square$

Proposition 13.27 (Liouville's Theorem)

> **If an analytic function on $\mathbb{C}$ grows polynomially $|f(z)| \leqslant c|z|^n$ as $|z| \to \infty$, then f *is* a polynomial of degree at most n. In particular, if f is bounded then it is constant.**

Proof If $f : \mathbb{C} \to \mathbb{C}$ were analytic on $\mathbb{C}$, and grows polynomially, then its maximum value on a disk of radius r is $c_r \leqslant cr^n$. So the mth Taylor coefficient vanishes for $m > n$,

$$|a_m| \leqslant c_r/r^m \leqslant cr^{n-m} \to 0 \quad \text{as } r \to \infty.$$

This also applies to vector-valued analytic functions $F : \mathbb{C} \to X$. For any functional $\phi \in X^*$, $\phi \circ F : \mathbb{C} \to \mathbb{C}$ is also analytic. If F grows polynomially, then so does $\phi \circ F$

$$|\phi \circ F(z)| \leqslant \|\phi\| \|F(z)\| \leqslant \|\phi\| c|z|^n,$$

which implies that $\phi \circ F(z)$ is a polynomial $a_0 + a_1 z + \cdots + a_n z^n$. In fact, by Example 12.3(3), $a_n = \phi \circ F^{(n)}(0)/n!$, so that

$$\phi \circ F(z) = \phi \circ (F(0) + F'(0)z + \cdots + F^{(n)}(0)z^n/n!).$$

As ϕ is arbitrary, we deduce that $F(z)$ is a polynomial in z. $\square$

Theorem 13.28 (Laurent Series)

> **If $f : \mathbb{C} \to \mathbb{C}$ is analytic in a ring $B_R(0) \setminus \overline{B_r(0)}$, and $r < \rho(T^{-1})^{-1} \leqslant \rho(T) < R$, then**
>
> $$f(T) := \frac{1}{2\pi i} \oint f(z)(z - T)^{-1} \, dz = \sum_{n=-\infty}^{\infty} a_n T^n,$$
>
> **where $a_n = \frac{1}{2\pi i} \oint f(z)z^{-1-n} \, dz$, for $n \in \mathbb{Z}$. The residue of f in $\overline{B_r(0)}$ is a_{-1}.**

The path of integration is here understood to be just within the boundary of the ring, going counter-clockwise around a circle of radius just smaller than R, and clockwise around a circle just larger than r. Note that R is allowed to be infinite, in which case substitute R in the proof with any value larger than $\rho(T)$.

Proof A Laurent series can be thought of as the sum of two separate power series, $\sum_{n=0}^{\infty} a_n T^n + \sum_{n=1}^{\infty} a_{-n} T^{-n}$, one in T and the other in T^{-1}. If R and R' are the respective radii of convergence, then absolute convergence occurs only when $\rho(T) < R$ and $\rho(T^{-1}) < R'$.

For z on the bigger circle, $\rho(T/z) = \rho(T)/|z| < 1$ if the radius is close enough to R, so just like the proof of the Taylor series,

$$\frac{1}{2\pi i} \oint_1 f(z)(z-T)^{-1} \, dz = \sum_{n=0}^{\infty} a_n T^n.$$

For z on the smaller circle, $\rho(zT^{-1}) = |z|\rho(T^{-1}) < 1$ when its radius is close enough to r, so

$$(z-T)^{-1} = -(1-zT^{-1})^{-1} T^{-1} = -\sum_{n=0}^{\infty} z^n T^{-n-1},$$

and (along an counter-clockwise path)

$$\frac{1}{2\pi i} \oint_2 f(z)(z-T)^{-1} \, dz = -\sum_{n=1}^{\infty} \frac{1}{2\pi i} \oint f(z) z^{n-1} \, dz \, T^{-n} = -\sum_{n=1}^{\infty} a_{-n} T^{-n}.$$

Combining the two integrals and series gives Laurent's expansion. Note that the second series vanishes when f is analytic within $B_r(0)$, by Cauchy's theorem, so it is consistent with Taylor's theorem.

Since the Laurent series converges uniformly strictly within the annulus, we obtain

$$\frac{1}{2\pi i} \oint f(z) \, dz = \frac{1}{2\pi i} \sum_{n=-\infty}^{\infty} \oint a_n z^n \, dz = a_{-1}.$$

□

These two theorems of course also apply, by translating, to disks and rings with center z_0; the resulting series will then be $\sum_n a_n (T - z_0)^n$.

Proposition 13.29

> **The zeros of a non-zero analytic function, defined on an open connected subset of $\mathbb{C}$, are isolated.**

Proof Suppose an interior zero w of $f : \Omega \to \mathbb{C}$ is a limit point of other zeros, $z_n \to w$ $(z_n \neq w)$. Then f can be written as a power series $f(z) = \sum_k a_k(z - w)^k$ in some neighborhood of w. If a_K is the first non-zero coefficient, then

$$0 = f(z_n) = (z_n - w)^K (a_K + a_{K+1}(z_n - w) + \cdots),$$

$$\therefore\ 0 = a_K + a_{K+1}(z_n - w) + \cdots \to a_K \text{ as } z_n \to w.$$

This contradiction determines that f is locally zero in Ω. Hence it is zero in Ω (Exercise 5.8(9)).   □

Examples 13.30

1. The Fourier series $\sum_{n=-\infty}^{\infty} a_n e^{in\theta}$ is a Laurent series with $T = e^{i\theta}$.
2. ▶ For polynomials (and circular paths as in the theorems),

$$p(T) = \frac{1}{2\pi i} \oint p(z)(z - T)^{-1}\, dz.$$

For example,

$$1 = \frac{1}{2\pi i} \oint (z - T)^{-1}\, dz, \quad T = \frac{1}{2\pi i} \oint z(z - T)^{-1}\, dz,$$

$$T^{-1} = \frac{1}{2\pi i} \oint \frac{1}{z}(z - T)^{-1}\, dz.$$

Proof for T^{-1}: We can use Laurent's expansion on a path $z(\theta) = re^{i\theta}$, since $1/z$ is analytic everywhere except at 0,

$$a_n = \frac{1}{2\pi i} \oint \frac{1}{z^{n+2}}\, dz = \frac{1}{2\pi} \int_0^{2\pi} \frac{1}{r^{n+1}} e^{-i(n+1)\theta}\, d\theta = 0$$

unless $n = -1$, in which case $a_{-1} = 1$. So $\sum_n a_n T^n = T^{-1}$.
3. ▶ We can finally show $e^{\log(1+T)} = 1 + T$ for $\rho(T) < 1$.
   Proof: Let $f(z) := e^{\log(1+z)}$ for $|z| < 1$; then $f'(z) = e^{\log(1+z)}/(1 + z)$ and $f''(z) = 0$ (check!). So the non-zero coefficients of its Taylor series are $a_0 = f(0) = e^0 = 1$ and $a_1 = f'(0) = 1$. Hence $f(T) = 1 + T$.
4. *Binomial theorem*: $(1 + T)^p := e^{p\log(1+T)} = 1 + pT + \binom{p}{2}T^2 + \cdots$ provided $\rho(T) < 1$, $p \in \mathbb{C}$, and $\binom{p}{n} := \frac{p(p-1)\cdots(p-n+1)}{n!}$.
   Proof: Define the analytic function $f(z) := (1+z)^p = e^{p\log(1+z)}$ inside the unit disk $B_{\mathbb{C}}$. Its derivatives are, by induction,

$$f^{(n)}(z) = p(p-1)\cdots(p-n+1)e^{(p-n+1)\log(1+z)}(1+z)^{-1}$$

$$= p(p-1)\cdots(p-n+1)(1+z)^{p-n},$$

so its power series coefficients are $a_n = f^{(n)}(0)/n! = \binom{p}{n}$.

5. ▶ There are versions of these series expansions valid for a vector-valued function $F : \mathbb{C} \to X$, where X is a Banach space and F is analytic inside a ring, $r < |z| < R$,

$$F(z) = \frac{1}{2\pi i} \oint F(w)(w - z)^{-1}\, dw = \sum_{n \in \mathbb{Z}} A_n z^n,$$

where $A_n := \frac{1}{2\pi i} \oint F(w) w^{-1-n}\, dw \in X$.

Proof For any $\phi \in X^*$, the map $\phi \circ F : \mathbb{C} \to \mathbb{C}$, being the composition of differentiable functions, is analytic on the ring $B_R(0) \setminus \overline{B_r(0)}$, so it has a Laurent expansion $\phi \circ F(z) = \frac{1}{2\pi i} \oint \phi \circ F(w)(w - z)^{-1}\, dw = \sum_n b_n z^n$ for $r < |z| < R$ and $b_n = \phi A_n$. But ϕ is linear and continuous, so it can be extracted out of the integrals and series,

$$\phi \circ F(z) = \phi \left(\frac{1}{2\pi i} \oint F(w)(w - z)^{-1}\, dw \right) = \phi \sum_{n \in \mathbb{Z}} A_n z^n,$$

and as ϕ is arbitrary, the result follows.

Exercises 13.31

1. Let $T := \begin{pmatrix} 0 & 1 \\ 0 & 0 \end{pmatrix}$; verify directly that $T = \frac{1}{2\pi i} \oint z(z - T)^{-1}\, dz$ by calculating the integral in a circular path around the origin.
2. Show that there are no analytic functions in $\mathbb{C}$ which grow at a fractional power rate $|z|^{m/n}$ $(m/n \notin \mathbb{N})$.
3. Show that the Laurent series for $\cot T$, valid for $\rho(T) < \pi$, $\rho(T^{-1}) > 0$, is

$$\cot T = T^{-1} - \tfrac{1}{3}T - \tfrac{1}{45}T^3 - \tfrac{2}{945}T^5 - \cdots,$$

and find its residue at 0. (Hint: $\cot z = (1 - z^2/2 + z^4/24 + \cdots)/z(1 - z^2/6 + \cdots)$.)
4. If an identity between analytic functions, $f(z) = g(z)$, holds in a complex disk $B_r(0)$, then it holds for any T with $\rho(T) < r$.
5. Justify the identity $n \log(1 + T) = \log(1 + T)^n$, hence deduce the assertion $\lim_{n \to \infty} (1 + T/n)^n = e^T$.
6. A function on $\mathbb{C}$ has a pole a of order N if, and only if, it has a Laurent series expansion $\sum_{n=-N}^{\infty} a_n(z - a)^n$ about a.
7. ⋆ Two analytic functions on an open connected subset of $\mathbb{C}$ must be identically equal if they are equal on an interior disk. (Consider the interior of the set for which $f = g$.)
8. Suppose f is analytic on the extended complex plane, except for isolated points, i.e., $f(1/z)$ is also analytic at 0.

   a. Show that f has a finite number of zeros and poles (except when $f = 0$),
   b. Using polynomials p, q whose roots are these zeros and poles, respectively, deduce that f is a rational function p/q.

Remarks 13.32

1. A subalgebra must have the same unity as the algebra—it is not enough that it has a unity. For example, $\mathcal{C}$ (Exercise 13.10(2)) contains the set $\{\,(0, a) : a \in \mathbb{C}\,\}$ which is closed under addition and multiplication and has its own unity $(0, 1)$, different from $\mathcal{C}$'s unity $(1, 0)$; it is an algebra, but not a subalgebra of $\mathcal{C}$. Instead, the set $\{\,(a, 0) : a \in \mathbb{C}\,\}$ *is* a subalgebra of $\mathcal{C}$.

2. The axiom $\Phi 1 = 1$ of an algebra morphism does not follow from the other properties of Φ. For example, the map $\Phi : \mathbb{C} \to \mathcal{C}$ defined by $\Phi(z) := (0, z)$ satisfies all the properties of a Banach algebra morphism, except that $\Phi(1) = (0, 1) \neq (1, 0)$. But continuity of characters follows from their other properties (see the proof of Proposition 14.35).

3. ⋆ The proof of the embedding of $\mathcal{X}$ into $B(\mathcal{X})$ does not make essential use of the axiom $\|1\| = 1$, or of $\|ax\| \leqslant \|a\|\|x\|$. If instead, $\|1\| = c$ and $\|ax\| \leqslant c'\|a\|\|x\|$, one gets

$$\|a\| = \|L_a 1\| \leqslant c\|L_a\|, \qquad \|L_a\| \leqslant c'\|a\|.$$

Thus $\mathcal{X}$ has an equivalent norm defined by $\|\!|a|\!\| := \|L_a\|$, with $\|\!|1|\!\| = \|I\| = 1$ and

$$\|\!|xy|\!\| = \|L_{xy}\| = \|L_x L_y\| \leqslant \|L_x\|\|L_y\| = \|\!|x|\!\|\,\|\!|y|\!\|.$$

4. In the Banach algebra $B(X)$, one can define $\rho_x(T) := \limsup_n \|T^n x\|^{\frac{1}{n}}$; so $0 \leqslant \rho_x(T) \leqslant \rho(T)$. The series $\sum_n a_n T^n x$ converges absolutely when $\rho_x(T)$ is less than the radius of convergence.

Chapter 14
Spectral Theory

14.1 The Spectrum of T

A moment's reflection shows that, by Cauchy's residue theorem, the path of integration in $f(T) = \frac{1}{2\pi i} \oint f(z)(z - T)^{-1} \, dz$ can be modified, as long as f and $(z - T)^{-1}$ remain analytic over the swept area. We are thus led to study the region where $z - T$ is not invertible, called the *spectrum* of T.

Definition 14.1

> The **spectrum** of an element T in a Banach algebra is defined as the set
>
> $$\sigma(T) := \{ \lambda \in \mathbb{C} : T - \lambda \text{ is not invertible} \}.$$
>
> Its complement $\mathbb{C} \setminus \sigma(T)$ is called the *resolvent* of T.

Examples 14.2

1. $\sigma(z) = \{z\}$ (since $z - \lambda$ is not invertible only when $\lambda = z$).
2. ▶ Recall that a square matrix A is non-invertible $\Leftrightarrow$ A is not 1–1 $\Leftrightarrow$ $\det A = 0$.
   The spectrum of an $n \times n$ matrix consists of its *eigenvalues*, i.e., the roots of the *characteristic* polynomial equation $\det(T - \lambda) = 0$ of degree n.
   For example, the spectra of the 2×2 matrices $\begin{pmatrix} 0 & 1 \\ 0 & 0 \end{pmatrix}$, $\begin{pmatrix} 0 & 0 \\ 1 & 0 \end{pmatrix}$, $\begin{pmatrix} 0 & 1 \\ 1 & 0 \end{pmatrix}$, and $\begin{pmatrix} a & 0 \\ 0 & b \end{pmatrix}$, are $\{0\}$, $\{0\}$, $\{-1, 1\}$, and $\{a, b\}$ respectively.
   Note that it is possible to have different elements with the same spectrum. The spectrum is a sort of 'shadow' of T—it yields important information about T, but need not identify it.
3. ▶ The spectrum of a sequence $x = (a_n)_{n \in \mathbb{N}} \in \ell^\infty$ is $\sigma(x) = \overline{\operatorname{im} x} = \overline{\{ a_n : n \in \mathbb{N} \}}$.

© The Author(s), under exclusive license to Springer Nature Switzerland AG 2024
J. Muscat, *Functional Analysis*, https://doi.org/10.1007/978-3-031-27537-1_14

Proof: The inverse of $x - \lambda = (a_n - \lambda)_{n \in \mathbb{N}}$ is bounded iff $|a_n - \lambda| \geqslant c > 0$ for all n, hence $\lambda \notin \sigma(x) \Leftrightarrow \lambda$ is an exterior point of $\{a_n\}_{n \in \mathbb{N}}$.

4. A *spectral value* of an operator $T \in B(X)$ is a complex number λ for which the equation $(T - \lambda)x = y$ is not well-posed; one sometimes sees in practice that as one varies a parameter λ of a model, some specific values have unstable solutions that 'resonate'.

5. (a) ▶ Translations, 'rotations' (in the sense of multiplication by $e^{i\theta}$) and scaling of T have corresponding actions on its spectrum:

$$\sigma(T + z) = \sigma(T) + z, \quad \sigma(zT) = z\sigma(T),$$

since $(T + z) - \lambda = T - (\lambda - z)$, so $\lambda \in \sigma(T + z) \Leftrightarrow \lambda - z \in \sigma(T)$; for $z \neq 0$, $(zT) - \lambda = z(T - \lambda/z)$, so $\lambda \in \sigma(zT) \Leftrightarrow \lambda/z \in \sigma(T)$.

   (b) If T is invertible, then $\sigma(T^{-1}) = \sigma(T)^{-1} := \{\lambda^{-1} : \lambda \in \sigma(T)\}$, since $T^{-1} - \lambda = -\lambda T^{-1}(T - \lambda^{-1})$, so $\lambda \in \sigma(T^{-1}) \Leftrightarrow \lambda^{-1} \in \sigma(T)$ (note that $\lambda \neq 0$).

   (c) The matrices $S := \begin{pmatrix} 0 & 1 \\ 0 & 0 \end{pmatrix}$ and $T := \begin{pmatrix} 0 & 0 \\ 1 & 0 \end{pmatrix}$ show that there is no simple relation between $\sigma(S + T)$ or $\sigma(ST)$ and $\sigma(S)$ and $\sigma(T)$ in general.

6. (a) $\sigma(ST) = \sigma(TS) \cup \{0\}$ OR $\sigma(ST) = \sigma(TS) \smallsetminus \{0\}$.

   (b) In particular, $\sigma(S^{-1}TS) = \sigma(T)$.

Proof: (a) For $\lambda \neq 0$ and $ST - \lambda$ invertible, $(TS - \lambda) = -\lambda^{-1}(1 - \lambda^{-1}TS)^{-1} = \lambda^{-1}(T(ST - \lambda)^{-1}S - 1)$, using Woodbury's formula. Thus, $\sigma(TS) \subseteq \sigma(ST) \cup \{0\}$; indeed, reversing the roles of S and T shows $\sigma(TS) \cup \{0\} = \sigma(ST) \cup \{0\}$.

Application: Quadratic Forms

Extracting the spectrum of matrices features prominently as one of the most useful applications of mathematics. It is used to find eigenfunctions of partial differential equations, in pattern recognition, stability analysis, etc.

Quadratic forms are expressions of degree 2 in a number of variables, such as

$$q(x, y, z) = ax^2 + by^2 + cz^2 + dxy + eyz + fzx = (x \; y \; z) \begin{pmatrix} a & d/2 & f/2 \\ d/2 & b & e/2 \\ f/2 & e/2 & c \end{pmatrix} \begin{pmatrix} x \\ y \\ z \end{pmatrix}.$$

They are found in the equations of conics and quadrics, the fundamental forms of surface geometry, the inertia tensor and stress tensor of mechanics, the integral forms of number theory, the covariances of statistics, etc. A quadratic form can always be written as $q(x) = x^\top Ax$, with A a symmetric matrix. We will see later that when the coefficients are real, such matrices have real eigenvalues, $\lambda_1, \ldots, \lambda_n$, and there exists an orthogonal matrix P such that $P^{-1}AP = D$, where D consists solely of the eigenvalues on the main diagonal. So the orthogonal transformation $x \mapsto \tilde{x} := P^{-1}x$ gives a simplified but equivalent quadratic form

$$q(x) = x^\top Ax = \tilde{x}^\top P^\top AP\tilde{x} = \tilde{x}^\top D\tilde{x} = \lambda_1 \tilde{x}_1^2 + \cdots + \lambda_n \tilde{x}_n^2 =: \tilde{q}(\tilde{x}).$$

These eigenvalues are intrinsic to the quadratic form, in the sense that any rotation of the variables gives a quadratic form with the same spectrum, and so represent real information about it rather than about the choice of variables. Not surprisingly these values were discovered before the connection with linear algebra became clear, and called by a variety of names such as "principal curvatures", "principal moments", "principal component variances", etc., in the different contexts. For example, a conic that satisfies the equation $ax^2 + bxy + cy^2 = 1$ can also be represented by the equation $\lambda \tilde{x}^2 + \mu \tilde{y}^2 = 1$, where $(\tilde{x}, \tilde{y})$ are obtained by a rotation/reflection of (x, y). Hence we can conclude that there result four conic types having this equation, depending on the signs of λ, μ: ellipses, hyperbolas, parallel lines, or the empty set.

The Spectral Radius

Determining the exact spectral values of an element is usually a non-trivial problem. The fundamental theorem for the general case is:

Theorem 14.3

> **The spectrum of T is a non-empty compact subset of $\mathbb{C}$. The largest extent of $\sigma(T)$, called the *spectral radius* of T, is**
>
> $$\max\{ |\lambda| : \lambda \in \sigma(T) \} = \rho(T) = \lim_{n \to \infty} \|T^n\|^{\frac{1}{n}}.$$

Proof $\sigma(T)$ *is compact*: If $|\lambda| > \rho(T)$, then $\rho(T/\lambda) = \rho(T)/|\lambda| < 1$, so $T - \lambda = -\lambda(1 - T/\lambda)$ is invertible (Theorem 13.20). Spectral values are therefore bounded by $\rho(T)$.

The resolvent set is none other than $f^{-1}\mathcal{G}(\mathcal{X})$ where $f(z) := T - z$, and $\mathcal{G}(\mathcal{X})$ is the set of invertible elements of $\mathcal{X}$. Since $\mathcal{G}(\mathcal{X})$ is open in $\mathcal{X}$ and f is continuous, it follows that the resolvent is open (Theorem 3.8), and the spectrum is closed in $\mathbb{C}$. More concretely, if $T - \lambda$ is invertible, and z is close enough to λ, then $|z - \lambda| = \|(T - z) - (T - \lambda)\|$ implies that $T - z$ is also invertible (Theorem 13.21).

The spectrum $\sigma(T)$, being a closed and bounded subset of $\mathbb{C}$, is compact (Corollary 6.20).

$\sigma(T)$ *is non-empty*: Applying Taylor's Theorem (13.26), with $f(z) := 1$, and a circular path centered at the origin with radius larger than $\rho(T)$, gives

$$1 = \frac{1}{2\pi i} \oint (z - T)^{-1} \, \mathrm{d}z.$$

But the map $z \mapsto (z-T)^{-1}$ is analytic on $\mathbb{C} \setminus \sigma(T)$ by (13.3). This would contradict Cauchy's theorem (Theorem 12.16) were the spectrum empty.

The spectral radius is $\rho(T)$: Let r_σ be the largest extent of $\sigma(T)$, and consider the function $f : z \mapsto (z - T)^{-1}$; it is analytic on $\mathbb{C} \setminus \sigma(T)$, in particular on $\mathbb{C} \setminus \overline{B_{r_\sigma}(0)}$. So it has a Laurent series $\sum_n A_n z^n$, valid for all $|z| > r_\sigma$ (Example 13.30(5)). On the other hand, we know that

$$(z - T)^{-1} = \frac{1}{z}(1 - T/z)^{-1} = \sum_{n=0}^{\infty} \frac{T^n}{z^{1+n}} \quad \text{for } |z| > \rho(T).$$

The two series must be identical, $\sum_{n=-\infty}^{\infty} A_n z^n = \sum_{n=0}^{\infty} T^n/z^{n+1}$, and remain valid for all $|z| > r_\sigma$. But the second series diverges when $\rho(T) > \liminf_n |z^{-n}|^{-1/n} = |z|$ by the Cauchy-Hadamard theorem, so there can be no $z \in \mathbb{C}$ such that $r_\sigma < |z| < \rho(T)$, in other words, $r_\sigma = \rho(T)$.   □

This result might appear unexpected because the formula $r_\sigma(T) = \lim_{n \to \infty} \|T^n\|^{1/n}$ for a matrix T seems to relate its eigenvalues, which are determined by a unique algebraic equation, with the norm, which can be changed. However, $\rho(T)$ does not depend on which equivalent norm is used to calculate it, and in finite dimensions, all norms are equivalent.

Corollary 14.4 (Fundamental Theorem of Algebra)

> **Every non-constant polynomial in $\mathbb{C}$ has a root.**

Proof The roots of the polynomial equation $z^n + a_{n-1}z^{n-1} + \cdots + a_0 = 0$ are precisely the spectral values of the matrix

$$\begin{pmatrix} 0 \cdots\cdots 0 & -a_0 \\ 1 \; 0 \cdots 0 & -a_1 \\ 0 \; \ddots \;\; \vdots & \vdots \\ \vdots \;\; \ddots \;\; 0 & \vdots \\ 0 \cdots 0 \; 1 & -a_{n-1} \end{pmatrix}.$$

□

Examples 14.5

1. The *smallest* extent of $\sigma(T)$ is $\rho(T^{-1})^{-1}$ when T is invertible (otherwise it is 0). Thus the condition $r < \rho(T^{-1})^{-1} \leqslant \rho(T) < R$ for a Laurent series expansion to exist (Theorem 13.28) can be restated as "the spectrum of T lies inside the ring with radii r and R".
2. ▶ *Every Banach division algebra is isomorphic to $\mathbb{C}$* (Gelfand-Mazur theorem).

Proof: A division algebra is defined as one in which the only non-invertible element is 0. Hence $T - \lambda$ is not invertible precisely when $T = \lambda \in \mathbb{C}1$. But $\sigma(T)$ is non-empty, so this must be the case for some λ.

3. ▶ *Every Banach algebra, except $\mathbb{C}$, has non-zero topological divisors of zero.*
   Proof: Suppose that the only topological divisor of zero is 0. Since the spectrum $\sigma(T)$ of every T has a non-empty boundary (Proposition 5.4), there is a $T - \lambda$ which is a topological divisor of zero, so $T = \lambda \in \mathbb{C}1$.

4. ▶ *Every commutative Banach algebra, except $\mathbb{C}$, has non-trivial ideals.*
   Proof: Suppose the only ideals are $\{0\}$ and $\mathcal{X}$. Then the ideal generated by $T \neq 0$, namely $\mathcal{X}T$ (in a commutative algebra), must equal $\mathcal{X}$. It follows that $ST = 1$ for some $S \in \mathcal{X}$, and T is invertible. But the only Banach division algebra is $\mathbb{C}$.

5. A morphism $J : \mathcal{X} \to \mathcal{Y}$ may only decrease the spectrum of an element, since a non-invertible element in $\mathcal{X}$ may become invertible in $\mathcal{Y}$, but an invertible in $\mathcal{X}$ cannot become non-invertible in $\mathcal{Y}$. If J is an embedding, the boundary of the spectrum in $\mathcal{X}$, consisting of topological divisors of zero, is preserved in $\mathcal{Y}$. The spectrum may decrease but its boundary (and the spectral radius) does not.

6. Recall the commutant algebra $\mathcal{Y} := \mathcal{A}'' \subseteq \mathcal{X}$ with which the elements of $\mathcal{A}$ commute. By part (c) of Exercise 13.10(14), for any $T \in \mathcal{Y}$, if $T - \lambda$ is invertible in $\mathcal{X}$ then its inverse is in $\mathcal{Y}$, so $\sigma_{\mathcal{Y}}(T) = \sigma(T)$.

Little else can be said about spectra of general elements of an algebra. The following proposition shows that the spectrum $\sigma(T)$ depends somewhat 'continuously' on T:

Proposition 14.6

> **If $T_n \to T$, then**
>
> $$\forall \epsilon > 0, \ \exists N, \quad n \geqslant N \implies \sigma(T_n) \subseteq \sigma(T) + B_\epsilon(0).$$

Proof Let U be any open subset of $\mathbb{C}$ containing $\sigma(T)$, for example $\sigma(T) + B_\epsilon(0)$. It is claimed that for all $z \notin U$, $\|(T - z)^{-1}\| \leqslant c$. When $|z| \geqslant r > \|T\|$,

$$\|(T - z)^{-1}\| = \Big\| \sum_{n=0}^{\infty} \frac{T^n}{z^{n+1}} \Big\| \leqslant \sum_{n=0}^{\infty} \frac{\|T\|^n}{r^{n+1}} = \frac{1}{r - \|T\|},$$

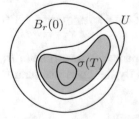

while on the remaining closed and bounded set $\overline{B_r(0)} \setminus U$, the continuous function $z \mapsto \|(T-z)^{-1}\|$ is bounded (Corollary 6.16). If $\|T-S\| < \frac{1}{c}$, then when $z \notin U$, $\|(T-z)^{-1}(T-S)\| < 1$. This implies that

$$S - z = (T-z) - (T-S) = (T-z)(1 - (T-z)^{-1}(T-S))$$

is invertible (Theorem 13.20). Thus $\sigma(S) \subseteq U$, and we have shown that any open set that contains $\sigma(T)$ also contains $\sigma(S)$ for S close enough to T.

For example, if $U := \sigma(T) + B_\epsilon(0)$ and T_n is close enough to T, then $\sigma(T_n) \subseteq U$.

$\square$

Exercises 14.7

1. The spectrum of $(z_1, \ldots, z_n) \in \mathbb{C}^n$ is $\{z_1, \ldots, z_n\}$.
2. The spectrum of $f \in C[0, 1]$ is $\sigma(f) = \operatorname{im}(f)$.
3. Verify directly that for a matrix A with eigenvalue λ, $A - \lambda$ is a divisor of zero.
4. Prove that $\sigma(T^2) = \sigma(T)^2 = \{\lambda^2 : \lambda \in \sigma(T)\}$ as follows, by considering $T^2 - \lambda^2 = (T-\lambda)(T+\lambda)$:

   (a) If $\lambda^2 \notin \sigma(T^2)$ then $T - \lambda$ is invertible.
   (b) If $\pm\lambda \notin \sigma(T)$ then $T^2 - \lambda^2$ is invertible.

   (We will see later a broad generalization of this (Theorem 14.26)).
5. Show that $\sigma(LR) = \{1\}$, but $\sigma(RL) = \{0, 1\}$, where L and R are the shift operators.
6. Show that $ST - TS = z \neq 0$ for $S, T \in \mathcal{X}$ implies $\sigma(ST)$ is unbounded, which is impossible. (Hint: $\lambda \in \sigma(TS) \Rightarrow \lambda + z \in \sigma(TS)$.)
7. The spectrum of $(S, T) \in \mathcal{X} \times \mathcal{Y}$ is $\sigma(S) \cup \sigma(T)$.
8. If $T \in B(X)$ and $S \in B(Y)$, let $T \odot S : X \times Y \to X \times Y$ be defined by $T \odot S(x, y) := (Tx, Sy)$. Then $\sigma(T \odot S) = \sigma(T) \cup \sigma(S)$.
9. If λ is a boundary point of the spectrum, then $T - \lambda$ is at the boundary of $\mathcal{G}(\mathcal{X})$, and so is a topological divisor of zero (Proposition 13.23). Moreover, if $T - \mu$ is invertible, then

$$\|(T-\mu)^{-1}\| \geqslant 1/d(\mu, \sigma(T)).$$

10. Recall the Hausdorff distance between subsets (Exercise 2.20(10)). Show that if $S \to T$, then $d(\sigma(S), \sigma(T)) \to 0$.

14.2   The Spectrum of an Operator

An operator T on a Banach space X is invertible in $B(X)$ when T has a continuous linear inverse $T^{-1} \in B(X)$. By the open mapping theorem, this is automatically true once T is bijective. So an operator $T \in B(X)$ is *not* invertible when one of the following cases holds:

```
                                          T not 1-1
                          ┌────────────────────────────────────────────
 T not invertible in B(X) │
─────────────────────────┤                                  im T = X
                          │ T is 1-1 but not onto ┌──────────────────────
                          └───────────────────────┤
                                                  └──────────────────────
                                                            im T ≠ X
```

- T is not 1–1 (i.e., $\ker T \neq 0$). In this case, T is a left divisor of zero as $TS = 0$ for any non-zero $S \in B(X)$ with $\operatorname{im} S \subseteq \ker T$.
- T is 1–1, but not onto, yet it is "almost" onto, in the sense that its image is dense, $\overline{\operatorname{im} T} = X$. Here, it cannot be the case that $\|Tx\| \geqslant c\|x\|$ for all x and some $c > 0$, otherwise $\operatorname{im} T$ would be closed (Example 8.16(3)) and T onto. This means that one can decrease $\|Tx\|$ but keep $\|x\|$ fixed, i.e., there are unit vectors x_n such that $Tx_n \to 0$. By taking any unit operators with $\operatorname{im} S_n = [\![x_n]\!]$, we get $TS_n \to 0$, so T is a topological left divisor of zero.
- T is 1–1, and its image is not even dense in X. In this case, by Proposition 11.20, there exists a vector x_0 and a functional ϕ such that $\phi x_0 \neq 0$ and $\phi[\overline{\operatorname{im} T}] = 0$. Then $Sx := x_0 \phi$ defines a non-zero operator with kernel containing $\operatorname{im} T$, so $ST = 0$, and T is a right divisor of zero.

The spectrum of an operator $T \in B(X)$ thus consists of λ in:

○ the **point spectrum** $\sigma_p(T)$, when $T - \lambda$ is not 1–1, i.e., $Tx = \lambda x$ for some $x \neq 0$; we say that λ is an **eigenvalue** and x an **eigenvector** of λ (note that a non-zero multiple of an eigenvector is another eigenvector, so they are often taken to be of unit length); the subspace $\ker(T - \lambda)$ of eigenvectors of λ (together with the zero vector) is called its *eigenspace*.

○ the **continuous spectrum** $\sigma_c(T)$, when $T - \lambda$ is 1–1, not onto, but $\overline{\operatorname{im}(T - \lambda)} = X$.

○ the **residual spectrum** $\sigma_r(T)$, when $T - \lambda$ is 1–1, and $\overline{\operatorname{im}(T - \lambda)} \neq X$.

In finite dimensions, a matrix is 1–1 iff it is onto, so only eigenvalues make up the spectrum. The direct way of finding eigenvalues and their corresponding eigenvectors is to solve $(T - \lambda)x = 0$; in finite dimensions, this implies the 'characteristic' polynomial equation $\det(T - \lambda) = 0$. What are the additional 'continuous' and 'residual' spectral values in infinite dimensions? Let us take the right shift operator to illustrate what can happen:

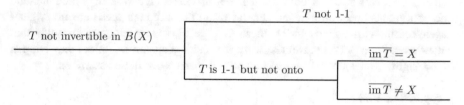

In the first example, although the vector is not an exact eigenvector of 1, it is very close to satisfying the equation $(A - 1)x \approx 0$ (when x is scaled to unit length). We call such spectral values, *approximate eigenvalues*, which include the continuous spectrum. In the second example, although there are no 'right'-eigenvectors with eigenvalue 0, there is a left- or row eigenvector with that eigenvalue. Such 'left eigenvalues' form the residual spectrum, unless they happen to be 'right eigenvalues'. The next few propositions prove these assertions.

Proposition 14.8

Eigenvectors of distinct eigenvalues are linearly independent.

Proof Let $v_i \neq 0$ be eigenvectors associated with the distinct eigenvalues λ_i, $i = 1, 2, \ldots$, so that $(T - \lambda)v_i = (\lambda_i - \lambda)v_i$. The sum $\sum_{i=1}^{n} \alpha_i v_i = 0$ implies

$$0 = (T - \lambda_2) \cdots (T - \lambda_n) \sum_{i=1}^{n} \alpha_i v_i$$

$$= (T - \lambda_2) \cdots (T - \lambda_{n-1}) \sum_{i=1}^{n-1} \alpha_i (\lambda_i - \lambda_n) v_i$$

$$= \cdots = \alpha_1 (\lambda_1 - \lambda_2) \cdots (\lambda_1 - \lambda_n) v_1$$

forcing $\alpha_1 = 0$. Since the argument can be repeated for any other index i, we have $\alpha_i = 0$. $\qquad\square$

In general, it is a hard task to find the point spectrum of most operators. So any result that gives us approximate alternatives are welcome.

Proposition 14.9 (Gershgorin's Theorem)

If $T = [T_{i,j}]$ is an operator on c_0, then the disks $\bigcup_{n \in \mathbb{N}} \overline{B_{r_n}(T_{n,n})}$, where $r_n := \sum_{j \neq n} |T_{n,j}|$, cover all the eigenvalues.

Proof Let $x = (a_j)_{j \in \mathbb{N}}$ be an eigenvector of T and let $|a_n|$ be its largest coefficient. Then rearranging $Tx = \lambda x$ we get

$$\lambda a_n = \sum_{j \in \mathbb{N}} T_{n,j} a_j = T_{n,n} a_n + \sum_{j \neq n} T_{n,j} a_j,$$

$$\therefore \ |\lambda - T_{n,n}||a_n| \leqslant \sum_{j \neq n} |T_{n,j}||a_j| \leqslant r_n |a_n|.$$

as required. Note that the row sum $\sum_j |T_{i,j}|$ converges since the dual space of c_0 is ℓ^1.   □

Proposition 14.10

> If λ is a limit of eigenvalues, or is in $\sigma_c(T)$, or is a boundary point of $\sigma(T)$, then λ is an *approximate eigenvalue*, meaning there are unit vectors x_n, such that
>
> $$(T - \lambda)x_n \to 0 \text{ as } n \to \infty.$$

Proof If $\lambda_n \to \lambda$ and $Tx_n = \lambda_n x_n$ with $\|x_n\| = 1$, then

$$(T - \lambda)x_n = (\lambda_n - \lambda)x_n \to 0.$$

λ is an approximate eigenvalue exactly when $T - \lambda$ is a topological left divisor of zero, because suppose there are unit operators S_n with $(T - \lambda)S_n \to 0$. Let x_n be vectors such that $\|S_n x_n\| = 1$ and $\|x_n\| \leqslant 2$ (possible since $\|S_n\| = 1$); then $(T - \lambda)S_n x_n \to 0$, and λ is an approximate eigenvalue.

Conversely, given $(T - \lambda)x_n \to 0$ with x_n unit vectors, let $S_n := x_n\phi$ for any $\phi \in X^*$ with unit norm. Then $\|S_n\| = 1$ and $(T - \lambda)S_n = (T - \lambda)x_n\phi \to 0$ as $n \to \infty$.

This includes the case when λ is at the boundary of $\sigma(T)$ (Proposition 13.23), and when $\lambda \in \sigma_c(T)$ as we have just seen at the beginning of this section.   □

Examples 14.11

1. ▶ The spectrum of the left-shift operator $L(a_n) := (a_{n+1})_{n\in\mathbb{N}}$, on ℓ^∞ is the unit closed disk.
   Proof: The norm of L is 1, so $\sigma(L) \subseteq \overline{B_1(0)}$. To find its eigenvalues, we need to solve $Lx = \lambda x$ for some non-zero $x = (a_n)_{n\in\mathbb{N}} \in \ell^\infty$, i.e.,

$$\forall n \in \mathbb{N}, \quad a_{n+1} = \lambda a_n, \quad |a_n| \leqslant c.$$

This recurrence relation gives $a_n = \lambda^n a_0$, satisfying $|a_0||\lambda|^n = |a_n| \leqslant c$. Thus the only possible candidates for eigenvalues are $|\lambda| \leqslant 1$. In fact, for any such λ, the sequence $(1, \lambda, \lambda^2, \ldots)$ is an eigenvector in ℓ^∞. Hence $\sigma(L) = \overline{B_\mathbb{C}}$, and all spectral points are eigenvalues.

2. ▶ The spectrum of the left-shift operator on ℓ^1 is the unit closed disk.
   Proof: The same analysis as in Example 1 applies: $\rho(L) \leqslant \|L\| = 1$, and $a_n = \lambda^n a_0$. This time, the condition $x \in \ell^1$ is $\sum_n |a_n| = |a_0| \sum_n |\lambda|^n < \infty$. This is only possible when $|\lambda| < 1$. Once again, but only for $|\lambda| < 1$, the sequence $(1, \lambda, \lambda^2, \ldots)$ is an eigenvector in ℓ^1. Still, since it is closed, bounded by 1,

and contains $B_{\mathbb{C}}$, the spectrum must be the closed disk. The spectral values in the interior are eigenvalues, and those on the circular perimeter are approximate eigenvalues.

3. Let $T : \ell^2 \to \ell^2$ be the multiplier operator $T(a_n) := (b_n a_n)_{n \in \mathbb{N}}$ where b_n are bounded. Its eigenvalues are b_n, and its spectrum is $\overline{\{b_1, b_2, \ldots\}}$.

   Proof: For eigenvalues, $T(a_n) = (b_n a_n)_{n \in \mathbb{N}} = \lambda(a_n)_{n \in \mathbb{N}}$, so $(b_n - \lambda)a_n = 0$ for all n. This implies $\lambda = b_n$ for some n, otherwise $(a_n)_{n \in \mathbb{N}} = \mathbf{0}$. In fact, $T e_n = b_n e_n$, so b_n is indeed an eigenvalue. Now, suppose λ is not a limit point of $\{b_1, b_2, \ldots\}$; there is then a minimum positive distance between them, i.e., $|\lambda - b_n| \geqslant d > 0$. So the equation $(T - \lambda)(a_n) = (c_n)$ can be inverted, $a_n = c_n/(b_n - \lambda)$, with $|a_n| \leqslant |c_n|/d$; $\|(T - \lambda)^{-1}\| \leqslant 1/d$. The spectrum therefore must include the eigenvalues and their limit points, but nothing else.

4. Let $T : L^\infty[0, 1] \to L^\infty[0, 1]$ be defined by $Tf(s) := \int_{1-s}^1 f(t)\, dt$. Then T is linear, and continuous with $\|T\| \leqslant 1$ since

$$\|Tf\|_{L^\infty} = \sup_{s \in [0,1]} \left| \int_{1-s}^1 f(t)\, dt \right| \leqslant \|f\|_{L^\infty} \sup_{s \in [0,1]} \int_{1-s}^1 dt = \|f\|_{L^\infty}.$$

   For eigenvalues, we need to solve $\int_{1-s}^1 f(t)\, dt = \lambda f(s)$. Differentiating twice gives $f''(s) + \frac{1}{\lambda^2} f(s) = 0$ with boundary conditions $f(0) = 0 = f'(1)$. Thus the eigenvectors (or "eigenfunctions") are $f(t) = \sin(t/\lambda)$ with eigenvalues $\lambda = 2/k\pi$, k odd. The spectrum must also include 0, because it is their limit point, but at this stage we cannot conclude anything further about the spectrum.

5. If $S : X \to Y$, $T : Y \to X$ are operators, then ST and TS share the same non-zero eigenvalues.

   Proof: If $STx = \lambda x$ ($x \neq 0$), then $TS(Tx) = T(ST)x = \lambda(Tx)$, so either $Tx = 0$, in which case $\lambda = 0$, or Tx is an eigenvector of TS with the same eigenvalue λ; similarly, every non-zero eigenvalue of TS is also an eigenvalue of ST. (Compare with Example 14.2(6d).)

6. Real eigenvalues of real operators have real eigenvectors.

   Proof: If X is a real Banach space, then $T \in B(X)$ is not guaranteed to have a spectral element, but it will have when considered as an operator on the complex space $X + iX$. Nevertheless if the eigenvalue is real, with eigenvector $u + iv$, then u and v are also eigenvectors (unless 0),

$$T(u + iv) = \lambda(u + iv) \Rightarrow Tu = \lambda u, Tv = \lambda v.$$

The Spectrum of the Adjoint

Let us prove our previous assertion that the residual spectrum consists of the 'left eigenvalues' of T, that is, the eigenvalues of $T^\top$:

Proposition 14.12

$$\sigma(T^\top) = \sigma(T), \qquad \sigma_r(T) = \sigma_p(T^\top)\setminus\sigma_p(T), \qquad \sigma_c(T^\top) \subseteq \sigma_c(T)$$

Proof (i) $T - \lambda$ is invertible in $B(X)$, if and only if, its adjoint is invertible (Exercise 11.34(7)),

$$(T^\top - \lambda)^{-1} = (T - \lambda)^{-1\top}.$$

So $\lambda \notin \sigma(T) \Leftrightarrow \lambda \notin \sigma(T^\top)$.

(ii) By definition, $\lambda \in \sigma_p(T^\top)$ when there is a $\phi \neq 0$ in X^* such that

$$\phi \circ (T - \lambda) = (T^\top - \lambda)\phi = 0.$$

This implies there is an $x \in X$, $\phi x \neq 0$, such that $x \notin \overline{\text{im}(T - \lambda)}$. In turn, if $x \in X\setminus\overline{\text{im}(T - \lambda)}$ exists, then there is a $\phi \neq 0$ such that $\phi(T - \lambda) = 0$ (Proposition 11.20), and we have proved

$$\lambda \in \sigma_p(T^\top) \Leftrightarrow \overline{\text{im}(T - \lambda)} \neq X.$$

This condition is certainly satisfied when λ is a residual spectral value of $\sigma(T)$, but not when it is in the continuous spectrum of T, so

$$\lambda \in \sigma_r(T) \Rightarrow \lambda \in \sigma_p(T^\top) \Rightarrow \lambda \notin \sigma_c(T).$$

(iii) When $A^\top$ is 1–1 but $\overline{\text{im}\,A^\top} = X^*$, then we can infer, by Proposition 11.32, that (a) $(\ker A)^\perp \supseteq \overline{\text{im}\,A^\top} = X^*$, so A is 1–1; and (b) $(\text{im}\,A)^\perp = \ker A^\top = 0$, so $\overline{\text{im}\,A} = X$. Applying this to $A := T - \lambda$ when $\lambda \in \sigma_c(T^\top)$, we find that $T - \lambda$ is 1–1 and has a dense image, that is, $\lambda \in \sigma_c(T)$. $\qquad\square$

Examples 14.13

1. When $T^{\top\top} = T$ (e.g., on a Hilbert space) then $\sigma_r(T^\top) = \sigma_p(T)\setminus\sigma_p(T^\top)$ as well as $\sigma_c(T^\top) = \sigma_c(T)$.
2. In c_0 or ℓ^2, the left-shift and right-shift operators have

$$\sigma_p(L) = B_{\mathbb{C}}, \quad \sigma_r(L) = \varnothing, \quad \sigma_c(L) = \mathbb{S}^1,$$
$$\sigma_p(R) = \varnothing, \quad \sigma_r(R) = B_{\mathbb{C}}, \quad \sigma_c(R) = \mathbb{S}^1.$$

Proof: That $\sigma_p(L^\top) = \varnothing$ has already been shown since $L^\top$ is the right shift on ℓ^1; in the same way can be proved $\sigma_p(L) = B_{\mathbb{C}}$. Applying this proposition, we find that $\sigma_r(L) \subseteq \sigma_p(L^\top) = \varnothing$, leaving $\sigma_c(L) = \mathbb{S}^1$. Similarly for R, $\sigma_r(R) \subseteq$

$\sigma_p(R^\top) \subseteq \sigma_r(R)$ since $\sigma_p(R) = \varnothing$ (prove!), hence $\sigma_r(R) = \sigma_p(R^\top) = B_\mathbb{C}$ and $\sigma_c(R) = \mathbb{S}^1$.

3. The analogous results for the Hilbert adjoint T^* are similar:[1]

$$\sigma(T^*) = \sigma(T)^*, \quad \sigma_r(T^*) = \sigma_p(T)^* \setminus \sigma_p(T^*), \quad \sigma_c(T^*) = \sigma_c(T)^*$$

Proof: Let $A := T - \lambda$, then A is invertible iff A^* is; and $\ker A^* = (\operatorname{im} A)^\perp$, so

$$\lambda \in \sigma_r(T) \Rightarrow \ker A^* = (\operatorname{im} A)^\perp \neq 0 \Rightarrow \bar{\lambda} \in \sigma_p(T^*),$$

$$\bar{\lambda} \in \sigma_p(T^*) \Rightarrow (\operatorname{im} A)^\perp = \ker A^* \neq 0 \Rightarrow \lambda \notin \sigma_c(T).$$

If $\bar{\lambda} \in \sigma_c(T^*)$, then $\ker A = (\operatorname{im} A^*)^\perp = 0$ and $\overline{\operatorname{im} A} = (\ker A^*)^\perp = 0^\perp = H$, so $\lambda \in \sigma_c(T)$.

4. In finite dimensions, the 'left eigenvalues' are the same as the 'right eigenvalues' because both A and $A^\top$ satisfy the same characteristic polynomial equation; but the 'left eigenvectors' are usually different from the 'right eigenvectors'.

Exercises 14.14

1. Show that the right-shift operator R (on ℓ^∞ or ℓ^1) has no eigenvalues.
2. The right-shift operator $R \in B(\ell^1)$ and its adjoint $L \in B(\ell^\infty)$ have spectra

$$\sigma(L) = \sigma_p(L) = \bar{B}_\mathbb{C} = \sigma_r(R) = \sigma(R).$$

3. The spectrum of the left-shift operator L on $\ell^1(\mathbb{Z})$ is the circle $\mathbb{S}^1$. This is an example of the hollowing out of a spectrum when the algebra increases, in this case when ℓ^1 is embedded in $\ell^1(\mathbb{Z})$.
4. The operator $T(a_0, a_1, \ldots) := (a_0, 0, a_1, a_2, \ldots)$, on c_0, has a single eigenvalue 1, but its adjoint has $\sigma_p(T^\top) = B_\mathbb{C} \cup \{1\}$. Deduce that $\sigma_p(T) = \{1\}$, $\sigma_r(T) = B_\mathbb{C}$, and $\sigma_c(T) = \mathbb{S}^1 \setminus \{1\}$.
   But the same operator on ℓ^1 has a single eigenvalue 1 and no continuous spectrum.
5. The operator $T(a_0, a_1, \ldots) := (a_0, 0, a_1, a_2/2, a_3/3, \ldots)$, on c_0, has a single eigenvalue 1, and its adjoint has two eigenvalues, 1 and 0.
6. The spectrum of the multiplier operator $Tx := ax$, on ℓ^2, has no residual spectrum.
7. The spectrum of $x\phi \in B(X)$, where $x \in X$ and $\phi \in X^*$, consists of the eigenvalues ϕx and 0 (unless X is 1-dimensional).
8. Let $T : X \to Y$, $S : Y \to X$ be operators and consider $R \in B(X \times Y)$ defined by $R(x, y) := (Sy, Tx)$; the 'matrix' form of R looks like $\begin{pmatrix} 0 & S \\ T & 0 \end{pmatrix}$. Then non-zero eigenvalues of R come in pairs $\pm\lambda$. (Hint: consider $(x, -y)$.)

---

[1] To avoid ambiguity with the closure $\bar{F}$ of a set $F \subseteq \mathbb{C}$, we use F^* to denote the set of conjugate numbers $\{ \bar{z} : z \in F \}$.

9. Let $T : C[0, 1] \to C[0, 1]$ be defined by $Tf(t) := tf(t)$. Show that T is linear and continuous, find its norm and show that its spectrum is the line $[0, 1]$ in $\mathbb{C}$, consisting of only the residual part.

   More generally the spectrum of $Tf := gf$ in $C[0, 1]$, where $g \in C[0, 1]$, is im g.

   The reader is encouraged to explore the spectrum of this operator in other spaces, such as $L^1[0, 1]$ or $L^2[0, 1]$.

10. Find the eigenvalues of $Tf(s) := \int_0^1 s^2 t^2 f(t)\, dt$ on $C[0, 1]$.

11. ⋆ Let $V : C[0, 1] \to C[0, 1]$ be the Volterra operator $Vf(t) := \int_0^t f$. Show that

$$V^{n+1} f(t) = \frac{1}{n!} \int_0^t (t - s)^n f(s)\, ds,$$

    and that $\|V^n\| \leq 1/n!$. Deduce, using the spectral radius formula, that its spectrum is just $\{0\}$. Show that 0 is not an eigenvalue (hint: differentiate) but a residual boundary spectral value.

12. The spectrum of an isometry T lies in $\bar{B}_{\mathbb{C}}$. Any eigenvalues or approximate eigenvalues lie in $e^{i\mathbb{R}}$. If T is an invertible isometry, then $\sigma(T) \subseteq e^{i\mathbb{R}}$, otherwise the spectrum must be the whole closed unit disk (e.g., the right-shift operator). (Hint: $T - \lambda = T(1 - \lambda S)$.)

13. Show that the set $\{T \in B(X) : T$ is 1–1 and has a closed image $\}$ is open in $B(X)$. (Hint: Proposition 11.3.)

14.3   Spectra of Compact Operators

Ascents and Descents

For any operator, the eigenspace associated with an eigenvalue λ is $\ker(T - \lambda)$. But this is not the whole story: for example, $T := \begin{pmatrix} 0 & 1 \\ 0 & 0 \end{pmatrix}$ has just one eigenvalue, and a *one*-dimensional eigenspace generated by $\begin{pmatrix} 1 \\ 0 \end{pmatrix}$; the vector $v := \begin{pmatrix} 0 \\ 1 \end{pmatrix}$ is mapped by T to $\begin{pmatrix} 1 \\ 0 \end{pmatrix}$, and only a second application of T kills it off. We can think of it as a "generalized" eigenvector, with $(T - \lambda)^2 v = 0$. In general, one can consider the spaces of vectors that vanish when $(T - \lambda)^n$ is applied to them. Two nested sequences of spaces can be formed (here shown for $\lambda = 0$),

- an *ascending sequence*

$$0 \subseteq \ker T \subseteq \ker T^2 \subseteq \cdots \subseteq \ker T^n \subseteq \cdots \subseteq \bigcup_n \ker T^n,$$

- a *descending sequence*

$$X \supseteq \operatorname{im} T \supseteq \operatorname{im} T^2 \supseteq \cdots \supseteq \operatorname{im} T^n \supseteq \cdots \supseteq \bigcap_n \operatorname{im} T^n.$$

As usual whenever we are dealing with infinite processes, it would be interesting to study operators with *finite* ascents or descents.

Suppose there is an n such that $\ker T^n = \ker T^{n+1}$, i.e., for all x,

$$T^n x = 0 \;\Leftrightarrow\; T^{n+1} x = 0.$$

Substituting Tx instead of x gives

$$T^{n+1} x = 0 \;\Leftrightarrow\; T^{n+2} x = 0$$

and $\ker T^{n+2} = \ker T^{n+1} = \ker T^n$. By induction, all the subsequent spaces in the ascending sequence are identical, $\ker T^{n+k} = \ker T^n$. Operators with this property are said to have a *finite ascent* up to n, $0 \subset \ker T \subset \cdots \subset \ker T^n$.

Similarly, if $\operatorname{im} T^m = \operatorname{im} T^{m+1}$ then for any $x \in \operatorname{im} T^{m+1}$,

$$x = T^{m+1} y = T(T^m y) = T(T^{m+1} z) = T^{m+2} z \in \operatorname{im} T^{m+2}.$$

By induction, $\operatorname{im} T^{m+k} = \operatorname{im} T^m$. Operators with this property are said to have a *finite descent* down to m.

Proposition 14.15

An operator T has

  (i)   **finite ascent up to at most n** $\Leftrightarrow$ $\forall k \in \mathbb{N}$, $\operatorname{im} T^n \cap \ker T^k = 0$,
 (ii)   **finite descent down to at most m** $\Leftrightarrow$ $\forall k \in \mathbb{N}$, $X = \ker T^m + \operatorname{im} T^k$,
(iii)   **finite ascent up to n and descent down to m implies $m = n$ and**

$$X = \ker T^n \oplus \operatorname{im} T^n.$$

Proof (i) Let T have finite ascent and let $x \in \operatorname{im} T^n \cap \ker T^k$, that is, $x = T^n y$ and $T^k x = 0$. Then $T^{n+k} y = 0$ and $y \in \ker T^{n+k} = \ker T^n$; so $x = T^n y = 0$. For the converse, if $\operatorname{im} T^n \cap \ker T = 0$, then $T^{n+1} x = 0 \Rightarrow T^n x \in \operatorname{im} T^n \cap \ker T = 0$, and T has finite ascent up to at most n.

(ii) Let $x \in X$, then $T^m x = T^{m+1} y = \cdots = T^{m+k} z$, assuming finite descent to m. So $T^m (x - T^k z) = 0$ and $x = T^k z + (x - T^k z) \in \operatorname{im} T^k + \ker T^m$. Conversely, if $X = \operatorname{im} T + \ker T^m$, then for any $x = Ty + z$, we have $T^m x = T^{m+1} y$ and $\operatorname{im} T^m = \operatorname{im} T^{m+1}$.

(iii) Suppose $\operatorname{im} T^n = \operatorname{im} T^{n+1}$, but $\ker T^n \subset \ker T^{n+1}$. Then there is an x_1 such that $T^{n+1}x_1 = 0$ but

$$0 \neq T^n x_1 = T^{n+1} x_2 = T^{n+2} x_3 = \cdots$$

so $x_k \in \ker T^{n+k} \setminus \ker T^{n+k-1}$, and T has an infinite ascent. This shows that a finite ascent cannot be longer than the descent.

Next suppose the ascent goes up to $\ker T^n = \ker T^{n+1}$ but the descent goes down to $\operatorname{im} T^m = \operatorname{im} T^{m+1}$ with $m \geqslant n$. Then for any $x \in X$, there is a y such that

$$T^m x = T^{m+1} y \;\Rightarrow\; T^m(x - Ty) = 0$$

$$\Rightarrow\; x - Ty \in \ker T^m = \ker T^n$$

$$\Rightarrow\; T^n x = T^{n+1} y$$

so a finite descent cannot be longer than the ascent.

Combining the results of (i) and (ii) gives $X = \ker T^n \oplus \operatorname{im} T^n$. $\qquad\square$

Proposition 14.16 (Fredholm Alternative)

> **A Fredholm operator T with**
>
>   (i) **finite ascent, satisfies** $\operatorname{index}(T) \leqslant 0$,
> (ii) **finite descent, satisfies** $\operatorname{index}(T) \geqslant 0$,
> (iii) **finite ascent and descent, satisfies** $\operatorname{index}(T) = 0$ **and**
>
> $$T \text{ is 1–1} \;\Leftrightarrow\; T \text{ is onto.}$$

Ivar Fredholm(1866–1927) Fredholm studied p.d.e.s under Mittag-Leffler in 1893 at the new University of Stockholm; he saw the connection between Volterra's equation and potential theory, especially in 1899 while working on Dirichlet's problem; in 1903 he analyzed the theory of general integral equations $f(x) - \lambda \int_a^b k(x, y) f(y) \, dy = g(x)$ covering much that was then known about boundary value problems (mostly self-adjoint), proved the Fredholm alternative and defined the Fredholm determinant $\det(1 - K) = e^{-\sum_n \frac{1}{n} \operatorname{tr} K^n}$. He was then 'distracted' by actuarial science and government.

Proof Recall that the codimension of a closed subspace $Y \subseteq X$ is defined as $\dim(X/Y)$, that Fredholm operators have finite-dimensional kernels and finite codimensional images, and $\text{index}(T) = \dim \ker T - \text{codim im } T$ (Definition 11.12). For T with finite ascent to n, by the index theorem,

$$0 \leqslant \text{codim im } T^k = \dim \ker T^k - \text{index}(T^k)$$
$$= \dim \ker T^n - k \, \text{index}(T), \text{ for } k \geqslant n.$$

Since k can be arbitrarily large, it must be the case that $\text{index}(T) \leqslant 0$.

For Fredholm operators with finite descent to m,

$$0 \leqslant \dim \ker T^k = \text{codim im } T^k + \text{index}(T^k)$$
$$= \text{codim im } T^m + k \, \text{index}(T), \text{ for } k \geqslant m.$$

This time, we must have $\text{index}(T) \geqslant 0$.

A special case is when $m = n = 0$, known as the *Fredholm alternative*: $\ker T = 0$ if, and only if, $\text{im } T = X$, i.e., T is 1–1 $\Leftrightarrow$ T is onto; in other words, T is either invertible or it is neither 1–1 nor onto.   □

Examples 14.17

1. The spaces $M := \text{im } T^m$ and $N := \ker T^n$ are both T-invariant and such that $T|_M$ is an isomorphism while $T|_N$ is nilpotent.
2. For matrices, the Fredholm alternative boils down to the statement that either $A\mathbf{x} = \mathbf{b}$ has a unique solution or $A\mathbf{x} = \mathbf{0}$ has non-trivial solutions.
3. The Fredholm alternative only applies to (Fredholm) operators with finite ascent and descent; e.g., the right-shift operator is 1–1 but not onto.
4. If T is Fredholm with finite ascent and descent, then $\dim \ker T = \dim \ker T^\top$ (Exercise 11.34(10)).

The Spectrum of a Compact Operator

The spectra of operators are usually hard to determine, with those of compact operators often being the most tractable. The following two results are peaks in the landscape of Operator Theory.

Proposition 14.18

Let $T : X \to X$ be compact on a Banach space X, then $I + T$ is a Fredholm operator with finite ascent and descent.

Proof $I + T$ is Fredholm by Proposition 11.15.

Suppose $S := I + T$ has infinite ascent, so $\ker S^{n-1} \subset \ker S^n$. By Riesz's lemma (Proposition 8.22), choose unit vectors $x_n \in \ker S^n$ with $\|x_n + \ker S^{n-1}\| \geqslant \frac{1}{2}$. Then for $m < n$,

$$\|Tx_n - Tx_m\| = \|(x_n - x_m) - S(x_n - x_m)\| \geqslant \tfrac{1}{2}$$

since $S^{n-1}(x_m + S(x_n - x_m)) = 0$. So $(Tx_n)_{n \in \mathbb{N}}$ has no Cauchy subsequence, contradicting the compactness of T.

Suppose S has infinite descent, with $\operatorname{im} S^{n-1} \supset \operatorname{im} S^n$. One can choose unit vectors $x_n \in \operatorname{im} S^n$ with $\|x_n + \operatorname{im} S^{n+1}\| \geqslant \frac{1}{2}$. Then for $m > n$,

$$\|Tx_n - Tx_m\| = \|(x_n - x_m) - S(x_n - x_m)\| \geqslant \tfrac{1}{2}$$

since $x_m + S(x_n - x_m) \in \operatorname{im} S^{n+1}$. Again this would contradict the hypothesis. □

It follows from the propositions and examples above, that the index of S vanishes and $\dim \ker(S^\top) = \dim \ker S$.

Theorem 14.19 (Riesz-Schauder)

> **If $T \in B(X)$ is compact, then**
>
> (i) **its spectrum $\sigma(T)$ is a countable set, whose only possible limit point may be 0,**
> (ii) **each non-zero $\lambda \in \sigma(T)$ is an eigenvalue with a finite-dimensional eigenspace $\ker(T - \lambda)$,**
> (iii) **$T^\top$ and T have the same non-zero eigenvalues and eigenspace dimensions.**

Proof For $\lambda \neq 0$, $T - \lambda = \lambda(I - T/\lambda)$ is a Fredholm operator with finite ascent and descent, so its kernel is finite dimensional and it satisfies the Fredholm alternative, namely it is either invertible ($\lambda \notin \sigma(T)$) or not 1–1 (λ is an eigenvalue). $T - \lambda$ has index 0, so $T^\top$ has the same number of eigenvectors of λ as T,

$$\dim \ker(T^\top - \lambda) = \dim \operatorname{im}(T - \lambda)^\perp = \operatorname{codim} \operatorname{im}(T - \lambda) = \dim \ker(T - \lambda).$$

Consider those eigenvalues λ for which $|\lambda| \geqslant \epsilon > 0$. Taking any list of them, λ_n (distinct), choose a unit eigenvector e_n for each, such that $\|e_n + [\![e_1, \ldots, e_{n-1}]\!]\| \geqslant \frac{1}{2}$ (Propositions 8.22 and 14.8). Hence, taking $n > m$, say,

$$\|Te_n - Te_m\| = \|\lambda_n e_n - \lambda_m e_m\| = |\lambda_n| \|e_n - \frac{\lambda_m}{\lambda_n} e_m\| \geqslant \frac{1}{2}|\lambda_n| \geqslant \frac{\epsilon}{2}.$$

Now the bounded set $\{e_1, e_2, \ldots\}$ is mapped to $\{Te_1, Te_2, \ldots\}$. If the first set is infinite, the latter set would have no Cauchy subsequence, contradicting the compactness of T. So the number of such eigenvectors, and corresponding eigenvalues, is finite. The rest of the eigenvalues must be within ϵ of 0. By taking $\epsilon = 1/n \to 0$, it follows that the number of non-zero eigenvalues is countable. □

To clarify, in finite dimensions, the set of eigenvalues is finite and need not include 0, but in infinite dimensions, 0 must be part of the spectrum (else $I = T^{-1}T$ is compact). If there is an infinite sequence of non-zero eigenvalues, then $\lambda_n \to 0$, and 0 is an approximate eigenvalue. What remains to complete the theory is to find the form of T on each generalized eigenspace.

Proposition 14.20 (Jordan Canonical Form)

On each **finite-dimensional space** $\ker(T - \lambda)^n$ $(\lambda \neq 0)$ **of a compact operator** T **on a Banach space** X, **there is a matrix of** T **consisting of blocks on the main diagonal, each of the type**

$$\begin{pmatrix} \lambda & 1 & 0 & \cdots & 0 \\ 0 & \lambda & & & \vdots \\ \vdots & & \ddots & & 0 \\ & & & & 1 \\ 0 & \cdots & & 0 & \lambda \end{pmatrix}$$

Proof The operator T can be split as $\lambda + (T - \lambda)$. The latter is nilpotent on the subspace $\ker(T - \lambda)^n$ (finite dimensional since $(T - \lambda)^n$ is Fredholm), while λI is diagonal. This is the claimed Jordan form, once it is shown that a nilpotent operator has the following form.

A nilpotent operator of index n on an n-dimensional space can be represented by a matrix of 0s except for 1s and 0s in the super-diagonal: Suppose A is a nilpotent operator of index n, $A^n = 0$; it has a descending sequence down to n, and an ascending sequence up to n, $0 \subset \ker A \subset \cdots \subset \ker A^n$. For each non-zero vector $A^{n-1}u \in \operatorname{im} A^{n-1}$ there is a sequence of vectors $e_1 := A^{n-1}u$, $e_2 := A^{n-2}u$, ..., $e_n := u$. They are linearly independent because $e_i \in \ker A^i \smallsetminus \ker A^{i-1}$, so to have $e_m \in [\![e_1, \ldots, e_{m-1}]\!] \subseteq \ker A^{m-1}$ is impossible. Since $Ae_i = e_{i-1}$ and $Ae_1 = 0$, the matrix of A restricted to the space generated by these vectors is

$$\begin{pmatrix} 0 & 1 & 0 & \cdots & 0 \\ \vdots & & \ddots & & 0 \\ \vdots & & & & 1 \\ 0 & \cdots & & & 0 \end{pmatrix}.$$

A remains nilpotent on the rest of the space $\ker A^n / [\![e_1, \ldots, e_n]\!]$, with perhaps a lower index. The same argument can be repeated to yield other sets of independent vectors. As $X = \ker A^n$ is finite-dimensional, this process ends with a finite basis for X and the matrix of A with respect to it consists of such blocks placed on the diagonal. $\square$

Examples 14.21

1. The total number of λs in a Jordan matrix, called its *algebraic multiplicity*, is the dimension of $\ker(T - \lambda)^n$, the largest generalized eigenspace of λ. The number of Jordan blocks associated with λ is $\dim \ker(T - \lambda)$, called the *geometric multiplicity* of λ. The size of the largest Jordan block is sometimes called its (Jordan) *index*. For example, the matrix below has an eigenvalue 2 with algebraic multiplicity 4, geometric multiplicity 2, and index 3; the other eigenvalue 3 has algebraic multiplicity 2, geometric multiplicity 1, and index 2.

$$\begin{pmatrix} 2 & & & & & \\ & 2 & 1 & & & \\ & & 2 & 1 & & \\ & & & 2 & & \\ & & & & 3 & 1 \\ & & & & & 3 \end{pmatrix}$$

2. Using the appropriate basis on each eigenspace, $E^{-1}TE = J$; rewritten as $TE = EJ$, this shows exactly which vectors are the eigenvectors and generalized eigenvectors of T. Written as $E^{-1}T = JE^{-1}$, this shows which rows are the left eigenvectors of T.

3. The set of $n \times n$ matrices with distinct eigenvalues is dense and open in $B(\mathbb{C}^n)$.
   Proof: Suppose a matrix A has the Jordan-form matrix $A = D + C$ where D is diagonal with the eigenvalues $\lambda_1, \ldots, \lambda_r$ and C is nilpotent. Alter each eigenvalue slightly so λ_i' are all distinct and let $A' := D' + C$; then $\|A' - A\| = \|D' - D\| = \max_i |\lambda_i' - \lambda_i| < \epsilon$.
   Because of this, the Jordan canonical form of a numerical matrix is impossible to calculate, due to the limited accuracy of the matrix coefficients; small changes in the coefficients result in a diagonal Jordan matrix with distinct eigenvalues.

Exercises 14.22

In these exercises, let K be a compact operator on a Banach space X.

1. When T is 1–1, the ascending sequence of spaces are all 0.
   When T is onto, the descending sequence of spaces are all X.
2. For the matrix $\left(\begin{smallmatrix} 0 & 1 \\ 0 & 0 \end{smallmatrix}\right)$, the ascending and descending sequences are the same.
3. The left-shift operator L is onto and has an infinite ascending sequence; its adjoint R is 1–1 and has an infinite descending sequence.
   The operator $f(t) \mapsto tf(t)$ acting on $C[0, 1]$, is 1–1, and also has an infinite descending sequence, e.g., each of the functions $1, t, t^2, \ldots$ belongs to a different image space.
4. If T has a finite descent then $T^\top$ has a finite ascent, of the same order.

5. (a) Suppose that $\ker T^n \subseteq \operatorname{im} T$ for some n. Show that $Tx = 0 \Rightarrow x = T^2 z$ so

$$\ker T^{n-1} \subseteq \operatorname{im} T^2, \ldots, \ker T \subseteq \operatorname{im} T^n.$$

   (b) Suppose $\ker T \subseteq \operatorname{im} T^n$ for some n, then $x \in \ker T^2 \Rightarrow x - Ty \in \ker T$ for some y, so

$$\ker T^2 \subseteq \operatorname{im} T^{n-1}, \ldots, \ker T^n \subseteq \operatorname{im} T.$$

6. There is an eigenvalue at the spectral radius of a compact operator, except possibly when it is 0.
7. In ℓ^1, the multiplier map $M(a_n) := (c_n a_n)_{n \in \mathbb{N}}$ is compact when $c_n \to 0$; its eigenvalues are c_n. 0 is part of the continuous spectrum, unless it is an eigenvalue.
   For example, take $c_n := 1/n$ (and $c_0 := 1$), and the shift operators L and R; then ML is also compact but has no eigenvalues except 0; RM is compact with no eigenvalues at all but 0 is part of the residual spectrum.
8. *The original Fredholm alternative*: For $\lambda \neq 0$, either $(K - \lambda)x = y$ has a unique solution for each y or $K^\top y = \lambda y$ has a non-trivial solution.
9. The minimal polynomial of each Jordan block is $(z - \lambda)^n$.
10. *Cayley-Hamilton theorem*: If p is the characteristic polynomial of a matrix T, then $p(T) = 0$. (Hint: Consider the characteristic polynomial of each Jordan block.)

14.4 The Functional Calculus

The previous definition of $f(T)$ in Taylor's theorem can be extended to functions that are analytic on the spectrum of T, since, by Cauchy's theorem, the path of integration can be swept over analytic regions of f and $(z - T)^{-1}$.

Definition 14.23

For any function $f : \mathbb{C} \to \mathbb{C}$ which is analytic in a neighborhood of $\sigma(T)$, let

$$f(T) := \frac{1}{2\pi i} \oint f(z)(z - T)^{-1} \, dz,$$

where the path of integration is taken along simple closed curves enclosing $\sigma(T)$ in a direction which keeps $\sigma(T)$ to its left.

Note that the integral is defined since $f(z)$ and $\|(z - T)^{-1}\|$ are continuous in z on the selected compact path; hence

$$\|f(T)\| \leqslant \frac{1}{2\pi} \int |f(z)| \|(z - T)^{-1}\| \, ds < \infty.$$

Examples 14.24

1. ▶ If $TS = SR$ then $f(T)S = Sf(R)$ when f is analytic on a neighborhood of $\sigma(T) \cup \sigma(R)$, since

$$S(z - R) = (z - T)S$$

$$\therefore \ (z - T)^{-1}S = S(z - R)^{-1}$$

$$\therefore \ f(T)S = \oint f(z)(z - T)^{-1}S \, dz = \oint f(z)S(z - R)^{-1} \, dz = Sf(R).$$

In particular

(a) $f(S^{-1}TS) = S^{-1}f(T)S$; for example, $e^{S^{-1}TS} = S^{-1}e^T S$.
(b) $ST = TS$ implies $f(T)S = Sf(T)$ and $f(T)g(S) = g(S)f(T)$.

2. If $f \in C^\omega(\sigma(T))$ is zero on $\sigma(T)$, it does not follow that $f(T) = 0$, because $f(T)$ is defined in terms of a path-integral just *outside* $\sigma(T)$. For example, $T := \left(\begin{smallmatrix} 0 & 1 \\ 0 & 0 \end{smallmatrix}\right)$ has $\sigma(T) = \{0\}$, and $f(z) := z$ vanishes there, yet $f(T) = T \neq 0$.

3. ⋆ f is differentiable (and continuous) at T: for H sufficiently small, $f(T + H)$ is defined since $\sigma(T + H) \subseteq \sigma(T) + B_\epsilon(0)$ (Proposition 14.6), and

$$f(T + H) = f(T) + \frac{1}{2\pi i} \oint f(w)(w - T)^{-1}H(w - T)^{-1} \, dw + o(H).$$

The next theorem proves that all algebraic properties of a complex function are mirrored by properties of $f(T)$.

Theorem 14.25 (The Functional Calculus)

Given $T \in \mathcal{X}$, the map $f \mapsto f(T)$, $C^\omega(\sigma(T)) \to \mathcal{X}$, satisfies

$$(f + g)(T) = f(T) + g(T), \quad (\lambda f)(T) = \lambda f(T),$$

$$(fg)(T) = f(T)g(T), \quad 1(T) = 1,$$

$$f \circ g(T) = f(g(T)),$$

$$f_n \to f \text{ in } C(\sigma(T) + B_\epsilon(0)) \ (\exists \epsilon > 0) \Rightarrow f_n(T) \to f(T) \text{ in } \mathcal{X}.$$

Proof We have already seen part of this theorem in action for power series. In particular, the cases $1 = \frac{1}{2\pi i} \oint (z - T)^{-1} dz$ and $T^{-1} = \frac{1}{2\pi i} \oint z^{-1}(z - T)^{-1} dz$ were covered (Example 13.30(2)).

(i) $(f + \lambda g)(T) = f(T) + \lambda g(T)$ expresses the linearity property of the integral.
(ii) $(fg)(T) = f(T)g(T)$: We require the identity

$$(z - w)(z - T)^{-1}(w - T)^{-1} = (w - T)^{-1} - (z - T)^{-1},$$

which follows easily from $z - w = (z - T) - (w - T)$. In the following analysis, consider two paths around $\sigma(T)$, one (with variable z) *nested* inside another (with variable w).

$$f(T)g(T) = \frac{1}{(2\pi i)^2} \oint \oint f(z)g(w)(z - T)^{-1}(w - T)^{-1} \, dz \, dw$$

$$= \frac{1}{(2\pi i)^2} \oint \oint f(z)g(w) \left(\frac{(w - T)^{-1}}{z - w} + \frac{(z - T)^{-1}}{w - z} \right) dz \, dw$$

$$= \frac{1}{2\pi i} \oint g(w)(w - T)^{-1} \frac{1}{2\pi i} \oint f(z)(z - w)^{-1} \, dz \, dw$$

$$+ \frac{1}{2\pi i} \oint f(z)(z - T)^{-1} \frac{1}{2\pi i} \oint g(w)(w - z)^{-1} \, dw \, dz$$

$$= \frac{1}{2\pi i} \oint f(z)(z - T)^{-1} g(z) \, dz,$$

$$= (fg)(T)$$

where we have changed the order of integration in the third line, and used the fact that $(w - z)^{-1}$ leaves a residue when integrated on the outer path, but not when integrated on the inner path (because the singularity at w would then be outside the path of integration).
In particular, note that if f is invertible on a neighborhood of $\sigma(T)$,

$$f(T)^{-1} = \frac{1}{2\pi i} \oint f(z)^{-1}(z - T)^{-1} \, dz. \tag{14.1}$$

(iii) $f(g(T)) := \frac{1}{2\pi i} \oint f(z)(z - g(T))^{-1} \, dz$, where the right part of the integrand is $(z - g(T))^{-1} = \frac{1}{2\pi i} \oint (z - g(w))^{-1}(w - T)^{-1} \, dw$ by (14.1). Combining the two and using Cauchy's integral formula (Proposition 12.19), we get

$$f(g(T)) = \frac{1}{(2\pi i)^2} \oint \oint f(z)(z - g(w))^{-1}(w - T)^{-1} \, dw \, dz,$$

$$= \frac{1}{(2\pi i)^2} \oint \oint f(z)(z - g(w))^{-1} \, dz \, (w - T)^{-1} \, dw,$$

$$= \frac{1}{2\pi i} \oint f \circ g(w)(w - T)^{-1} \, dw,$$

$$= f \circ g(T).$$

Note that f has to be analytic on $\sigma(g(T))$ and $g[\sigma(T)]$ for $f(g(T))$ and $f \circ g(T)$ to be defined, but the two sets are equal by the next theorem (which only uses part (ii) of this theorem).

(iv) The mapping is continuous, since $\|(z - T)^{-1}\|$ is bounded by some constant c on the compact path enclosing the open set $U := \sigma(T) + B_\epsilon(0)$:

$$\|f(T) - g(T)\| \leqslant \frac{1}{2\pi} \oint |f(z) - g(z)| \|(z - T)^{-1}\| \, ds$$

$$\leqslant c\|f - g\|_{C(U)}.$$

$\square$

Theorem 14.26 (Spectral Mapping Theorem)

> **The spectrum of $f(T)$ is equal to the set $\{\, f(\lambda) : \lambda \in \sigma(T) \,\}$, that is,**
>
> $$\sigma(f(T)) = f[\sigma(T)]$$

Proof For any f analytic in a neighborhood of $\sigma(T)$:

(i) $\lambda \notin f[\sigma(T)] \Rightarrow \lambda \notin \sigma(f(T))$: Let $\lambda \neq f(z)$ for all $z \in \sigma(T)$; since $f[\sigma(T)]$ is a closed set, there is a minimum distance between λ and $f[\sigma(T)]$. So $(f(z) - \lambda)^{-1}$ is analytic on $\sigma(T) + B_\epsilon(0)$ if ϵ is small enough, and by the functional calculus $(f(T) - \lambda)^{-1}$ exists. Thus $f(T) - \lambda$ is invertible.

(ii) $f(T) - f(\lambda)$ *invertible* $\Rightarrow T - \lambda$ *invertible*: if $f(T) - f(\lambda)$ has an inverse S, we see from rewriting $f(z) - f(\lambda) = (z - \lambda)F(z)$, and the functional calculus, that

$$(T - \lambda)F(T)S = 1 = SF(T)(T - \lambda)$$

which implies that the factor $T - \lambda$ itself is invertible. This is justified once it is shown that $F(z)$ is analytic about $\sigma(T)$; this is apparent when $z \neq \lambda$, but even so,

$$f(z) = f(\lambda) + f'(\lambda)(z - \lambda) + \tfrac{1}{2} f''(\lambda)(z - \lambda)^2 + o((z - \lambda)^2),$$

$$\Rightarrow F(z) = \frac{f(z) - f(\lambda)}{z - \lambda} = f'(\lambda) + \tfrac{1}{2} f''(\lambda)(z - \lambda) + o(z - \lambda),$$

meaning F is analytic at λ. $\square$

Examples 14.27

1. $\log T$ can be defined whenever there is a path, or "branch", connecting 0 to ∞ without meeting $\sigma(T)$, because in this case, $\log z$ can be defined and is analytic on $\sigma(T)$. But note that $\log z$, and consequently $\log T$, depends on the actual branch used. Examples include, of course, invertible $n \times n$ complex matrices. When defined, $e^{\log T} = T$. Such elements must be in $\mathcal{G}_1$ (Proposition 13.24).

2. Similarly one can define $T^a := e^{a \log T}$ (again not uniquely); then $(T^{1/n})^n = T$ $(n = 1, 2, \ldots)$, and $T^{a+b} = T^a T^b$ (at least for a, b real). By the spectral mapping theorem, $\rho(T^a) = \rho(T)^a$ for $a \geqslant 0$.

3. If T satisfies a polynomial $p(T) = 0$, then $\sigma(T)$ consists of the roots of the minimal polynomial of T (Example 13.3(13)).
   Proof: The spectral theorem shows that $p[\sigma(T)] = 0$, i.e., that the spectrum consists of roots of p. Conversely, if λ is a root of the minimal polynomial, $p(\lambda) = 0$, then $p(z) = (z - \lambda)^n q(z)$, so $0 = p(T) = (T - \lambda)^n q(T)$, where $q(T) \neq 0$ and thus $T - \lambda$ is not invertible.

4. ▶ If λ is an eigenvalue of $T \in B(X)$ then $f(\lambda)$ is an eigenvalue of $f(T)$, with the same eigenvector.
   Proof: When $Tx = \lambda x$, then $(z - T)x = (z - \lambda)x$ and $(z - T)^{-1}x = (z - \lambda)^{-1}x$ $(z \notin \sigma(T))$, so

$$f(T)x = \frac{1}{2\pi i} \oint f(z)(z - T)^{-1}x \, dz = \frac{1}{2\pi i} \oint f(z)(z - \lambda)^{-1}x \, dz = f(\lambda)x.$$

Conversely suppose $f(T) - f(\lambda)$ is not 1–1. Take an open neighborhood $U \supset \sigma(T)$ in which f is analytic. Then, either f is constant on U, or else there are only a finite number of $\lambda_i \in \sigma(T)$ satisfying $f(\lambda_i) = f(\lambda)$. So, for $z \in U$, $f(z) - f(\lambda) = (z - \lambda_1) \cdots (z - \lambda_k)g(z)$ (where multiple roots are repeated) with g analytic and non-zero on U, and consequently

$$f(T) - f(\lambda) = (T - \lambda_1) \cdots (T - \lambda_k)g(T).$$

But $f(T) - f(\lambda)$ is not 1–1, so there must be a λ_i such that $T - \lambda_i$ is not 1–1 ($g(T)$ is invertible), and $f(\lambda_i) = f(\lambda)$.

Proposition 14.28

> If $\sigma(T)$ **disconnects into two closed sets** $\sigma_1 \cup \sigma_2$, **each surrounded by simple closed paths in open neighborhoods of them, then**
>
> (i) $T = TP_1 + TP_2$, **with** P_1, P_2 **(called** *spectral idempotents*) **such that**
> $1 = P_1 + P_2$, $P_i P_j = \delta_{ij}$,
> (ii) **In the reduced algebras** $P_1 \mathcal{X} P_1$, $P_2 \mathcal{X} P_2$ **respectively,**
>
> $$\sigma(TP_1) = \sigma_1, \quad \sigma(TP_2) = \sigma_2.$$

Proof The disjoint closed sets σ_1 and σ_2 can be separated by disjoint open sets U_1, U_2 (Exercise 5.8(5)). Consider the functions χ_i $(i = 1, 2)$ which take the constant value 1 on one open set $U_i \supset \sigma_i$, and 0 on the other. They are analytic on $U_1 \cup U_2$, so we can define

$$P_i := \chi_i(T) = \frac{1}{2\pi i} \oint_{\sigma_i} (z - T)^{-1} \, dz.$$

The path of integration is the union of the two paths surrounding σ_1 and σ_2, but one of the two integrals vanishes.

P_i are idempotents, $P_1 P_2 = 0$, and $P_1 + P_2 = 1$, because $\chi_i^2 = \chi_i$, $\chi_1 \chi_2 = 0$ and $\chi_1 + \chi_2 = 1$ on $U_1 \cup U_2 \supset \sigma(T)$.

Let $f_i(z) := z \chi_i(z)$; then $f_i(T) = TP_i$ and $\sigma(f_i(T)) = f_i[\sigma(T)] = \sigma_i \cup \{0\}$. However, if we restrict to the reduced algebra $P_i \mathcal{X} P_i$, with unity P_i, this changes slightly. Since $z - \lambda$ is invertible in $C^\omega(\sigma_i)$ if, and only if, $\lambda \notin \sigma_i$, it follows that there exists an S such that $S(T - \lambda)P_i = P_i = (T - \lambda)SP_i$ whenever $\lambda \notin \sigma_i$; this means that $(T - \lambda)P_i$ is invertible in $P_i \mathcal{X} P_i$. Thus, $\sigma(TP_i) = \sigma_i$ in this algebra. $\square$

Examples 14.29

1. ▶ When the algebra is $B(X)$, P_i are projections, and the spectral decomposition of an operator T into TP_1 and TP_2 also gives a decomposition of $X = X_1 \oplus X_2$ where $X_i = \mathrm{im}\, P_i$ are T-invariant, and $\sigma(T|_{X_i}) = \sigma_i$. (Proposition 11.5)
2. If 0 is an isolated point of $\sigma(T)$, with spectral idempotent P, then there is a Laurent expansion

$$(z - T)^{-1} P = P z^{-1} + TP z^{-2} + T^2 P z^{-3} + \cdots.$$

3. If $0 \notin \sigma_1$, then $P_1 = T \left(\frac{1}{2\pi i} \oint_{\{\lambda\}} \frac{(z-T)^{-1}}{z} \, dz \right)$. For example, when T is a compact operator and $\lambda \neq 0$ is an isolated point of $\sigma(T)$, then the projection P_λ is also compact, confirming that the eigenspace of λ is finite-dimensional.

Exercises 14.30

1. The non-trivial *idempotents* have spectrum $\{0, 1\}$, and the *nilpotents* have spectrum $\{0\}$. What can the spectrum of a *cyclic* element be?

2. If f takes the value 0 inside $\sigma(T)$ then $f(T)$ is not invertible.

3. Use the spectral mapping theorem to show that if $e^T = 1$ then $\sigma(T) \subset 2\pi i \mathbb{Z}$. If P is an idempotent, then $e^{2\pi i P} = 1$.

4. If J is a Banach algebra morphism, then $f(J(T)) = J(f(T))$ (recall $\sigma(J(T)) \subseteq \sigma(T)$).

5. Show directly that the matrix $\begin{pmatrix} 0 & 1 \\ 0 & 0 \end{pmatrix}$ has no square root at all.

   The shift operators on ℓ^2, say, cannot have a square root because their spectrum encloses 0 (even on $\ell^1(\mathbb{Z})$ when L and R are invertible). Prove this directly by showing the contradictions

   (a) if $T^2 = L$, then T must be onto and $\ker T = \ker L = [\![e_0]\!]$, so $e_0 = \alpha T e_0 = \mathbf{0}$;

   (b) if $T^2 = R$, then T is 1–1, and $\operatorname{im} T = \operatorname{im} R$, so $TRx = RTx = (0, 0, b_0, \ldots)$.

6. A simple linear electronic circuit with feedback can be modeled as an operator, transforming an input signal $x = (x_n)_{n \in \mathbb{N}}$ to an output signal $y = (y_n)_{n \in \mathbb{N}}$ such that

$$y_n = b x_n - a_1 y_{n-1} - \cdots - a_r y_{n-r},$$

where b, a_i are parameters determined by the circuit. Equivalently,

$$(1 + a_1 R + \cdots + a_r R^r) y = bx,$$

where R is the right-shift operator. To avoid the once-familiar feedback loop instability, it is desired that the values y_n do not grow of their own accord, meaning that $1 + a_1 R + \cdots + a_r R^r$ has a continuous inverse. This is the case when the roots of the polynomial $1 + a_1 z + \cdots + a_r z^r$ all have magnitude greater than 1.

14.5   The Gelfand Transform

Quasinilpotents and the Radical

How much can an operator T be modified and still retain the same spectrum? That is, when is $\sigma(T+Q) = \sigma(T)$? If this is to hold for all T, including invertible ones, then $I + T^{-1}Q$ would need to be invertible for all such T, thus $\rho(T^{-1}Q) = 0 = \rho(Q)$. The next definition explores these ideas.

Definition 14.31

The **quasinilpotents** are those elements $Q \in \mathcal{X}$ with $\rho(Q) = 0$.

The (Jacobson) **radical** of $\mathcal{X}$ is

$$\operatorname{rad} \mathcal{X} := \{ Q \in \mathcal{X} : \forall T \in \mathcal{X}, \ \rho(TQ) = 0 \}.$$

A Banach algebra with a trivial radical is called *semi-primitive* or *$\mathcal{J}$-semi-simple*.

The next proposition shows that the radical is a closed ideal, which can be factored out to leave a semi-primitive Banach algebra.

Examples 14.32

1. The prime examples of quasinilpotents are the *nilpotents*, defined as those elements which satisfy $Q^n = 0$ for some n, so $\rho(Q) \leqslant \|Q^n\|^{1/n} = 0$; e.g., $\begin{pmatrix} 0 & 1 \\ 0 & 0 \end{pmatrix}$.
   But the right/left shift operators are not quasinilpotent, even though their matrices resemble nilpotent ones.
2. Every operator $Tf(s) := \int_0^s k(s,t) f(t) \, dy$ on $C[0,1]$, where $k \in L^\infty[0,1]^2$, is a quasinilpotent.
   Proof: $|Tf(s)| \leqslant \int_0^s |k(s,t)||f(t)| \, dt \leqslant \|k\|\|f\|s$. By induction one can conclude $|T^n f(s)| \leqslant \|k\|^n \|f\| s^n / n!$,

   $$|T^{n+1} f(s)| = \left| \int_0^s k(s,t) T^n f(t) \, dt \right|$$

   $$\leqslant \int_0^s \|k\|^{n+1} \|f\| t^n / n! \, dt$$

   $$\leqslant \|k\|^{n+1} \|f\| s^{n+1} / (n+1)!$$

   so $\|T^n\| \leqslant \|k\|^n / n!$ and $\rho(T) \leqslant \|T^n\|^{1/n} \leqslant \|k\| / \sqrt[n]{n!} \to 0$.
3. The sum and product of quasinilpotents need not be quasinilpotents, e.g., $\begin{pmatrix} 0 & 1 \\ 0 & 0 \end{pmatrix}$ and $\begin{pmatrix} 0 & 0 \\ 1 & 0 \end{pmatrix}$.
4. The quasinilpotents are topological divisors of zero since their spectrum is a boundary point. Idempotents (except 0 and 1) are divisors of zero but not quasinilpotents.

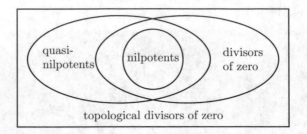

5. ▶ For any $T \in \mathcal{X}$, $Q \in \mathrm{rad}\,\mathcal{X}$, $\sigma(T + Q) = \sigma(T)$.
   Proof: For any invertible S, the sum $S + Q = S(1 + S^{-1}Q)$ is also invertible, since $\rho(S^{-1}Q) = 0$ (Theorem 13.20). Thus

$$\lambda \notin \sigma(T+Q) \ \Leftrightarrow \ T+Q-\lambda \text{ is invertible } \ \Leftrightarrow \ T-\lambda \text{ is invertible } \ \Leftrightarrow \ \lambda \notin \sigma(T).$$

6. Radical elements are obviously quasinilpotents, $\rho(Q) = \rho(1Q) = 0$.
7. It is enough to show that $1 \notin \sigma(TQ)$ for all T, in order that $Q \in \mathrm{rad}\,\mathcal{X}$.
   Proof: For any $\lambda \neq 0$, $1 \notin \sigma(TQ/\lambda) = \sigma(TQ)/\lambda \Rightarrow \lambda \notin \sigma(TQ)$.
8. $B(X)$ has nilpotents (except for $X = \mathbb{C}$) but only a trivial radical.
   Proof: For any $Q \neq 0$, an operator T can be found such that $1 - TQ$ is non-invertible, so $1 \in \sigma(TQ)$. One such operator is $T := x\phi$, where $Qx \neq 0$, $\phi \in X^*$, $\phi Qx = 1$; then $(1 - TQ)x = x - x\phi Qx = 0$ but $x \neq 0$.

Proposition 14.33

The radical is a closed ideal.

Proof *The radical is contained in every maximal left-ideal*: Recall that a maximal left-ideal is closed and that every proper left-ideal can be enlarged to a maximal left-ideal (Example 13.5(7,8)). Let $Q \in \mathrm{rad}\,\mathcal{X}$, and let $\mathcal{M}$ be a maximal left-ideal. Then $\mathcal{M} + \mathcal{X}Q$ is a left-ideal which contains $\mathcal{M}$. Either

(a) $\mathcal{M} + \mathcal{X}Q = \mathcal{X}$, in which case $1 = R + TQ$ for some $R \in \mathcal{M}$, $T \in \mathcal{X}$, so that $R = 1 - TQ$ is invertible, contradicting $R \in \mathcal{M}$ (Example 13.5(5)); or else,
(b) $\mathcal{M} + \mathcal{X}Q = \mathcal{M}$, in which case $Q = 0 + 1Q \in \mathcal{M}$.

Thus $\mathrm{rad}\,\mathcal{X} \subseteq \mathcal{M}$ as required; an analogous argument shows that $\mathrm{rad}\,\mathcal{X}$ is contained in every maximal right-ideal.

The radical is the intersection of the maximal left-ideals: Let P be an element that is contained in every maximal left-ideal. For any $T \in \mathcal{X}$, the left-ideal $\mathcal{X}(1 - TP)$ cannot be proper, otherwise it would lie inside some maximal left-ideal $\mathcal{M}$, forcing $P \in \mathcal{M}$, and $TP \in \mathcal{M}$, and so $1 = TP + (1 - TP) \in \mathcal{M}$, a contradiction. Hence $\mathcal{X}(1 - TP) = \mathcal{X}$, and there is an S such that $S(1 - TP) = 1$.

To show $1 - TP$ is invertible we need to prove $(1 - TP)S = 1$ as well. To this end one can substitute $-ST$ for T in the above argument, to conclude that there is an $R \in \mathcal{X}$ such that

$$1 = R(1 + STP) = R(S + 1 - S(1 - TP)) = RS.$$

But $RS = 1 = S(1 - TP)$ implies $1 - TP = S^{-1}$ is invertible. With $1 \notin \sigma(TP)$ for any T, P must be in the radical.

The radical is a closed ideal: Being the intersection of closed sets, rad $\mathcal{X}$ is also closed (Proposition 2.18). For any $S, T \in \mathcal{X}$ and $Q, Q' \in \text{rad}\,\mathcal{X}$,

(a) $\rho(STQ) = 0 = \rho(SQT)$, so $TQ, QT \in \text{rad}\,\mathcal{X}$,
(b) $\sigma(T(Q + Q')) = \sigma(TQ) = \{0\}$ from Example 5 above ($TQ' \in \text{rad}\,\mathcal{X}$), so $Q + Q' \in \text{rad}\,\mathcal{X}$,
(c) $\rho(T(\lambda Q)) = |\lambda|\rho(TQ) = 0$, so $\lambda Q \in \text{rad}\,\mathcal{X}$,

and rad $\mathcal{X}$ is an ideal.   □

The State Space

The spectrum of an element $T \in \mathcal{X}$ is a subset that gives us important information about T. However, it is not well behaved under addition or multiplication of elements. There exists a set that contains the spectrum which is much better behaved. Consider a functional ϕ on $\mathcal{X}$, then $\phi(T - \lambda) = \phi(T) - \lambda$ if we insist that $\phi(1) = 1$; moreover, as it turns out, ϕ maps non-invertible elements to 0, and thus $\lambda = \phi(T)$ for $\lambda \in \sigma(T)$, if we restrict the functionals to the following definition:

Definition 14.34

The **state space** of a Banach algebra $\mathcal{X}$ is the set of functionals

$$\mathcal{S}(\mathcal{X}) := \{\, \phi \in \mathcal{X}^* : \phi 1 = 1 = \|\phi\| \,\}.$$

We often write $\mathcal{S}$ for $\mathcal{S}(\mathcal{X})$ and $\mathcal{S}(T) := \{\, \phi T \in \mathbb{C} : \phi \in \mathcal{S}(\mathcal{X}) \,\}$, for example, $\mathcal{S}(1) = \{1\}$.

Proposition 14.35

The state space $\mathcal{S}(\mathcal{X})$ is a convex set containing the character space $\Delta(\mathcal{X})$. For any $T \in \mathcal{X}$, $\mathcal{S}(T)$ is a compact convex subset of $\mathbb{C}$, and

$$\Delta(T) \subseteq \sigma(T) \subseteq \mathcal{S}(T).$$

Moreover, $\mathcal{S}(T_1 + T_2) \subseteq \mathcal{S}(T_1) + \mathcal{S}(T_2)$, $\mathcal{S}(\lambda T) = \lambda\mathcal{S}(T)$.

Proof (i) $S(\mathcal{X})$ *and* $S(T)$ *are convex*: For $\phi, \psi \in S$ and $0 \leqslant t \leqslant 1$,

$$(t\phi + (1 - t)\psi)1 = t + 1 - t = 1, \text{ and}$$

$$\|t\phi + (1 - t)\psi\| \leqslant t\|\phi\| + (1 - t)\|\psi\| = 1.$$

It follows from $t\phi T + (1 - t)\psi T = (t\phi + (1 - t)\psi)T \in S(T)$ that $S(T)$ is convex.

$S(T)$ *is compact*: $S(T)$ is bounded since $|\phi T| \leqslant \|T\|$ for any $\phi \in S$. Now recall that every bounded sequence in $\mathcal{X}^*$ has a weak*-convergent subsequence (Theorem 11.42 for $\mathcal{X}$ separable). So whenever $\phi_n T \in S(T)$ converges to a limit point z, there is a subsequence of ϕ_n that converges in the weak* sense, $\phi_{n_i} \rightharpoonup \phi \in \mathcal{X}^*$, implying

(a) $\phi_{n_i} T \to \phi T = z$ and $1 = \phi_{n_i} 1 \to \phi 1$,
(b) $\|\phi\| \leqslant \liminf_i \|\phi_{n_i}\| = 1$ (Corollary 11.37).

Hence $\phi \in S$ and $z \in S(T)$, that is, $S(T)$ is closed and bounded.

(ii) $\sigma(T) \subseteq S(T)$: If $R \in \mathcal{X}$ is not invertible, then $1 \notin [\![R]\!]$; indeed $d(1, [\![R]\!]) = 1$ as $[\![R]\!]$ contains no invertible elements (Theorem 13.20). So by the Hahn-Banach theorem, there is a $\phi \in \mathcal{X}^*$ satisfying $\phi 1 = 1 = \|\phi\|$ and $\phi R = 0$ (Proposition 11.20). In particular, for $R = T - \lambda$, where $\lambda \in \sigma(T)$, there is a $\phi \in S$ such that

$$0 = \phi(T - \lambda) = \phi T - \lambda,$$

so $\lambda = \phi T \in S(T)$.

(iii) $\Delta(T) \subseteq \sigma(T)$: Recall that any character $\psi \in \Delta$ maps invertible elements to invertible complex numbers (Example 13.7(1)), including $\psi 1 = 1$. So for any $\lambda \notin \sigma(T)$, $\psi T - \lambda = \psi(T - \lambda) \neq 0$, and $\lambda \notin \Delta(T)$. Equivalently, $\Delta(T) \subseteq \sigma(T)$ and $|\psi T| \leqslant \rho(T) \leqslant \|T\|$. This means that ψ is automatically continuous with $\|\psi\| = 1$, and so $\Delta \subseteq S$.

(iv) For any $\phi \in S$,

$$\phi(T_1 + T_2) = \phi(T_1) + \phi(T_2) \in S(T_1) + S(T_2),$$

$$\phi(\lambda T) = \lambda \phi T \implies S(\lambda T) = \lambda S(T).$$

$\square$

Examples 14.36

1. ▶ The state space of ℓ^1 consists of bounded sequences with $b_0 = 1$, $|b_n| \leqslant 1$. The characters of ℓ^1 are of the type $(z^n)_{n \in \mathbb{N}}$, for some $|z| \leqslant 1$.
   Proof: Let $\phi \in S(\ell^1) \subseteq \ell^{1*} \equiv \ell^\infty$ (Proposition 9.6); then $\phi \sim (b_n)_{n \in \mathbb{N}}$ and the requirements $\phi 1 = 1 = \|\phi\|$ become $b_0 = 1$, $|b_n| \leqslant 1$. In particular for any

$\psi \in \Delta$, and any $x \in \ell^1$,

$$x = (a_0, a_1, \ldots) = \sum_{n=0}^{\infty} a_n e_n = \sum_{n=0}^{\infty} a_n (\underbrace{e_1 * \cdots * e_1}_{n}),$$

$$\therefore \psi x = \sum_{n=0}^{\infty} a_n \psi(e_1 * \cdots * e_1) = \sum_{n=0}^{\infty} a_n z^n = (z^n) \cdot (a_n), \quad (z := \psi e_1),$$

where the multiplicative property $\psi(e_1 * \cdots * e_1) = (\psi e_1)^n$ is used.

2. The characters of $\ell^1(\mathbb{Z})$ are $\psi_\theta \sim (e^{in\theta})_{n \in \mathbb{Z}}$.
   The proof is the same as above except $|z^{-1}| \leq 1$ holds additionally, that is, $z = e^{i\theta} \in \mathbb{S}^1$ for some $0 \leq \theta < 2\pi$.

3. For $L^1(\mathbb{S}^1)$, the characters are $\psi_n(f) = \int_0^{2\pi} e^{in\theta} f(\theta) \, d\theta$, $(n \in \mathbb{Z})$.
   Proof: Let $\psi \in \Delta \subseteq L^1(\mathbb{S}^1)^* \equiv L^\infty(\mathbb{S}^1)$, so $\psi(f) = \int_0^{2\pi} h(\theta) f(\theta) \, d\theta$ for some $h \in L^\infty(\mathbb{S}^1)$. Recall that $L^1(A)$ does not contain a unity for convolution (Example 13.3(5)); nevertheless, one can be added artificially, so Δ exists and its characters act on $L^1(A)$. Again we require

   (a) $1 = \|\psi\| = \|h\|_{L^\infty}$, so $|h(\theta)| \leq 1$ for almost all θ;
   (b) $\psi(f * g) = \psi(f)\psi(g)$, or equivalently,

$$\int_0^{2\pi} h(\theta) \int_0^{2\pi} f(\theta - \eta) g(\eta) \, d\eta \, d\theta = \int_0^{2\pi} h(\theta) f(\theta) \, d\theta \int_0^{2\pi} h(\eta) g(\eta) \, d\eta.$$

This implies that $h(\theta + \eta) = h(\theta)h(\eta)$ a.e.; we've met this identity before in our preliminary discussion on the exponential function in Sect. 13.2, where we concluded that $h(\theta) = h(1)^\theta = e^{z\theta}$, assuming h is continuous. That this can be taken to be the case follows from Corollary 9.31,

$$\left| \int \big(h(y + \epsilon) - h(y)\big) f(y) \, dy \right| = \left| \int h(y)\big(f(y - \epsilon) - f(y)\big) \, dy \right|$$

$$\leq \int |f(y - \epsilon) - f(y)| \, dy \to 0.$$

Moreover, $h(2\pi) = h(0) = 1$ implies that $h(1) = e^{in}$ for some $n \in \mathbb{Z}$.

4. For $L^1(\mathbb{R})$, the characters are $\psi_\xi(f) = \int_{\mathbb{R}} e^{it\xi} f(t) \, dt$, $(\xi \in \mathbb{R})$.
   Proof: Let $\psi \in \Delta \subseteq L^1(\mathbb{R})^* \equiv L^\infty(\mathbb{R})$; so $\psi(f) = \int hf$. As before, $|h(t)| \leq 1$ for all t, while the condition $\psi(f * g) = \psi(f)\psi(g)$ is equivalent to $h(t + s) = h(t)h(s)$ a.e., so $h(t) = h(1)^t$. To avoid $h(t)$ growing arbitrarily large as $t \to \pm\infty$, $|h(1)|$ must be 1, and $h(t) = e^{it\xi}$.

5. Repeating for $L^1(\mathbb{R}^+)$, $\Delta \cong \{ e^{-zt} : \text{Re } z \geq 0 \}$.

6. ⋆ For $C[0, 1]$, $\Delta = \{\delta_t \in C[0, 1]^* : \delta_t(f) = f(t),\ t \in [0, 1]\} \cong [0, 1]$.
   Proof: That δ_t are unit functionals should be obvious. In addition,

   $$\delta_t(fg) = (fg)(t) = f(t)g(t) = \delta_t(f)\delta_t(g), \text{ and } \delta_t(1) = 1.$$

   Note that for $s \neq t$, $\delta_s(f) \neq \delta_t(f)$ for some $f \in C[0, 1]$.
   For the converse, let ψ be a character of $C[0, 1]$. Define 'triangle' functions,
   $\tau_{n,i}(x)$, as in the accompanying plot; note that these functions overlap and sum
   to 1 everywhere, $\sum_{i=0}^{2^n} \tau_{n,i} = 1$.

   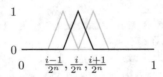

   Then $1 = \psi 1 = \sum_i \psi(\tau_{n,i})$ and at least one triangle function must give
   $\psi(\tau_{n,i_n}) \neq 0$. In fact, $\psi(\tau_{n,i}) = 0$ for $i \neq i_n - 1, i_n, i_n + 1$, since $\tau_{n,i}\tau_{n,i_n} = 0$.
   By taking larger values of n, and selected values of i_n, the nested intervals
   $[\frac{i_n-1}{2^n}, \frac{i_n+1}{2^n}]$ shrink to some point t. For any function $f \in C[0, 1]$,

   $$\psi f = \sum_i \psi(\tau_{n,i})\psi(f) = \psi\left(\sum_{i=i_n-1}^{i_n+1} \tau_{n,i} f\right) \to f(t), \quad \text{as } n \to \infty.$$

   The map $x \mapsto \delta_t$ is thus 1–1 and onto Δ. Furthermore $t_n \to t \Leftrightarrow \delta_{t_n} \rightharpoonup \delta_t$,
   since the latter means $f(t_n) \to f(t)$ for all $f \in C[0, 1]$, in particular for the
   identity function $f(x) := x$.

7. ⋆ The character space of the Banach algebra $\overline{\mathbb{C}[T_1, \ldots, T_n]}$ generated by
   commuting elements, is isomorphic to a compact subset of $\mathbb{C}^n$ (use the map
   $\psi \mapsto (\psi T_1, \ldots, \psi T_n)$).

8. ⋆ The character space is weakly closed: $\psi_n \in \Delta$ AND $\psi_n \rightharpoonup \psi \Rightarrow \psi \in \Delta$.
   Consequently, for a separable Banach algebra, Δ is a compact metric space.
   Proof: Taking the limits of $\psi_n(S + T) = \psi_n S + \psi_n T$, $\psi_n(\lambda T) = \lambda \psi_n T$,
   $\psi_n(ST) = (\psi_n S)(\psi_n T)$, and $\psi_n 1 = 1$, shows that ψ is an algebraic morphism.
   Also $|\psi_n T| \leq \|T\|$ becomes $|\psi T| \leq \|T\|$ in the limit $n \to \infty$, and ψ is
   continuous. For a separable Banach algebra, the unit ball in $\mathcal{X}^*$ is compact
   with respect to the weak*-metric (Theorem 11.42), and so is its weakly closed
   subset Δ.

The Gelfand Transform

To see why characters may be useful, consider the algebra ℓ^1 and its characters p_z.
A sequence such as $x = (\frac{1}{2}, \frac{1}{4}, \frac{1}{8}, \ldots)$ can be encoded as a complex power series in

terms of its characters, $p_z(x) = \sum_{i=0}^{\infty} z^i/2^{i+1} = (2-z)^{-1}$. Then the convolution product $x * \cdots * x$ can be evaluated using characters instead of working it out directly,

$$p_z(x * \cdots * x) = p_z(x)^n = \frac{1}{(2-z)^n} = \sum_{i=0}^{\infty} \frac{n(n+1)\cdots(n+i-1)}{i!\, 2^{n+i}} z^i.$$

For an example from probability theory, consider a random variable that outputs a natural number $i = 0, 1, 2, \ldots$, with probability $1/2^{i+1}$. The probability distribution of the sum of n such random outputs is $x * \cdots * x$, which can be read off from the coefficients of $p_z(x)^n$; e.g., the probability of getting a total of, say 2, after n trials is $n(n+1)/2^{n+3}$. Further, the mean of such a sum of random variables is given by differentiating $(2-z)^{-n}$ at $z = 1$, that is n. The key step that makes all of this work is to consider $p_z(x)$ as a function of z. Its generalization leads to:

Definition 14.37

> The **Gelfand transform** of T is the map $\widehat{T} : \Delta(\mathcal{X}) \to \sigma(T)$ defined by
>
> $$\widehat{T}(\psi) := \psi T.$$

The element T is transformed into a function on the compact space Δ. The algebraic structure is preserved, but the transform is generally neither 1–1 nor onto.

Proposition 14.38

> The Gelfand transform $\mathcal{G} : T \mapsto \widehat{T}$ is a Banach algebra morphism $\mathcal{X} \to C(\Delta)$,
>
> $$\widehat{S+T} = \widehat{S} + \widehat{T}, \qquad \widehat{\lambda T} = \lambda \widehat{T},$$
> $$\widehat{ST} = \widehat{S}\,\widehat{T}, \qquad \widehat{1} = 1, \qquad \|\widehat{T}\| \leqslant \|T\|.$$
>
> **Its kernel $\ker \mathcal{G}$ contains the quasinilpotents and the commutators. For any analytic function on the spectrum of T, $f \in C^{\omega}(\sigma(T))$,**
>
> $$\widehat{f(T)} = f \circ \widehat{T}.$$

Proof It is clear from

$$|\widehat{T}(\psi) - \widehat{T}(\phi)| = |\psi T - \phi T| \leqslant \|\psi - \phi\| \|T\|,$$

$$\text{and } |\widehat{T}(\psi)| = |\psi T| \leqslant \|T\|, \text{ for all } \psi, \phi \in \Delta,$$

that $\widehat{T}$ is a (continuous) Lipschitz and bounded function on Δ, with $\|\widehat{T}\|_C \leqslant \|T\|$.

Israel Gelfand(1913–2009) Gelfand studied functional analysis at the University of Moscow under Kolmogorov in 1935, specializing in commutative normed rings. During 1939-41 he studied Banach algebras, introducing his transform and proving the spectral radius formula, which gave much impetus to the subject; in 1943, with Naimark, he proved the embedding of special commutative $*$-algebras into $B(H)$; and then in 1948 he simplified the subject-matter with the introduction of the C^*-condition $\|x^*x\| = \|x\|^2$.

For any $\psi \in \Delta$, we have:

$$\widehat{1}(\psi) = \psi 1 = 1,$$

$$\widehat{\lambda T}(\psi) = \psi(\lambda T) = \lambda \psi T = \lambda \widehat{T}(\psi),$$

$$\widehat{S + T}(\psi) = \psi(S + T) = \psi S + \psi T = (\widehat{S} + \widehat{T})(\psi),$$

$$\widehat{ST}(\psi) = \psi(ST) = \psi S \psi T = \widehat{S}(\psi)\,\widehat{T}(\psi) = (\widehat{ST})(\psi).$$

Clearly, from $\widehat{T}(\psi) = \psi T$, $\widehat{T} = 0 \Leftrightarrow \Delta(T) = 0$. If Q is a quasinilpotent then $\Delta(Q) \subseteq \sigma(Q) = \{0\}$. Also, $\widehat{[S,T]} = \widehat{S}\widehat{T} - \widehat{T}\widehat{S} = 0$ since $C(\Delta)$ is commutative. Lastly, as $\psi(S^{-1}) = (\psi S)^{-1}$, for any $\psi \in \Delta$, $S \in \mathcal{X}$,

$$\widehat{f(T)}(\psi) = \psi f(T) = \psi \left(\frac{1}{2\pi i} \oint f(z)(z - T)^{-1}\, dz \right)$$

$$= \frac{1}{2\pi i} \oint f(z)(z - \psi T)^{-1}\, dz \qquad (\psi T \in \sigma(T))$$

$$= f(\psi T) = f \circ \widehat{T}(\psi).$$

$\square$

We cannot expect the Gelfand transform to be very useful for general algebras as it loses information by representing $\mathcal{X}$ as a subspace of the special commutative algebra $C(\Delta)$; for example, $\widehat{S^{-1}TS} = \widehat{S}^{-1}\widehat{T}\widehat{S} = \widehat{T}$. But for commutative Banach algebras the situation is much improved:

Theorem 14.39

For a commutative Banach algebra $\mathcal{X}$,

$$\operatorname{im} \widehat{T} = \Delta(T) = \sigma(T), \quad \|\widehat{T}\|_{C(\Delta)} = \rho(T), \quad \ker \mathcal{G} = \operatorname{rad} \mathcal{X}.$$

Proof *Any maximal ideal of a commutative Banach algebra is the kernel of some character*: Given a closed ideal $\mathcal{M}$, the mapping $\Phi(T) := T + \mathcal{M}$ is a Banach algebra morphism $\mathcal{X} \to \mathcal{X}/\mathcal{M}$ with $\mathcal{M} = \ker \Phi$ (Exercise 13.10(21)). By Exercise 13.10(20), when $\mathcal{M}$ is also maximal in $\mathcal{X}$, then $\mathcal{X}/\mathcal{M}$ has no non-trivial ideals, and so is isomorphic to $\mathbb{C}$ (Example 14.5(4)). Hence $\Phi : \mathcal{X} \to \mathcal{X}/\mathcal{M} \cong \mathbb{C}$ is a character.

But any non-invertible T belongs to some maximal ideal $\mathcal{M}$ (Example 13.5(8)); so there must be some $\psi \in \Delta$ such that $\mathcal{M} = \ker \psi$, implying $\psi T = 0$. Thus $T - \lambda$ is not invertible if, and only if, there is a $\psi \in \Delta$, with $\psi T - \lambda = \psi(T - \lambda) = 0$, i.e., $\lambda \in \Delta(T)$, and therefore $\Delta(T) = \sigma(T)$. (Note that this shows the existence of characters in a commutative Banach algebra.) Since the two sets are the same, they have the same greatest extent,

$$\|\widehat{T}\|_C = \max_{\psi \in \Delta} |\psi T| = \max_{\lambda \in \sigma(T)} |\lambda| = \rho(T).$$

The quasinilpotents are in the radical: If Q is a quasinilpotent, and $T \in \mathcal{X}$, then

$$\rho(TQ) = \lim_{n \to \infty} \|(TQ)^n\|^{1/n} = \lim_{n \to \infty} \|T^n Q^n\|^{1/n} \leqslant \rho(T)\rho(Q) = 0,$$

so Q is in the radical. Moreover, $\ker \mathcal{G} = \operatorname{rad} \mathcal{X}$ since

$$\widehat{T} = 0 \iff \Delta(T) = \{0\} \iff \sigma(T) = \{0\}.$$

$\square$

Proposition 14.40

A Banach algebra which satisfies, for some $c > 0$ and all T,

$$\|T\|^2 \leqslant c\|T^2\|,$$

can be embedded in the commutative semi-simple Banach algebra $C(\Delta)$, via the Gelfand map.

Proof By induction on n,

$$\|T\|^{2^n} \leqslant (c\|T^2\|)^{2^{n-1}} \leqslant \cdots \leqslant c^{2^n-1} \|T^{2^n}\|$$

from which can be concluded

$$\|T\| \leqslant \lim_{n \to \infty} c^{1-2^{-n}} \|T^{2^n}\|^{2^{-n}} = c\,\rho(T).$$

This inequality has various strong implications:

$\mathcal{X}$ is *semi-primitive*: 0 is clearly the only quasinilpotent.
$\mathcal{X}$ is *commutative*: For any $S, T \in \mathcal{X}$,

$$\|ST\| \leqslant c\,\rho(ST) = c\,\rho(TS) \leqslant c\|TS\|.$$

Hence, the analytic function $F(z) := e^{-zT} S e^{zT}$ is bounded,

$$\forall z \in \mathbb{C}, \qquad \|F(z)\| \leqslant c\|S e^{zT} e^{-zT}\| = c\|S\|.$$

By Liouville's theorem, F must be constant, $e^{-zT} S e^{zT} = S$, that is, $e^{zT} S = S e^{zT}$. Comparing the second terms of their power series expansions,

$$(1 + zT + o(z))S = S(1 + zT + o(z)),$$

gives $TS = ST$.

The Gelfand map is an embedding: $\mathcal{G}$ has the trivial kernel rad $\mathcal{X} = \{0\}$, and is thus an algebra isomorphism onto $\widehat{\mathcal{X}} \subseteq C(\Delta)$. Moreover, $\|T\| \leqslant c\,\rho(T) = c\|\widehat{T}\|_C$, so $\mathcal{G}^{-1}$ is continuous. $\square$

Exercises 14.41

1. In $\mathbb{C}$, as well as $\mathbb{C}^n$, ℓ^∞ and $C[0, 1]$, the only quasinilpotent is 0.
2. Quasinilpotents are preserved by Banach algebra morphisms.
3. A quasinilpotent upper triangular matrix must have 0s on the main diagonal, so is nilpotent. Deduce, using the Jordan canonical form and Theorem 13.8, that every quasinilpotent of a finite-dimensional Banach algebra is nilpotent.
4. $(Q, R) \in \mathcal{X} \times \mathcal{Y}$ is quasinilpotent (or radical) when both Q and R are.
5. The operator $V : \ell^\infty \to \ell^\infty$ defined by $V(a_n) := (0, a_0, a_1/2, a_2/3, \ldots)$ is quasinilpotent.
6. Prove directly that the Volterra operator $f \mapsto \int_0^x f$, on $C[0, 1]$, is a quasinilpotent.
7. A quasinilpotent for which $\|(z - T)^{-1}\| \leqslant \dfrac{c}{|z|^n}$ for all z in a neighborhood of 0, must in fact be a nilpotent. (Hint: use $\|T^n\| \leqslant \frac{1}{2\pi} \int |z|^n \|(z - T)^{-1}\|\, dz \leqslant \epsilon c$.)
8. $\rho(TQS) = 0$ for any $S, T \in \mathcal{X}$, $Q \in \operatorname{rad} \mathcal{X}$. (Hint: Example 14.2(6).)
9. If $\psi \in \Delta$ and $f \in C^\omega(\sigma(T))$, then $\psi(f(T)) = f(\psi T)$.
10. $\Delta(T)$ has better properties than $\sigma(T)$, and may yield useful information about it (unless $\Delta = \varnothing$):

$$\Delta(S + T) \subseteq \Delta(S) + \Delta(T), \qquad \Delta(ST) \subseteq \Delta(S)\Delta(T)$$

11. For $\mathbb{C}^n$, $\Delta = \{\delta_1, \ldots, \delta_n\}$ where $\delta_i(z_1, \ldots, z_n) := z_i$ are the dual basis. The same is true for the space c_0, $\Delta = \{\delta_i \in c_0^* : \delta_i(a_0, a_1, \ldots) = a_i\}$.
12. For $B(\mathbb{C}^n)$, $\Delta = \varnothing$. (Hint: Consider products of $\left(\begin{smallmatrix} 1 & 0 \\ 0 & 0 \end{smallmatrix}\right)$, $\left(\begin{smallmatrix} 0 & 1 \\ 0 & 0 \end{smallmatrix}\right)$, etc.)

13. For characters of the group algebra $\mathbb{C}^G$, $\psi(e_{h^{-1}gh}) = \psi(e_g)$ and $|\psi(e_g)| = 1$.
14. ▶ The invertible elements of a commutative $\mathcal{X}$ correspond to the invertible elements of $\widehat{\mathcal{X}}$.
15. The Gelfand transform on $\mathbb{C}^n$, mapping $\mathbb{C}^n \to C(\Delta) \cong \mathbb{C}^n$, is the identity map. The same is true for $C[0, 1]$, so $\sigma(f) = \mathrm{im}\, f$ for $f \in C[0, 1]$.
16. ▶ The Gelfand transform gathers together various classical transforms under one theoretical umbrella:

(a) *Generating functions*: $\mathcal{G} : \ell^1 \to C(\bar{B}_\mathbb{C})$, maps a sequence $x = (a_n)_{n \in \mathbb{N}}$ to a power series on the unit closed disk in $\mathbb{C}$,

$$(a_n)_{n \in \mathbb{N}} \mapsto \sum_{n=0}^{\infty} a_n z^n.$$

(b) $\mathcal{G} : \ell^1(\mathbb{Z}) \to C(\mathbb{S}^1)$ is similar, $\widehat{x}(\theta) := \sum_{n=-\infty}^{\infty} a_n e^{in\theta}$.

It follows that $\sigma(x) = \{\widehat{x}(\theta) : 0 \leqslant \theta < 2\pi\}$, and the sequence x is invertible in $\ell^1(\mathbb{Z})$ (in the convolution sense) exactly when $\sum_n a_n e^{in\theta} \neq 0$ for all θ. This is essentially *Wiener's theorem*: If $f \in C(\mathbb{S}^1)$ is nowhere 0 and $\widehat{f} \in \ell^1(\mathbb{Z})$ then the Fourier coefficients of $1/f$ are also in $\ell^1(\mathbb{Z})$.

(c) *Fourier coefficients*: $L^1(\mathbb{S}^1) \to C_b(\mathbb{Z}) \equiv \ell^\infty(\mathbb{Z})$,

$$\widehat{f}(n) := \int_0^{2\pi} e^{-in\theta} f(\theta) \, d\theta.$$

(d) *Fourier transform*: $L^1(\mathbb{R}) \to C(\mathbb{R})$,

$$\widehat{f}(\xi) := \int_{-\infty}^{\infty} e^{-it\xi} f(t) \, dt.$$

(e) *Laplace transform*: $L^1(\mathbb{R}^+) \to C(\mathbb{R}^+ + i\mathbb{R})$,

$$\mathcal{L}f(s) := \int_0^{\infty} e^{-sx} f(x) \, dx, \quad \mathrm{Re}\, s \geqslant 0.$$

In all these cases, $\widehat{f * g} = \widehat{f}\,\widehat{g}$.

17. ⋆ In any Banach algebra, if $ST = TS$ then $\sigma(S + T) \subseteq \sigma(S) + \sigma(T)$ and $\sigma(ST) \subseteq \sigma(S)\sigma(T)$. (Hint: Consider the commutant algebra $\{S, T\}''$ of Exercise 13.10(14) and Example 14.5(6).)
18. In a commutative Banach algebra, $e^{S+T} = e^S e^T$, and $De^T = e^T$. The set of exponentials $e^\mathcal{X}$ is a connected group, so $e^\mathcal{X} = \mathcal{E} = \mathcal{G}_1$ (Proposition 13.24).
19. A Banach algebra which satisfies $\|T^2\| = \|T\|^2$ is isometrically isomorphic to a subalgebra of $C(\Delta)$: the condition is equivalent to $\|T\| = \rho(T) = \|\widehat{T}\|$.

20. Conversely to the proposition, a Banach algebra that can be embedded in some $C(K)$ (K compact) satisfies $\|T\|^2 \leqslant c\|T^2\|$.

Remarks 14.42

1. Given a compact set $K \subset \mathbb{C}$, is there an element T with spectrum $\sigma(T) = K$? Of course, this is false in the Banach algebra $\mathbb{C}$, where all spectra consist of single points, and in $B(\mathbb{C}^n)$, where the spectra are finite sets of points. But in ℓ^∞ there are elements with any given compact set K for spectrum (Example 14.2(3)).

2. The distinction between σ_p, σ_c and σ_r is not of purely mathematical interest. In quantum mechanics, a solution of Schrödinger's time-independent equation $H\psi = E\psi$ gives energy-eigenvalues with eigenfunctions that are "localized" (since $\psi \in L^2(\mathbb{R}^3)$), whereas the continuous spectrum corresponds to "free" states.

3. Among the operators in Sect. 14.2, one can find examples without point, continuous or residual spectra (and any combination thereof, except all empty). Note also that the spectra of these examples are misleadingly not hard to compute in contrast to generic operators.

4. There are various definitions of spectra of T that are subsets of $\sigma(T)$. The *singular spectrum* is the set of λ such that $T - \lambda$ is a topological divisor of zero. The *essential spectrum* consists of λ such that $T - \lambda$ is not Fredholm.

5. Recalling $\rho_x(T) := \limsup_n \|T^n x\|^{\frac{1}{n}}$, defined for $T \in B(X)$ and $x \in X$ (Remark 13.32(4)), suppose a closed subset of the spectrum of T is isolated from the rest of the spectrum by a disk, $\sigma_1 \subset B_r(a)$. If $\rho_x(T - a) < r$ then $x \in X_1$ since

$$P_1 x = \frac{1}{2\pi i} \oint_{\sigma_1} (z - T)^{-1} x \, \mathrm{d}z = \sum_n a_n (T - a)^n x = x.$$

Chapter 15
C^*-Algebras

$B(H)$ is a special Banach algebra when H is a Hilbert space because there is an *adjoint* operation that pairs up operators together. Its properties can be generalized to Banach algebras as follows.

Definition 15.1

A C^* -**algebra** is a (unital) Banach algebra with an **involution** map $* : \mathcal{X} \to \mathcal{X}$ having the properties:

$$T^{**} = T, \quad (T + S)^* = T^* + S^*, \quad (\lambda T)^* = \bar{\lambda}\, T^*,$$

$$(ST)^* = T^* S^*, \quad \|T^* T\| = \|T\|^2.$$

A $*$-*morphism* is defined as a Banach algebra morphism Φ which also preserves the involution $\Phi(T^*) = (\Phi T)^*$.

Easy Consequences
1. $0^* = 0$, $1^* = 1$, $z^* = \bar{z}$ (by expanding $(0+1)^*$, $(1*1)^*$, and $(z1)^*$).
2. $\|T\| = \|T^*\|$ (since $\|T\|^2 = \|T^* T\| \leqslant \|T^*\|\|T\|$, and so $\|T\| \leqslant \|T^*\| \leqslant \|T^{**}\|$); the involution map is thus continuous and bijective. But it is neither linear $((iT)^* = -iT^*)$, nor differentiable (since $(T + H)^* = T^* + H^*$).
3. $\|TT^*\| = \|T\|^2$.
4. $(T^*)^{-1} = (T^{-1})^*$ when T is invertible.
5. $\rho(T^*) = \rho(T)$, $\sigma(T^*) = \sigma(T)^*$ (since $(T^* - \bar{\lambda})^{-1} = (T - \lambda)^{-1*}$).

One might expect that $\|T^*\| = \|T\|$ be taken as an axiom, and indeed Banach algebras with involutions satisfying this weaker axiom are studied and called *Banach $*$-algebras*. C^*-algebras resemble $\mathbb{C}$ and $B(H)$ more closely: the chosen

© The Author(s), under exclusive license to Springer Nature Switzerland AG 2024
J. Muscat, *Functional Analysis*, https://doi.org/10.1007/978-3-031-27537-1_15

axiom, which is the analogue of the familiar $\bar{z}z = |z|^2$, is much stronger and can only be satisfied by a unique norm, if at all (Example 15.10(5)).

Examples 15.2

1. The simplest example is $\mathbb{C}$ with conjugacy. $\mathbb{C}^n$ has an involution

$$(z_1, \ldots, z_n)^* := (\bar{z}_1, \ldots, \bar{z}_n).$$

This example extends to ℓ^∞.

2. $C[0, 1]$ with conjugate functions $\bar{f} : t \mapsto \overline{f(t)}$.

3. $B(H)$ with the adjoint operator, where H is a Hilbert space (Proposition 10.19). We will see later (Theorem 15.53) that every C^*-algebra can be embedded into $B(H)$ for some Hilbert space H.

4. $B(H)$ contains the closed $*$-*subalgebra*

$$\mathbb{C} \oplus \mathcal{K} := \{ a + T : a \in \mathbb{C},\ T \in B(H) \text{ compact} \}$$

5. If $\mathcal{X}$ and $\mathcal{Y}$ are C^*-algebras then so is $\mathcal{X} \times \mathcal{Y}$ with $(S, T)^* := (S^*, T^*)$ (Example 13.3(7)).

6. $\diamond$ $\ell^1(\mathbb{Z})$ has an involution $(a_n)_{n \in \mathbb{N}}^* := (\bar{a}_{-n})_{n \in \mathbb{N}}$, that satisfies $\|x^*\| = \|x\|$ but not $\|x^* * x\| = \|x\|^2$. However, it can be given a new norm, $\|x\| := \|L_x\|$ where $L_x y := x * y$ for $y \in \ell^2$, and $L : x \mapsto L_x$ embeds $\ell^1(\mathbb{Z})$ as a commutative C^*-subalgebra of $B(\ell^2)$. Similarly for $L^1(\mathbb{R})$.

7. $\diamond$ The *group algebra* $\mathbb{C}^G$ has an involution making it a $*$-algebra, but not a C^*-algebra,

$$x^* = \left(\sum_{g \in G} a_g e_g \right)^* := \sum_{g \in G} \bar{a}_g e_{g^{-1}}.$$

Exercises 15.3

1. *Polarization identity*: If ω is a primitive root of unity, $\omega^n = 1$, then

$$T^* S = \frac{1}{n} \sum_{i=1}^n \omega^i (S + \omega^i T)^* (S + \omega^i T),$$

$$S^* S + T^* T = \frac{1}{n} \sum_{i=1}^n (S + \omega^i T)^* (S + \omega^i T).$$

2. For any *real* polynomial (or power series) in T, $p(T)^* = p(T^*)$.

3. If T is a nilpotent, a quasinilpotent, a divisor of zero, or a topological divisor of zero, then so is T^*, respectively. If T^*T is a nilpotent, then so is TT^*; but find an example in $B(\ell^2)$ where T^*T is invertible yet TT^* isn't.

4. If T^*T and TT^* are both invertible then so is T,

$$T^{-1} = (T^*T)^{-1}T^* = T^*(TT^*)^{-1}.$$

5. T^* has the same condition number as T; that of T^*T is squared (Sect. 8.3).
6. The inner automorphism $T \mapsto S^{-1}TS$ is a $*$-automorphism exactly when SS^* belongs to the center $\mathcal{X}'$ (in which case $S^*S = SS^*$).
7. $\star$ A $*$-isomorphism $B(H_1) \to B(H_2)$ is of the type $T \mapsto UTU^{-1}$ where $U : H_1 \to H_2$ is a Hilbert-space isomorphism.
8. A $*$-*ideal* is an ideal that is closed under involution. Examples include the kernel of any $*$-morphism, rad $\mathcal{X}$, and the set of compact operators of $B(H)$.
9. If $\mathcal{A} \subseteq \mathcal{X}$ is closed under involution ($\mathcal{A}^* = \mathcal{A}$), then so is its commutant $\mathcal{A}'$ (which is thus a C^*-subalgebra) (Exercise 13.10(14)).
10. $\star$ Suppose $\mathcal{X}$ has no unity but otherwise satisfies all the axioms of a C^*-algebra. Show that the embedding $L : \mathcal{X} \to B(\mathcal{X})$ (Theorem 13.8) is still isometric, and that $L\mathcal{X} \oplus [\![I]\!]$ with the adjoint operation $(L_a + \lambda)^* := L_{a^*} + \bar{\lambda}$ is a unital C^*-algebra.

15.1 Normal Elements

It is a well-known fact in Linear Algebra that real symmetric matrices are diagonalizable with real eigenvalues and orthogonal eigenvectors. This makes them particularly useful and simple to work with, e.g., if $T = PDP^{-1}$ then $f(T) = Pf(D)P^{-1}$ can easily be calculated when D is diagonal. However, these matrices do not exhaust the set of diagonalizable matrices via orthogonal eigenvectors: for example, matrices such as $\left(\begin{smallmatrix} 1 & -1 \\ 1 & 1 \end{smallmatrix}\right)$ may be diagonalizable with complex eigenvalues. As we shall see later, diagonalization is closely related to the commutativity of T with T^*.

Definition 15.4

An element T is called **normal** when $T^*T = TT^*$, **unitary** when $T^* = T^{-1}$, and **self-adjoint** when $T^* = T$.

Examples 15.5

1. It is clear that self-adjoint and unitary elements are normal.
2. Any $z \in \mathbb{C}$ is normal; it is self-adjoint only when $z \in \mathbb{R}$; it is unitary only when $|z| = 1$.
3. A diagonal matrix is normal; it is self-adjoint when it is real, and unitary when each diagonal element is of unit length $|a_{ii}| = 1$.

More generally, diagonalizable matrices, of the type $T = UDU^*$ where U is unitary and D is diagonal, are normal: $T^*T = UD^*U^*UDU^* = UD^*DU^* = UDD^*U^* = TT^*$.

4. The operator $Tf(s) := \int_0^1 k(s,t)f(t)\,dt$ on $L^2[0,1]$ is normal when (Example 9.28(2c))

$$\int_0^1 \overline{k(u,s)}k(u,t)\,du = \int_0^1 k(s,u)\overline{k(t,u)}\,du \quad \text{a.e.}(s,t)$$

5. When T is normal, a polynomial in T and T^* looks like

$$p(T,T^*) = \sum_{n=1}^{N}\sum_{m=1}^{M} a_{n,m}T^n T^{*m}.$$

The set of such polynomials $\mathbb{C}[T,T^*]$ is a commutative $*$-subalgebra. The character space of its closure $\overline{\mathbb{C}[T,T^*]}$ is denoted by Δ_T.

6. A unitary matrix is a square matrix whose column vectors are orthonormal. A self-adjoint matrix is a square matrix $[a_{i,j}]$ such that $a_{j,i} = \overline{a}_{i,j}$, e.g., $\left(\begin{smallmatrix} 1 & i \\ -i & 0 \end{smallmatrix}\right)$. *Proof*: If $\boldsymbol{u}_i$ denotes the ith column of U, then $U^*U = I$ implies

$$\langle \boldsymbol{u}_i, \boldsymbol{u}_j \rangle = \boldsymbol{u}_i^* \boldsymbol{u}_j = \delta_{ij}.$$

7. The unitary operators of $B(H)$ are the Hilbert-space automorphisms of H (Proposition 10.22).

8. ▶ If T is normal, then so are T^*, $T + z$, zT, T^n, and T^{-1} when it exists. But the addition and product of normal elements need not be normal, e.g., $\left(\begin{smallmatrix} 1 & 0 \\ 0 & 2 \end{smallmatrix}\right)$ and $\left(\begin{smallmatrix} i & i \\ i & i \end{smallmatrix}\right)$.
   Proof for T^{-1}: Taking the inverse of $TT^* = T^*T$ together with $(T^{-1})^* = (T^*)^{-1}$ gives the normality of T^{-1}.

9. ▶ If T_n are normal and $T_n \to T$, then T is also normal, i.e., the set of normal elements is closed (as are the sets of self-adjoint and unitary elements).
   Proof: The limit as $n \to \infty$ of $T_n^* T_n = T_n T_n^*$ is $T^*T = TT^*$ since the adjoint is continuous. Similarly take the limit of $T_n^* = T_n$ or $T_n^* = T_n^{-1}$ to prove the other statements.

10. ▶ If S, T are self-adjoint, then so are $S + T$, λT ($\lambda \in \mathbb{R}$), $p(T)$ for any real polynomial p, and T^{-1} if it exists. But ST is self-adjoint iff $ST = TS$.

11. ▶ If T is self-adjoint, then e^{iT} is unitary; in fact, letting $U^t := e^{itT}$, $t \in \mathbb{R}$, gives a one-parameter group of unitary elements (Exercise 13.25(9) for definition).

The analogy of self-adjoint elements with real numbers and unitary elements with unit complex numbers raises the issue of which propositions about complex numbers generalize to C^*-algebras.

Proposition 15.6

Every element T can be written uniquely as $A + iB$ with A and B self-adjoint, called the real and imaginary parts of T, respectively. Then T is normal iff $AB = BA$.

The real and imaginary parts of T are denoted $\operatorname{Re} T$ and $\operatorname{Im} T$.

Proof Simply check that $A := (T + T^*)/2$ and $B := (T - T^*)/2i$ are self-adjoint. The sum $A + iB$ is obviously T. Uniqueness follows from the fact that if $A + iB = 0$ for A, B self-adjoint then $A = 0 = B$ since

$$0 = (A + iB) + (A + iB)^* = A + iB + A - iB = 2A.$$

$$\therefore \quad T^*T = (A - iB)(A + iB) = (A^2 + B^2) + i[A, B],$$

$$TT^* = (A + iB)(A - iB) = (A^2 + B^2) - i[A, B],$$

so $T^*T = TT^*$ precisely when $[A, B] = 0$.   □

Proposition 15.7

The set of unitary elements $\mathcal{U}(\mathcal{X})$ is a closed subgroup of $\mathcal{G}(\mathcal{X})$,

$$U, V \text{ unitary} \ \Rightarrow \ UV, \ U^{-1} \text{ unitary.}$$

Unitary elements have unit norm, $\|U\| = 1$.

Proof If U_n are unitary and $U_n \to T$, then by continuity of the involution, $U_n^* \to T^*$. Also, the equations $U_n^*U_n = 1 = U_nU_n^*$ become $T^*T = 1 = TT^*$ in the limit, that is, $T^{-1} = T^*$.

For any $U, V \in \mathcal{U}(\mathcal{X})$, UV and $U^*(= U^{-1})$ are also unitary, and U is of unit norm:

$$(UV)^* = V^*U^* = V^{-1}U^{-1} = (UV)^{-1}$$

$$U^{**} = U = (U^{-1})^{-1} = (U^*)^{-1},$$

$$\|U\|^2 = \|U^*U\| = \|1\| = 1.$$

                                                                                      □

The next theorem starts to unravel the close connection between normal elements and their spectra.

Proposition 15.8

> **For T normal, $\rho(T) = \|T\|$, and $\mathcal{S}(T)$ is the closed convex hull of $\sigma(T)$.**

Proof (i) For any normal element T, $\|T^2\| = \|T\|^2$ since

$$\|T\|^4 = \|T^*T\|^2 = \|(T^*T)^*(T^*T)\| = \|(T^2)^*T^2\| = \|T^2\|^2.$$

But T^2 itself is normal, so the doubling game can be repeated to get, by induction, $\|T^{2^k}\| = \|T\|^{2^k}$ and

$$\rho(T) = \lim_{n \to \infty} \|T^n\|^{1/n} = \lim_{k \to \infty} \|T^{2^k}\|^{2^{-k}} = \|T\|.$$

(ii) As $\mathcal{S}(T)$ is a closed convex set that contains $\sigma(T)$ (Proposition 14.35), it must also contain the convex hull of the latter. Notice that, by (i), $\sigma(T)$ reaches out to the boundary of $\mathcal{S}(T)$.

Conversely, suppose λ is not in the closed convex hull of $\sigma(T)$. There must be a straight line through λ not intersecting $\sigma(T)$ (why? Hint: join λ to its closest point in $\sigma(T)$). So the spectrum can be enclosed by a ball $\overline{B_r(z)}$ that does not meet the line (Exercise 6.22(8)).

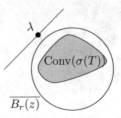

For any $\phi \in \mathcal{S}$,

$$|\phi T - z| = |\phi(T - z)| \leqslant \|T - z\| = \rho(T - z) \leqslant r < |\lambda - z|$$

so $\lambda \neq \phi T$. It follows that $\mathcal{S}(T)$ has the same points as the closed convex hull of $\sigma(T)$.   □

Proposition 15.9 (Fuglede's Theorem)

> **If T is normal and $ST = TS$ then $ST^* = T^*S$.**

Proof From $f(T)S = Sf(T)$ (Example 14.24(1b)), we have $e^{-\bar{z}T}Se^{\bar{z}T} = S$. Writing $\bar{z}T = A + iB$ and noting that $\bar{z}T$ is normal, so $AB = BA$, we find

$$F(z) := e^{-zT^*}Se^{zT^*} = e^{-A+iB}Se^{A-iB}$$

$$= e^{2iB}e^{-\bar{z}T}Se^{\bar{z}T}e^{-2iB}$$

$$= e^{2iB}Se^{-2iB}$$

$$\therefore \|F(z)\| \leqslant \|S\| \qquad\qquad \text{by Example 15.5(11).}$$

As F is a bounded analytic function of z, by Liouville's theorem it is constant, $F(z) = F(0) = S$, i.e., $e^{zT^*}S = Se^{zT^*}$. Comparing the second term of their power series gives $T^*S = ST^*$. $\qquad\square$

Examples 15.10

1. If $T = A + iB$, where A, B are self-adjoint, then $T^* = A - iB$, so

    (a) T is unitary if, and only if, $AB = BA$ and $A^2 + B^2 = 1$;
    (b) T is self-adjoint if, and only if, $B = 0$.

2. $\mathcal{X}$ is commutative if, and only if, every element is normal.
    Proof: If every element $A + iB$ is normal, then $AB = BA$, i.e., any two self-adjoint elements commute. But then $TS = (A + iB)(C + iD) = ST$. The converse is obvious.

3. (a) For T normal, $\|T^n\| = \|T\|^n$, since $\|T\| = \rho(T) \leqslant \|T^n\|^{1/n} \leqslant \|T\|$.
    (b) For any T, $\|T\|^{2n} = \|(T^*T)^n\|$ and $\|T\| = \sqrt{\rho(T^*T)}$.

4. ▶ 0 is the only normal quasinilpotent and the only radical element, that is, every C^*-algebra is semi-primitive. More generally, if T is normal with $\sigma(T) = \{z\}$, then $T = z$.
    Proof: If Q is a normal quasinilpotent, then $\|Q\| = \rho(Q) = 0$, so $Q = 0$. If P is a radical element, then $\|P\|^2 = \|P^*P\| = \rho(P^*P) = 0$.

5. Every C^*-algebra has a unique norm satisfying $\|T^*T\| = \|T\|^2$.
    Proof: Suppose there is a second C^*-norm. Then $\|T\| = \rho(T^*T)^{\frac{1}{2}} = \|T\|$.

Exercises 15.11

1. What are the normal, self-adjoint and unitary elements of ℓ^∞ and $C[0, 1]$?
2. Diagonal 'matrices' in ℓ^2 are multiplier operators, $(a_n)_{n\in\mathbb{N}} \mapsto (b_n a_n)_{n\in\mathbb{N}}$. Show they are normal, self-adjoint when $b_n \in \mathbb{R}$, and unitary when $|b_n| = 1$, for all n.
    Generalize for multiplier operators on $L^2(\mathbb{R})$, $Tf := gf$, $(g \in C_b(\mathbb{R}))$.
3. Triangular matrices, such as $\left(\begin{smallmatrix} 1 & 1 \\ 0 & 2 \end{smallmatrix}\right)$, are not normal (unless diagonal). A real diagonalizable matrix, such as $\left(\begin{smallmatrix} 0 & 1 \\ -1 & 0 \end{smallmatrix}\right)$, need not be self-adjoint.

4. For *any* T, $\alpha T + \beta T^*$ is normal when $|\alpha| = |\beta|$.

5. A $*$-morphism preserves normal, self-adjoint, and unitary elements.

6. If P_i are normal idempotents with $P_i P_j = \delta_{ij} P_i$ as well as $P_1 + \cdots + P_n = 1$, then $z_1 P_1 + \cdots + z_n P_n$ is normal (unitary when $|z_i| = 1$) and for any polynomial p,

$$p(z_1 P_1 + \cdots + z_n P_n) = p(z_1) P_1 + \cdots + p(z_n) P_n.$$

7. If S and T are commuting normal elements, then ST is also normal.

8. The shift-operators on $\ell^2(\mathbb{Z})$ are unitary, with $\sigma(R) = \sigma(L) = \mathbb{S}^1$ (but on ℓ^2, they are not even normal).

9. Translations $T_a f(t) := f(t - a)$ and stretches $S_a f(t) := a^{\frac{1}{2}} f(at)$ $(a > 0)$, acting on $L^2(\mathbb{R})$, are unitary.

10. If U is unitary then for any T, $\|UT\| = \|T\| = \|TU\|$.

11. If $U \in \mathcal{X}$ is unitary, then $T \mapsto U^*TU$ is an inner $*$-automorphism of $\mathcal{X}$.

12. If T is an invertible normal element, then T^*T^{-1} is unitary.
    For example, the Cayley transformation $U := (i - T^*)(i + T)^{-1}$ maps T to a unitary element if $i + T$ is invertible. Compare with the transformation $z \mapsto (i - \bar{z})/(i + z)$, which takes $\mathbb{R}$ to the unit circle $(0 \mapsto 1, 1 \mapsto i, \infty \mapsto -1)$. (Note that not every unitary operator U is of this form, only those such that $-1 \notin \sigma(U)$.)

13. $\mathcal{U}(\mathcal{X})$ need not be a normal subgroup of $\mathcal{G}(\mathcal{X})$; when does $T^{-1}\mathcal{U}T \subseteq \mathcal{U}$ hold?

14. The operator $Tf(s) := \int k(s, t) f(t) \, dt$ on $L^2(\mathbb{R})$ $(k \in L^2(\mathbb{R}^2))$ is self-adjoint when $k(s, t) = \overline{k(t, s)}$ a.e. (Hint: Examples 10.23(3), 9.28(2)).

15. For any $T \in \mathcal{X}$, the elements $T + T^*$, T^*T and TT^* are self-adjoint.

16. The real and imaginary parts of T satisfy $\|\operatorname{Re} T\| \leqslant \|T\|$, $\|\operatorname{Im} T\| \leqslant \|T\|$.

17. Find the real and imaginary parts of ST when S and T are self-adjoint.

18. For S, T normal, $\rho(S + T) \leqslant \rho(S) + \rho(T)$, and $\rho(ST) \leqslant \rho(S)\rho(T)$.

19. When T is normal, then $\|T\|e^{i\theta}$ is a spectral value for some θ.

20. Let $Q \neq 0$ be a quasinilpotent, then $1 + Q$ is not normal. More generally, if T is normal and $TQ = QT$, then $T + Q$ is not normal.

21. If $A^*B = 0 = AB^*$, then $\|A + B\| = \max(\|A\|, \|B\|)$.
    (Hint: Show, by induction, $\|A + B\|^{2n} = \|(A^*A)^n + (B^*B)^n\|$.)

22. If T^*T is an idempotent then so is TT^*.

23. A commutative C^*-algebra is isometrically embedded in some $C(K)$ (Exercise 14.41(19)).

24. Let $\Phi : \mathcal{X} \to \mathcal{Y}$ be a $*$-morphism between C^*-algebras with $\mathcal{X}$ commutative. Then $\Phi(T)$ is normal in $\mathcal{Y}$ for any $T \in \mathcal{X}$, and Φ is continuous with $\|\Phi\| \leqslant 1$ (Hint: $\sigma(\Phi(T)) \subseteq \sigma(T)$).

15.2 Normal Operators in $B(H)$

Let us see what special properties normal elements have for the most important C^*-algebra, the space of operators $B(H)$ when H is a Hilbert space.

Proposition 15.12

> **For a normal operator $T \in B(H)$,**
>
> (i) $\|T^*x\| = \|Tx\|$,
> (ii) $\ker T^2 = \ker T = \ker T^*$,
> (iii) **im T is dense in X** $\Leftrightarrow$ **T is 1–1,**
> (iv) **T is invertible in $B(H)$** $\Leftrightarrow$ $\exists c > 0, \forall x \in H, \quad c\|x\| \leqslant \|Tx\|$.

Proof (i) follows from

$$\|T^*x\|^2 = \langle T^*x, T^*x \rangle = \langle x, TT^*x \rangle = \langle x, T^*Tx \rangle = \langle Tx, Tx \rangle = \|Tx\|^2.$$

(ii) $\ker T = \ker T^*$ is due to $T^*x = 0 \Leftrightarrow \|T^*x\| = \|Tx\| = 0 \Leftrightarrow Tx = 0$, using (i). $\ker T^2 = \ker T$, i.e., $T^2x = 0 \Leftrightarrow Tx = 0$ follows from

$$\|Tx\|^2 = \langle x, T^*Tx \rangle \leqslant \|x\|\|T^*Tx\| = \|x\|\|T^2x\|.$$

(iii) Recall that $(\text{im } T)^\perp = \ker T^*$ (Proposition 10.20). Hence, by (ii), T is 1–1 if, and only if $(\text{im } T)^\perp = \ker T = 0$ iff $\overline{\text{im } T} = H$.

(iv) If T has a continuous inverse, then $\|x\| = \|T^{-1}Tx\| \leqslant \|T^{-1}\|\|Tx\|$. Conversely, if the given inequality is true for all $x \in H$, then T is 1–1 and the image of T is closed (Examples 8.16(3)). By (iii), $\text{im } T = H$ and T is bijective. Its inverse is continuous:

$$\forall x \in H, \quad c\|T^{-1}x\| \leqslant \|TT^{-1}x\| = \|x\|.$$

$\square$

Proposition 15.13

> **For a normal operator $T \in B(H)$,**
>
> (i) $Tv = \lambda v \Leftrightarrow T^*v = \bar{\lambda}v$, **and eigenvectors of distinct eigenvalues of T are orthogonal,**
> (ii) $\sigma(T)$ **contains no residual spectrum,** $\sigma_r(T) = \varnothing$,
> (iii) **isolated points of $\sigma(T)$ are eigenvalues.**

Proof (i) is a direct application of $\ker(T - \lambda) = \ker(T^* - \bar{\lambda})$, as $T - \lambda$ is normal. Note that the eigenvectors of T and T^* are identical. For eigenvalues λ and μ with corresponding eigenvectors x and y, we have

$$\lambda \langle y, x \rangle = \langle y, Tx \rangle = \langle T^* y, x \rangle = \langle \bar{\mu} y, x \rangle = \mu \langle y, x \rangle,$$

implying either $\lambda = \mu$ or $\langle y, x \rangle = 0$.

(ii) Let $\lambda \in \sigma(T)$; either $T - \lambda$ is not 1–1, in which case λ is an eigenvalue (point spectrum); or it is 1–1, in which case its image is dense in H by the previous proposition, and λ forms part of the continuous spectrum.

(iii) If $\{\lambda\}$ is an isolated point of $\sigma(T)$, form the projection

$$P := \frac{1}{2\pi i} \oint_{\{\lambda\}} (z - T)^{-1} \, dz$$

onto a space $X_\lambda \neq 0$ (Example 14.29(1)). Then $\sigma(T|_{X_\lambda}) = \{\lambda\}$, and since $T|_{X_\lambda}$ is normal as well, $\|T|_{X_\lambda} - \lambda\| = \rho(T|_{X_\lambda} - \lambda) = 0$, i.e., $Tx = \lambda x$ for any $x \in X_\lambda$. □

Examples 15.14

1. ▶ A projection $P \in B(H)$ is normal $\Leftrightarrow$ self-adjoint $\Leftrightarrow$ orthogonal $\Leftrightarrow$ $\|P\| = 0$ or 1.
   Proof: If P is orthogonal, then $(x - Px) \perp Px$ (Theorem 10.12), so

   $$\langle x, Px \rangle = \langle (I - P)x + Px, Px \rangle = \|Px\|^2 \in \mathbb{R}$$

   hence $\langle x, Px \rangle = \langle Px, x \rangle = \langle x, P^*x \rangle$ for all $x \in H$, and $P = P^*$ (Example 10.7(3)).
   If $\|P\| = 1$, let $x \in (\ker P)^\perp$, so that $x \perp (x - Px)$. Then $\|Px\|^2 = \|x\|^2 + \|Px - x\|^2$, yet $\|Px\| \leqslant \|x\|$, so $x = Px \in \operatorname{im} P$ and $\ker P \perp \operatorname{im} P$. The other implications should be obvious.
2. All spectral values of a normal operator are approximate eigenvalues (either eigenvalues or part of the continuous spectrum) and there are no proper generalized eigenvectors (Sect. 14.3). Note that a normal operator need not have any eigenvalues, e.g., $Tf(t) := tf(t)$ on $L^2[0, 1]$.

Exercises 15.15

1. ▶ Conversely to the proposition, an operator which satisfies $\|T^* x\| = \|Tx\|$, for all x, is normal.
2. When T is a normal operator, $\ker T$ and $\overline{\operatorname{im} T}$ are both T- and T^*-invariant.
3. Suppose $T_n x \to Tx$ for all $x \in H$, where T_n are normal operators in $B(H)$. Then T is normal if, and only if, $\forall x$, $T_n^* x \to T^* x$. (Note: T is an operator by Corollary 11.37.)
4. The eigenvalues of self-adjoint operators are real, and those of unitary operators satisfy $|\lambda| = 1$.

5. A normal operator on a separable Hilbert space can have at most a countable number of distinct eigenvalues.
6. Suppose H has an orthonormal basis of eigenvectors of an operator $T \in B(H)$. Prove that T is normal. (Hint: show $T^* e_n = \bar{\lambda}_n e_n$.)
7. If $Tx = \lambda x$, $T^* y = \mu y$, then $\mu = \bar{\lambda}$ or $\langle y, x \rangle = 0$ (T not necessarily normal).
8. *An Ergodic Theorem*: Consider the Cesáro sum

$$T_n := (I + T + \cdots + T^{n-1})/n.$$

If $\rho(T) < 1$ then $T_n = (I - T^n)(I - T)^{-1}/n \to 0$ as $n \to \infty$. Now let T be a normal operator with $\rho(T) = 1$.

(a) For $Tx = x$ (i.e., $x \in \ker(T - I)$), we get $T_n x = x$;
(b) For $x = y - Ty \in \operatorname{im}(T - I)$ we get $T_n x = (y - T^n y)/n \to 0$;
(c) For any $x \in H$, $T_n x \to x_0 \in \ker(T - I)$, the closest fixed point of T.

If T is not normal then T_n may diverge, e.g., $T = \begin{pmatrix} 1 & a \\ 0 & 1 \end{pmatrix}$ gives $T_n = \begin{pmatrix} 1 & (n-1)a/2 \\ 0 & 1 \end{pmatrix}$. As an application of this theorem in discrete dynamical systems, let $Tf(x) := f \circ g(x)$ where g is a volume-preserving mapping $\mathbb{R}^N \to \mathbb{R}^N$, that is, its Jacobian determinant is everywhere 1; so $T^n f = f \circ g^n$ and $\|T^n\| = 1$; then the average of such 'positions' converges to f_0, where $f_0 \circ g(x) = f_0(x)$. To take a concrete example, let g be a rotation of $\mathbb{R}^2$, then f_0 is rotationally symmetric.

15.3   The Numerical Range

To help us further with analyzing the spectra of normal operators, we turn to the state space of $B(H)$ (Definition 14.34). An example of a state is the map $T \mapsto \langle x, Tx \rangle$, when x is a fixed unit vector; it is linear on T, has norm 1, and maps I to 1. Furthermore, that value of λ which makes λx closest to Tx, i.e., minimizes $\|Tx - \lambda x\|$ can be obtained from Theorem 10.12: it satisfies $(Tx - \lambda x) \perp x$, or equivalently, $\lambda = \langle x, Tx \rangle$. This number is sometimes called the *mean value* of T at x, or the *Rayleigh coefficient*, and denoted by $\langle T \rangle_x$. We are thus led to the following definition:

Definition 15.16

The **numerical range** of an operator $T \in B(H)$ is the set

$$W(T) := \{ \langle x, Tx \rangle : \|x\| = 1 \}.$$

The extent of $W(T)$ is called the *numerical radius*, $\|T\|_N := \sup_{\|x\|=1} |\langle x, Tx \rangle|$.

Examples 15.17

1. $\langle I \rangle_x = 1$, $\langle T + S \rangle_x = \langle T \rangle_x + \langle S \rangle_x$, $\langle \lambda T \rangle_x = \lambda \langle T \rangle_x$, $\langle T^* \rangle_x = \overline{\langle T \rangle}_x$.
   These are easily verified, e.g.,

   $$\langle x, T^* x \rangle = \langle Tx, x \rangle = \overline{\langle x, Tx \rangle}$$

2. ▶ For operators on a complex Hilbert space,

   (a) $W(I) = \{1\}$, indeed for $z \in \mathbb{C}$, $W(T) = \{z\} \Leftrightarrow T = z$,
   (b) $W(T + z) = W(T) + z$ (translations), and $W(\lambda T) = \lambda W(T)$,
   (c) $W(S + T) \subseteq W(S) + W(T)$,
   (d) $W(T^*) = W(T)^*$,
   (e) $W(U^{-1} T U) = W(T)$ when U is unitary.

3. The above properties show that $\| \cdot \|_N$ is a norm; it is equivalent to the operator norm:

   $$\tfrac{1}{2} \| T \| \leqslant \| T \|_N \leqslant \| T \|$$

   Proof: For unit x, $|\langle x, Tx \rangle| \leqslant \| Tx \| \leqslant \| T \|$ by the Cauchy-Schwarz inequality, so $\| T \|_N \leqslant \| T \|$. Conversely, note that in general, $|\langle v, Tv \rangle| \leqslant \| T \|_N \| v \|^2$, so for any unit vectors x, y, and using the parallelogram law,

   $$|\langle x, Ty \rangle| = \tfrac{1}{4} \big| \langle x + y, T(x + y) \rangle - \langle x - y, T(x - y) \rangle$$
   $$- i \langle x + iy, T(x + iy) \rangle + i \langle x - iy, T(x - iy) \rangle \big|$$
   $$\leqslant \tfrac{1}{4} \| T \|_N (\| x + y \|^2 + \| x - y \|^2 + \| x + iy \|^2 + \| x - iy \|^2)$$
   $$= \| T \|_N (\| x \|^2 + \| y \|^2) = 2 \| T \|_N$$

   So maximizing over x, y, $\| T \| = \sup_{\| x \| = 1 = \| y \|} |\langle x, Ty \rangle| \leqslant 2 \| T \|_N$.

4. (a) $W(T)$ includes the point and residual spectra of T.
   Proof: If $Tx = \lambda x$ for x a unit vector, then $\langle T \rangle_x = \langle x, Tx \rangle = \lambda$.
   $\sigma_r(T) \subseteq \sigma_p(T^*)^* \subseteq W(T^*)^* = W(T)$; more concretely, for $\lambda \in \sigma_r(T)$, there is a unit vector $x \in \operatorname{im}(T - \lambda)^\perp$, so $\langle x, (T - \lambda)x \rangle = 0$, i.e., $\lambda = \langle x, Tx \rangle$.

   (b) If $\lambda \in W(T)$ has magnitude $\| T \|$, then it is an eigenvalue of T.
   Proof: Given $\lambda = \langle x, Tx \rangle$ with x unit and $|\lambda| = \| T \|$, then

   $$\| Tx - \lambda x \|^2 = \| Tx \|^2 - 2 \operatorname{Re} \bar{\lambda} \langle x, Tx \rangle + |\lambda|^2$$
   $$= \| Tx \|^2 - |\lambda|^2$$
   $$\leqslant \| T \|^2 - |\lambda|^2 = 0$$

   so $Tx = \lambda x$.

(c) If $\lambda \in \overline{W(T)}$ has magnitude $\|T\|$, then it is an approximate eigenvalue.
Proof: Given $\langle x_n, T x_n \rangle \to \lambda$, $|\lambda| = \|T\|$, then the same argument as above shows that $(T - \lambda)x_n \to 0$.

5. In finite dimensions, $W(T)$ is closed in $\mathbb{C}$.
   Proof: If $\lambda_n \to \lambda$ with $\lambda_n = \langle x_n, T x_n \rangle$, x_n unit, then there is a convergent subsequence $x_n \to x$, also unit, so taking limits, $\lambda = \langle x, T x \rangle$.

6. Although the quadratic form $x \mapsto \langle x, T x \rangle$ is unique to T, i.e., $\langle x, T x \rangle = \langle x, S x \rangle$ for all x if, and only if, $T = S$ (Example 10.7(3)), the numerical range $W(T)$ does not identify T in general, e.g., $W(P) = [0, 1]$ for any non-zero orthogonal projection.

7. For a fixed unit $x \in H$, one can define two semi-inner-products on $B(H)$,

   (a) $\langle S, T \rangle := \langle S x, T x \rangle = \langle S^* T \rangle_x$ (with associated semi-norm $\|T\|_x := \|T x\|$), and

   (b) the *covariance* semi-inner-product

$$\mathrm{Cov}(S, T) := \langle S - \langle S \rangle_x, T - \langle T \rangle_x \rangle = \langle S^* T \rangle_x - \overline{\langle S \rangle}_x \langle T \rangle_x,$$

   with the associated semi-norm called the *standard deviation*

$$\sigma_T^2 := \mathrm{Cov}(T, T) = \|T x\|^2 - |\langle T \rangle_x|^2.$$

The *uncertainty principle* states that

$$|\mathrm{Cov}(S, T)| \leqslant \sigma_S \sigma_T$$

(essentially the Cauchy-Schwarz inequality—Exercise 10.10(16)).
The *correlation* is the normalized inner product $\mathrm{Cov}(S, T)/\sigma_S \sigma_T$; T and S are called *independent* when they are orthogonal, $\mathrm{Cov}(S, T) = 0$, which is equivalent to $\langle S, T \rangle = \overline{\langle S \rangle}_x \langle T \rangle_x$.
These definitions are usually applied to $L^2(A)$, where x corresponds to a function $p \in L^2(A)$, with $|p(s)|^2$ interpreted as a probability distribution, and the operators are multiplication by functions $T p := f p$, that is,

the mean $\langle f \rangle_p = \int_A f(s) |p(s)|^2 \, ds,$ the rms $\|f\|_p = \sqrt{\int_A |f|^2 |p|^2},$

the covariance $\mathrm{Cov}(f, g) = \int_A (f - \langle f \rangle)(g - \langle g \rangle)|p|^2.$

We can now elucidate the connection between the spectrum of an operator, its numerical range, and its state space values, hinted at in the examples above.

Proposition 15.18 (Hausdorff-Toeplitz)

> $\overline{W(T)}$ **is a convex compact subset of** $\mathbb{C}$, **such that**
>
> $$\sigma(T) \subseteq \overline{W(T)} \subseteq \mathcal{S}(T).$$
>
> **When** T **is normal,** $\overline{W(T)} = \mathcal{S}(T) = \overline{\text{Conv}}(\sigma(T)).$

Proof The inclusion $W(T) \subseteq \mathcal{S}(T)$ is obvious: for any unit vector x, the functional $\phi(T) := \langle x, Tx \rangle$ is linear in T, maps I to 1, and $|\phi(T)| = |\langle x, Tx \rangle| \leqslant \|T\|$, so $\|\phi\| = 1$ and $\phi \in \mathcal{S}$. As $\mathcal{S}(T)$ is compact (Proposition 14.35), so must be its closed subset $\overline{W(T)}$.

The main part of the proof is to show the other inclusion $\sigma(T) \subseteq \overline{W(T)}$: for $\|x\| = 1$, $\lambda \in \mathbb{C}$,

$$\alpha := d(\lambda, W(T)) \leqslant |\langle x, Tx \rangle - \lambda| = |\langle x, (T - \lambda)x \rangle| \leqslant \|(T - \lambda)x\|,$$

so for any $x \in H$,

$$\alpha \|x\| \leqslant \|(T - \lambda)x\|.$$

When $\lambda \notin \overline{W(T)}$, α is strictly positive, and the inequality shows that $T - \lambda$ is 1–1 with a closed image (Example 8.16(3)). Moreover, since $W(T^*) = W(T)^*$ and $d(\bar{\lambda}, W(T)^*) = d(\lambda, W(T))$,

$$\alpha \|x\| \leqslant \|(T^* - \bar{\lambda})x\|.$$

This implies that $(T - \lambda)^*$ is 1–1, hence $T - \lambda$ is surjective (Proposition 10.20). Thus $T - \lambda$ has an inverse, which is continuous (Proposition 8.15), and $\lambda \notin \sigma(T)$.

$W(T)$ *is convex:* Given λ, μ in $W(T)$ ($\lambda \neq \mu$), let x, y be unit vectors such that $\langle x, Tx \rangle = \lambda$, $\langle y, Ty \rangle = \mu$. Any vector $v := \alpha e^{i\phi_1} x + \beta e^{i\phi_2} y$ ($\alpha, \beta, \phi_1, \phi_2 \in \mathbb{R}$) has norm

$$\|v\|^2 = \alpha^2 + 2\alpha\beta \, \text{Re} \, e^{i(\phi_2 - \phi_1)} \langle x, y \rangle + \beta^2 = 1 + \sin 2\theta \, \text{Re}(e^{i\phi} \langle x, y \rangle),$$

for $\alpha = \cos\theta$, $\beta = \sin\theta$, $\phi := \phi_2 - \phi_1$. Then $\langle v, Tv \rangle$ works out to

$$\langle \alpha e^{i\phi_1} x + \beta e^{i\phi_2} y, \alpha e^{i\phi_1} Tx + \beta e^{i\phi_2} Ty \rangle$$

$$= \alpha^2 \lambda + \alpha\beta(e^{i\phi} \langle x, Ty \rangle + e^{-i\phi} \langle y, Tx \rangle) + \beta^2 \mu$$

$$= \lambda \cos^2\theta + \sin 2\theta(w \cos\phi + z \sin\phi) + \mu \sin^2\theta$$

$$= \tfrac{\lambda + \mu}{2} + \tfrac{\lambda - \mu}{2} \cos 2\theta + (w \cos\phi + z \sin\phi) \sin 2\theta$$

where $w := \frac{1}{2}(\langle x, Ty\rangle + \langle y, Tx\rangle)$, $z := \frac{i}{2}(\langle x, Ty\rangle - \langle y, Tx\rangle)$. But $w_\phi := w\cos\phi +$ $z\sin\phi$ traces out an ellipse as ϕ varies. By choosing the correct value of ϕ, w_ϕ can be made to point in any direction in the complex plane, including that of $\lambda - \mu$. With this choice, $\langle v, Tv\rangle/\|v\|^2$ gives a line segment as θ varies, a line that contains λ and μ (at $\theta = 0, \pi/2$). Thus $W(T)$, and its closure $\overline{W(T)}$, are convex sets.

Since $\overline{W(T)}$ is convex, the closed convex hull of $\sigma(T)$ lies in it. Proposition 15.8 shows $\mathcal{S}(T) = \overline{\text{Conv}}(\sigma(T)) \subseteq \overline{W(T)}$ when T is normal. $\qquad\square$

Two follow-up results are given next: one is a sharp refinement due to T. H. Hildebrandt, and another allows us to identify the self-adjoint operators among the normal ones from their spectrum.

Proposition 15.19

The closed convex hull of the spectrum of an operator $T \in B(H)$ is equal to

$$\overline{\text{Conv}}(\sigma(T)) = \bigcap_{S\in\mathcal{G}} \overline{W(S^{-1}TS)}.$$

Proof We need a lemma that will be reproved later for any C^*-algebra. Recall the geometric series $(1 - A)^{-1} = \sum_{n\in\mathbb{N}} A^n$ valid for $\rho(A) < 1$. This result, applied for the operator A^*A, has a variant:

$$R := 1 + A^*A + (A^2)^*(A^2) + (A^3)^*(A^3) + \cdots$$

converges for $\rho(A) < 1$ and is self-adjoint and invertible. Convergence is justified by the root test, $\limsup_n \|(A^n)^*(A^n)\|^{1/n} = \lim_{n\to\infty} \|A^n\|^{2/n} = \rho(A)^2 < 1$. For any unit vector x,

$$\langle x, Rx\rangle = \|x\|^2 + \|Ax\|^2 + \|A^2x\|^2 + \cdots \geqslant 1,$$

so $\sigma(R) \subseteq \overline{W(R)} \subseteq [1, \infty[$. We can deduce (i) R is invertible with $\sigma(1 - R^{-1}) = 1 - \sigma(R)^{-1} \subseteq [0, 1[$; and (ii) the operators $R^{\pm\frac{1}{2}}$ can be defined by the functional calculus, since there is a branch line from the origin that does not meet its spectrum. Note that, since the norm and spectral radius of a self-adjoint agree,

$$1 - R^{-1} = R^{-\frac{1}{2}}(R - 1)R^{-\frac{1}{2}} = R^{-\frac{1}{2}}A^*RAR^{-\frac{1}{2}},$$

$$\therefore \ \|R^{\frac{1}{2}}AR^{-\frac{1}{2}}\|^2 = \|1 - R^{-1}\| = \rho(1 - R^{-1}) < 1$$

Now we can start the main proof. For any invertible operator S,

$$\sigma(T) = \sigma(S^{-1}TS) \subseteq \overline{W(S^{-1}TS)}$$

and since the intersection of closed convex subsets remains so, we can conclude that

$$\overline{\text{Conv}}(\sigma(T)) \subseteq \bigcap_{S \in \mathcal{G}} \overline{W(S^{-1}TS)}.$$

Conversely, suppose $\lambda \notin \overline{\text{Conv}}(\sigma(T))$. As in the proof of Proposition 15.8, there is a disk $B_r(z)$ which covers the latter but separates it from λ. By considering $\frac{T-z}{r}$, the lemma above shows there is an operator $S = R^{-1/2}$ such that $\|S^{-1}(T - z)S\| < r$, i.e., $\overline{W(S^{-1}(T - z)S)} \subseteq B_r(0)$, so $W(S^{-1}TS) \subseteq B_r(z)$. The conclusion is that $\lambda \notin \overline{W(S^{-1}TS)}$ and hence is not in the intersection.

□

Proposition 15.20

A normal operator T is self-adjoint $\Leftrightarrow$ $W(T)$ is real $\Leftrightarrow$ $\sigma(T)$ is real.

Proof When $T \in B(H)$ is self-adjoint, $\langle x, Tx \rangle = \langle Tx, x \rangle = \overline{\langle x, Tx \rangle}$ for all $x \in H$, which implies $W(T) \subseteq \mathbb{R}$. Conversely, if $\langle x, Tx \rangle \in \mathbb{R}$ for all vectors x, then

$$\langle Tx, x \rangle = \langle x, Tx \rangle = \langle T^*x, x \rangle$$

which can only hold when $T^* = T$ (Example 10.7(3)).

The spectrum $\sigma(T)$ is real iff $\overline{W(T)}$ is real since the latter is the convex closure of the former.

□

Exercises 15.21

1. Show that, for the shift operators on ℓ^2, $\overline{W(L)} = \overline{B_1(0)} = \overline{W(R)}$.
2. Let T be a square matrix $\begin{pmatrix} A & B \\ C & D \end{pmatrix}$ with respect to an orthonormal basis, where A, D are square sub-matrices.

   (a) $W(A) \cup W(D) \subseteq W(T)$.
   (b) If $B = C = 0$, then $\overline{W(T)}$ is the closed convex hull of $W(A) \cup W(D)$.

3. Write a program that samples $W(T)$ for 2×2 matrices (by plotting the points x^*Tx for a large number of unit complex 2-vectors x), and test it on random matrices. Verify, and then prove, that $\overline{W(T)}$ for

   (a) $T := \begin{pmatrix} a & 0 \\ 0 & b \end{pmatrix}$ is the line joining a to b;
   (b) $T := \begin{pmatrix} a & 1 \\ 0 & a \end{pmatrix}$ is the closed disk $\overline{B_{\frac{1}{2}}(a)}$ (although its spectrum is $\{a\}$);
   (c) $\star T := \begin{pmatrix} a & b \\ c & d \end{pmatrix}$ is generically an ellipse with its interior.

4. Go through the proof of Hildebrandt's proposition for $T = \begin{pmatrix} 1 & 1 \\ 0 & 1 \end{pmatrix}$; verify that $\frac{T-1}{r} = \begin{pmatrix} 0 & 1/r \\ 0 & 0 \end{pmatrix}$, $S = R^{-\frac{1}{2}} = \begin{pmatrix} 1 & 0 \\ 0 & \alpha \end{pmatrix}$, $S^{-1}TS = \begin{pmatrix} 1 & \alpha \\ 0 & 1 \end{pmatrix}$, $W(S^{-1}TS) = \overline{B_{\alpha/2}(1)}$.

5. Let T be a square matrix with positive coefficients. If $x = (a_1, \ldots, a_n) \in \mathbb{C}^n$ and $x_+ := (|a_1|, \ldots, |a_n|)$, then

$$|\langle x, Tx \rangle| \leqslant \langle x_+, Tx_+ \rangle$$

so that the largest extent of $W(T)$ occurs at a positive real number.

6. An operator is called *real* when it is an operator acting on a real Banach space X, extended to act linearly on the complex space $X + iX$. Show that a real operator on a Hilbert space has a numerical range that is symmetric about the real axis.

7. The classical proofs of some of the statements above do not use the convexity properties of the numerical range. For a self-adjoint operator T,

   (a) $\sigma(T)$ is real. Prove this by letting $\lambda := \alpha + i\beta$ with $\beta \neq 0$, and showing

   $$\|(T - \lambda)x\|^2 = \|(T - \alpha)x\|^2 + \beta^2\|x\|^2 \geqslant |\beta|^2\|x\|^2.$$

   (b) $\overline{W(T)}$ is the smallest interval containing $\sigma(T)$. Show this by taking $\sigma(T) \subseteq [a, b]$, letting $c := (a + b)/2$, and proving that for any unit vector x,

   $$|\langle x, Tx \rangle - c| = |\langle x, (T - c)x \rangle| \leqslant b - c = c - a.$$

8. For any $T \in B(H_1, H_2)$, $\overline{W(T^*T)} = [a, b]$, where $a \geqslant 0$ and $b = \|T\|^2$.

9. If $\lambda \notin \overline{W(T)}$, then $\|(\lambda - T)^{-1}\| \leqslant 1/d(\lambda, W(T))$.

10. A *coercive* operator $T \in B(H)$ satisfies $|\langle x, Tx \rangle| \geqslant c > 0$ for all unit $x \in H$. Show that it has a continuous inverse. An *elliptic* operator is one which satisfies $\langle x, Tx \rangle \geqslant c > 0$, a special case of a coercive self-adjoint operator.

11. Let $\phi : B(H) \to \mathbb{C}$ be defined by $T \mapsto \langle x, Ty \rangle$ for some fixed unit $x, y \in H$; show that $\phi \in S \Leftrightarrow x = y$.

12. (a) $\mathrm{Cov}(I, T) = 0$, $\mathrm{Cov}(S, T + \lambda) = \mathrm{Cov}(S, T)$, $\sigma_{T+\lambda} = \sigma_T$;

    (b) For every λ and unit x, $\sigma_T \leqslant \|(T - \lambda)x\|$, so $\sigma_T \leqslant \frac{1}{2} \mathrm{diam}\, \sigma(T)$ for T normal;

    (c) $\sigma_T = 0 \Leftrightarrow x$ is an eigenvector of T, with eigenvalue $\langle T \rangle_x$.

    (d) If S, T are self-adjoint operators, let $A := \frac{1}{i}[S, T]$ and $h := \frac{1}{2}\langle A \rangle_x = \mathrm{Cov}(S, T)$, then

    $$\sigma_S \sigma_T \geqslant h.$$

15.4   The Spectral Theorem for Compact Normal Operators

As seen before, multiplier operators such as diagonal matrices are normal. In fact, all normal operators are of this type in an appropriate basis; we show this first in the simple case of compact normal operators, in a theorem due to David Hilbert and Erhard Schmidt.

Theorem 15.22 (Spectral Theorem for Compact Normal Operators)

If T is a compact normal operator on a Hilbert space, then

$$Tx = \sum_{n=0}^{\infty} \lambda_n \langle e_n, x \rangle e_n,$$

where e_n are the eigenvectors of T with corresponding non-zero eigenvalues λ_n.

The statement is written supposing an infinite number of eigenvectors; otherwise the sum is finite.

Proof Let T be a compact normal operator. We show that H has an orthonormal basis of eigenvectors.

(a) The fact that T is compact implies that $\sigma(T) \smallsetminus \{0\}$ consists of a countable set of eigenvalues, and each generalized eigenspace $X_\lambda := \ker(T - \lambda)^{k_\lambda}$ is finite-dimensional (Theorem 14.19).
(b) The fact that $T - \lambda$ is normal implies, firstly, that $X_\lambda = \ker(T - \lambda)$ consists of eigenvectors, and secondly, that X_λ are orthogonal to each other (Propositions 15.12, 15.13).

Note that the eigenvalues decrease to 0 (unless there are a finite number of them). This is part of Theorem 14.19, but its proof in the context of a Hilbert space is simpler: As T is compact, for any infinite set of orthonormal eigenvectors e_n, Te_n ($= \lambda_n e_n$) has a Cauchy subsequence, so

$$|\lambda_n|^2 + |\lambda_m|^2 = \|\lambda_n e_n - \lambda_m e_m\|^2 = \|Te_n - Te_m\|^2 \to 0, \text{ as } n, m \to \infty$$

implying both $\lambda_n \to 0$ and that each eigenspace $\ker(T - \lambda)$ is finite-dimensional.

Thus a countable number of orthonormal eigenvectors e_n (a finite number from each X_λ) account for all the non-zero eigenvalues, and form an orthonormal basis for the closed space $M := [\![e_1, e_2, \dots]\!]$ generated by them. $M^\perp$ is T-invariant since $x \in M^\perp$ implies that for all n, $\langle e_n, x \rangle = 0$, and as $T^* e_n = \bar{\lambda}_n e_n$,

$$\langle e_n, Tx \rangle = \langle T^* e_n, x \rangle = \lambda_n \langle e_n, x \rangle = 0.$$

Thus T can be restricted to $M^\perp$, when it remains compact (Exercises 11.17(5), 6.9(5)) and normal, yet without non-zero eigenvalues, because those are all accounted for by the eigenvectors in M. Its spectrum must therefore be 0, implying $T|_{M^\perp} = 0$, i.e., $M^\perp = \ker T$. Unless $M^\perp = 0$, there is an orthonormal basis of eigenvectors e_i for it, and collectively with e_n, form a basis for $H = M \oplus M^\perp$,

$$x = \sum_n \langle e_n, x \rangle e_n + \sum_i \langle e_i, x \rangle e_i.$$

Finally, since T is linear and continuous, and $T e_i = 0$, we find that

$$Tx = T\left(\sum_n \langle e_n, x \rangle e_n \right) = \sum_n \langle e_n, x \rangle T e_n = \sum_n \langle e_n, x \rangle \lambda_n e_n.$$

$\square$

Corollary 15.23

A normal complex matrix is diagonalizable.

The Singular Value Decomposition

There is a remarkable extension of diagonalization applicable to any compact operator between Hilbert spaces, including rectangular matrices. The analogue of the eigenvalues are called singular values, although they are not closely related, for several reasons. The bases for the domain and codomain are different, even if the spaces happen to be the same. Thus the singular values of a square matrices are usually not the same as its eigenvalues, except when the matrix is diagonalizable by an orthonormal basis.

Theorem 15.24 (Singular Value Decomposition)

If $T : X \to Y$ is a compact operator between Hilbert spaces, then there are isometry operators $U : Y \to Y$ and $V : X \to X$ such that $T = U \Sigma V^*$ with Σ diagonal.

The numbers σ_n comprising the diagonal of Σ are called the *singular values* of T; and u_n, v_n which make up U and V are called its *singular vectors* (u_n are also called the *principal components* of T).

Proof T^*T and TT^* are compact self-adjoint operators, on X and Y respectively. They share the same non-zero eigenvalues (Example 14.11(5)), which are positive, since if $T^*Tv = \lambda v$, $\|v\| = 1$, then

$$\lambda = \langle v, T^*Tv \rangle = \|Tv\|^2 > 0.$$

By the spectral theorem there is an orthonormal set of eigenvectors $v_n \in X$ of T^*T with eigenvalues $\lambda_n = \sigma_n^2 > 0$. It turns out that the vectors $Tv_n \in Y$ are also orthogonal,

$$\langle Tv_m, Tv_n \rangle = \langle v_m, T^*Tv_n \rangle = \sigma_n^2 \delta_{nm},$$

so $u_n := Tv_n/\sigma_n$ form an orthonormal set in Y. Note that, by the above,

$$Tv_n = \sigma_n u_n, \qquad T^* u_n = \sigma_n v_n.$$

In fact, v_n form an orthonormal basis for $(\ker T^*T)^{\perp} = (\ker T)^{\perp} = \overline{\operatorname{im} T^*}$, and similarly u_n is an orthonormal basis for $\overline{\operatorname{im} T}$ (Exercise 10.24(12) and Proposition 10.20).

It follows that for any $x \in X$ and $y \in Y$,

$$x = Px + \sum_n \langle v_n, x \rangle v_n, \qquad Tx = \sum_n \sigma_n \langle v_n, x \rangle u_n, \qquad T^*y = \sum_n \sigma_n \langle u_n, y \rangle v_n$$

where $P \in B(X)$ is the orthogonal projection onto $\ker T$. Indeed a stronger statement is true:

$$T = \sum_n \sigma_n u_n v_n^*$$

That is, the convergence is in norm, not just pointwise, the reason being

$$\left\| (T - \sum_{n=1}^N \sigma_n u_n v_n^*) x \right\|^2 = \left\| \sum_{n=N+1}^{\infty} \sigma_n \langle v_n, x \rangle u_n \right\|^2$$

$$= \sum_{n=N+1}^{\infty} \sigma_n^2 |\langle v_n, x \rangle|^2 \leqslant (\max_{n > N} \sigma_n^2) \|x\|^2$$

and $\max_{n>N} \sigma_n \to 0$ as $N \to \infty$ since $\sigma_n \to 0$ as $n \to \infty$.

Let U be that operator representing a change of basis in $\overline{\operatorname{im} T}$ from u_n to some arbitrary basis (leaving the perpendicular space $\ker T^*$ invariant), V a similar change of basis in $\overline{\operatorname{im} T^*}$ from v_n. Then the 'matrix' of T with respect to v_n and u_n is $\Sigma := U^*TV$; as $Tv_n := \sigma_n u_n$ and $Tx := 0$ for $x \in \ker T$, Σ is diagonal. $\square$

Examples 15.25

1. The spectral theorem is often stated as: If a compact normal operator has "matrix" T with respect to a given orthonormal basis $\tilde{e}_n$, then $T = UDU^{-1}$, where D is diagonal and U is the unitary change-of-basis operator that maps $(\tilde{e}_n)$ to (e_n), the orthonormal basis of eigenvectors of T.

2. $\|T\| = \sigma_{max}$, the largest singular value.
   Proof: If $x = \sum_n \alpha_n v_n + Px$ (as in the proof) is a unit vector, then $Tx = \sum_n \alpha_n \sigma_n u_n$, so

$$\|Tx\|^2 = \sum_n |\sigma_n|^2 |\alpha_n|^2 \leqslant \sigma_{max}^2.$$

   Moreover there is an index k such that $Tv_k = \sigma_{max} u_k$, so that $\|Tv_k\| = \sigma_{max}$.

3. The converse of the spectral theorem is true, i.e., defining the operator

$$Tx := \sum_{n=0}^{\infty} \lambda_n \langle e_n, x \rangle e_n$$

   in terms of an orthonormal basis gives a normal operator, assuming λ_n bounded, because $\|Tx\|^2 = \sum_n |\lambda_n|^2 |\langle e_n, x \rangle|^2 = \|T^*x\|^2$. If $\lambda_n \to 0$, then T is compact because it is the limit of finite-rank operators (prove!).

4. Given a compact normal operator in $B(H)$, and any function $f \in C(\sigma(T))$, with $f(0) = 0$, one can define the compact operator $f(T)$ by the formula

$$f(T)x := \sum_{n=0}^{\infty} f(\lambda_n) \langle e_n, x \rangle e_n.$$

   For example,

   (a) $\sqrt{T}$ is compact when T is a self-adjoint compact operator with non-negative eigenvalues,

   (b) for any $\lambda \neq 0$, there is a projection $P_\lambda := f_\lambda(T)$, where f_λ is a continuous function which takes the value 1 around λ and 0 around all other eigenvalues.

5. The projections P_n to the eigenspaces X_{λ_n} of T commute and are orthogonal, so $E_n := P_1 + \cdots + P_n$ is a projection onto $X_{\lambda_1} + \cdots + X_{\lambda_n}$ (Exercise 8.19(12)). The spectral decomposition can be rewritten as $Tx = \sum_n \lambda_n \delta E_n x$, where $\delta E_n := E_n - E_{n-1} = P_n$. This can be seen as a breakup of $T = \frac{1}{2\pi i} \int_{\sigma(T)} z(z - T)^{-1} dz$ into integrals on the disconnected components of the spectrum.

6. If $T \in B(X)$ is compact normal, then the singular values of T are the absolute values of its non-zero eigenvalues.

Proof: Clearly, if $Tx = \lambda x$ then $T^*Tx = |\lambda|^2 x$. Conversely, if $T^*Tx = \mu x$ ($\mu \neq 0$) then

$$0 = (T^*T - \mu)x = \sum_n (|\lambda_n|^2 - \mu)\langle e_n, x\rangle e_n.$$

The e_n are the eigenvectors of T with non-zero eigenvalues λ_n, so $\langle e_n, x\rangle \neq 0$ for some n, and $\mu = |\lambda_n|^2$.

Application: Feature Extraction

According to SVD, any matrix T can be approximated by $\sum_i \sigma_i u_i v_i^*$, which is a useful way of representing the information content of T. Typically, data from variables $x_1, \ldots, x_m$ is organized in the form of an $m \times n$ matrix T with the rows representing the different variables and the columns the normalized instances; the resulting matrix U associated with the largest singular values are linear combinations of the variables x_i that account for the most variability in the data.

To take a visual example, consider the numerical digits as images of 16×16 gray-level pixels; the $m = 256$ variables are the pixel values and the n column vectors represent each 'training set' digit image, of which there would typically be well over a hundred. An SVD results in three matrices: U, Σ, V. The orthogonal matrix U consists of $m \times m$ rows and columns; Σ is an $m \times n$ rectangular matrix with the singular values on the main diagonal, and V is an $n \times n$ orthogonal matrix. In the digits example, the training set t_i and the first few columns u_i are shown:

Note that u_1 is a sort of "average" of all the data vectors, u_2 is the most significant correction, and so on—they are the *features* of the data. One can truncate the U, Σ, and V matrices to sizes of $m \times k, k \times k$, and $k \times n$ respectively.

After the training is over, and the U matrix extracted, it can be used for many tasks. One is to compress the data effectively. Given a new data instance x, one can find its 'coordinates' relative to the basis u_i by taking $\alpha_i := \langle u_i, x\rangle, i = 1, \ldots, k$. The N largest coefficients α_i can be retained, and the vector rebuilt as $\sum_{|\alpha_i|>\epsilon} \alpha_i u_i$. In the table below, one can see the convergence of an image as more terms are added. With as few as a hundred coefficients, instead of 256 pixel values, the image is practically indistinguishable from the original (to the right).

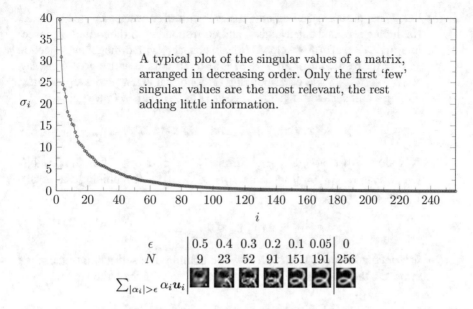

A typical plot of the singular values of a matrix, arranged in decreasing order. Only the first 'few' singular values are the most relevant, the rest adding little information.

ϵ	0.5	0.4	0.3	0.2	0.1	0.05	0
N	9	23	52	91	151	191	256

$\sum_{|\alpha_i| > \epsilon} \alpha_i u_i$

Another use of SVD is to test whether a new data point is similar to the training set. In the following example, a number of training images representing the digit 3 are placed as columns of a matrix, its SVD extracted, and the first few columns of U stored as U_k.

Then given a new test image, its coefficients α_i are computed, and the reconstructed image can be compared with the original as

$$r := x - \sum_{i=1}^{k} \langle u_i, x \rangle u_i = (I - U_k U_k^*)x$$

If the 'residual' $\|r\|$ is below a chosen threshold, then the test image is classified as similar to the test data. For example, the residual of the image with $k = 10$ is 3.45, while the residual of is 1.59; the former might be rejected as a 'three', while the latter is accepted.

Exercises 15.26

1. Find the singular values and vectors of $\begin{pmatrix} 2 & 3 \\ 0 & 2 \end{pmatrix}$ and $\begin{pmatrix} 1 & 1 & 0 \\ 0 & 1 & 1 \end{pmatrix}$.

2. If S and T are commuting self-adjoint compact operators, then they are simultaneously diagonalizable. (Hint: consider $S + iT$.)

3. (a) Let T be an $n \times n$ self-adjoint matrix, with eigenvalues $\lambda_1 \leqslant \cdots \leqslant \lambda_n$ (including repeated eigenvalues), and corresponding orthonormal eigenvectors $v_1, \ldots, v_n$. If M is a closed linear subspace, with orthogonal projection P, then the restriction of PTP to M is also self-adjoint with eigenvalues, say, $\mu_1 \leqslant \cdots \leqslant \mu_m$, and corresponding orthonormal eigenvectors $u_1, \ldots, u_m$. Taking a unit vector $x \in [\![u_1, \ldots, u_i]\!] \cap [\![v_i, \ldots, v_n]\!] \neq 0$, we get

$$\mu_1 \leqslant \langle x, Tx \rangle \leqslant \mu_i \quad \text{and} \quad \lambda_i \leqslant \langle x, Tx \rangle \leqslant \lambda_n.$$

It follows that $\lambda_i \leqslant \mu_i$. Similarly, take $x \in [\![u_i, \ldots, u_m]\!] \cap [\![v_1, \ldots, v_{i+n-m}]\!] \neq 0$ to deduce $\mu_i \leqslant \lambda_{i+n-m}$. Combining the results we get

$$\lambda_i \leqslant \mu_i \leqslant \lambda_{n-m+i}.$$

(b) *Interlacing theorem*: If the kth row and column of a self-adjoint matrix are removed, the new eigenvalues μ_i are interlaced with the old ones λ_i:

$$\lambda_1 \leqslant \mu_1 \leqslant \lambda_2 \leqslant \cdots \leqslant \lambda_{n-1} \leqslant \mu_{n-1} \leqslant \lambda_n.$$

4. *Picard's criterion*: Suppose $T \in B(X, Y)$ is a compact operator on Hilbert spaces X, Y, having singular values σ_n and singular vectors v_n, u_n. In solving $Tx = y$, we find for all n,

$$\langle u_n, y \rangle = \sigma_n \langle v_n, x \rangle.$$

A necessary condition is $\langle u_n, y \rangle / \sigma_n \in \ell^2$ as well as $y \in (\ker T^*)^\perp$. Thus the coefficients of y must 'diminish faster' than σ_n.

5. *Truncated Singular Value Decomposition (TSVD)*: The series solution

$$x = \sum_n \frac{\langle u_n, y \rangle}{\sigma_n} v_n$$

of $T^*Tx = T^*y$ need not converge in general. Even if it does, any small errors in $\langle u_n, y \rangle$ are magnified as $\sigma_n \to 0$. In practice, the series is *truncated* at some stage to avoid this. The cutoff point is best taken when the error in y becomes appreciable compared to σ_n. Use the Tikhonov regularization method (Sect. 10.4) to derive another way of doing this (for the right choice of α),

$$x = \sum_n \frac{\sigma_n^2}{\sigma_n^2 + \alpha} \frac{\langle u_n, y \rangle}{\sigma_n} v_n.$$

But any other weighting $\sum_n w_n \frac{\langle u_n, y \rangle}{\sigma_n} v_n$ where w_n vanishes sufficiently rapidly as $\sigma_n \to 0$, is just as valid.

6. It is instructive to compare with the case of solving the equation $(T - \lambda)x = y$ where T is compact in $B(H)$ and $0 \neq \lambda \in \sigma(T)$ (the case $\lambda \notin \sigma(T)$ is trivial). It has a solution $\Leftrightarrow y \in \ker(T - \lambda)^{\perp}$. That solution of minimum norm is then

$$x = \sum_n \frac{\langle e_n, y \rangle}{\lambda_n - \lambda} e_n - y_0/\lambda,$$

where the sum is taken over $\lambda_n \neq \lambda, 0$, and y_0 is the projection of y to $\ker T$. There is no issue of convergence of the series as $|\lambda_n - \lambda| \geqslant c > 0$.

7. ⋆ If T is a compact normal operator, then the iteration $v_{n+1} := T v_n / \| T v_n \|$ (starting from a generic vector v_0) converges to an eigenvector of the largest eigenvalue, if this is unique and strictly positive. What happens otherwise?

15.5 Ideals of Compact Operators

Another way of looking at the spectral theorem (or even the singular value decomposition), is the following:

Proposition 15.27

A compact operator on a separable Hilbert space can be approximated in norm by a square matrix.

A compact normal operator on a separable complex Hilbert space can be approximated in norm by a diagonalizable matrix.

Proof An operator $T \in B(H)$ takes the matrix form, in terms of a countable orthonormal basis e_i of H,

$$\begin{pmatrix} P_n T P_n & P_n T (I - P_n) \\ \hline (I - P_n) T P_n & (I - P_n) T (I - P_n) \end{pmatrix}$$

where P_n is the self-adjoint/orthogonal projection onto $[\![e_1, \ldots, e_n]\!]$ (Example 15.14(1)). Note that for any vector $x \in H$, $P_n x \to x$ as $n \to \infty$ (Theorem 10.31). The claim is that when T is compact, the finite square matrices $P_n T P_n$ converge to T. This is the same as claiming that the other three sub-matrices vanish as $n \to \infty$.

$(I - P_n)T \to 0$: Suppose, for contradiction, that there is a subsequence of P_n and unit vectors x_n such that $\|(I - P_n)Tx_n\| \geqslant c > 0$. Since T is compact, there is a convergent subsequence $Tx_n \to x$, hence

$$(I - P_n)Tx_n = (I - P_n)x + (I - P_n)(Tx_n - x) \to 0$$

leads to an impossibility.

$(I - P_n)TP_n \to 0$ and $(I - P_n)T(I - P_n) \to 0$ now follow from $\|P_n\| = 1 = \|I - P_n\|$. Finally, $T(I - P_n) \to 0$ is also true and follows from $(I - P_n)T^* \to 0$, since T^* is also a compact operator (Proposition 11.33).

For a compact normal operator, the orthonormal basis e_i can be chosen to consist of the eigenvectors of T by the Spectral Theorem, in which case $P_n T P_n$ is a diagonal matrix

$$P_n T P_n = \sum_{i=1}^{n} \lambda_i e_i e_i^*.$$

$\square$

Proposition 15.28

The compact operators of finite rank acting on a Hilbert space H form a simple $*$-ideal $\mathcal{K}_F(H)$, which is contained in every non-zero ideal of $B(H)$.

Its closure in $B(H)$ is the $*$-ideal of compact operators $\mathcal{K}(H)$.

Proof The facts that the sum of compact operators, the product of a compact operator with any other operator, and the adjoint of a compact operator, are compact have already been proved earlier (Propositions 11.9 and 11.33), so $\mathcal{K}(H)$ is a $*$-ideal in $B(H)$.

Similarly, it is not difficult to show that the sum of two finite-rank operators, and the product (left or right) of a finite-rank operator with any other operator, are again finite-rank. The details are left to the reader.

Let $\mathcal{I}$ be an ideal in $B(H)$ which contains a non-zero operator S. There exist non-zero vectors v, w such that $Sv = w$. For any vectors x, y, define the operator $E_{xy} := xy^*$, so that

$$E_{xw} S E_{vy} = x(w^* Sv)y^* = \|w\|^2 E_{xy}$$

and $E_{xy} \in \mathcal{I}$. Hence, any finite-rank operator, which is a sum of such operators, $T = \sum_{n=1}^{N} u_n v_n^*$, is also in $\mathcal{I}$, that is, we have proved $\mathcal{K}_F(H) \subseteq \mathcal{I}$.

In particular $\mathcal{K}_F(H)$ contains no non-zero ideals; we say it is *simple*. That the closure of $\mathcal{K}_F(H)$ is $\mathcal{K}(H)$ is essentially the content of the previous proposition: More precisely, recall that the image of a compact operator is separable, so $M := \overline{\operatorname{im} T}$ has a countable basis (e_i). Let P_n be the orthogonal projection onto $[\![e_1, \ldots, e_n]\!]$. Then, as in the proof of the previous proposition, the finite-rank operators $P_n T$ converge to T.   □

Examples 15.29

1. The ideal of compact operators, being the closure $\mathcal{K}(H) = \overline{\mathcal{K}_F(H)}$, is contained in every closed ideal of $B(H)$.
2. The algebra of square matrices $B(\mathbb{C}^n) = \mathcal{K}_F(\mathbb{C}^n) = \mathcal{K}(\mathbb{C}^n)$ is simple.
3. ▶ The above argument cannot be extended to show, more generally, that compact operators on a Banach space can be approximated by finite-rank operators. Spaces for which this is true are said to have the "approximation property"; even separable spaces may fail to have this property [41].

Hilbert-Schmidt Operators

Definition 15.30

The **trace** of an operator T on a Hilbert space with an orthonormal basis e_n, is, when finite,

$$\operatorname{tr}(T) := \sum_{n \in \mathbb{N}} \langle e_n, T e_n \rangle.$$

A **Hilbert-Schmidt** operator is one such that $\operatorname{tr}(T^*T) = \sum_{n \in \mathbb{N}} \|T e_n\|^2$ is finite.

As defined, the trace of an operator can depend on the choice of orthonormal basis. But for a Hilbert-Schmidt operator, $\operatorname{tr}(T^*T)$ *is* well-defined as the proof of the next proposition shows:

Proposition 15.31

If the right-hand traces exist,

$$\operatorname{tr}(S + T) = \operatorname{tr}(S) + \operatorname{tr}(T), \quad \operatorname{tr}(\lambda T) = \lambda \operatorname{tr}(T), \quad \operatorname{tr}(T^*) = \overline{\operatorname{tr}(T)}.$$

If S, T are Hilbert-Schmidt, then $\operatorname{tr}(ST) = \operatorname{tr}(TS)$.

Proof The identities $\text{tr}(S + T) = \text{tr}(S) + \text{tr}(T)$ and $\text{tr}(\lambda T) = \lambda \, \text{tr}(T)$ follow easily from the linearity of the inner product and summation, while

$$\text{tr}(T^*) = \sum_{n \in \mathbb{N}} \langle e_n, T^* e_n \rangle = \sum_{n \in \mathbb{N}} \overline{\langle T^* e_n, e_n \rangle} = \sum_{n \in \mathbb{N}} \overline{\langle e_n, T e_n \rangle} = \overline{\text{tr}(T)}.$$

Let e_n and $\tilde{e}_m$ be orthonormal bases for the Hilbert space H; then $T e_n = \sum_m \langle \tilde{e}_m, T e_n \rangle \tilde{e}_m$ and $ST e_n = \sum_m \langle \tilde{e}_m, T e_n \rangle S \tilde{e}_m$, so

$$\text{tr}(ST) = \sum_n \langle e_n, ST e_n \rangle = \sum_{n,m} \langle \tilde{e}_m, T e_n \rangle \langle e_n, S \tilde{e}_m \rangle = \sum_m \langle \tilde{e}_m, T S \tilde{e}_m \rangle, \quad (15.1)$$

exchanging the order of summation. This would be justified if the convergence is absolute, which is the case when S^* and T are Hilbert-Schmidt,

$$\sum_{n,m} |\langle \tilde{e}_m, T e_n \rangle \langle e_n, S \tilde{e}_m \rangle| \leqslant \sqrt{\sum_{n,m} |\langle \tilde{e}_m, T e_n \rangle|^2} \sqrt{\sum_{n,m} |\langle e_n, S \tilde{e}_m \rangle|^2}$$

$$= \sqrt{\sum_n \|T e_n\|^2 \sum_n \|S^* e_n\|^2}, \quad (15.2)$$

applying the Cauchy-Schwarz inequality and Parseval's identity. So, putting $S = T^*$ and $\tilde{e}_n = e_n$ in (15.1) shows that $\text{tr}(T^*T) = \text{tr}(TT^*)$, when T is Hilbert-Schmidt, i.e., T^* is also Hilbert-Schmidt. This, in turn, implies that when S and T are Hilbert-Schmidt, (15.2) and (15.1) are satisfied, so $\text{tr}(TS) = \text{tr}(ST)$ (in particular $\text{tr}(T^*T)$) is independent of the orthonormal basis. $\qquad\qquad\qquad\square$

Theorem 15.32

> **The Hilbert-Schmidt operators of $B(H)$ form a Hilbert space $\mathcal{HS}$, with inner product**
>
> $$\langle S, T \rangle_F := \text{tr}(S^*T) = \sum_{n \in \mathbb{N}} \langle S e_n, T e_n \rangle,$$
>
> **which is a $*$-ideal of compact operators, and**
>
> $$\|T\| \leqslant \|T\|_F, \qquad \|ST\|_F \leqslant \|S\| \|T\|_F.$$

Proof Let e_n be an orthonormal basis for H. First note that $\|T\|_F := \sqrt{\langle T, T \rangle_F} = \sqrt{\text{tr}(T^*T)}$ is finite for Hilbert-Schmidt operators.

(i) We have remarked in the preceding proposition that if $T \in \mathcal{HS}$ then $T^* \in \mathcal{HS}$, and

$$\|T^*\|_F = \sqrt{\text{tr}(TT^*)} = \sqrt{\text{tr}(T^*T)} = \|T\|_F.$$

The product $\langle S, T \rangle := \text{tr}(S^*T)$ is finite and independent of the choice of orthonormal basis when $S, T \in \mathcal{HS}$, by (15.1) and (15.2). Moreover, both of the following traces are finite,

$$\text{tr}((S + T)^*(S + T)) = \text{tr}(S^*S) + \text{tr}(S^*T) + \text{tr}(T^*S) + \text{tr}(T^*T)$$

$$\text{tr}((\lambda T)^*(\lambda T)) = |\lambda|^2 \, \text{tr}(T^*T),$$

so that $\mathcal{HS}$ is a vector space.
Linearity and 'symmetry' of the product follow from

$$\langle S, T_1 + T_2 \rangle = \text{tr}(S^*T_1 + S^*T_2) = \text{tr}(S^*T_1) + \text{tr}(S^*T_2) = \langle S, T_1 \rangle + \langle S, T_2 \rangle,$$

$$\langle S, \lambda T \rangle = \text{tr}(S^*\lambda T) = \lambda \, \text{tr}(S^*T) = \lambda \langle S, T \rangle,$$

$$\langle T, S \rangle = \text{tr}(T^*S) = \text{tr}(S^*T)^* = \overline{\text{tr}(S^*T)} = \overline{\langle S, T \rangle}.$$

That $\|T\| \leqslant \|T\|_F$ (and hence $\|T\|_F = 0 \Rightarrow T = 0$) follows from

$$\|Tx\| = \|\sum_n \langle e_n, x \rangle T e_n\| \leqslant \sum_n |\langle e_n, x \rangle| \|T e_n\|$$

$$\leqslant \sqrt{\sum_n |\langle e_n, x \rangle|^2} \sqrt{\sum_n \|T e_n\|^2} = \|x\| \|T\|_F.$$

$\langle \cdot, \cdot \rangle$ is therefore a legitimate inner product on $\mathcal{HS}$.
Finally, $\mathcal{HS}$ is an ideal of $B(H)$, since for any $S \in B(H)$ and $T \in \mathcal{HS}$,

$$\|ST\|_F^2 = \sum_{n \in \mathbb{N}} \|ST e_n\|^2 \leqslant \sum_{n \in \mathbb{N}} \|S\|^2 \|T e_n\|^2 = \|S\|^2 \|T\|_F^2,$$

and $\quad \|TS\|_F = \|(TS)^*\|_F \leqslant \|S^*\| \|T^*\|_F = \|S\| \|T\|_F.$

(ii) *Hilbert-Schmidt operators are compact*: Given $T \in \mathcal{HS}$, define the finite-rank operator T_N by $T_N e_n := \begin{cases} T e_n, & \text{if } n \leqslant N \\ 0, & \text{if } n > N \end{cases}$.

$$\|T - T_N\|^2 \leqslant \|T - T_N\|_F^2 = \sum_{n \in \mathbb{N}} \|(T - T_N)e_n\|^2$$

$$= \sum_{n=N+1}^{\infty} \|T e_n\|^2 \to 0 \text{ as } N \to \infty.$$

T is thus the limit of finite-rank operators, making it compact (Proposition 11.9).

(iii) *The space $\mathcal{HS}$ is complete in the $\mathcal{HS}$-norm:* Let $(T_n)_{n\in\mathbb{N}}$ be an $\mathcal{HS}$-Cauchy sequence

$$\|T_n - T_m\|_F^2 = \sum_{i\in\mathbb{N}} \|(T_n - T_m)e_i\|^2 \to 0 \quad \text{as } n, m \to \infty,$$

then it is a Cauchy sequence in the operator norm, and thus $T_n \to T$ in $B(H)$. But writing the Cauchy condition in a slightly different way, the sequences $x_n := (\|(T_n - T)e_i\|)_{i\in\mathbb{N}}$ form a Cauchy sequence in ℓ^2,

$$\|x_n - x_m\|_{\ell^2}^2 = \sum_{i\in\mathbb{N}} \left| \|(T_n - T)e_i\| - \|(T_m - T)e_i\| \right|^2$$

$$\leqslant \sum_{i\in\mathbb{N}} \|(T_n - T_m)e_i\|^2 \to 0,$$

as $n, m \to \infty$; so x_n converges to some sequence $(a_i)_{i\in\mathbb{N}} \in \ell^2$. Combining $T_n e_i \to T e_i$ with $\|T_n e_i - T e_i\| \to a_i$ for all i, each a_i must be 0, and

$$\|T_n - T\|_F^2 = \sum_i \|(T_n - T)e_i\|^2 = \|x_n\|_{\ell^2}^2 \to 0 \quad \text{as } n \to \infty,$$

so $T_n \to T$ in $\mathcal{HS}$, and $T \in \mathcal{HS}$ since $\|T\|_F \leqslant \|T - T_n\|_F + \|T_n\|_F < \infty$. Note that the space $\mathcal{HS}$ is not necessarily complete in the operator norm.   □

Having established a theory of Hilbert-Schmidt operators, we populate it with some important examples on $L^2(\mathbb{R})$:

Theorem 15.33

If $k \in L^2(\mathbb{R}^2)$, then the operator on $L^2(\mathbb{R})$

$$Tf(s) := \int k(s, t) f(t) \, dt,$$

is Hilbert-Schmidt with $\|T\|_F = \|k\|_{L^2}$.

Proof Let $e_n(t)$ be any orthonormal basis for $L^2(\mathbb{R})$. Then any function of t in $L^2(\mathbb{R})$ can be written as a sum of these basis functions. Analogously any function of two variables t, s in $L^2(\mathbb{R}^2)$ can be written as a sum (convergent in $L^2(\mathbb{R}^2)$)

$$k(s, t) = \sum_{m,n} \alpha_{n,m} \overline{e_n(t)} e_m(s),$$

(by first approximating k by simple functions $\sum_i a_i 1_{E_i}(t, s) = \sum_{m,n} a_{m,n} 1_{E_m}(t)$ $1_{E'_n}(s)$.) Write $\overline{e_n} \otimes e_m$ for the basis functions $(t, s) \mapsto \overline{e_n}(t)e_m(s)$. They are orthonormal, since

$$\langle \overline{e_n} \otimes e_m, \overline{e_{n'}} \otimes e_{m'} \rangle = \iint e_n(t)\overline{e_m}(s)\overline{e_{n'}}(t)e_{m'}(s)\,dt\,ds$$

$$= \langle e_{n'}, e_n \rangle \langle e_{m'}, e_m \rangle = \delta_{n'n}\delta_{m'm}.$$

By Parseval's identity $\|k\|_{L^2}^2 = \iint |k(t, s)|^2\,dt\,ds = \sum_{m,n} |\alpha_{n,m}|^2$. Clearly,

$$\langle e_m, Te_n \rangle = \iint \overline{e_m(t)}k(t, s)e_n(s)\,dt\,ds = \langle \overline{e_n} \otimes e_m, k \rangle_{L^2(\mathbb{R}^2)} = \alpha_{n,m},$$

so

$$\|T\|_F^2 = \sum_n \|Te_n\|^2 = \sum_{n,m} |\langle e_m, Te_n \rangle|^2 = \sum_{n,m} |\alpha_{n,m}|^2 = \|k\|_{L^2}^2.$$

$\square$

Examples 15.34

1. ▶ For square matrices $T = [T_{i,j}]$, $S = [S_{i,j}]$,

$$\operatorname{tr} T = \sum_i T_{ii}, \quad \|T\|_F = \sqrt{\sum_{i,j} |T_{i,j}|^2}, \quad \langle S, T \rangle_F = \sum_{ij} \bar{S}_{i,j} T_{i,j}.$$

2. ▶ More generally, for any Hilbert-Schmidt operator on any Hilbert space,

$$\|T\|_F^2 = \sum_{i \in \mathbb{N}} \|Te_i\|^2 = \sum_{i,j} |\langle e_j, Te_i \rangle|^2 = \sum_{i,j} \sigma_i^2 |\langle e_j, \tilde{e}_i \rangle|^2 = \sum_{i \in \mathbb{N}} \sigma_i^2,$$

where σ_i are the singular values of T; Parseval's identity is used on orthonormal bases such that $Te_i = \sigma_i \tilde{e}_i$.
It is evident that $\|T\|_F \geqslant \max_n |\sigma_n| = \|T\|$. It also follows that

$$|\langle S, T \rangle_F| \leqslant \|\sigma_S\| \|\sigma_T\|.$$

The fact that $T = \sum_n \lambda_n e_n e_n^*$ for a Hilbert-Schmidt normal operator is one of Hilbert's major theorems.

3. Find the eigenvalues and eigenfunctions of the integral operator on $L^2[0, 1]$ with kernel $k(s, t) := \begin{cases} t(1 - s), & 0 \leqslant t \leqslant s \leqslant 1 \\ s(1 - t), & 0 \leqslant s \leqslant t \leqslant 1 \end{cases}$.

Solution: The operator is Hilbert-Schmidt since $|k(s, t)| \leqslant 1$ on the bounded domain $[0, 1]$. The eigenvalue equation is

$$\int_0^s t(1-s)f(t)\,dt + \int_s^1 s(1-t)f(t)\,dt = \lambda f(s).$$

The eigenfunctions can be assumed to be differentiable, essentially because they are integrals. Differentiating gives

$$s(1-s)f(s) - \int_0^s tf(t)\,dt - s(1-s)f(s) + \int_s^1 (1-t)f(t)\,dy = \lambda f'(s),$$

and again,

$$- sf(s) - (1-s)f(s) = \lambda f''(s),$$

$$f''(s) + \frac{1}{\lambda}f(s) = 0, \quad f(0) = 0 = f(1).$$

The solutions of this differential equation are the eigenfunctions $f_n(t) = \sin(n\pi t)$ with eigenvalues $\lambda_n = 1/(n^2\pi^2)$.

4. A traceless operator in $B(\mathbb{C}^n)$ has a matrix with a zero diagonal, with respect to some orthonormal basis.
   Proof: Let A be an $n \times n$ matrix with tr $A = 0$. The proof is by induction on n. Since the numerical range of A is convex,

$$0 = \frac{1}{n}\,\mathrm{tr}\,A = \frac{1}{n}\sum_{i=1}^n \lambda_i \in W(A)$$

where λ_i are the eigenvalues of A. So there is a unit vector u such that $\langle u, Au \rangle = 0$. The matrix restricted to $u^\perp$, $\tilde{A} := A|_{u^\perp}$, is still traceless

$$0 = \mathrm{tr}\,A = \mathrm{tr}\,\tilde{A} + \langle u, Au \rangle = \mathrm{tr}\,\tilde{A}.$$

Therefore, by induction, there is an orthonormal basis $e_1, \ldots, e_{n-1}$ of $u^\perp$ in which $\tilde{A}$ has zero diagonal, i.e., $\langle e_i, Ae_i \rangle = 0$. This basis, together with u is the required basis for the whole n-dimensional space.

5. ▶ There is a correspondence between various ideals of compact operators and the sequence spaces of their singular values (σ_n):

Finite-rank operators	$\mathcal{K}_F(H)$	$(\sigma_n) \in c_{00}$
Trace-class operators	$\mathrm{Tr}(H)$	$(\sigma_n) \in \ell^1$
Hilbert-Schmidt operators	$\mathcal{HS}(H)$	$(\sigma_n) \in \ell^2$
Compact operators	$\mathcal{K}(H)$	$(\sigma_n) \in c_0$
Bounded operators	$B(H)$	$(\sigma_n) \in \ell^\infty$

where the set of *trace-class* operators has been added to complete the picture (Exercise 15.51(11)). More generally, the *Schatten-von Neumann* class of operators C_p corresponds to $(\sigma_n)_{n\in\mathbb{N}} \in \ell^p$. The analogy goes deeper than this: $\mathcal{K}(H)^* \cong \mathrm{Tr}(H)$ and $\mathrm{Tr}(H)^* \cong B(H)$ (via the functionals $T \mapsto \mathrm{tr}(ST)$).

Exercises 15.35

1. (a) $\langle S^*, T^* \rangle_F = \langle T, S \rangle_F$,
   (b) $\langle RT^*, S \rangle_F = \langle R, ST \rangle_F = \langle S^*R, T \rangle_F$.

2. The closest number to an $n \times n$ matrix T (in the $\mathcal{HS}$-norm) is $\mathrm{tr}(T)/n$. (Hint: $\lambda - T \perp I$.)

3. The map $x \mapsto M_x$, where $M_x y := x y$, embeds ℓ^2 into $\mathcal{HS}(\ell^2)$ (isometrically). More generally, if $x_n \in H$ satisfy $\sum_n \|x_n\|^2 < \infty$, then $T := \sum_n x_n e_n^*$ is Hilbert-Schmidt with $\|T\|_F^2 = \sum_n \|x_n\|^2$.

4. Let A be a normal matrix and P any orthogonal (self-adjoint) projection of rank r. Using the eigenbasis of A, $\langle P, AP \rangle = \langle P, A \rangle = \sum_i \alpha_i \lambda_i$ where λ_i are the eigenvalues of A, $\alpha_i = P_{ii}$, and $\sum_i \alpha_i = \langle I, P \rangle = \mathrm{tr}(P) = r$. It follows that the largest value of $\langle P, AP \rangle$, as P is varied, is the largest sum of r eigenvalues of A.

5. The Volterra operator on $L^2[0, 1]$, $Vf(t) := \int_0^t f$ is Hilbert-Schmidt with singular values $(n + \frac{1}{2})\pi, n \in \mathbb{N}$, and $\|V\|_F = \frac{1}{\sqrt{2}}$, $\|V\| = \frac{2}{\pi}$.

6. If $k(s, t) = k(s - t)$ for a real function $k(t) \in L^2[0, 1]$ (Example 9.28(3)), then $Tf := k * f$ is Hilbert-Schmidt, with eigenvalues $\widehat{k}(n)$.

7. Find the eigenfunctions and eigenvalues of the $\mathcal{HS}$-compact self-adjoint operators $Tf(s) := \int_0^1 k(s, t) f(t) \, dt$ (on $L^2[0, 1]$), where

   (a) $k(s, t) := s + t$,

   (b) $k(s, t) := \begin{cases} 1, & 1 - s \leqslant t \leqslant 1 \\ 0, & 0 \leqslant t \leqslant 1 - s \end{cases}$,

   (c) $k(s, t) := \min(s, t)$; deduce that $\sum_{n\in\mathbb{N}} \frac{1}{(2n+1)^4} = \frac{\pi^4}{96}$ and $\sum_{n=1}^{\infty} \frac{1}{n^4} = \frac{\pi^4}{90}$.

8. In the original Fredholm theory, it was proved under certain hypotheses that the equation

$$f(t) + \int_a^b k(t, s) f(s) \, ds = g(t)$$

either has a unique solution, or else the same equation with $g = 0$ admits a finite number of linearly independent solutions. Show this for $f, g \in L^2(\mathbb{R})$, $k \in L^2(\mathbb{R}^2)$, using Proposition 14.18.

15.6   Representation Theorems

We return to a general unital C^*-algebra $\mathcal{X}$ and recover some of the previous propositions that were proved in the special case of $B(H)$ to the general setting. We will see that the functional calculus can be widened considerably for normal elements, allowing us to define square roots and absolute values of certain operators. The final aim and fitting end to the chapter is to prove that a C^* algebra is embedded in $B(H)$ for some Hilbert space H.

Proposition 15.36

For any $\phi \in \mathcal{S}(\mathcal{X}), T \in \mathcal{X},$

$$\phi T^* = \overline{\phi T}, \quad \widehat{T^*} = \widehat{T}^*.$$

Proof If A is self-adjoint and $t \in \mathbb{R}$, then

$$\|A + it\|^2 = \|(A + it)^*(A + it)\|$$
$$= \|A^2 + t^2\| \leqslant \|A\|^2 + t^2$$

(As a matter of fact, equality holds as the accompanying diagram shows.)

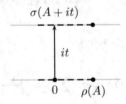

Writing $\phi A = a + ib$, we find

$$|b + t| \leqslant |a + ib + it| = |\phi(A + it)| \leqslant \|A + it\| \leqslant \sqrt{\|A\|^2 + t^2}$$

$$\therefore \qquad \forall t \in \mathbb{R}, \quad (2t + b)b \leqslant \|A\|^2$$

so $b = 0$ and $\phi A \in \mathbb{R}$. More generally, for any $T = A + iB \in \mathcal{X}$, with A, B self-adjoint,

$$\phi T^* = \phi(A - iB) = \phi A - i\phi B = \overline{\phi A + i\phi B} = \overline{\phi T}.$$

In particular, every $\psi \in \Delta$ is automatically a $*$-morphism, and

$$\widehat{T^*}(\psi) = \psi T^* = \overline{\psi T} = \widehat{T}(\psi)^*.$$

$\square$

Note that, for A self-adjoint, $\Delta(A) \subseteq \sigma(A) \subseteq \mathcal{S}(A) \subset \mathbb{R}$. The proposition above opens the path to dramatic results: not only are the commutative unital C^*-algebras completely characterised as $C(K)$ where K is a compact space, but even for non-commutative algebras, it allows an extension of the functional calculus to normal elements.

Corollary 15.37 (Gelfand-Naimark Theorem)

Every commutative unital C^*-algebra is isometrically $*$-isomorphic to $C(\Delta)$ via the Gelfand map.

Proof Recall that the Gelfand map $\mathcal{G}$ is a Banach algebra morphism (Proposition 14.38). In a commutative C^*-algebra, every element T is normal, so $\|\widehat{T}\|_C = \rho(T) = \|T\|$ (Theorem 14.39); furthermore $\widehat{T^*} = \widehat{T}^*$, and the Gelfand transform is an isometric $*$-embedding. Moreover,

$$\|T^2\| = \rho(T^2) = \rho(T)^2 = \|T\|^2,$$

so by Proposition 14.40 and Exercise 14.41(19), $\mathcal{X}$ is isomorphic to $C(\Delta)$. $\qquad\square$

Theorem 15.38 (The Functional Calculus for Normal Elements)

When T is normal, $\mathcal{T} := \overline{\mathbb{C}[T, T^*]}$ is a commutative closed $*$-subalgebra of $\mathcal{X}$, isometrically $*$-isomorphic to $C(\sigma(T))$.

The identity $\widehat{f(T)} = f \circ \widehat{T}$ defines a normal element $f(T)$ whenever $f \in C(\sigma(T))$; then $\sigma(f(T)) = f[\sigma(T)]$.

Proof $\mathcal{T}$ *is a commutative closed $*$-subalgebra of* $\mathcal{X}$: Since T is normal, $T^n(T^*)^m = (T^*)^m T^n$ (by induction), so it should be obvious that (i) any polynomial in T and T^* can be written uniquely in the form $\sum_{n,m} a_{n,m} T^n (T^*)^m$, (ii) the product (and addition) of two polynomials in T and T^* is another polynomial, (iii) this product commutes, and (iv) the involute of a polynomial $p(T, T^*)$ remains in $\mathcal{T}$,

$$p(T, T^*)^* = \left(\sum_{n,m} a_{n,m} T^n (T^*)^m \right)^* = \sum_{n,m} \overline{a_{n,m}} \, T^m (T^*)^n \in \mathbb{C}[T, T^*].$$

$\mathbb{C}[T, T^*]$ is thus a commutative $*$-subalgebra. The closure of such a subalgebra in $\mathcal{X}$ remains a commutative $*$-subalgebra (Prove!). Note that $\mathcal{T} = \overline{[\![T^n T^{*m}]\!]}_{(n,m) \in \mathbb{N}^2}$ is obviously separable.

The spectrum of $S \in \mathcal{Y}$, with respect to a closed $*$-subalgebra $\mathcal{Y} \subseteq \mathcal{X}$, is $\sigma(S)$: Clearly, if S (or $S - \lambda$) is invertible in $\mathcal{Y}$, it remains so in $\mathcal{X}$. Conversely, if S is invertible in $\mathcal{X}$, then so are S^*, S^*S and SS^*. But S^*S is self-adjoint, with a real spectrum (in $\mathcal{Y}$ and $\mathcal{X}$), hence $S^*S + i/n$ is invertible in $\mathcal{Y}$. As $\mathcal{Y}$ is closed and $(S^*S + i/n)^{-1} \to (S^*S)^{-1}$ in $\mathcal{X}$, as $n \to \infty$, we can deduce $(S^*S)^{-1} \in \mathcal{Y}$. Similarly $(SS^*)^{-1} \in \mathcal{Y}$, implying S is invertible in $\mathcal{Y}$ (Exercise 15.3(4)).

$\widehat{T} : \Delta_T \to \sigma(T)$ *is a homeomorphism*: (Δ_T is the character space of $\mathcal{T}$.) $\widehat{T}$ is 1–1 since suppose $\widehat{T}(\psi_1) = \widehat{T}(\psi_2)$ for some $\psi_1, \psi_2 \in \Delta_T$, i.e., $\psi_1 T = \psi_2 T$. Then

$$\psi_1 T^* = \overline{\psi_1 T} = \overline{\psi_2 T} = \psi_2 T^*$$

$$\therefore \psi_1 p(T, T^*) = \psi_1 \left(\sum_{n,m} a_{n,m} T^n (T^m)^* \right)$$

$$= \sum_{n,m} a_{n,m} (\psi_2 T)^n \overline{(\psi_2 T)^m} = \psi_2 p(T, T^*)$$

for any polynomial p; finally, by continuity of ψ_1 and ψ_2, $\psi_1 S = \psi_2 S$ for all $S \in \mathcal{T}$, proving $\psi_1 = \psi_2$. That $\widehat{T}$ is onto was proved in Theorem 14.39. It is continuous because

$$\psi_n \rightharpoonup \psi \implies \widehat{T}(\psi_n) = \psi_n T \to \psi T = \widehat{T}(\psi).$$

So $\widehat{T}$ is a homeomorphism since Δ_T is a compact metric space (Proposition 6.17 and Example 14.36(8)). Hence any $z \in \sigma(T)$ corresponds uniquely to some $\psi \in \Delta_T$ via $z = \widehat{T}(\psi) = \psi T$.

The Gelfand transform $\mathcal{G} : \mathcal{T} \to C(\Delta_T) \cong C(\sigma(T))$ *is an isometric* $*$-*isomorphism*: Since $\mathcal{T}$ is a commutative C^*-algebra, it is $*$-isometric to $C(\Delta_T)$.

The continuous function calculus: The correspondence between elements in $\mathcal{T}$ and functions in $C(\Delta_T)$ allows us to extend the analytic function calculus established earlier. For any *continuous* function $f \in C(\sigma(T))$, the composition $f \circ \widehat{T} : \Delta_T \to \mathbb{C}$ corresponds to some (normal) element in $\mathcal{T}$ which is denoted by $f(T)$. By this definition, $\widehat{f(T)} = f \circ \widehat{T}$. The following identities are true because they mirror the same properties in $C(\Delta_T)$,

$$(f + g)(T) = f(T) + g(T), \quad (\lambda f)(T) = \lambda f(T),$$

$$(fg)(T) = f(T)g(T), \quad \bar{f}(T) = f(T)^*.$$

Finally $\|f(T)\| = \|f\|_C$ is due to $\mathcal{G}$ being an isometry and $g \circ f(T) = g(f(T))$ follows after

$$\sigma(f(T)) = \text{im } \widehat{f(T)} = \text{im } f \circ \widehat{T} = f \text{ im } \widehat{T} = f(\sigma(T)).$$

$\square$

Examples 15.39

1. To take a simple example, consider a 2×2 diagonalizable matrix T with distinct eigenvalues λ_i and corresponding orthonormal eigenvectors v_i, $i = 1, 2$. Its character space Δ_T consists of the two morphisms $\psi_i S := \langle v_i, S v_i \rangle$ for $S \in \mathcal{T}$. The Gelfand transform takes T to (λ_1, λ_2); any other matrix $f(T)$ is simultaneously 'diagonalized' to $(f(\lambda_1), f(\lambda_2))$.

2. ▶ For any elements $S_1, S_2 \in \mathcal{T}$,

$$\sigma(S_1 + S_2) \subseteq \sigma(S_1) + \sigma(S_2), \quad \sigma(S_1 S_2) \subseteq \sigma(S_1)\sigma(S_2).$$

   Proof: As $\mathcal{T}$ is commutative, Theorem 14.39 shows that $\sigma(S) = \Delta_T S$ for any $S \in \mathcal{T}$. Hence the statements follow from Exercise 14.41(10)).

3. If S, T are commuting normal elements, and $f \in C(\sigma(S))$, $g \in C(\sigma(T))$, then $f(S)g(T) = g(T)f(S)$.
   Proof: For polynomials p and q in z and z^*, $p(S, S^*)q(T, T^*) = q(T, T^*)p(S, S^*)$ since they are sums of terms of the form

$$aS^n S^{*m} T^i T^{*j} = aT^i T^{j*} S^n S^{*m}$$

   by an application of Fuglede's theorem. Taking the limit of polynomials converging to f, g (by the Stone-Weierstrass theorem) gives the required result.

4. The self-adjoint elements of $\mathcal{T}$ correspond to the real-valued functions $f \in C(\Delta_T)$ and form a real Banach algebra, while the unitary elements correspond to functions with unit absolute value, $|f| = 1$.

Proposition 15.40

> **For T normal,**
>
> $$T \text{ is unitary } \Leftrightarrow \sigma(T) \subseteq e^{i\mathbb{R}},$$
>
> $$T \text{ is self-adjoint } \Leftrightarrow \sigma(T) \subseteq \mathbb{R}.$$

Proof (i) The spectrum of a unitary element U must lie in the unit closed ball since $\|U\| = 1$. Now, $U - \lambda = U(1 - \lambda U^*)$ and $\|\lambda U^*\| = |\lambda| \|U^*\| = |\lambda|$; so $|\lambda| < 1$ implies $1 - \lambda U^*$, and thus $U - \lambda$, are invertible (Theorem 13.20).
(Equivalently, if $\lambda \in \sigma(U)$ then $\lambda^{-1} \in \sigma(U^{-1}) = \sigma(U^*) = \sigma(U)^*$ and so both $|\lambda|$ and $1/|\lambda|$ are less than 1.)

(ii) We have already seen that $S(T) \subset \mathbb{R}$ when T is self-adjoint, and $S(T)$ includes $\sigma(T)$. Alternatively, e^{iT} is unitary (Example 15.5(11)) and the spectral mapping theorem gives $e^{i\sigma(T)} = \sigma(e^{iT}) \subseteq e^{i\mathbb{R}}$. But $|e^{i(a+ib)}| = e^{-b}$ is 1 only when $b = 0$, from which follows that $\sigma(T) \subseteq \mathbb{R}$.

(iii) For the converses, let T be normal with $\sigma(T) \subset \mathbb{R}$. Writing it as $A + iB$ with A, B commuting self-adjoint, we see that $iB = T - A$, so

$$\sigma(iB) \subseteq \sigma(T) + \sigma(-A) \subset \mathbb{R}, \qquad \text{by Example 2 above,}$$

yet $\sigma(iB) = i\sigma(B) \subset i\mathbb{R}$. Thus $\sigma(B) = \{0\}$, $B = 0$, and $T = A$ is self-adjoint. (Alternatively, we can work with $\mathcal{S}$: if T is normal and $\sigma(T)$ is real, then $\mathcal{S}(T) \subset \mathbb{R}$; for any $\phi \in \mathcal{S}$, $\phi(T - T^*) = \phi T - \overline{\phi T} = 0$, hence $T - T^* = 0$.)

(iv) If T is normal with $\sigma(T) \subseteq e^{i\mathbb{R}}$, then

$$\sigma(T^*T) \subseteq \sigma(T^*)\sigma(T) = \sigma(T)^*\sigma(T) \subseteq e^{i\mathbb{R}}.$$

As T^*T is self-adjoint and has a real spectrum, that leaves only ± 1 as possible spectral values. But $1 + T^*T$ is invertible, otherwise there is a $\psi \in \Delta_T$ such that

$$-1 = \psi(T^*T) = \psi T^* \psi T = |\psi T|^2,$$

a contradiction. So $\sigma(T^*T) = \{1\}$, $1 = T^*T = TT^*$ and T is unitary.   □

Exercises 15.41

1. Find an example of an operator T having a real spectrum, without T being self-adjoint.
2. If J is a $*$-morphism and T is normal, then $J(f(T)) = f(J(T))$ (first prove, for any polynomial p, $J(p(T, T^*)) = p(J(T), J(T)^*)$).
3. ▶ In a C^*-algebra, $\mathcal{S}(T) = 0 \Rightarrow T = 0$ (write $T = A + iB$). We say that $\mathcal{S}(\mathcal{X})$ *separates points* of $\mathcal{X}$: if $T \neq S$, then there is a $\phi \in \mathcal{S}$ such that $\phi T \neq \phi S$.
4. Suppose a C^*-algebra has two involutions, $*$ and $\star$ (with the same norm). Show that $T^* = T^\star$ for all T—the involution is unique. (Hint: $\phi(T^*) = \overline{\phi T} = \phi(T^\star)$.)
5. Every normal cyclic element is unitary. In particular, the normal elements of a finite subgroup of $\mathcal{G}(\mathcal{X})$ are unitary.
6. The Fourier transform $\mathcal{F} : L^2(\mathbb{R}) \to L^2(\mathbb{R})$ is unitary; in fact it is cyclic $\mathcal{F}^4 = 1$, so that it has four eigenvalues $\pm 1, \pm i$. Verify that the following are eigenfunctions: $e^{-\pi t^2}$, $te^{-\pi t^2}$, $(4\pi t^2 - 1)e^{-\pi t^2}$, $(4\pi t^3 - 3t)e^{-\pi t^2}$.
7. A normal T such that $\|T\| = 1 = \|T^{-1}\|$ is unitary.
8. Normal idempotents are self-adjoint. A normal element T with $\sigma(T) \subseteq \{0, 1\}$ is an idempotent, e.g., when T is normal and $T^{n+1} = T^n$ for some integer n.
9. Suppose M is a closed subspace of a Hilbert space which is invariant under a group of unitary operators. Show that $M^\perp$ is also invariant.
10. If T_n are self-adjoint operators and $T_n \rightharpoonup T$ then T is self-adjoint.

15.7 Positive Self-Adjoint Elements

For T, S self-adjoint, let $T \leqslant S$ be defined to mean $\sigma(S - T) \subseteq [0, \infty[$. Equivalently, since $\mathcal{S}(S - T)$ is the closed convex hull of $\sigma(S - T)$ (Proposition 15.8),

$$T \leqslant S \Leftrightarrow \forall \phi \in \mathcal{S}(\mathcal{X}), \ \phi T \leqslant \phi S.$$

Proposition 15.42

> **The self-adjoint elements form an ordered real Banach space, such that**
>
> $$T \leqslant S \text{ AND } R \leqslant Q \ \Rightarrow \ T + R \leqslant S + Q,$$
> $$T \leqslant S \ \Rightarrow \ R^* T R \leqslant R^* S R \quad \forall R \in \mathcal{X}.$$

Proof First note that, by the definition, $T \leqslant S \Leftrightarrow 0 \leqslant S - T \Leftrightarrow T - S \leqslant 0$ ($\Leftrightarrow -S \leqslant -T$), so we might as well consider $A := S - T \geqslant 0$ and $B := Q - R \geqslant 0$ in proving some of the assertions.

(i) It is trivially true that self-adjoint elements form a real vector subspace

$$(S + T)^* = S^* + T^* = S + T, \qquad (\lambda T)^* = \bar{\lambda} T^* = \lambda T, \quad \forall \lambda \in \mathbb{R}.$$

If $T_n \to T$ with $T_n^* = T_n$, then in the limit, $T^* = T$, so the subspace is closed.

(ii) That $T \leqslant T$ is immediate from $\sigma(0) = \{0\}$. For anti-symmetry, note that

$$0 \leqslant A \leqslant 0 \Rightarrow \sigma(A) = \{0\} \Rightarrow \|A\| = \rho(A) = 0 \Rightarrow A = 0,$$
$$\text{so} \quad S \leqslant T \leqslant S \Rightarrow T = S.$$

(iii) To facilitate the rest of the proof, we demonstrate

$$a \leqslant T \leqslant b \Leftrightarrow \sigma(T) \subseteq [a, b] \tag{15.3}$$

in two parts,

$$a \leqslant T \Leftrightarrow \sigma(T) - a = \sigma(T - a) \subseteq [0, \infty[\Leftrightarrow \sigma(T) \subseteq [a, \infty[$$
$$T \leqslant b \Leftrightarrow \sigma(T) - b = \sigma(T - b) \subseteq \ ]-\infty, 0] \Leftrightarrow \sigma(T) \subseteq \ ]-\infty, b].$$

In particular, note that $T \leqslant \rho(T) = \|T\|$ and that if $-b \leqslant T \leqslant b$ then $\rho(T) \leqslant b$.

(iv) $A, B \geqslant 0 \Rightarrow A + B \geqslant 0$: In general,

$$C + D \leqslant \|C + D\| \leqslant \|C\| + \|D\| = \rho(C) + \rho(D).$$

Let $a := \rho(A)$ and $b := \rho(B)$, then $0 \leqslant A \leqslant a$ can be rewritten as $0 \leqslant a - A \leqslant a$ and hence $\rho(a - A) \leqslant a$. Similarly $\rho(b - B) \leqslant b$, so $(a - A) + (b - B) \leqslant a + b$, or equivalently, $A + B \geqslant 0$.

(v) A special case of this shows transitivity of the order relation,

$$T \leqslant S \leqslant R \implies 0 \leqslant (R - S) + (S - T) = R - T \implies T \leqslant R$$

(vi) We are not at this stage able to prove the full product-inequality rule as claimed in the proposition. The proof is deferred to the next proposition. Here we show only the simple case when R is scalar, i.e., if $\lambda \geqslant 0$ and $A = S - T \geqslant 0$, then $\sigma(\lambda A) = \lambda \sigma(A) \subseteq \mathbb{R}^+$, meaning $\lambda T \leqslant \lambda S$.   □

The continuous functional calculus allows us to extend the domain of all continuous real functions $f : \mathbb{R} \to \mathbb{R}$ to the set of self-adjoint elements. Two functions in particular stand out:

(i) the positive square root $\sqrt{A}$ when $A \geqslant 0$, satisfying $(\sqrt{A})^2 = A = \sqrt{A^2}$,

(ii) A_+ for all A self-adjoint, from the function $t_+ := \begin{cases} t, & \text{when } t \geqslant 0 \\ 0, & \text{when } t < 0 \end{cases}$; similarly

   A_- from $t_- := (-t)_+$. Their sum then gives $|A|$, which corresponds to the function $t \mapsto |t|$.

Examples 15.43

1. (a) If $-T \leqslant S \leqslant T$ then $\|S\| \leqslant \|T\|$.
   (b) If $0 < a \leqslant T \leqslant b$ then T is invertible and $b^{-1} \leqslant T^{-1} \leqslant a^{-1}$.
   (c) If $ST \geqslant 0$ then $TS \geqslant 0$.
   (d) If $S, T \geqslant 0$ and ST is self-adjoint, then $ST \geqslant 0$. In particular, $T \geqslant 0 \implies T^n \geqslant 0$.
   (e) If $S_n \leqslant T_n$ and $S_n \to S$, $T_n \to T$, then $S \leqslant T$.
   Proof:

   (a) $-\|T\| \leqslant S \leqslant \|T\|$, so $\sigma(S) \subseteq [-\|T\|, \|T\|]$ and $\|S\| = \rho(S) \leqslant \|T\|$.
   (b) $\sigma(T) \subseteq [a, b]$ does not include 0; $\sigma(T^{-1}) = \sigma(T)^{-1} \subseteq [b^{-1}, a^{-1}]$.
   (c) $\sigma(TS)$ is the same as $\sigma(ST)$ except possibly for the inclusion or exclusion of 0. In any case $\sigma(ST) \subseteq \mathbb{R}^+ \Leftrightarrow \sigma(TS) \subseteq \mathbb{R}^+$.
   (d) Recall that ST is self-adjoint exactly when $ST = TS$. So, by Exercise 14.41(17)), $\sigma(ST) \subseteq \sigma(S)\sigma(T) \subseteq \mathbb{R}^+$.
   (e) Let $A_n := T_n - S_n \geqslant 0$ and $A_n \to A := T - S$. Then $0 \leqslant \phi A_n \to \phi A$ for any $\phi \in S$, so $S(A) \subseteq [0, \infty]$.

2. The set of positive elements is a closed convex 'cone' (meaning $T \geqslant 0$ AND $\lambda \geqslant 0 \implies \lambda T \geqslant 0$), with non-empty interior in the real Banach space of self-adjoints. *Proof*: The only non-trivial statement is that the cone contains an open set of self-adjoints, namely the unit ball around 1: If A is self-adjoint and $\|A\| < 1$ then $-1 \leqslant A \leqslant 1$, so $1 + A \geqslant 0$.

3. Positive continuous functions $f : \mathbb{R} \to \mathbb{R}^+$ give positive elements $f(A) \geqslant 0$ for A self-adjoint. For example, A_+, A_-, $|A|$, A^2, and e^A are all positive. More generally, for any normal operator T and $f \in C(\mathbb{C}, \mathbb{R}^+)$, $f(T) \geqslant 0$.
   Proof: By the functional calculus, $\sigma(f(T)) = f[\sigma(T)] \subseteq [0, \infty[$.
4. Every self-adjoint element decomposes into two positive elements

   (a) $A = A_+ - A_-$, $|A| = A_+ + A_-$,
   (b) $A_+ A_- = 0$, $A_\pm |A| = A_\pm^2$, $A_\pm A = \pm A_\pm^2$, and A_+, A_-, A and $|A|$ all commute with each other,
   (c) $-A_- \leqslant A \leqslant A_+ \leqslant |A| \leqslant \|A\|$.

   Proof: The identities $t = t_+ - t_-$, $|t| = t_+ + t_-$, $t_+ t_- = 0$, $t_\pm |t| = t_\pm^2$, $t_\pm t = \pm t_\pm^2$ imply (a) and (b). Moreover, $A_+ A_- = A_+ \geqslant 0$, $|A| - A_+ = A_+ - A = A_- \geqslant 0$. Finally, $\sigma(|A|) = \{ |\lambda| : \lambda \in \sigma(A) \}$ is bounded above by $\rho(A) = \|A\|$.
5. By the spectral mapping theorem, the spectral values of $\sqrt{A}$ are the positive square roots of those of $A \geqslant 0$. Overall there may be an infinite number of square roots of A, e.g., for any $z \in \mathbb{C}$, $\left(\begin{smallmatrix} z & 1+z \\ 1-z & -z \end{smallmatrix} \right)^2 = \left(\begin{smallmatrix} 1 & 0 \\ 0 & 1 \end{smallmatrix} \right)$.
6. If S, T are invertible positive self-adjoints, then

   (a) $(T^{\frac{1}{2}})^{-1} = T^{-\frac{1}{2}}$,
   (b) $0 \leqslant S \leqslant T \Rightarrow T^{-1} \leqslant S^{-1}$,
   (c) $0 \leqslant S \leqslant T \Rightarrow \sqrt{S} \leqslant \sqrt{T}$.

   Proof: (a) follows from the same identity that holds in $C(\mathbb{R}^+)$. For (b), (c), note that $0 \leqslant T^{-\frac{1}{2}} S T^{-\frac{1}{2}} \leqslant 1$ (Proposition 15.42); using example 1(b) then gives $T^{\frac{1}{2}} S^{-1} T^{\frac{1}{2}} \geqslant 1$, and hence $S^{-1} \geqslant T^{-\frac{1}{2}} T^{-\frac{1}{2}} = T^{-1}$. Also, $\|S^{\frac{1}{2}} T^{-\frac{1}{2}}\|^2 = \|T^{-\frac{1}{2}} S T^{-\frac{1}{2}}\| \leqslant 1$, from which follows $T^{-\frac{1}{4}} S^{\frac{1}{2}} T^{-\frac{1}{4}} \leqslant 1$ and $S^{\frac{1}{2}} \leqslant T^{\frac{1}{2}}$.
7. If $\rho(T) < 1$ then

$$S := 1 + T^* T + (T^2)^* (T^2) + (T^3)^* (T^3) + \cdots$$

converges and is positive invertible; $\|1 - S^{-1}\| = \|S^{\frac{1}{2}} T S^{-\frac{1}{2}}\| < 1$.
   Proof: Applying the root test, $\|(T^n)^* (T^n)\|^{\frac{1}{n}} = \|T^n\|^{\frac{2}{n}} \to \rho(T)^2 < 1$. Assuming the next proposition that each term is non-negative, $(T^n)^* (T^n) \geqslant 0$, it follows that $S \geqslant 1$, and thus invertible. Moreover, $0 < S^{-1} \leqslant 1$ and $0 \leqslant 1 - S^{-1} < 1$, so

$$1 - S^{-1} = S^{-\frac{1}{2}} (S - 1) S^{-\frac{1}{2}} = S^{-\frac{1}{2}} T^* S T S^{-\frac{1}{2}}$$

$$\therefore \|S^{\frac{1}{2}} T S^{-\frac{1}{2}}\|^2 = \|(S^{\frac{1}{2}} T S^{-\frac{1}{2}})^* (S^{\frac{1}{2}} T S^{-\frac{1}{2}})\| = \|1 - S^{-1}\| < 1$$

Proposition 15.44

> **For any $T \in \mathcal{X}$ and $\phi \in S(\mathcal{X})$,**
>
> (i) $T^*T \geqslant 0$,
> (ii) $T \geqslant 0 \Leftrightarrow T = R^*R$, **for some** $R \in \mathcal{X}$,
> (iii) $\langle S, T \rangle := \phi(S^*T)$ **gives a semi-inner product,**
> (iv) $|\phi(S^*T)|^2 \leqslant \phi(S^*S)\phi(T^*T)$, $|\phi T|^2 \leqslant \phi(T^*T)$,
> (v) $|\phi(S^*TS)| \leqslant \phi(S^*S)\|T\|$.

Proof (i) T^*T is certainly self-adjoint, and can be decomposed as $T^*T = A - B$ where $A, B \geqslant 0$, $AB = BA = 0$ (Example 4b above). Now

$$(TB)^*(TB) = BT^*TB = B(A - B)B = -B^3 \leqslant 0$$

and hence $(TB)(TB)^* \leqslant 0$ (Examples 15.43(1c)). Writing $TB = C + iD$, with C, D self-adjoint, we find

$$0 \leqslant 2(C^2 + D^2) = (TB)^*(TB) + (TB)(TB)^* \leqslant 0$$

$$\therefore \qquad 0 \leqslant C^2 = -D^2 \leqslant 0$$

$$\therefore \qquad C = 0 = D$$

so $TB = 0$. But then, $0 = (TB)^*(TB) = -B^3$ forces $B = 0$ and $T^*T = A \geqslant 0$. This allows us to conclude part (vi) of the proof of Proposition 15.42. If $T \leqslant S$ let $A := S - T \geqslant 0$, so for any $R \in \mathcal{X}$, $R^*AR = (\sqrt{A}R)^*(\sqrt{A}R) \geqslant 0$, i.e., $R^*TR \leqslant R^*SR$.

(ii) Conversely, if T is positive, let $R := \sqrt{T} \geqslant 0$, so $R^*R = R^2 = T$.

(iii) The product satisfies the following inner-product axioms,

$$\langle S, \lambda T_1 + \mu T_2 \rangle = \phi(\lambda ST_1 + \mu ST_2) = \lambda \langle S, T_1 \rangle + \mu \langle S, T_2 \rangle,$$

$$\langle T, S \rangle = \phi(T^*S) = \phi(S^*T)^* = \overline{\langle S, T \rangle},$$

$$\langle T, T \rangle = \phi(T^*T) \geqslant 0 \qquad \text{since } T^*T \geqslant 0.$$

However, it need not be definite, i.e., $\phi(T^*T) = 0$ may be possible without $T = 0$.

(iv) This is the Cauchy-Schwarz inequality, which is valid even for semi-definite inner products (Exercise 10.10(16)). In particular, taking $S = 1$ gives the second inequality.

(v) As ϕ preserves inequalities,

$$T^*T \leqslant \|T^*T\| = \|T\|^2 \Rightarrow S^*T^*TS \leqslant \|T\|^2 S^*S$$

$$\Rightarrow \phi(S^*T^*TS) \leqslant \phi(S^*S)\|T\|^2.$$

$$\therefore \quad |\phi(S^*(TS))|^2 \leqslant \phi(S^*S)\phi(S^*T^*TS) \quad \text{by (iv)},$$

$$\leqslant \phi(S^*S)^2 \|T\|^2$$

$\square$

Proposition 15.45

> If $J : \mathcal{X} \to \mathcal{Y}$ is an algebraic $*$-morphism between C^*-algebras, then it is
> continuous with $\|J\| = 1$, and preserves $\leqslant$.
>
> If J is also injective, then it is isometric.

By an algebraic $*$-morphism is meant a map which preserves $+$, $\cdot$, 1, and $*$.

Proof If $A \geqslant 0$, then $A = R^*R$ and $J(A) = J(R)^*J(R) \geqslant 0$. Thus J preserves the order of self-adjoint elements,

$$S \leqslant T \Rightarrow J(T - S) \geqslant 0 \Rightarrow J(S) \leqslant J(T).$$

Now for any T (noting that $J(1) = 1$),

$$0 \leqslant T^*T \leqslant \|T\|^2,$$

$$\therefore \quad 0 \leqslant J(T^*T) \leqslant \|T\|^2,$$

$$\therefore \quad \|J(T)\| = \|J(T)^*J(T)\|^{\frac{1}{2}} = \|J(T^*T)\|^{\frac{1}{2}} \leqslant \|T\|.$$

If J is 1–1, then one can form the 'inverse' $J^{-1} : \text{im } J \to \mathcal{X}$. It is automatically an algebraic $*$-morphism (check!), for example, for any $S \in \text{im } J$,

$$J^{-1}(S^*) = J^{-1}(JT)^* = J^{-1}J(T^*) = T^* = (J^{-1}(JT))^* = (J^{-1}S)^*,$$

and so $\|J^{-1}(S)\| \leqslant \|S\|$. Thus $\|T\| \leqslant \|J(T)\| \leqslant \|T\|$ as required.

(Alternatively, defining $\|\!|T|\!\| := \|J(T)\|_{\mathcal{Y}}$ gives a C^*-norm on $\mathcal{X}$. But there can only be one C^*-norm (Exercise 15.10(5)), so J is an isometry and im J is closed.)

$\square$

Examples 15.46

1. Characters are *extremal* points of $\mathcal{S}(\mathcal{X})$: If $\psi \in \Delta$ lies between $\phi_1, \phi_2 \in \mathcal{S}$, then $\psi = \phi_1 = \phi_2$.

Proof: By convexity of $\mathcal{S}$, we can assume $\psi = \frac{\phi_1 + \phi_2}{2}$. Then

$$|\phi_1(T)|^2 + |\phi_2(T)|^2 \leqslant \phi_1(T^*T) + \phi_2(T^*T)$$
$$= 2\psi(T^*T) = 2|\psi(T)|^2$$
$$= \tfrac{1}{2}|\phi_1(T) + \phi_2(T)|^2$$

from which follows that $\phi_1(T) = \phi_2(T)$.

2. Consider $\phi \in \mathcal{X}^*$ which preserves inequalities, $0 \leqslant A \Rightarrow 0 \leqslant \phi A$; it satisfies Proposition 15.44 except that $|\phi T|^2 \leqslant \phi 1 \phi(T^*T) \leqslant (\phi 1)^2 \|T\|^2$. Such *positive* functionals, as they are called, are positive multiples of states.

Proof: The proofs of 15.44 (iii)–(v) are still valid: the only assumption that needs justification is $\phi(T^*) = \phi(T)^*$. For any self-adjoint A, $A = A_+ - A_-$, so $\phi A = \phi A_+ - \phi A_- \in \mathbb{R}$. So

$$\phi T = \phi(A + iB) = \phi A + +i\phi B,$$
$$\phi T^* = \phi(A - iB) = \phi A - i\phi B = (\phi T)^*.$$

Consider $\hat{\phi} := \phi/(\phi 1)$; obviously $\hat{\phi} \in \mathcal{X}^*$ and $\hat{\phi} 1 = 1$. By the proposition, $|\hat{\phi} T| = |\phi T|/(\phi 1) \leqslant \|T\|$; combined with $\hat{\phi} 1 = 1$, we find $\|\hat{\phi}\| = 1$. Thus $\hat{\phi} \in \mathcal{S}(\mathcal{X})$ and $\phi = (\phi 1)\hat{\phi}$. Note that $\phi 1 \geqslant 0$ since $1 \geqslant 0$.

Exercises 15.47

1. $0 \leqslant 1$ (as self-adjoint elements), and the order relation of $\mathbb{R}$ is subsumed in that of the self-adjoint elements. Similarly, in $C[0, 1]$, $f \leqslant g \Leftrightarrow \forall t, \; f(t) \leqslant g(t)$.
2. $\begin{pmatrix} 0 & 1-i \\ 1+i & -1 \end{pmatrix} \leqslant \begin{pmatrix} 1 & 1 \\ 1 & 0 \end{pmatrix}$. Note that $T \leqslant S$ does not mean "$\sigma(T) \leqslant \sigma(S)$" in general.
3. (a) A diagonal matrix is positive when all its diagonal coefficients are real and positive.
   (b) If the coefficients of a real symmetric matrix are positive, it does *not* follow that it is positive: $\forall i, j, \; A_{ij} \geqslant 0 \nRightarrow A \geqslant 0$.
   (c) But if a real symmetric matrix is dominated by its positive diagonal, meaning $A_{ii} \geqslant \sum_{j \neq i} |A_{ij}|$, then $A \geqslant 0$ (Gershgorin's theorem, Proposition 14.9).
4. Show $\mathrm{Re}(T) \geqslant 0 \Leftrightarrow \mathrm{Re}\,\mathcal{S}(T) \subseteq \mathbb{R}^+$; $T \geqslant 0 \Leftrightarrow \mathcal{S}(T) \subseteq \mathbb{R}^+$.
5. The similarity between self-adjoints and real numbers is striking. But not every property about inequalities of real numbers carries through to self-adjoints:

   (a) Not every two self-adjoints S and T are comparable, e.g., $T := \begin{pmatrix} 1 & 0 \\ 0 & -1 \end{pmatrix}$ satisfies neither $T \leqslant 0$ nor $T \geqslant 0$;
   (b) $0 \leqslant S \leqslant T$ does not imply $S^2 \leqslant T^2$ (unless S, T commute), e.g., $S := \begin{pmatrix} 2 & 1 \\ 1 & 1 \end{pmatrix}$, $T := \begin{pmatrix} 3 & 1 \\ 1 & 1 \end{pmatrix}$.

6. In $B(H)$, $S \leqslant T \Leftrightarrow \langle x, Sx \rangle \leqslant \langle x, Tx \rangle$ for all $x \in H$. In particular, $S^*S \leqslant T^*T \Leftrightarrow \|Sx\| \leqslant \|Tx\|$ for all $x \in H$ (e.g., $T^*T \geqslant 0$); deduce

   (a) If T is compact then so is S,
   (b) If T is Hilbert-Schmidt, then so is S,
   (c) For self-adjoint projections in $B(H)$, $P \leqslant Q$ when im $P \subseteq$ im Q,
   (d) The 'ellipsoid' associated with S^*S, namely $B_S := \{x : \|Sx\| \leqslant 1\}$ satisfies $B_T \subseteq B_S$.

7. Prove directly $S \leqslant T \Rightarrow R^*SR \leqslant R^*TR$ for all R, in $B(H)$.
8. In $B(H)$, if $T \geqslant 0$ then $\langle\!\langle x, y \rangle\!\rangle := \langle x, Ty \rangle$ is "almost" an inner product on H, except that it need not be definite; it still satisfies the Cauchy-Schwarz inequality though,

$$|\langle x, Ty \rangle|^2 \leqslant \langle x, Tx \rangle \langle y, Ty \rangle.$$

   Conversely, every inner product $\langle\!\langle , \rangle\!\rangle$ on H that is bounded, in the sense that $|\langle\!\langle x, y \rangle\!\rangle| \leqslant c\|x\|\|y\|$, is of this type. Use Exercise 10.17(1) to deduce that, for all $x \in H$,

$$\|Tx\| \leqslant \sqrt{\|T\|}\sqrt{\langle x, Tx \rangle}.$$

   In particular, $\langle x, Tx \rangle = 0 \Leftrightarrow Tx = 0$.
9. If $f : \mathbb{R} \to \mathbb{R}$ is increasing and $a \leqslant T \leqslant b$ then $f(a) \leqslant f(T) \leqslant f(b)$.
10. To calculate $f(A)$ for a positive self-adjoint matrix A, first diagonalize it as $A = PDP^{-1}$, then work out $f(A) = Pf(D)P^{-1}$. For example,

$$\begin{pmatrix} 0 & 1 \\ 1 & 0 \end{pmatrix}_{\pm} = \frac{1}{2}\begin{pmatrix} 1 & \pm 1 \\ \pm 1 & 1 \end{pmatrix}, \qquad \sqrt{\begin{pmatrix} 5 & 4 \\ 4 & 5 \end{pmatrix}} = \begin{pmatrix} 2 & 1 \\ 1 & 2 \end{pmatrix}.$$

11. There exists $A^{\alpha} \geqslant 0$ for $\alpha > 0$ when $A \geqslant 0$, for which $(A^{\alpha})^{1/\alpha} = A$.
12. If $-1 \leqslant A \leqslant 1$ then $A + i\sqrt{1 - A^2}$ is unitary. Hence any $T \in \mathcal{X}$ is the linear combination of at most four unitary elements. (Hint: $A = (U + U^*)/2$.)
13. Solve the equation $TAT = B$ for the unknown $T \geqslant 0$, given $A, B \geqslant 0$ invertible (Hint: $A^{\frac{1}{2}}TATA^{\frac{1}{2}} = (A^{\frac{1}{2}}TA^{\frac{1}{2}})^2$.)
14. If $J : \mathcal{X} \to \mathcal{Y}$ is an algebraic $*$-morphism, then

$$\mathcal{X}/\ker J \cong \operatorname{im} J \Leftrightarrow \operatorname{im} J \text{ is closed.}$$

Polar Decomposition

An important application of the use of square roots of positive self-adjoint elements is the following generalization of the polar decomposition of complex numbers to $B(H)$:

Proposition 15.48 (Polar Decomposition)

> **Every operator $T \in B(H)$ has a decomposition $T = U|T|$, in which $|T| :=$ $\sqrt{T^*T} \geqslant 0$ and $U : \overline{\operatorname{im}|T|} \to \overline{\operatorname{im} T}$ is an isometry.**

Proof T^*T is non-negative, so its square root $R := \sqrt{T^*T} \geqslant 0$ can be defined. R reduces to the previous definition of $|T|$ when T is normal, so it is common to write $|T|$ for R. Then $\||T|x\| = \|Tx\|$ for all $x \in H$, as

$$\langle |T|x, |T|y \rangle = \langle x, |T|^2 y \rangle = \langle x, T^*Ty \rangle = \langle Tx, Ty \rangle. \tag{15.4}$$

Let $U : \operatorname{im}|T| \to \operatorname{im} T$ be defined by $U(|T|x) := Tx$; it is well-defined by (15.4),

$$|T|(x - y) = 0 \ \Leftrightarrow \ T(x - y) = 0,$$

and isometric, so can be extended isometrically to $\overline{\operatorname{im}|T|} \to \overline{\operatorname{im} T}$ (Example 8.9(5)). It can be extended further to the whole of the Hilbert space H by letting $Ux = 0$ whenever x belongs to the orthogonal space $\ker|T|$, in which case it is called a *partial isometry*. $\square$

Examples 15.49

1. If the SVD of a compact operator is given by $T = U\Sigma V^*$, then its polar decomposition is $T = (UV^*)(V\Sigma V^*)$, since $T^*T = V\Sigma^2 V^*$ and $|T| = V\Sigma V^*$.
2. When T is normal and U is extended to a partial isometry, $T = |T|U$ is also true: $\ker|T| = \ker T$ by (15.4) and since $\ker T^* = \ker T$ (Proposition 15.12),

$$\overline{\operatorname{im}|T|} = (\ker|T|)^\perp = (\ker T)^\perp = (\ker T^*)^\perp = \overline{\operatorname{im} T}.$$

In fact,

$$\begin{aligned} &\text{for } x \in \ker|T|, &&|T|Ux = 0 = Tx, \\ &\text{for } x = |T|y \in \operatorname{im}|T|, &&|T|Ux = |T|U|T|y = |T|Ty = T|T|y = Tx, \end{aligned}$$

and by extension $|T|Ux = Tx$ for $x \in \overline{\operatorname{im}|T|}$ as well.

3. If T is invertible, then it implies, in succession, that T^*, T^*T, and $|T|$ are invertible; thus U is an onto isometry on H, hence unitary, and can be written as an exponential. Then $T = e^{i\Theta}|T|$ for some self-adjoint operator Θ, analogous to the polar decomposition of complex numbers.

Proposition 15.50

Every unitary operator in $B(H)$ is of the type e^{iT} with $T \in B(H)$ self-adjoint.

The group of invertible operators $\mathcal{G}(H) \subseteq B(H)$ is connected and generated by the exponentials.

Proof (i) The polar decomposition of any self-adjoint operator $B \in B(H)$ is $B = V|B|$ where

$$Vx := \begin{cases} x, & x \in \ker B_- \\ -x, & x \in (\ker B_-)^\perp = \overline{\operatorname{im} B_-} \end{cases}$$

since $B_+x \in \ker B_-$ ($B_-B_+ = 0$). Note that $V^2 = I$. Hence

$$V|B|x = VB_+x + VB_-x = B_+x - B_-x = Bx.$$

Let U be any unitary operator on H. It equals $U = A + iB$ where A, B are commuting self-adjoint operators such that $A^2 + B^2 = I$. It follows that A commutes with B_- (Example 15.39(3)) and thus preserves $\ker B_-$ and $\overline{\operatorname{im} B_-}$ (Exercise 8.6(10)). Accordingly, if $B = V|B|$ is the polar decomposition of B, as above, then V commutes with A: for all $x = u + v \in \ker B_- \oplus \overline{\operatorname{im} B_-}$,

$$VAx = VA(u + v) = Au - Av = A(u - v) = AVx.$$

The function $\arccos : [-1, 1] \to [0, \pi]$ is a continuous function, and $-1 \leqslant A \leqslant 1$, so we can define $C := \arccos A \in B(H)$, and this commutes with V. Let $T := VC$, so that $T^2 = V^2C^2 = C^2$. Hence,

$$\begin{aligned} e^{iT} &= (I - \tfrac{1}{2!}T^2 + \cdots) + iT(I - \tfrac{1}{3!}T^2 \cdots) \\ &= (I - \tfrac{1}{2!}C^2 + \cdots) + iV(C - \tfrac{1}{3!}C^3 + \cdots) \\ &= \cos C + iV \sin C \\ &= A + iV|B| \qquad (\sin \circ \arccos(A) = \sqrt{1 - A^2} = |B|) \\ &= U. \end{aligned}$$

(ii) Consider the polar decomposition of an invertible operator $T = U|T|$, where U is unitary and $|T|$ is invertible. By the above, $U = e^{iA}$, while $|T|$ has a logarithm, $|T| = e^B$ (Example 14.27(1)). Hence $T = e^{iA}e^B$ lies in the connected component of I (Proposition 13.24), which must therefore equal $\mathcal{G}(H)$.   □

Exercises 15.51

1. Examples of polar decompositions are $\begin{pmatrix} 1 & 0 \\ 0 & -2 \end{pmatrix} = e^{i\pi\begin{pmatrix} 0 & 0 \\ 0 & 1 \end{pmatrix}}\begin{pmatrix} 1 & 0 \\ 0 & 2 \end{pmatrix}$, and $\begin{pmatrix} 1 & 1 \\ 0 & 1 \end{pmatrix} \approx \begin{pmatrix} 0.89 & 0.45 \\ -0.45 & 0.89 \end{pmatrix}\begin{pmatrix} 0.89 & 0.45 \\ 0.45 & 1.34 \end{pmatrix}$.

2. If T is a compact operator in $B(H)$ with singular values σ_n and singular vectors $e_n, \tilde{e}_n$, then $|T|e_n = \sigma_n e_n$ and $U : e_n \mapsto \tilde{e}_n$.

3. The polar decomposition of the right-shift operator in ℓ^2 is trivial: $|R| = I$. What is it for the left-shift operator?

4. $T^* = |T|U^*$, $|T| = U^*T = T^*U$, and $|T^*| = UT^* = TU^*$, since U^*U is a projection onto $\operatorname{im}|T|$ and UU^* is a projection onto $\operatorname{im} T$. $\||T|\| = \|T\|$.

5. (a) T is normal $\Leftrightarrow |T^*| = |T|$,
   (b) T is positive self-adjoint $\Leftrightarrow T = |T|$,
   (c) T is unitary $\Leftrightarrow |T| = I$ AND T is invertible.

6. If $|S| = |T|$ and T is invertible then ST^{-1} is unitary.

7. When T is compact normal, with polar decomposition $T = |T|U = U|T|$, then U and $|T|$ are simultaneously diagonalizable, $U = P^{-1}e^{i\Theta}P$, $|T| = P^{-1}DP$, so that $T = P^{-1}De^{i\Theta}P$.

8. Adapt the proof of the Polar Decomposition theorem to show that if $T^*T \leqslant S^*S$ then the map $U : \operatorname{im} S \to \operatorname{im} T$, $Sx \mapsto Tx$, is a well-defined operator with $\|U\| \leqslant 1$ and $T = US$.

9. Every ideal in $B(H)$ is a $*$-ideal since

$$T \in \mathcal{I} \Rightarrow |T| = U^*T \in \mathcal{I} \Rightarrow T^* = |T|U^* \in \mathcal{I}.$$

10. Every invertible element T of a C^*-algebra can be written uniquely as $T = U|T|$ where U is unitary.

11. *Trace-class Operators:* Let $\operatorname{Tr} := \{T \in B(H) : \operatorname{tr}|T| < \infty\}$ with norm $\|T\|_{\operatorname{Tr}} := \operatorname{tr}|T|$ (Proposition 15.31 and Example 15.34(5)).

    (a) $\|T\|_{\operatorname{Tr}} = \||T|^{\frac{1}{2}}\|_F^2$, and $T \in \operatorname{Tr} \Leftrightarrow |T|^{\frac{1}{2}} \in \mathcal{HS}$,
    (b) $\operatorname{tr}(T)$ is independent of the orthonormal basis,
    (c) $|\operatorname{tr}(ST)| \leqslant \|S\|\|T\|_{\operatorname{Tr}}$; in particular $\|T\|_F \leqslant \|T\|_{\operatorname{Tr}}$,
    (d) Tr is a closed $*$-ideal in $B(H)$,
    (e) $T \in \operatorname{Tr} \Leftrightarrow T = AB$ where $A, B \in \mathcal{HS}$,
    (f) $\operatorname{tr}|T| = \sum_n \sigma_n$, where σ_n are the singular values of T (repeated according to their multiplicities). $\operatorname{tr} T = \sum_n \lambda_n$ holds when T is normal and λ_n are its eigenvalues.

15.8 Spectral Theorem for Normal Operators

There is one further extension of the functional calculus of the C^*-algebra $B(H)$: when T is a normal operator, $f(T)$ may be defined even for bounded measurable functions.

Let 1_Ω be the characteristic function defined on a bounded open subset $\Omega \subseteq \mathbb{C}$. To find an operator that corresponds to 1_Ω, we will be needing the following lemma:

Monotone Convergence Theorem for Self-Adjoint Operators: *If $A_n \geqslant 0$ is a decreasing sequence of commuting self-adjoint operators in $B(H)$ then A_n converges strongly to some operator $A \geqslant 0$.*

Proof It is easy to show that when $0 \leqslant S \leqslant T$ commute,

$$S^2 \leqslant S^2 + (T - S)^2 = T^2 - 2S(T - S) \leqslant T^2.$$

From this it follows that A_n^2 is also a decreasing sequence, as is $\|A_n x\|$ by Example 15.43(6c). Also $\|A_n x - A_m x\|^2 \leqslant \big| \|A_m x\|^2 - \|A_n x\|^2 \big| \to 0$ as $n, m \to \infty$, since $A_n A_m \geqslant A_n^2$ for $n \geqslant m$, so $(A_n x)$ is a Cauchy sequence in H. Now apply the corollary of the uniform boundedness theorem (Corollary 11.37). $\qquad \square$

It follows easily from this that an *increasing* sequence of bounded self-adjoint operators $A_n \leqslant c$ converges strongly to some operator $A \leqslant c$.

There exist increasing sequences of positive continuous functions $f_n : \mathbb{C} \to \mathbb{R}^+$ which converge pointwise to 1_Ω; for example, take $f_n(z) := \min(1, n\, d(z, \Omega^c))$. Using the continuous functional calculus defined in Theorem 15.38, $f_n(T)$ exist as positive self-adjoint operators on H with norm equal to $\|f_n(T)\| = \|f_n\|_C = 1$.

We can therefore define $1_\Omega(T)x := \lim_{n \to \infty} f_n(T)x$ for all $x \in H$. This definition can be extended to closed subsets F of $\mathbb{C}$: there are nested open sets U_n such that $F = \bigcap_n U_n$, so $1_F(T)$ can be defined by $1_F(T)x := \lim_{n \to \infty} 1_{U_n}(T)x$ by the monotone convergence theorem above. Some properties of $1_\Omega(T)$ are:

1. $1_\Omega(T)$ is an orthogonal projection; so $1_\Omega(T) \geqslant 0$.
   Proof: Write $A_n := f_n(T)$ and $A := 1_\Omega(T)$. Then

$$\langle Ay, x \rangle = \lim_{n \to \infty} \langle A_n y, x \rangle = \lim_{n \to \infty} \langle y, A_n x \rangle = \langle y, Ax \rangle,$$

$$\|(A_n^2 - A^2)x\| = \|(A_n + A)(A_n - A)x\| \leqslant (1 + \|A\|)\|(A_n - A)x\| \to 0.$$

   Thus $1_\Omega(T)^2 = 1_\Omega(T)$ is self-adjoint, and hence othogonal (Example 15.14(1)).

2. (a) If U, V are disjoint open sets, then $1_U(T) + 1_V(T) = 1_{U \cup V}(T)$,
   (b) $1_{U \cap V}(T) = 1_U(T)1_V(T)$.
   Proof: If $f_n(z) \to 1_U(z)$ and $g_n(z) \to 1_V(z)$ for $z \in \mathbb{C}$ then $f_n(z) + g_n(z) \to 1_U(z) + 1_V b(z) = 1_{U \cup V}(z)$. So by the continuous functional calculus and the strong convergence of f_n and g_n, it follows that $f_n(T)x + g_n(T)x \to 1_{U \cup V}(T)x$ for any $x \in H$.

John von Neumann(1903–1957) Originally from Budapest, von Neumann studied in Berlin, under Weyl and Polya, but graduated at 23 years under Fejér in Budapest with a thesis on ordinal numbers. A young party-going genius, in 1926-30 he defined Hilbert spaces axiomatically as foundation for the brand new quantum mechanics and generalized the spectral theorem to unbounded self-adjoint operators. In the 1930s he went to the Princeton Institute, proved the ergodic theorem, and studied rings of operators and group representations; only turbulent fluid dynamics proved too hard (it remains unsolved today); in 1944 he started game theory, proving the mini-max theorem, then on to computers and automata theory.

Similarly, the second statement results from $f_n(z)g_n(z) \to 1_U(z)1_V(z) = 1_{U \cap V}(z)$.

3. $1_\varnothing(T) = 0$, $1_{\sigma(T)}(T) = I$ (since if $\sigma(T) \subseteq U$ and $f_n \to 1_U$, then $f_n|_{\sigma(T)} = 1$ for n large enough).

The projections $1_E(T)$ for Borel sets E are defined by the same procedure and are said to be the *spectral measure* associated with T. We gloss over the details of the exact definition (see [10]).

One can now follow the same steps of creating the space of simple functions through to $L^1(\mathbb{C})$, but starting from the projections $1_E(T)$ as 'indicator functions'. The end result is a functional calculus in which $f(T)$ is defined for any complex-valued $f \in L^\infty(\sigma(T))$: If f is approximated by $\sum_i a_i 1_{U_i}$, then $f(T)$ is approximately $\sum_i a_i 1_{U_i}(T)$. Indeed, $f(T)$ is still meaningful even if $f \in L^1(\sigma(T))$ but need not be a "bounded" (i.e., continuous) operator.

Theorem 15.52 (von Neumann's Spectral Theorem)

For any normal operator T and $f \in L^\infty(\sigma(T))$, there is a spectral measure E_λ such that

$$f(T) = \int_{\sigma(T)} f(\lambda) \, dE_\lambda$$

Proof For any $x, y \in H$, define $\mu_{x,y}(U) := \langle x, 1_U(T)y \rangle$ for any open bounded subset $U \subseteq \mathbb{C}$. By the properties proved above, $\mu_{x,y}$ can be extended to a measure with support equal to $\sigma(T)$. (It is not a Lebesgue measure on $\mathbb{C}$ as it is not translation invariant, but Borel sets are $\mu_{x,y}$-measurable.) It has the additional properties:

$$\mu_{x,y_1+y_2} = \mu_{x,y_1} + \mu_{x,y_2}, \quad \mu_{x,\lambda y} = \lambda \mu_{x,y}, \quad \mu_{y,x} = \overline{\mu_{x,y}}, \quad 0 \leqslant \mu_{x,x} \leqslant \|x\|^2.$$

It follows that for any $f \in L^\infty(\sigma(T))$, $\langle\langle x, y \rangle\rangle := \int_{\sigma(T)} f \, d\mu_{x,y}$ is a semi-inner-product which is bounded in the sense $|\langle\langle x, y \rangle\rangle| \leqslant \|f\|_{L^\infty} \|x\| \|y\|$. Thus, by Exercise 15.47(8), $\langle\langle x, y \rangle\rangle = \langle x, Sy \rangle$ for some continuous operator S which we henceforth call $f(T)$,

$$\langle x, f(T)y \rangle = \int_{\sigma(T)} f \, d\mu_{x,y}.$$

$f(T)$ *agrees with the earlier definition for* $f \in C(\sigma(T))$: Any such f is uniformly continuous, so for δ small enough $f B_\delta(z) \subseteq B_\epsilon(f(z))$, independently of $z \in \sigma(T)$. Let B_i be squares, with centers λ_i and diameter less than δ, which partition $\sigma(T)$; one can find slightly smaller closed squares $A_i \subset B_i$ and slightly larger open squares $C_i \supset B_i$, such that $\sum_i \mu_{x,y}(C_i \setminus A_i) < \epsilon$. Moreover, one can find continuous functions h_i such that $1_{A_i} \leqslant h_i \leqslant 1_{C_i}$ and $\sum_i h_i = 1$; for example, let $h_i(s, t) := h(s)h(t)$ where $h(t) = \min(1, r \, d(t, I^c))$ is a continuous real function with support equal to I and taking the value 1 just inside it. Then (writing $\mu = \mu_{x,y}$)

$$\langle x, f(T)y \rangle = \sum_i \langle x, f h_i(T)y \rangle \approx f(\lambda_i)\langle x, h_i(T)y \rangle \approx \sum_i f(\lambda_i)\mu(B_i).$$

More rigorously, (it is enough to consider real-valued functions)

$$\langle x, f h_i(T)y \rangle \leqslant (f(\lambda_i) + \epsilon)\mu(C_i)$$

$$= (f(\lambda_i) + \epsilon)\mu(B_i) + (f(\lambda_i) + \epsilon)(\mu(C_i) - \mu(B_i))$$

$$-\langle x, f h_i(T)y \rangle \leqslant -f(\lambda_i)\mu(B_i) + \epsilon\mu(B_i) + (f(\lambda_i) - \epsilon)(\mu(B_i) - \mu(A_i))$$

$$\therefore |\langle x, f h_i(T)y \rangle - f(\lambda_i)\mu(B_i)| \leqslant \epsilon\mu(B_i) + |f(\lambda_i) + \epsilon|(\mu(C_i) - \mu(A_i))$$

$$\therefore |\langle x, f(T)y \rangle - \sum_i f(\lambda_i)\mu(B_i)| = \Big|\sum_i \langle x, f h_i(T)y \rangle - \sum_i f(\lambda_i)\mu(B_i)\Big|$$

$$\leqslant \sum_i |\langle x, f h_i(T)y \rangle - f(\lambda_i)\mu(B_i)|$$

$$\leqslant \sum_i (|f(\lambda_i)| + \epsilon)(\mu(C_i) - \mu(A_i)) + \epsilon\mu(B_i)$$

$$\leqslant (\|f\|_C + \epsilon)\epsilon + \epsilon$$

Hence, in the limit $\epsilon \to 0$, $\langle x, f(T)y \rangle = \int_{\sigma(T)} f \, d\mu_{x,y}$.
The map $f \mapsto f(T)$ *is a* ∗-*morphism from* $L^\infty(\sigma(T))$ *to* $B(H)$: Linearity is immediate,

$$(f + \lambda g)(T) = \int_{\sigma(T)} (f + \lambda g) \, d\mu_{x,y} = f(T) + \lambda g(T).$$

$\bar{f}(T) = f(T)^*$ since

$$\langle x, \bar{f}(T)y \rangle = \int \bar{f}\,\mathrm{d}\mu_{x,y} = \overline{\int f\,\mathrm{d}\mu_{x,y}} = \overline{\langle y, f(T)x \rangle} = \langle f(T)x, y \rangle = \langle x, f(T)^*y \rangle.$$

$fg(T) = f(T)g(T)$ follows from

$$\int_{\sigma(T)} \mathrm{d}\mu_{x,f(T)y} = \langle x, f(T)y \rangle = \int_{\sigma(T)} f\,\mathrm{d}\mu_{x,y}$$

$$\Rightarrow \quad \langle x, fg(T)y \rangle = \int fg\,\mathrm{d}\mu_{x,y} = \int f\,\mathrm{d}\mu_{x,g(T)y} = \langle x, f(T)g(T)y \rangle.$$

$\square$

In particular, $T = \int_{\sigma(T)} \lambda\,\mathrm{d}E_\lambda$. This result, and the next one, are often claimed to be the pinnacle of the subject of functional analysis.

Embedding in $B(H)$

Theorem 15.53 (Gelfand-Naimark)

Every C^*-algebra is isometrically embedded in $B(H)$, for some Hilbert space H.

Proof We have already seen that every Banach algebra $\mathcal{X}$ is embedded in $B(\mathcal{X})$ (Theorem 13.8); as in the proof of that theorem, we will again denote elements of $\mathcal{X}$ by lower-case letters. The main difficulty is that there is no natural inner product defined on $\mathcal{X}$ or $B(\mathcal{X})$. Rather there are many semi-inner-products, one for each $\phi \in S$, $\langle x, y \rangle_\phi := \phi(x^*y)$.

Let $\mathcal{M}_\phi := \{ x : \phi(x^*x) = 0 \}$; it is a closed left-ideal, since for any $a \in \mathcal{X}$ and $x \in \mathcal{M}$, then $ax \in \mathcal{M}_\phi$

$$0 \leqslant \phi(x^*a^*ax) \leqslant \phi(x^*x)\|a\|^2 = 0.$$

This allows us to turn $\mathcal{X}/\mathcal{M}_\phi$ into an inner product space, which can be completed to a Hilbert space H_ϕ (Examples 10.7(2) and 13.10(21)). The inner product on $\mathcal{X}/\mathcal{M}_\phi$ is given by

$$\langle x + \mathcal{M}_\phi, y + \mathcal{M}_\phi \rangle := \phi(x^*y).$$

The ∗-morphism $L : \mathcal{X} \to B(H_\phi)$: For any $a \in \mathcal{X}$, consider the linear map defined by $L_a(x + \mathcal{M}_\phi) := ax + \mathcal{M}_\phi$ on $\mathcal{X}/\mathcal{M}_\phi$; this is well-defined since $a\mathcal{M}_\phi \subseteq \mathcal{M}_\phi$. It is continuous with $\|L_a\| \leqslant \|a\|$ since,

$$\|L_a(x+\mathcal{M}_\phi)\| = \|ax+\mathcal{M}_\phi\| = \sqrt{\phi(x^*a^*ax)} \leqslant \sqrt{\phi(x^*x)}\|a\|=\|a\|\,\|x + \mathcal{M}_\phi\|.$$

This map extends uniquely to one in $B(H_\phi)$ (Example 8.9(5)).

Clearly L_a is linear in a, $L_{ab} = L_a L_b$, and $L_1 = I$, but it also preserves the involution $L_{a^*} = L_a^*$,

$$\langle x + \mathcal{M}_\phi, L_a(y + \mathcal{M}_\phi)\rangle = \phi(x^*ay) = \phi((a^*x)^*y) = \langle L_{a^*}x + \mathcal{M}_\phi, y + \mathcal{M}_\phi\rangle.$$

It remains a ∗-morphism when extended to $B(H_\phi)$, by continuity of the adjoint.

The final Hilbert space: However L need not be 1–1. To remedy this deficiency, let $H := \prod_{\phi \in S} H_\phi$ be the Hilbert space of "sequences" $\boldsymbol{x} := (x_\phi)_{\phi \in S}$ such that $x_\phi \in H_\phi$ and $\sum_{\phi \in S} \langle x_\phi, x_\phi\rangle_{H_\phi} < \infty$; it has the inner product

$$\langle \boldsymbol{x}, \boldsymbol{y}\rangle := \sum_{\phi \in S} \langle x_\phi, y_\phi\rangle_{H_\phi}.$$

It is straightforward to show that H is indeed a Hilbert space, by analogy with ℓ^2.

Let $J_a\boldsymbol{x} := (L_a x_\phi)_{\phi \in S}$, so that $J_a : H \to H$ is obviously linear, and also continuous since

$$\|J_a\boldsymbol{x}\|^2 = \sum_\phi \|L_a x_\phi\|^2 \leqslant \|a\|^2 \sum_\phi \|x_\phi\|^2 = \|a\|^2\|\boldsymbol{x}\|^2.$$

The mapping $a \mapsto J_a$, $\mathcal{X} \mapsto B(H)$ is an algebraic ∗-morphism,

$$\langle \boldsymbol{y}, J_a\boldsymbol{x}\rangle = \sum_\phi \langle y_\phi, L_a x_\phi\rangle = \sum_\phi \langle L_a^* y_\phi, x_\phi\rangle = \langle J_{a^*}\boldsymbol{y}, \boldsymbol{x}\rangle.$$

Moreover it is 1–1, for if $J_a = 0$ then $L_a x_\phi = 0$ for any x_ϕ and $\phi \in S$, in particular $a + \mathcal{M}_\phi = L_a 1 = 0$. But this means that for all $\phi \in S$, $a \in \mathcal{M}_\phi$, i.e., $\phi(a^*a) = 0$, and this can only hold when $\sigma(a^*a) \subseteq S(a^*a) = 0$, so $\|a\|^2 = \|a^*a\| = 0$ and $a = 0$.

Since every such ∗-morphism between C^*-algebras is isometric, the theorem is proved. $\qquad\square$

Note that in the above *GNS construction*, when $\mathcal{X}$ is represented in $B(H)$, every state $\phi \in S(\mathcal{X})$ is associated with a unit vector $\boldsymbol{x} \in H$, such that $\phi y = \langle \boldsymbol{x}, J_y\boldsymbol{x}\rangle$. *Proof*: The vector in question is $\boldsymbol{x} := (x_\psi)_{\psi \in S}$ where $x_\phi = 1 + \mathcal{M}_\phi$ and $x_\psi = 0$ otherwise. For every $y \in \mathcal{X}$,

$$\phi(y) = \langle 1+\mathcal{M}_\phi, y+\mathcal{M}_\phi\rangle_{H_\phi} = \langle x_\phi, L_y x_\phi\rangle_{H_\phi} = \sum_\psi \langle x_\psi, L_y x_\psi\rangle_{H_\psi}=\langle \boldsymbol{x}, J_y\boldsymbol{x}\rangle_H.$$

Remarks 15.54

1. The Banach algebra axiom $\|1\| = 1$ is redundant for C^*-algebras as it follows from $\|T^*T\| = \|T\|^2$ (assuming $\mathcal{X} \neq 0$).
2. The use of $A < B$ is best avoided: it may either mean $A \leqslant B$ but $A \neq B$ or that $\sigma(B - A) \subset \,]0, \infty[$. However the use of $A > 0$ in the latter sense is standard.

Hints to Selected Problems

2.2 (1) Writing $s := a - c$, $t := c - b$, and substituting into $|s + t| \leqslant |s| + |t|$ gives the triangle inequality.

2.3 (2) (a) $\exists a \in A, \exists b \in B, \ d(a, b) \leqslant 2$, (b) $\forall \epsilon > 0, \exists a \in A, \exists b \in B, \ d(a, b) \leqslant \epsilon$.

2.14 (7) (i) The two sets have, respectively, the shapes of a diamond, and a square with a smaller concentric square removed (the outer boundary is included but the inner one is not). (ii) The shapes are the same but intersected with the first quadrant.

(9) For example, $\mathbb{R} \setminus \{a\}$.

2.20 (2) The complement of the set is $\{x \in \mathbb{Q} : x^2 > 2\}$ since $\sqrt{2}$ is irrational. To prove the set is open, one needs to find a small enough ϵ such that

$$2 < (x - \epsilon)^2 = x^2 - 2\epsilon x + \epsilon^2.$$

(6) Try the graph of the exponential function and the x-axis in $\mathbb{R}^2$.

(7) The Cantor set is the intersection of all of these closed intervals.

(8) First show the set $\{x \in [0, 1] : \frac{n_1 + \cdots + n_k}{k} \leqslant 5\}$ for fixed k is closed.

(11) The answer to the first question is of course no: all points on a circle are equally close to the center; the second is also false e.g., in $\mathbb{Z}$; it is true however in $\mathbb{R}^2$ because the line joining an interior point to x contains closer points. What properties does the metric space need to have for this statement to be true?

(14) No. Take the subsets $A := [-1, 1]$ and $B := \mathbb{R} \setminus \{0\}$ in $\mathbb{R}$.

2.23 (6) Any ball $B_r(x)$ will contain a point a of the dense open set A. There will therefore be a small ball $B_\epsilon(a) \subseteq A \cap B_r(x)$ which contains a point $b \in B$.

(7) The complement of the Cantor set is open and dense.

(10) $\partial U = \bar{U} \setminus U$ contains no balls.

3.6 (1c) $n/a^n = n/(1 + \delta)^n \leqslant \frac{n}{n(n-1)\delta^2/2} \leqslant \frac{4}{\delta^2} \frac{1}{n} \to 0$.

(1e) $a_n := (1 + \frac{1}{n})^n = 2 + \frac{1}{2}(1 - \frac{1}{n}) + \frac{1}{3!}(1 - \frac{1}{n})(1 - \frac{2}{n}) + \cdots + \frac{1}{n^n}$. So $a_{n+1} > a_n$, yet $a_n < 2 + \frac{1}{2} + \frac{1}{3!} + \cdots < 2 + \frac{1}{2} + \frac{1}{4} + \cdots = 3$.

J. Muscat, *Functional Analysis*, https://doi.org/10.1007/978-3-031-27537-1

(1f) $a_n \to \infty$ means $\forall \epsilon > 0$, $\exists N$, $n \geqslant N \Rightarrow a_n > \epsilon$.

(2) The limits must satisfy $x = 2 + \sqrt{x}$ and $x = 1 + 1/x$ respectively.

(3) Eventually, $|a_n| < c < 1$, so $|a_n|^n < c^n$.

3.13 (3) See Proposition 7.9

(4) If $t_n \to t$ then $t \neq 0$ and $|t_n| \geqslant c > 0$, so that $|1/t_n - 1/t| = |t - t_n|/|t_n||t| \to 0$.

(10a) The map $f(t) = (\cos t, \sin t)$ is a continuous bijective map from $[0, 2\pi[$ to the circle. The inverse map is discontinuous at $(-1, 0)$.

(10b) Take f to be a constant function, and $t_n = n$.

(10c) Take $f(t) := t^2$ and $U :=]-1, 1[$. Examples of open mappings on $\mathbb{R}$ are polynomials which have no local maxima/minima.

(11) $(f^{-1}F)^{\complement} = f^{-1}F^{\complement}$ is open. The identity map $[0, 1[\to [0, 2]$ is a continuous open mapping whose image is not closed.

(17) $d(x, A)/(d(x, A) + d(x, B))$.

(18) The map $t \mapsto \frac{t-a}{b-a}$ $(a \neq b)$ is a homeomorphism between (i) $]a, b[$ and $]0, 1[$, (ii) $[a, b]$ and $[0, 1]$, (iii) $]a, b]$ and $]0, 1]$, (iv) $[a, b[$ and $[0, 1[$. Translations make (v) $]-\infty, a[$ homeomorphic to $]-\infty, 0[$, and (vi) $]-\infty, a]$ with $]-\infty, 0]$, (vii) $[a, \infty[$ with $[0, \infty[$, and (viii) $]a, \infty[$ with $]0, \infty[$. Reflections $t \mapsto -t$ (followed by a translation) then show that the intervals in (iii) and (iv) are homeomorphic, as well as (v) and (viii), and (vi) with (vii). Finally $]0, 1]$ is homeomorphic to $[0, \infty[$ via the map $t \mapsto \frac{1}{t} - 1$; this same map shows $]0, 1[$ is homeomorphic to $]0, \infty[$, and this in turn, is homeomorphic to $\mathbb{R}$ via $t \mapsto t - \frac{1}{t}$.

(19) Points $\{x\}$ are open in $\mathbb{N}$ but not in $\mathbb{Q}$.

4.11 (1) The difference between the nth and mth terms of decimal approximations is at most $10^{-\min(m,n)}$.

(4) The finite number of values have a minimum distance ϵ between them.

(5) Note that $\sum_n 1/n \to \infty$.

(6) $$|d(x_n, y_n) - d(x_m, y_m)| \leqslant |d(x_n, y_n) - d(y_n, x_m)|$$
$$+ |d(x_m, y_n) + d(x_m, y_m)|$$
$$\leqslant d(x_n, x_m) + d(y_n, y_m)$$

(7) For example, the continuous function $f(t) := 1/t$, defined on $]0, 1] \to [1, \infty[$, maps the Cauchy sequence $(1/n)$ to the unbounded sequence (n).

(9) $\sqrt{n+1} - \sqrt{n} = \sqrt{n}((1 + 1/n)^{1/2} - 1) = \frac{1}{2\sqrt{n}} + \cdots$

(11) If $\{x_n\}$ are the values of a Cauchy sequence, and x is a boundary point, then there is a subsequence $x_m \to x$ (by Proposition 3.4). If all points are isolated, then the sequence is eventually constant.

(14) Any Cauchy sequence in a discrete metric space must eventually be constant.

(15) The intersection of the balls can contain at most one point, since $r_n \to 0$. In fact, if $x_n \to x$, then $x \in B_{r_n}[x_n]$ for all n, since the balls are nested.

(16) First show that $f(n) = f(1 + \cdots + 1) = nf(1)$, then $f(m/n) = \frac{m}{n}f(1)$.

4.18 (1b) $\qquad |(x_2 - x_1)y_2 + x_1(y_2 - y_1)| \leqslant (|y_2||x_1 - x_2| + |x_1||y_2 - y_1|)$

$$\leqslant (|x_1 - x_2| + |y_2 - y_1|),$$

$\qquad |(x_1 + x_2)(x_1 - x_2) + (y_2 - y_1)| \leqslant 2|x_1 - x_2| + |y_1 - y_2|.$

(5) Let $f : X \to Y$ be an equivalence; then every Cauchy sequence $(x_n)_{n \in \mathbb{N}}$ in X corresponds to a Cauchy sequence in Y, by uniform continuity and Proposition 4.13. Since equivalences are homeomorphisms, $(x_n)_{n \in \mathbb{N}}$ converges precisely when $(f(x_n))_{n \in \mathbb{N}}$ does. So X is complete $\Leftrightarrow$ Y is complete.

4.22 (2) Repeat the proof of Proposition 2.10, using $B_{r_n}(a_n)$ instead of $B_{r(x)}(x)$, where a_n is an approximation of x.

(4) Let X be an uncountable set with the discrete metric. Then $B_{1/2}(x)$, for each $x \in X$, form an uncountable collection of disjoint sets.

5.8 (1) Take $X \setminus \{x_1\}$ and $X \setminus \{x_2\}$ as the open sets; alternatively take small enough balls. For (b), take $X \setminus F_1$ and $X \setminus F_2$.

(2) To show that every subset of $\mathbb{Q}$ with at least two points, is disconnected, use the same idea with some other irrational.

(5) Consider the open sets $f^{-1}\{0\}$ and $f^{-1}\{1\}$.

(11) Suppose $f(a) < f(y)$; $f(x) > f(y)$ is impossible else there is some $z \in [a, x]$ such that $f(z) = f(y)$.

5.13 (2) The metric space is the union of the path images, whose intersection contains the fixed point.

(5) Use Theorem 5.10 with $A_y := X \times \{y\}$ and $B := \{x_0\} \times Y$.

(6) Without loss of generality, take $x = 0$; then $\mathbb{R}^2 \setminus \{x\}$ is connected using the unit circle and radial lines te for $t > 0$ and unit vectors e.

(8) Otherwise, the interior and exterior of the set would disconnect a component.

(10a) If a component C has a boundary point $a \notin C$, then $C \cup B_\epsilon(a)$ would be a strictly larger connected set.

6.4 (3) If B is bounded, so $B \subseteq B_r(x)$, then $\overline{B} \subseteq \overline{B_r(x)}$.

6.9 (3) From some N onwards, $x_n \in B_\epsilon(x_N)$; cover the rest of the values x_m with $B_\epsilon(x_m)$.

(4) Let $B \subseteq \bigcup_{i=1}^{N} B_{\epsilon/2}(x_i)$, then $\overline{B} \subseteq \bigcup_{i=1}^{N} \overline{B_{\epsilon/2}(x_i)} \subseteq \bigcup_{i=1}^{N} B_\epsilon(x_i)$ (Theorem 2.19).

6.22 (7) Suppose $d(K, F) = 0$, then there are asymptotic sequences $a_n \in K, b_n \in F$; $(a_n)_{n \in \mathbb{N}}$ has a convergent subsequence, and therefore $(b_n)_{n \in \mathbb{N}}$ converges to the same limit. But then $K \cap F \neq \emptyset$.

(8) After showing $K \subset B_r(r, 0)$, use the fact that there is a point $a \in K$ which has maximum distance from $(r, 0)$ less than r.

(14) The unit sphere is a closed subset of the cube $[-1, 1]^n$.

(17) $X \times Y$ is complete and totally bounded by Proposition 4.7 and Exercise 6.9(1).

6.28 (2) If $f_n \to f$ with $f_n \in C(X, \mathbb{R})$, then $f_n(x) \to f(x)$ in $\mathbb{C}$, and taking the imaginary parts shows that $f(x) \in \mathbb{R}$.

(5) $f(y) - f_n(y) \leqslant f(y) - f_N(y) \leqslant |f(y) - f(x)| + |f(x) - f_N(x)| + |f_N(x) - f_N(y)| < \epsilon$ where N depends on x, and $|x - y| < \delta$, small enough but independent of x (Proposition 6.17). So $f - \epsilon \leqslant f_n \leqslant f$ on $B_\delta(x)$ for $n \geqslant N$. By compactness, one N will suffice.

(6) Convert any binary sequence (of 0s and 1s) into a "tent" function in $C(\mathbb{R}^+)$; there are uncountably many such functions and their distance from each other is at least 1.

(10) $(t + |t|)/2 \approx t(t + 1)/2$.

7.8 (3) Balls look like circles, squares and diamonds in the 2-norm, ∞-norm, and 1-norm respectively.

(6) Let $A := \{|a_n|\}$, $B := \{|b_n|\}$. Then from Review 7.2(13), $\sup |\lambda a_n| = \sup |\lambda| A = |\lambda| \sup A$, $\sup |a_n + b_n| \leqslant \sup(A + B) \leqslant \sup A + \sup B$, and if $\sup A = 0$, then $0 \leqslant |a_n| \leqslant 0$ implying $a_n = 0$ for all n.

(9) The functions $f_n := 1_{[0,1/n]}$ converge to 0 in $L^1[0, 1]$ but not in $L^\infty[0, 1]$. The inequality $\|x\|_{\ell^\infty} \leqslant \|x\|_{\ell^1}$ remains true for sequences, so convergence in ℓ^1 implies that in ℓ^∞.

(10) For $r > \|x\|$, $x \in rC$, so $\lambda x \in \lambda rC = |\lambda| rC$, i.e., $\|\lambda x\| \leqslant |\lambda| \|x\|$; but then $\|x\| \leqslant \frac{1}{\lambda} \|\lambda x\|$. If $s > \|y\|$, then $x + y \in rC + sC = (r + s)C$, hence $\|x + y\| \leqslant \|x\| + \|y\|$.

7.15 (5) Let $x, y \in \bar{C}$; then there are points $a, b \in C$ within ϵ of x and y. So any point on the line $tx + (1 - t)y$ is also close to a point on the line $ta + (1 - t)b$ which lies in C because

$$\|tx + (1 - t)y - ta - (1 - t)b\| \leqslant t\|x - a\| + (1 - t)\|y - b\| < \epsilon.$$

(7) A convex set C is the union of line segments that start from a fixed point $x_0 \in C$, then use Theorem 5.10.

(8) If $\lambda a_n \to x$, $a_n \in A$, then $a_n \to x/\lambda$ (for $\lambda \neq 0$) and $x/\lambda \in \bar{A}$. Conversely, if $x \in \lambda \bar{A}$, i.e., $x = \lambda a$ with $a_n \to a$, then $\lambda a_n \to \lambda a = x$ and $x \in \overline{\lambda A}$.

Similarly, when $a_n \to a$, $a_n \in A$, and $b_n \to b$, $b_n \in B$, then $a_n + b_n \to a + b$, so $a + b \in \overline{A + B}$. An example in $\mathbb{R}$ is $A := \{n + 1/n : n = 2, 3, \dots\}$ and $B := \{-n : n = 1, 2, \dots\}$.

7.21 (1c) $\sum_{i=N}^{N+k} x_i = \sum_{i=0}^{N+k} x_i - \sum_{i=0}^{N-1} x_i \to 0$ as $N \to \infty$, since convergent sequences are Cauchy.

(3) The odd sub-sums $a_1 - (a_2 - a_3) - (a_4 - a_5) + \cdots$ are decreasing, and bounded below by the increasing even sub-sums $(a_1 - a_2) + (a_3 - a_4) + \cdots$.

7.23 (6) Applying the Cauchy test to $\sum_n \frac{1}{n^p}$: the series $\sum_n 2^n/(2^{np})$ converges only when $p - 1 < 0$; for $p = 1$, $\sum_n \frac{1}{n}$ diverges; $\sum_n \frac{1}{n \log_2 n}$ becomes $\sum_n \frac{2^n}{2^n n}$ which diverges; etc.

(12) For N large enough $\|x_1 + \cdots + x_N - x\| < \epsilon$ as well as $\sum_{n=N+1}^\infty \|x_n\| < \epsilon$. So for k large enough that $n_1, \dots, n_k$ include $1, \dots, N$,

$$\|x_{n_1} + \cdots + x_{n_k} - x\| \leqslant \|x_1 + \cdots + x_N - x\| + \sum \|x_{\text{extra}}\|$$

8.6 (3) im R is closed since for $Rx_n \to y$, the first components give $0 \to y_0$, so $y = (0, y_1, \ldots) = R(y_1, \ldots)$.

(4) $T = L^4 - I$.

(7) If $x_n \to x$ then $x_n - x \to 0$ and $Tx_n - Tx = T(x_n - x) \to T0 = 0$.

8.13 (1) Proof that im T is not closed: Let $v_n := (1, 1/2, \ldots, 1/n, 0, 0, 0, \ldots)$, then $Tv_n = (1, 1/4, \ldots, 1/n^2, 0, \ldots)$ converges to $(1, 1/4, \ldots) \in \ell^1$ as $n \to \infty$ since

$$\|(0, \ldots, 0, 1/(n+1)^2, \ldots)\|_{\ell^1} = \sum_{n=N+1}^{\infty} \frac{1}{n^2} \to 0$$

Yet, there is no sequence in ℓ^1 which maps to this sequence as $(1, 1/2, 1/3, \ldots) \notin \ell^1$.

(2) $e_n/n \to 0$ in ℓ^1 because $\|(0, \ldots, 0, 1/n, 0, \ldots)\|_{\ell^1} = 1/n \to 0$, but $e_n \nrightarrow 0$ since $\|(0, \ldots, 0, 1, 0, \ldots)\|_{\ell^1} = 1$.

(4) $\|T\| = 1$, $\|T_a\| = 1$, $\|T_g\| = 1$, $\|M_g\| = \|g\|_C$

(5) Proof for first matrix. Assuming, without loss of generality, that $|\mu| \leqslant |\lambda|$,

$$\left\| \begin{pmatrix} \lambda & 0 \\ 0 & \mu \end{pmatrix} \begin{pmatrix} x \\ y \end{pmatrix} \right\|^2 = \left\| \begin{pmatrix} \lambda x \\ \mu y \end{pmatrix} \right\|^2 = |\lambda|^2 |x|^2 + |\mu|^2 |y|^2 \leqslant |\lambda|^2 (|x|^2 + |y|^2)$$

so $\|Tx\| \leqslant |\lambda| \|x\|$. However for $x = \begin{pmatrix} 1 \\ 0 \end{pmatrix}$, $Tx = \lambda x$, so $|\lambda| \leqslant \|T\| \leqslant |\lambda|$.

(7) Choose unit x_n such that $\|T_n x_n\| \geqslant \|T_n\| - 1/2^n$.

8.19 (6) For $x = (a_i)$, take the supremum over i of

$$\left| T_{ii} a_i + \sum_{j \neq i} T_{ij} a_j \right| \geqslant (|T_{ii}||a_i| - \sum_{j \neq i} |T_{ij}| \|x\|)$$

$$\geqslant c\|x\| - (\sup_i |T_{ii}|)(\|x\| - |a_i|) \approx c\|x\|.$$

(8) If $J_X : X_1 \to X_2$ and $J_Y : Y_1 \to Y_2$ are the isomorphisms, then $J(T) := J_Y T J_X^{-1}$ gives the required isomorphism; note that $J^{-1}(S) = J_Y^{-1} S J_X$.

8.23 (3b) Show $y \mapsto (0, y) + X \times 0$ is an isometry.

(5) Let $\{a_n\}$ be dense in M and $\{b_n + M\}$ dense in X/M. Then $\{a_n + b_m\}$ is dense in X.

8.27 (5) See the Hilbert cube Exercise 9.10(3).

9.4 (2) The functionals on c are $y^\top$ $(y \in \ell^1)$ and Lim.

(6) $c_{00} \subset \ell_s^\infty$, so $\overline{\ell_s^\infty} = c_0$; $1/\log n$ does not belong to any ℓ_s^∞.

9.7 (1) Let $y_n := x_n - x \in \ell^1$; then $\sum_{i=N+1}^{\infty} |y_{ni}| \leqslant \sum_{i=N+1}^{\infty} |y_{1i}| < \epsilon$ for some N and all n. But $|y_{n1}| + \cdots + |y_{nN}| \to 0$ as $n \to \infty$, so $\sum_i |y_{ni}| < 2\epsilon$.

9.10 (4) It is required to show $\|x - a\|_{\ell^1} < \epsilon$ for $a = (a_0, \ldots, a_N, 0, \ldots) \in c_{00}$, N large enough.

9.17 (2) Try $|a_n|^{p'/p} e^{-i\theta_n}$.

(9b) Take $r \to \infty$ in $\|x\|_\infty \leqslant \|x\|_r \leqslant \|x\|_p^{p/r} \|x\|_\infty^{1 - p/r}$.

9.37 (2) Look at the dual spaces of $L^1[0, 1]$ and c_0 to see why they are not isomorphic.

(6) Write $\frac{\pi t^2}{\sigma^2} + 2\pi i t\xi = \frac{\pi}{\sigma^2}(t + i\sigma^2\xi)^2 + \pi\sigma^2\xi^2$ to simplify the integral.

10.10 (2) In Pythagoras' theorem, $\|y + z\|^2 = \|y\|^2$ exactly when $z = 0$.
Consider

$$\Big\|\sum_n x_n\Big\|^2 = \Big|\sum_{n,m}\langle x_n, x_m\rangle\Big| \leqslant \sum_{n,m}|\langle x_n, x_m\rangle|$$

$$\leqslant \sum_{n,m}\|x_n\|\|x_m\| = \Big(\sum_n\|x_n\|\Big)^2.$$

This can only be an equality when $|\langle x_n, x_m\rangle| = \|x_n\|\|x_m\|$ for each n, m.

(4) Writing $x = \sum_n a_n v_n$ and $y = \sum_m b_m v_m$ for a basis $v_1, \ldots, v_N$, we find

$$\langle x, y\rangle = \sum_{nm} a_n b_m \langle v_n, v_m\rangle$$

(9) $(1, 1, 0, \ldots)$ and $(1, -1, 0, \ldots)$ do not satisfy the parallelogram law; write these as step functions for L^1 and L^∞.

(11) $\int_{-\pi}^{\pi}\sin(t)\cos(t)\,dt = \frac{1}{2}\int_{-\pi}^{\pi}\sin(2t)\,dt = [-\cos 2t]_{-\pi}^{\pi} = 0$, and $\int_0^1 2t^3 - t\,dt = \frac{1}{2}[t^4 - t^2]_0^1 = 0$.

(14) Substitute $\lambda = \alpha + i\beta$, then find the minimum by differentiating in α, β to get $\lambda = -\overline{\langle x, y\rangle}$.

(15) $\|x_n - x_m\| \leqslant \|x_n + y_n - x_m - y_m\| \to 0$ since $\langle x_n - x_m, y_n - y_m\rangle = 0$.

(16) The 'inner product' remains continuous, so Z is closed.

10.15 (1) Answer $\frac{1}{14}(10x - 2y + 6z, -2x + 13y + 3z, 6x + 3y + 5z)$.

(2i) $Px \in M$ so $Px = \lambda y$, and $x - Px \in M^\perp$, so $\langle y, x - \lambda y\rangle = 0$. Expanding gives $\lambda = \langle y, x\rangle$.

(3) Consider $x \in M^\perp$, and $x = u + v$ where $u \in M$, $v \in N$; since $N \subseteq M^\perp$ it follows that $u = 0$.

(5) Any vector $x \in N$ can be written $x = u + v$ where $u \in M$, $v \in M^\perp$. Since $M \subseteq N$, then $v = x - u \in N$ as well.

(6) Let $x = u + v$, $u \in M$, $v \in M^\perp$; then $Tx = Tu + Tv$, $Tu = Au \in M$, $Tv = Bv \in M^\perp$.

$$\|T\|^2 = \sup\frac{\|Tu\|^2 + \|Tv\|^2}{\|u\|^2 + \|v\|^2}$$

by Pythagoras' theorem. But $\|Tu\| \leqslant \|A\|\|u\|$ and $\|Tv\| \leqslant \|B\|\|v\|$, so $\|T\|^2 \leqslant t\|A\|^2 + (1 - t)\|B\|^2$, where $t = \|u\|^2/(\|u\|^2 + \|v\|^2)$. Now take $t = 0$ or $t = 1$ depending on which is the maximum of the two.

(8b) Expand $d^2 \leqslant \|x - y\|^2 = 2 - 2\operatorname{Re}\langle x, y\rangle$ with $y = e^{i\theta}v$.

(9c) If $\|x - a\| = d = \|x - b\|$ is the shortest distance from x to M, then $\|tx + (1 - t)x - ta - (1 - t)b\| = d$.

(9d) The closest sequence would be $1 \notin c_0$.

(10) $\|Py_{n+1}\| \leqslant \|y_{n+1}\| \leqslant \|y_n\|$, so $\|y_n\|$ converges. But in general, as $Py \perp (y - Py)$, $\|y\|^2 = \|Py\|^2 + \|y - Py\|^2$, so $\|y_n - Py_n\| \to 0$, and similarly $Py_n - QPy_n \to 0$. In finite dimensions, the bounded sequence y_n has a convergent subsequence, $y_{n_i} \to y$, so $y = Py = QPy$, and y is in $\mathrm{im}\, P \cap \mathrm{im}\, Q$.

(11) $\sin t \approx 0.955 - 0.304t \approx -0.20 + 1.91t - 0.88t^2 + 0.093t^3$; $1 - t^3 \approx 1.13 \cos t - 0.43 \sin t$.

(12c) Answer: $\alpha = \frac{2}{\pi} \frac{3MR^2 - 5I}{R^4}$, $\beta = \frac{15}{2\pi} \frac{2I - MR^2}{R^5}$.

10.24 (1) Check that $\|x^*\|_{H^*}$ satisfies the parallelogram law, then use the polarization identity, noting that $(ix)^* = -ix^*$.

(2) $\tilde{\phi}$ corresponds to Px.

(3) The map $x \mapsto \langle\!\langle x, \rangle\!\rangle$ is a functional so corresponds to some vector Tx.

(6) $\|T\|^2 = \|T^*T\| \leqslant \|T^*\| \|T\|$, so $\|T\| \leqslant \|T^*\| \leqslant \|T^{**}\| = \|T\|$.

(7) For $x = (a_n)_{n \in \mathbb{N}}$, $y = (b_n)_{n \in \mathbb{N}}$, $z = (c_n)_{n \in \mathbb{N}}$,

$$\langle z, yx \rangle = \sum_n \bar{c}_n b_n a_n = \sum_n \overline{\bar{b}_n c_n} a_n = \langle \bar{y}z, x \rangle$$

(9) $\int_0^1 \overline{g(s)} Vf(s)\, ds = \int_0^1 \int_0^s \overline{g(s)} f(t)\, dt\, ds = \int_0^1 \int_t^1 \overline{g(s)} f(t)\, ds\, dt$.

(12) $T^*Tx = 0 \Rightarrow 0 = \langle x, T^*Tx \rangle = \langle Tx, Tx \rangle$.

(13) Fix a unit vector $u \in X$, $\lambda := \langle Tu, Tu \rangle > 0$, and let v be any orthogonal unit vector; then $\langle Tu, Tv \rangle = \langle u, v \rangle = 0$; similarly, $\langle T(u + v), T(u - v) \rangle = \langle u + v, u - v \rangle = 0$, so $\langle Tv, Tv \rangle = \lambda > 0$ constant. For vectors $x = \alpha u$, $y = \beta_1 u + \beta_2 v$, $\langle Tx, Ty \rangle = \bar{\alpha} \beta_1 \lambda = \lambda \langle x, y \rangle$.

10.26 (1) Answers: $(-5/2, -2/3, 7/6)$, $(-17, -5, 7)/3$.

(6) $T^\dagger T$ is the projection onto $\ker T^\perp$; $TT^\dagger$ is the projection onto $\mathrm{im}\, T$.

(8) $V^*Vf = V^*g$ is $\int_y^1 \int_0^x f(t)\, dt\, dx = \int_y^1 g(x)\, dx$.

(9) Answer: $r = 0.499$ m and $\kappa/m = 0.008\,\mathrm{m}^{-1}$ (the actual values used to generate the data were $r = 0.5$ m and $\kappa/m = 0.003\,m^{-1}$).

10.35 (1) Take the inner product of $\sum_n \alpha_n e_n = 0$ with e_m.

(3) $\langle (e_n, 0), (0, \tilde{e}_m) \rangle = \langle e_n, 0 \rangle + \langle 0, \tilde{e}_m \rangle = 0$; if x and y can be approximated by $x_N := \sum_{n=1}^N \alpha_n e_n$ and $y_M := \sum_{m=1}^M \beta_m \tilde{e}_m$ respectively, then

$$\|(x, y) - (x_N, y_M)\| = \|(x - x_N, y - y_M)\| = \sqrt{\|x - x_N\|^2 + \|y - y_M\|^2}$$

can be made small; note that $(x_N, y_M) = (x_N, 0) + (0, y_M) = \sum_{n=1}^N \alpha_n (e_n, 0) + \sum_{m=1}^M (0, \tilde{e}_m)$.

(4) $\langle e_n, x - x_* \rangle = 0$

(5) Suppose e_n and $U e_n$ are both orthonormal bases. Then, by Parseval's identity,

$$\langle Ux, Uy \rangle = \sum_{n,m} \bar{\alpha}_n \beta_m \langle U e_n, U e_m \rangle = \langle x, y \rangle.$$

U is onto because $y = \sum_n \alpha_n U e_n = U(\sum_n \alpha_n e_n)$.

Conversely, if $\{e_n\}$ is an orthonormal basis for H_1, and $y \in \{U e_n\}^\perp$, then $0 = \langle y, U e_n \rangle = \langle U^* y, e_n \rangle$ for all n, so $U^* y \in \{e_n\}^\perp = 0$ and $\|y\| = \|U^* y\| = 0$. The column vectors of the matrix of U are $U e_n$, so $\langle U e_n, U e_m \rangle = \langle e_n, e_m \rangle = \delta_{nm}$.

(6) Show $t = \frac{1}{2} + \frac{1}{\pi} \sum_{n \neq 0} \frac{1}{2\pi n} e^{2\pi i n t}$, then take $t = 1/4$. It is interesting to generate other series using other points and functions (e.g., $|t|$, $t/|t|$, $|\sin t|$).

(7) For f odd, $\alpha_{-n} = -\alpha_n$. In general, every f is the sum of an even and an odd function.

(9) For example, take $\lambda \binom{1}{0}$, $\frac{\lambda}{2} \binom{-1}{\pm\sqrt{3}}$. For the second part, substitute e_m instead of x, and deduce orthogonality; if $x \in \{e_n\}^\perp$, then $\|x\| = 0$.

11.7 (4b) Continuity of $\tilde{T}(Sx) := Tx$: For any $v \in \ker S$ and $y \in Y$, $\|\tilde{T} y\| = \|Tx\| = \|T(x+v)\| \leqslant c\|T\|\|x+v\|$, then use $\|x + \ker S\| \leqslant c\|Sx\|$.

(9) $|\alpha_n| = \|\alpha_n e_n\| \leqslant \|\sum_{i=1}^n \alpha_i e_i - \sum_{i=1}^{n-1} \alpha_i e_i\| \leqslant 2c\|x\|$.

11.17 (5) If $x_n \in M$ is bounded then $T|_M x_n = Tx_n$ has a Cauchy subsequence, which converges.

11.28 (4) The requirement is $\phi(x, y) = x + \lambda y$, $|x + \lambda y| \leqslant |x| + |y|$, so $|\lambda| \leqslant 1$.

(6) Consider unit functionals such that $\phi_i(x_i) = \|x_i\|$; let $y \in \bigcap_i \ker \phi_i$; then

$$\|x_i\| = |\phi_i(y - x_i)| \leqslant \|y - x_i\|.$$

(9) $^\perp \Phi = 0$, so $(^\perp \Phi)^\perp = \ell^{1*}$. Now in the correspondence of ℓ^{1*} with ℓ^∞, we get $[\![\Phi]\!] \equiv c_{00}$ and so $\overline{[\![\Phi]\!]} \equiv c_0$.

(10) $|\phi x| = |\phi(x+v)| \leqslant \|\phi\|\|x+v\|$ for any $v \in M$; in fact this approaches equality for certain $v \in M$, so $\|\psi\| = \|\phi\|$. Onto: for any $\psi \in (X/M)^*$, let $\phi x := \psi(x + M)$. Hint for the second part: the norm of $\|\phi + M^\perp\| = \inf_{\psi \in M^\perp} \|\phi + \psi\|$ is the same as $\|\phi|_M\|$.

11.34 (5) $(T^{\top\top} x^{**})\phi = x^{**}(T^\top \phi) = (T^\top \phi)x = \phi Tx$.

(7) If $T^\top$ is onto, then T is 1–1 by (1) and has a closed image; if $T^\top$ is also 1–1, then $\operatorname{im} T$ is dense, hence T is onto. If T is onto, use the open mapping theorem.

11.49 (1) For c_0, a functional is of the type $y^\top$ where $y = (b_n)_{n \in \mathbb{N}} \in \ell^1$. Now $y \cdot e_n = \sum_i b_i \delta_{ni} = b_n \to 0$ as $n \to \infty$ since $\ell^1 \subset c_0$.

(8b) $|\phi(T_n S_n - TS)x| \leqslant \|\phi\|\|T_n\|\|(S_n - S)x\| + |\phi(T_n S - TS)x| \to 0$.

(11) If $x \notin M$, there is a $\phi \in X^*$ such that $\phi x = 1$, $\phi M = 0$, so x is not a weak limit point of M. More generally, every closed convex set is weakly closed, because a hyperplane (so a functional) separates it from any point not in it.

12.12 (4) $\|o(h)\| = \|f(x+h) - f(x) - f'(x)h\|$

$$= \left\| \int_0^1 dt f(x+th) - f'(x)h \, dt \right\|$$

$$\leqslant \int_0^1 \|f'(x+th) - f'(x)\| \|h\| \, dt \leqslant \tfrac{1}{2} k \|h\|^2.$$

12.21 (6) Poles and residues are (a) $i : 1/2ie$, and $-i : ie/2$; (b) $1 : (e, e^{-1})/3$, and $\omega : (e^\omega, e^{-\omega})/3\omega^2$, and $\omega^2 : (e^{\omega^2}, e^{-\omega^2})/3\omega$; (c) $0 : 1$.

13.3 (11) If TR is invertible, then $P(ST) = 1 = (TR)Q$, so T is invertible.

13.10 (2) Each vector (a, b) corresponds to the matrix $\begin{pmatrix} a & 0 \\ b & a+b \end{pmatrix}$.

(4) $1, A, \ldots, A^N$ cannot be linearly independent, so $A^m = p(A)$ must be true for some polynomial p.

(10) This is a generalization of the convolution operation on ℓ^1. The proofs are very similar to that case; see Exercise 9.7(2).

(13) For any ϕ, $Tx\phi x = x\phi Tx$, i.e., $Tx = \lambda_x x$. So if x, y are linearly dependent then $Ty = \lambda_y y$, implying $Tx = \lambda_y x$ and $\lambda_y = \lambda_x$; if not, then $\lambda_y = \lambda_{x+y} = \lambda_x$.

(14d) If $S, T \in \mathcal{A}''$, then $TR = RT$ for any $R \in \mathcal{A}' \supseteq \mathcal{A}''$, including $R = S$.

(18) To show $\mathcal{I}_A \subseteq \mathcal{I}$, let $f \in \mathcal{I}_A$ and let K be a closed subset of $[0, 1] \setminus A$; then for any $x \in K$, one can find a function $g_x \in \mathcal{I}$ such that $g_x(x) > 1$ in a neighborhood of x. By compactness of K, a finite number of such functions "cover" K, so $g := g_{x_1} + \cdots + g_{x_n} \in \mathcal{I}$ is greater than 1 on K. Let $h(x) := \begin{cases} 1, & g(x) > 1 \\ g(x), & g(x) \leqslant 1 \end{cases}$, a continuous function with $h|_K = 1$ and belonging to $\mathcal{I}$ ($h = gk$). By making K larger, one can find a sequence of functions such that $h_n g \to g$, so $g \in \mathcal{I}$.

(19) To show $\|f + \mathcal{I}_A\| = \|f|_A\|$, it is required to find functions $g_n \in \mathcal{I}_A$ such that $\|f - g_n\| \to \|f|_A\|$. This can be done as follows: take $B := [0, 1] \setminus U$, where $U = A + B_\epsilon(0)$, and let h be a function such that $h|A = 0$, $h|B = 1$; so $fh \in \mathcal{I}_A$ yet $f - fh = 0$ on B, and $\|f - fh\| \to \|f|A\|$ as $\epsilon \to 0$.

(21) Multiplication is well-defined, for if $S - \tilde{S} \in \mathcal{I}$, $T - \tilde{T} \in \mathcal{I}$, then $ST - \tilde{S}\tilde{T} = (S - \tilde{S})T + \tilde{S}(T - \tilde{T}) \in \mathcal{I}$. Associativity and distributivity follow from those of $\mathcal{X}$. Suppose $\|S + A_n\| \to \|S + \mathcal{I}\|$, $\|T + B_n\| \to \|T + \mathcal{I}\|$, for some $A_n, B_n \in \mathcal{I}$, then

$$\|ST + \mathcal{I}\| \leqslant \|(S + A_n)(T + B_n)\| \leqslant \|S + A_n\| \|T + B_n\|$$

$$\to \|S + \mathcal{I}\| \|T + \mathcal{I}\|$$

Finally, $\|1 + \mathcal{I}\| \leqslant \|1 + 0\| = 1$ yet $\|1 + \mathcal{I}\| \neq 0$; but also in any normed algebra in which $\|ST\| \leqslant \|S\| \|T\|$ holds, $1 \leqslant \|1\|$, since $\|1\| = \|1^2\| \leqslant \|1\|^2$.

(26) $\mathcal{X}$ has a basis of two vectors, which can be taken to be $1 = \binom{1}{0}$ and $\binom{0}{1}$. Multiplication by 1 acts of course as the identity matrix; if $\binom{0}{1}\binom{1}{1} = \binom{\alpha}{\beta}$, then $\binom{0}{1}\binom{x}{y} = \binom{\alpha y}{x + \beta y} = \binom{0\ \alpha}{1\ \beta}\binom{x}{y}$.

13.19 (1) Answers (a) 0, (b) 1, (c) $\max(|a|, |b|)$, (d) $(a \neq 0)$

$$\begin{pmatrix} a & 1 \\ 0 & a \end{pmatrix}^n = \begin{pmatrix} a^n & na^{n-1} \\ 0 & a^n \end{pmatrix} = a^n \begin{pmatrix} 1 & n/a \\ 0 & 1 \end{pmatrix}.$$

Now $\begin{pmatrix} 1 & n/a \\ 0 & 1 \end{pmatrix}\binom{1}{0} = \binom{1}{0}$, so $1 \leqslant \left\| \begin{pmatrix} 1 & n/a \\ 0 & 1 \end{pmatrix} \right\| \leqslant \sqrt{2 + n^2/a^2}$ (Proposition 8.10). Taking the nth root gives $(2 + n^2/a^2)^{1/2n} \to 1$, so $\rho(T) = |a|$. Note how, in this case, $\|T^n\|$ first increases then decreases to 0. Only (c) has $\rho(T) = \|T\|$.

(3) Use the Cauchy inequality for $|x + ay| \leqslant \sqrt{1 + |a|^2}\sqrt{|x|^2 + |y|^2}$.

(7) Let R and S be the radii of convergence of $\sum_n a_n z^n$ and $\sum_n b_n z^n$. Then $\sum_n (a_n + b_n)z^n = \sum_n a_n z^n + \sum_n b_n z^n$ has radius of convergence at least $\min(R, S)$. $\sum_n a_n b_n z^n$ has radius of convergence RS since $\liminf |a_n b_n|^{-1/n} = \liminf |a_n|^{-1/n}|b_n|^{-1/n}$.

(8) $f + g$ and fg have coefficients $a_n + b_n$, $a_0 b_n + a_1 b_{n-1} + \cdots + a_n b_0$.

$$f \circ g(T) = a_0 + a_1 g(T) + a_2 g(T)^2 + \cdots$$
$$= (a_0 + a_1 b_0 + a_2 b_0^2 + \cdots) + (a_1 b_1 + 2a_2 b_1 + \cdots)T$$
$$+ (a_1 b_2 + a_2 b_1^2 + \cdots)T^2$$

(9) $\|f(T) - \sum_{n=0}^N a_n T^n\| = \|\sum_{n=N+1}^\infty a_n T^n\| \leqslant \sum_{n=N+1}^\infty |a_n|\|T\|^n \to 0$ when $\|T\| < r$.

(14) $\cos 0 = e^0 = 1$, but $\cos 2 = (\cos 1 - \sin 1)(\cos 1 + \sin 1) < 0$, so there is a number $0 < \beta < 2$, $\cos \beta = 0$. Since the conjugate of $e^{i\theta}$ is $e^{-i\theta}$, it follows that $|e^{i\theta}| = 1$, so $\sin \beta = 1$; hence $e^{i\beta} = i$ and $e^{4\beta i} = 1$.

(17) Expand $e^{\alpha_1 S} e^{\alpha_2 T} e^{\alpha_3 T} e^{\alpha_4 T}$ to second order, and equate with $e^{S+T} \approx 1 + (S + T) + (S + T)^2/2$, to get $\alpha_2 \alpha_3 = 1/2$; the two values can be chosen to be equal.

13.25 (2) $f(t)g(t) = 1 \Leftrightarrow f(t) = 1/g(t) \neq 0, \forall t \in [0, 1]$. g has a minimum distance to the origin Exercise 6.22(11), so $f = 1/g$ is also bounded.

(4) $\|T^{-1}\| = \sup_x \|T^{-1}x\|/\|x\| = \sup_y \|y\|/\|Ty\|$.

(8) $e^{(t+s)T} = e^{tT+sT} = e^{tT}e^{sT}$ since $(tT)(sT) = (sT)(tT)$.

$$\therefore \quad e^{(t+h)T} = e^{tT}e^{hT} = e^{tT}(1 + hT + o(h))$$

so the derivative at t is $e^{tT}T$.

(12) $SR = 0$ for $S(a_0, a_1, \ldots) := (a_0, 0, \ldots)$. But $\|RT\| = \|T\|$ for all T, so $RT_n \nrightarrow 0$ when T_n are unit elements.

13.31 (2) If $|f(z)| \leqslant c|z|^{m/n} \leqslant c|z|^k$, then f is still a polynomial.

(8) If a is a zero or pole of order $\pm N$, then $(z - a)^{\pm N} f(z)$ is analytic and non-zero at a. Thus qf/p is bounded analytic on $\mathbb{C}$, so must be constant.

14.7 (2) $f(t) - \lambda$ is not invertible precisely when $f(t_0) - \lambda = 0$ for some $t_0 \in [0, 1]$.

(4) $T^2 - z^2 = (T - z)(T + z)$, so $z^2 := \lambda \in \sigma(T^2) \Rightarrow \lambda = \pm z \in \sigma(T)$ (one of them). Conversely, if $T^2 - z^2$ has an inverse S, then $S(T + z)(T - z) = 1 = (T - z)(T + z)S$, so $T - z$ is invertible.

(7) $(S, T) - \lambda(1, 1) = (S - \lambda, T - \lambda)$ is not invertible iff $S - \lambda$ or $T - \lambda$ is not invertible.

(8) The map $T \odot S - \lambda : (x, y) \mapsto (Tx - \lambda x, Sy - \lambda y)$ is invertible exactly when $T - \lambda$ and $S - \lambda$ are invertible.

14.14 (1) $Rx = \lambda x$ means $a_n = \lambda a_{n+1}$, so $a_n = a_0/\lambda^n$; but also $0 = \lambda a_0$. There are no solutions to these algebraic equations.

(3) ℓ^1 is embedded in $\ell^1(\mathbb{Z})$, so $\sigma(L)$ decreases from the first case to the second. In fact, in $\ell^1(\mathbb{Z})$, there are no eigenvalues, because $\sum_{n=-\infty}^{\infty} |\lambda|^n$ cannot converge for any λ. Yet the boundary of $\sigma(T)$ in ℓ^1, consisting of generalized eigenvalues, is preserved in $\ell^1(\mathbb{Z})$.

(4) $T^\top x = (a_0, a_2, a_3, \ldots)$ on ℓ^1.

(5) $T^\top x = (a_0, a_2, a_3/2, \ldots)$ on ℓ^1.

(9) The operator $(T - \lambda)f(t) = (t - \lambda)f(t)$ is invertible only when $\lambda \notin [0, 1]$. There are no eigenvalues because $tf(t) = \lambda f(t)$ for all t implies $f = 0$. The image of $T - \lambda$ is a subset of $\{ g \in C[0, 1] : g(\lambda) = 0 \}$; as this set is closed and not $C[0, 1]$, all $\lambda \in [0, 1]$ are residual spectral values.

(11) Induction on n: Expand $V V^n f$ as a double integral and change the order of integration.

(12) $1 - |\lambda| \leqslant \|Tx_n - \lambda x_n\| \to 0$; $T - \lambda = T(1 - \lambda T^{-1})$, so $|\lambda| < 1 \Rightarrow \lambda \notin \sigma(T)$. The boundary of $\sigma(T)$ must be part of the circle.

(13) T is 1–1 with a closed image $\Leftrightarrow \|Tx\| \geqslant c\|x\|$, so

$$\|(T + H)x\| \geqslant \|Tx\| - \|Hx\| \geqslant (c - \|H\|)\|x\|$$

shows T is an interior point of the set.

14.22 (7) The eigenvalue equation for ML is $a_{n+1} = n\lambda a_n$, so $a_n = n!\lambda^n x_0 \to \infty$. For RM, $\{0\} = \sigma_p((RM)^\top) \subseteq \sigma_r(RM)$.

14.30 (3) $e^{\sigma(T)} = \sigma(e^T) = \sigma(1) = \{1\}$, so $\sigma(T) \subseteq 2\pi i\mathbb{Z}$. For an idempotent P,
$$e^{2\pi P} = 1 + P(2\pi i + \frac{(2\pi i)^2}{2!} + \cdots) = 1 + P(e^{2\pi i} - 1) = 1.$$

14.41 (11) $\mathbb{C}^n$ is generated by e_i, where $e_i e_j = 0$ when $i \neq j$, and $e_i e_i = 1$. So a character satisfies $\delta e_i \delta e_j = 0$ and $\delta e_i = \pm 1$. If $\delta e_1 = \pm 1$, say, then $\delta e_i = 0$ for $i \neq 1$. In fact $1 = \delta(1) = \sum_i \delta(e_i) = \delta(e_1)$.

(12) $B(\mathbb{C}^2)$ is generated by $\left(\begin{smallmatrix} 1 & 0 \\ 0 & 0 \end{smallmatrix}\right)$, $\left(\begin{smallmatrix} 0 & 1 \\ 0 & 0 \end{smallmatrix}\right)$, $\left(\begin{smallmatrix} 0 & 0 \\ 1 & 0 \end{smallmatrix}\right)$, and $\left(\begin{smallmatrix} 0 & 0 \\ 0 & 1 \end{smallmatrix}\right)$. A character δ maps them to $w_1, \ldots, w_4$, which must satisfy $w_2^2 = 0$, $w_3^2 = 0$, $w_2 w_3 = w_1$, $w_3 w_2 = w_4$, for which there are no non-zero solutions.

(15) $\widehat{x}$ acts on the n points in Δ as $(\delta_i x) = (x_i) = x$.

(17) In the commutative Banach algebra $\mathcal{Y} := \{\,S, T\,\}''$, the spectra remain the same, $\sigma_{\mathcal{Y}}(A) = \sigma(A)$, so $\Delta(A) = \sigma(A)$ and the inclusions follow from $\Delta(S + T) \subseteq \Delta(S) + \Delta(T)$ and $\Delta(ST) \subseteq \Delta(S)\Delta(T)$.

15.3 (4) T is left- and right-invertible: $TT^*R = 1 = R'T^*T$.

(7) Use Theorem 13.9; note that $L^*L = \alpha \in \mathbb{R}$, so $\alpha = \lambda^2$.

15.11 (5) What is meant is that if $T \in \mathcal{X}$ is normal, and J is a $*$-morphism, then $J(T) \in \mathcal{Y}$ is also normal, etc.

(9) The inverse of T_a is T_{-a}, which is the adjoint:

$$\int \overline{g(x)}\, T_a f(x)\, dx = \int \overline{g(x)}\, f(x - a)\, dx = \int \overline{g(t + a)}\, f(t)\, dt$$

$$= \int \overline{T_{-a} g(t)}\, f(t)\, dt.$$

(21) $\|(T^*T)^n\|^{1/2n} = \|(A^*A)^n + (B^*B)^n\|^{1/2n} \leqslant (\|A\|^{2n} + \|B\|^{2n})^{1/2n} \to \max(\|A\|, \|B\|)$

(22) If T^*T is idempotent, then $\sigma(TT^*) \subseteq \sigma(T^*T) \cup \{0\} \subseteq \{0, 1\}$. Hence $\sigma(TT^*TT^* - TT^*) = \{0\}$.

15.15 (1) $\langle x, TT^*x \rangle = \|T^*x\|^2 = \|Tx\|^2 = \langle x, T^*Tx \rangle$ and use Example 10.7(3).

(3) $\|T_n^*x - T^*x\| = \|(T_n - T)^*x\| = \|T_n x - Tx\| \to 0$. Conversely, take the limit of $\|T_n^*x\| = \|T_n x\|$ and use Exercise 1.

(4) $|\lambda|^2 \|x\|^2 = \|\lambda x\|^2 = \|Ux\|^2 = \|x\|^2$.

(5) Each distinct eigenvalue comes with an orthogonal eigenvector. In a separable space, there can only be a countable number of these.

(6) $\langle e_m, T^*e_n \rangle = \langle Te_m, e_n \rangle = \bar{\lambda}_n \delta_{nm}$, so $T^*e_n = \sum_m \langle e_m, T^*e_n \rangle e_m = \bar{\lambda}_n e_n$. Then show $\|T^*x\| = \|Tx\|$.

(8) For (b), note that $\sigma(I - T^n) = 1 - \sigma(T)^n \subseteq \overline{B_1(1)}$, so $\|I - T^n\| = \rho(I - T^n) \leqslant 2$. For (c) use $H = \ker(T^* - I) \oplus \ker(T^* - I)^\perp = \ker(T - I) \oplus \overline{\operatorname{im}(T - I)}$.

15.21 (2b) Let $M, M^\perp$ be the domains of A and D. For any $x = a + b \in M \oplus M^\perp$,

$$\langle x, Tx \rangle = \langle a + b, Ta + Tb \rangle = \langle a, Ta \rangle + \langle b, Tb \rangle,$$

$$\langle x, x \rangle = \langle a + b, a + b \rangle = \|a\|^2 + \|b\|^2.$$

As $\langle a, Ta \rangle = \|a\|^2 \lambda$ with $\lambda \in W(A)$, and similarly $\langle b, Tb \rangle = \|b\|^2 \mu$, $\mu \in W(D)$, the values of $\langle x, Tx \rangle / \|x\|^2$ includes the line between λ and μ. The collection of these lines is the convex hull of $W(A) \cup W(D)$.

(3b) For $T := \begin{pmatrix} a & 1 \\ 0 & a \end{pmatrix}$, let $x = \begin{pmatrix} \alpha \\ \beta \end{pmatrix}$, then $\langle x, Tx \rangle = |\alpha|^2 a + \bar{\alpha}\beta + |\beta|^2 a = a + \bar{\alpha}\beta$, because of the condition $1 = \|x\|^2 = |\alpha|^2 + |\beta|^2$. But $\bar{\alpha}\beta = \cos t \sin t\, e^{i\theta}$ takes the value of any complex number in the closed ball $\overline{B_{1/2}(0)}$. In general, the sum $\sum_{i=1}^n \alpha_i \alpha_{i+1}$ is a disk of radius $\cos \frac{\pi}{n+1}$ by Lagrange multipliers.

(12c) Let $\lambda := \langle T \rangle_x$, so $0 = \sigma_T^2 = \sigma_{T-\lambda}^2 = \|T - \lambda\|^2$.

15.26 (1) The singular values are (i) 4 with singular vectors proportional to $\binom{1}{2}$, $\binom{2}{1}$, and 1 with $\binom{2}{-1}$, $\binom{1}{-2}$; (ii) $\sqrt{3}$ with $\binom{2}{1}$, $\binom{1}{1}$, and 1 with $\binom{1}{0}$, $\binom{1}{-1}$.

(7) Let $S := T/\lambda$ where λ is the largest eigenvalue (in the sense of magnitude); it has the same eigenvectors e_n as T except with eigenvalues $\mu_n := \lambda_n/\lambda$. If $v_0 = \sum_n a_n e_n + y$, where $y \in \ker(T - \lambda)$ then $S^k v_0 = \sum_n \mu_n^k a_n e_n + y$. So

$$\|S^k v_0 - y\|^2 = \sum |\mu_n|^{2k} |a_n|^2 \leqslant c^{2k}\|v_0\|^2, \quad (0 \leqslant c < 1)$$

and $S^k v_0 \to y$ as $k \to \infty$. Hence $\frac{|\lambda|^k}{\lambda^k}\frac{T^k v_0}{\|T^k v_0\|} = \frac{S^k v_0}{\|S^k v_0\|} \to y/\|y\|$, and $v_{k+1} \approx \frac{\lambda^k y}{\|\lambda^k y\|}$; the sequence does not converge unless $\lambda = |\lambda|$ but behaves like $e^{ik\theta} y/\|y\|$.

15.35 (7) Answers: (b) eigenvalues $1/(n + \frac{1}{2})\pi$, eigenvectors $\sin(n + \frac{1}{2})\pi x$; (c) $1/(n + \frac{1}{2})^2\pi^2$, $\sin(n + \frac{1}{2})\pi x$; so

$$\sum_n \frac{1}{(n + \frac{1}{2})^4\pi^4} = \int_0^1 \int_0^1 \min(x, y)^2 \, dy \, dx = 1/6.$$

15.41 (3) If $\phi A = 0$ for all $\phi \in S$ and A is self-adjoint, then $\sigma(A) \subseteq S(A) = \{0\}$ and $A = 0$.

(7) $\sigma(T) \subseteq \overline{B_1(0)}$, and $\sigma(T)^{-1} = \sigma(T^{-1}) \subseteq \overline{B_1(0)}$.

(8) By the spectral mapping theorem $\{0\} = \sigma(P^2 - P) = \{\lambda^2 - \lambda : \lambda \in \sigma(P)\}$, so $\lambda = 0, 1$.

15.47 (4) Let $T = A + iB$ with A, B self-adjoint. Then $A \geqslant 0$ implies $S(T) \subseteq S(A) + iS(B) \subseteq \mathbb{R}^+ + i\mathbb{R}$. Conversely, if $0 > \lambda \in \sigma(A)$, then $\lambda = \phi A$ for some $\phi \in S$. If $\phi T = \phi A + i\phi B \geqslant 0$ for all ϕ, then $\phi B = 0$, so $B = 0$.

15.51 (4) $|T^*|^2 = T|T|U^* = TU^*U|T|U^* = (TU^*)^2$.

(10) $|T|$ is invertible, so let $U := T|T|^{-1}$; it is unitary, e.g., $UU^* = T|T|^{-2}T^*TT^{-1} = 1$.

(11) (b) $T = U|T| = S|T|^{\frac{1}{2}}$, where $:= U|T|^{\frac{1}{2}} \in \mathcal{HS}$, so $\text{tr}(T) = \text{tr}(S|T|^{\frac{1}{2}})$ is independent of the basis.

(c) $|\text{tr}(ST)| = |\text{tr}(SU|T|)| = |\langle U|T|^{\frac{1}{2}}, S^*|T|^{\frac{1}{2}}\rangle_{\mathcal{HS}}|$

$$\leqslant \|U|T|^{\frac{1}{2}}\|_{\mathcal{HS}}\|S^*|T|^{\frac{1}{2}}\|_{\mathcal{HS}} \leqslant \|S^*\|\||T|^{\frac{1}{2}}\|_{\mathcal{HS}}^2 = \|S\|\|T\|_{\text{Tr}}$$

(d) The norm axioms are satisfied because $\|T\|_{\text{Tr}} = \||T|^{\frac{1}{2}}\|_{\mathcal{HS}}^2$ and

$$\text{tr}|S + T| = \text{tr}\, U^*(S + T) = \langle U, S\rangle_{\mathcal{HS}} + \langle U, T\rangle_{\mathcal{HS}} \leqslant \|S\|_{\mathcal{HS}} + \|T\|_{\mathcal{HS}}.$$

Also,

$$\|T^*\|_{\mathrm{Tr}} = \operatorname{tr} U T^* = \operatorname{tr} T^* U = \operatorname{tr} |T|.$$

(e) $|T| = U^* T = CB$ where $C := U^* A \in \mathcal{HS}$. So $\operatorname{tr} |T| = \langle C^*, B \rangle \leqslant \|A\|_{\mathcal{HS}} \|B\|_{\mathcal{HS}}$.

(f) If e_n, e'_n are the singular vectors of T, then $|T|^2 e_n = |\lambda_n|^2 e_n$. Take the polar decomposition of $\langle e'_n, T e_n \rangle = e^{i\theta_n} |\langle e'_n, T e_n \rangle|$, and let $U e_n := e^{i\theta_n} e'_n$. Then

$$\sum_n |\langle e'_n, T e_n \rangle| = \sum_n \langle e_n, U^* T e_n \rangle = \operatorname{tr}(U^* T) \leqslant \|T\|_{\mathrm{Tr}}$$

If $T e_n = \lambda_n e'_n$, then $\|T\|_{\mathrm{Tr}} = \sum_n \langle e'_n, T e_n \rangle = \sum_n \lambda_n$.

Glossary of Symbols

$\to$	Converges to,	Definition 3.1
$\rightharpoonup$	Weak convergence,	Section 11.5
$\|\cdot\|_X$	Norm of space X,	Definition 7.3
$\langle\cdot,\cdot\rangle_X$	Inner product of space X,	Definition 10.1
1_E	Characteristic function on E,	Review 9.19(1)
$\sum_n$	A series of terms,	Definition 7.20
$[a_n]$	Equivalence class of sequence $(a_n)_{n\in\mathbb{N}}$,	Theorem 4.5
T^*	Hilbert adjoint of an operator T, or the involute of an algebra element,	Definitions 10.18 and 15.1
$T^\top$	Adjoint of an operator T,	Definition 11.29
$x^\top$	Dual of a sequence x,	Example 8.3(4)
$\widehat{T}$	Gelfand/Fourier transform of T,	Definitions 9.33, 14.37, and Exercise 9.37(5)
$\llbracket A \rrbracket$	Span of vectors in A,	Review 7.2(7)
A^c	Complement of set A,	Page 7
$\mathcal{A}'$	Commutant algebra of $\mathcal{A}$,	Exercise 13.10(14)
A°	Interior of set A,	Definition 2.7
$A^\perp$	Annihilator or orthogonal complement of A,	Proposition 10.9 and Definition 11.24
$^\perp A$	Pre-annihilator of A,	Definition 11.24
X^*	Dual space of X,	Definition 8.1
∂A	Boundary of set A,	Definition 2.7
$\bar{A}$	Closure of set A,	Definition 2.7
xy	Multiplication of sequences,	Exercise 9.4(3)
$x \cdot y$	Dot product of sequences,	Example 8.3(4)
$x * y$	Convolution of sequences or functions,	Exercise 9.7(2)
$A + B$	Addition of sets,	Review 7.2(12)
$A \oplus B$	Direct sum of subspaces,	Review 7.2(15)
$X \cong Y$	Isomorphic spaces,	Definition 8.14

$X \equiv Y$	Isometric spaces,	Definition 8.14
$X \subsetneqq Y$	X is embedded in Y,	Definition 8.14
X/M	Quotient space of X by M,	Proposition 8.20
$B(X)$	Space $B(X, X)$,	Proposition 8.8
$B(X, Y)$	Space of continuous linear operators $X \to Y$,	Definition 8.1
$B_r(x)$	Ball of radius r, center x,	Definition 2.4
$\overline{B_r(x)}$	Closed ball,	Example 2.17(3)
B_X	Unit ball of X,	ff. Proposition 7.6
c	Space of convergent sequences,	Proposition 9.2
c_0	Space of sequences that converge to zero,	Proposition 9.2
c_{00}	Space of sequences with a finite number of non-zero components,	Example 7.17(4)
$C(X)$	Space $C_b(X, \mathbb{C})$,	Theorem 6.23
$C_b(X, Y)$	Space of bounded continuous functions $f : X \to Y$,	Theorem 6.23
$C^n(\mathbb{R}, X)$	Space of n-times continuously differentiable functions,	Page 300
$C^\omega(A)$	Space of analytic functions on A,	Page 304
$\mathbb{C}[x, y]$	Space of polynomials in x, y,	Page 9
codim A	Codimension of subspace A,	ff. Proposition 8.20
d	Distance function,	Definition 2.1
D	Differentiation operator,	Proposition 12.2
$D(U, Y)$	Set of differentiable functions,	ff. Definition 12.1
D_1	"Taxicab" distance on $X \times Y$,	Example 2.2(6)
D_∞	Max distance on $X \times Y$,	Example 2.2(6)
Δ	Character space of $\mathcal{X}$,	Definition 13.6
δ_x	Dirac functional,	Example 8.3(10)
dim X	Dimension of space X,	Review 7.2(10)
$\mathbb{F}$	A field, usually $\mathbb{R}$ or $\mathbb{C}$,	Page 10
$\mathcal{G}(\mathcal{X})$	Group of invertibles of $\mathcal{X}$,	Theorem 13.21
I	Identity operator,	Page 8
im T	Image of a linear map T,	Page 8
index(T)	Index of an operator T,	Definition 11.12
ker T	Kernel space of a linear map T,	Proposition 8.4
L	Left-shift operator,	Example 8.3(7)
ℓ^p	Space of sequences with the p-norm,	Example 7.4(5)
$L^p(A)$	Space of functions on A with the p-norm,	Example 7.4(7)
$\lim_{n \to \infty}$	Limit as $n \to \infty$,	Proposition 3.2
μ	Lebesgue measure on $\mathbb{R}^n$,	Review 9.18(4)
M_a	Multiplication operator by a,	Exercise 9.4(4)
$\mathcal{S}(\mathcal{X})$	State space of an algebra,	Definition 14.34

R	Right-shift operator,	Example 8.6(3)
rad $\mathcal{X}$	Radical of an algebra,	Definition 14.31
$\rho(T)$	Spectral radius of T,	Proposition 13.12 and Theorem 14.3
$\sigma(T)$	Spectrum of T,	Definition 14.1
T_a	Translation by a,	Exercise 8.13(4)
$\mathrm{tr}(T)$	Trace of T,	Definition 15.30
$W(T)$	Numerical range of T,	Definition 15.16

Further Reading

Functional analysis impinges upon a wide range of mathematical branches, from linear algebra to differential equations, probability, number theory, and optimization, to name just a few, as well as such varied applications as financial investment/risk theory, bioinformatics, control engineering, quantum physics, etc.

As an example of how functional analysis techniques can be used to simplify classical theorems consider Picard's theorem for ordinary differential equations. The differential equation $y' = F(x, y)$, $y(a) = y_a$, is equivalent to the integral equation $y(x) = T(y) := y_a + \int_a^x F(s, y(s))\, ds$. It is not hard to show that if F is Lipschitz in y and continuous in x, then T is a contraction map on $C[a - h, a + h]$ for some $h > 0$, and the Banach fixed point theorem then implies that the equation has a unique solution locally.

However, the classical derivative operator is in many ways inadequate: its domain is not complete and it is unbounded on several norms of interest. But there is a way to extend differentiation to much larger spaces, namely *Sobolev spaces* and *Distributions*. The former are Banach spaces L_s^p of functions that have certain grades of integrability (p) and differentiability (s), while the latter are spaces of functionals that act on them with weak*-convergence. Distributions include all the familiar functions in L_{loc}^1, but also other 'singular' ones, such as Dirac's delta 'function' δ and $1/x^n$. Differentiation can be extended as a continuous operator on these spaces, e.g., $L_s^p \to L_{s-1}^p$. Moreover, distributions can be differentiated infinitely many times; for example, the derivative of the discontinuous Heaviside function $1_{\mathbb{R}^+}$ is δ. But, in general, 'singular' distributions cannot be multiplied together. A central result is the Sobolev inequality, $\|u\|_{L^q(\mathbb{R}^n)} \leqslant c_{n,p} \|Du\|_{L^p(\mathbb{R}^n)}$, for $n \geqslant 2$, $\frac{1}{q} = \frac{1}{p} - \frac{1}{n}$, which implies that the identity map $L_s^p(\mathbb{R}^n) \to L_t^q(\mathbb{R}^n)$, along the arrows in Fig. 1, is continuous. The study of operators on such generalized spaces is of fundamental importance: from extensions of the convolution and the Fourier transform, to pseudo-differential operators of the type $f(x, D)$, singular integrals, and various other transforms (see [12, 26, 28]).

© The Author(s), under exclusive license to Springer Nature Switzerland AG 2024   455
J. Muscat, *Functional Analysis*, https://doi.org/10.1007/978-3-031-27537-1

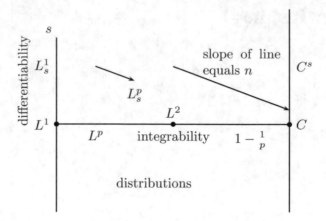

Fig. 1 Sobolev spaces

Although unbounded, classical differential operators are normal 'closed operators': these have a graph $\{(x, Tx) : x \in X\}$ which is closed in $X \times X$. Quite a lot of the spectral theory extends in modified form to them. For example their spectrum remains closed but not necessarily bounded. So, if one inverts in a point $\lambda \notin \sigma(T)$ then $(T - \lambda)^{-1}$ becomes a regular continuous operator, which can often be expressed as an integral operator, whose kernel is called its *Green's function*. Indeed, it turns out that 'elliptic' differential operators become Fredholm self-adjoint operators under this inversion. This immediately gives certain results, usually falling under the heading of Sturm-Liouville theory, such as that the spectrum of the Laplace operator $-\triangle$ on a compact shape in $\mathbb{R}^n$ is an unbounded sequence of isolated positive eigenvalues, called the "resonant frequencies" or "harmonics" of the shape. Deeper results include the Atiyah-Singer index theorem: the Fredholm index of an elliptic differential operator is equal to a certain topological invariant of the domain.

The concept of a Banach space can be generalized to a *topological vector space*, namely a vector space with a topology that makes its operations continuous. Many theorems continue to hold at least for "locally convex topological vector spaces", including the Hahn-Banach theorem, the open mapping theorem, and the uniform boundedness theorem. Other important results are Schauder's fixed point theorem, the Krein-Milman theorem, the analytic Fredholm index theorem, and the Hille-Yosida theorem.

Harmonic analysis is the study of general (but usually locally compact) group algebras, especially the Fourier transform. The central results are the Pontryagin duality theorem, which asserts that the character space of $L^1(G)$ is itself a group that is 'dual' to G, and the Peter-Weyl theorem. *von Neumann algebras* are *-algebras that arise as double commutators of C^*-algebras; equivalently, they are the weakly closed subspaces of $B(H)$. The spectral theorem holds for them. There is a lot of theory devoted to their structure, and a complete classification is still an open problem.

One must also include some outstanding conjectures: whether every operator on a separable Hilbert space has a non-trivial closed invariant subspace; whether every infinite-dimensional Banach space admits a quotient which is infinite-dimensional and separable; Selberg's conjecture about the first eigenvalue of a specific Laplace-Beltrami operator on Maass waveforms; the Hilbert-Pólya conjecture that the non-trivial zeros of the Riemann zeta function are the eigenvalues of some unbounded operator $\frac{1}{2} + iA$ with A self-adjoint; etc.

References

1. Bear H (2002) A primer of Lebesgue integration. Academic Press, California
2. Bollobás B (1999) Linear analysis: an introductory course. Cambridge University Press, Cambridge
3. Körner T (1989) Fourier analysis. Cambridge University Press, Cambridge
4. Kreyszig E (1978) Introductory functional analysis with applications. Wiley, New York
5. Riesz F, S-Nagy B (1990) Functional analysis. Dover Publications, New York
6. Schechter M (2002) Principles of functional analysis. AMS Bookstore, Boston
7. Shilov G (1996) Elementary functional analysis. Dover Publications, New York
8. Steele J (2004) The Cauchy-Schwarz master class. Cambridge University Press, Cambridge

More Advanced Books

 9. Bachman G, Narici L (2000) Functional analysis. Dover Publications, New York
10. Conway J (1990) A course in functional analysis. Springer, New York
11. Lax P (2002) Functional analysis. Wiley, New York
12. Lieb EH, Loss M (2001) Analysis. AMS Bookstore, California
13. Megginson R (1998) An introduction to Banach space theory. Birkhäuser, Berlin
14. Murphy GJ (1990) C^*-algebras and operator theory. Academic Press, New York
15. Palmer TW (2001) Banach algebras and the general theory of *-algebras, vol I and II. Cambridge University Press, Cambridge
16. Rudin W (1991) Functional analysis. McGraw-Hill, New York
17. Yoshida K (1995) Functional analysis. Springer, New York

Other References

18. Aster RC, Borchers B, Thurber CH (2018) Parameter estimation and inverse problems, 3rd edn. Elsevier, Amsterdam
19. Barnsley M (2000) Fractals everywhere. Morgan Kaufmann, California
20. Brown JW, Churchill RV (2013) Complex variables and applications, 9th edn. McGraw-Hill, New York
21. Devaney A (2012) Mathematical foundations of imaging, tomography and wavefield inversion. Cambridge University Press, Cambridge
22. Diestel J, Uhl J (1977) Vector measures. Mathematical Surveys, vol 15. AMS, Boston

23. Herman GT (2010) Fundamentals of computerized tomography: image reconstruction from projections, 2nd edn. Springer, New York
24. Jahnke HN (ed) (2003) A history of analysis, history of mathematics, vol 24. AMS, Boston
25. Peitgen HO, Jürgens H, Saupe D (2004) Chaos and Fractals: new frontiers of science. Springer, New York
26. Stein EM (1993) Harmonic analysis. Princeton University Press, Princeton
27. Walker JS (2008) A primer on wavelets and their scientific applications, 2nd edn. Chapman and Hall/CRC, Boca Raton
28. Ziemer WP (1989) Weakly differentiable functions. Springer, New York

Selected Historical Articles

29. Banach S (1932) Théorie des Opérations Linéaires. Subwencji Funduszu Kultury Narodowej, Warsaw
30. Carleson L (1966) On convergence and growth of partial sums of Fourier series. Acta Math 116(1–2):135–137
31. Enflo P (1973) A counterexample to the approximation problem in Banach spaces. Acta Math 130:1
32. Fréchet M (1906) Surquelques points du calcul fonctionnel. Ph.D. Thesis, Rendiconti del Circolo Matematico di Palermo, vol. 22
33. Fredholm EI (1903) Sur une classe d'equations fonctionnelles. Acta Math 27:365–390
34. Gowers WT, Maurey B (1993) The unconditional basic sequence problem. J Am Math Soc 6:851–874
35. Hausdorff F (1914) Grundzüge der Mengenlehre. von Veit, Leipzig
36. Hilbert D (1904–1910) Grundzüge einer allgemeinen Theorie der linearen Integralgleichungen, I-VI. Nachr Akad Wiss Göttingen Math-Phys Kl.
37. Lebesgue H (1902) Intégrale, longueure, aire, Ph.D. Thesis, Université de Paris
38. Lindenstrauss J, Tzafriri L (1971) On the complemented subspaces problem. Isr J Math 9:263–269
39. Phillips RS (1940) On linear transformations. Trans Am Math Soc 48:516–541
40. Riesz F (1918) Sur une espèce de géométrie analytiques des systèmes de fonctions sommables. C R Acad Sci Paris 144:1409–1411
41. Volterra V (1887) Sopra le funzioni che dipendono da altre funzioni; Sopra le funzioni dipendenti da linee. In: Rendiconti della Reale Academia dei Lincei, vol 3. Tipografia della R. Accademia dei Lincei, Roma
42. von Neumann J (1929) Zur Algebra der Funktionaloperationen und Theorie der normalen Operatoren. Math Ann 102:370–427

Index